Organoborane Chemistry

ORGANOMETALLIC CHEMISTRY

A Series of Monographs

BRIAN G. RAMSEY: Electronic Transitions in Organometalloids, 1969.
R. C. POLLER: The Chemistry of Organotin Compounds, 1970.
RUSSELL N. GRIMES: Carboranes, 1970.
PETER M. MAITLIS: The Organic Chemistry of Palladium, Volume I, Volume II – 1971.
DONALD S. MATTESON: Organometallic Reaction Mechanisms of the Nontransition Elements, 1974.
RICHARD F. HECK: Organotransition Metal Chemistry: A Mechanistic Approach, 1974.
P. W. JOLLY AND G. WILKE: The Organic Chemistry of Nickel, Volume I, Organonickel Complexes, 1974. Volume II, Organic Synthesis, in preparation.
P. C. WAILES, R. S. P. COUTTS, AND H. WEIGOLD: Organometallic Chemistry of Titanium, Zirconium, and Hafnium, 1974.
U. BELLUCO: Organometallic and Coordination Chemistry of Platinum, 1974.
P. S. BRATERMAN: Metal Carbonyl Spectra, 1974.
L. MALATESTA AND S. CENINI: Zerovalent Compounds of Metals, 1974.
THOMAS ONAK: Organoborane Chemistry, 1975.

In preparation

R. P. A. SNEEDEN: Organochromium Compounds

Organoborane Chemistry

Thomas Onak

Department of Chemistry
California State University
Los Angeles, California

Academic Press

New York San Francisco London 1975
A Subsidiary of Harcourt Brace Jovanovich, Publishers

ACADEMIC PRESS, INC.
111 Fifth Avenue, New York, New York 10003

United Kingdom Edition published by
ACADEMIC PRESS, INC. (LONDON) LTD.
24/28 Oval Road, London NW1

Library of Congress Cataloging in Publication Data

Onak, Thomas.
Organoborane chemistry.

(Organometallic chemistry series)
Bibliography: p.
Includes index.
1. Organoboron compounds. I. Title.
QD412.B105 547′.05′67 74-10198
ISBN 0–12–526550–6

PRINTED IN THE UNITED STATES OF AMERICA

Contents

Preface ix

Chapter 1 **Introduction and Nomenclature**

Text 1

Chapter 2 **Structure and Physical Properties of Organoboranes**

2-1. Boron–Carbon Bond Lengths 2
2-2. Boron–Carbon Bonding Characteristics 4
2-3. Nuclear Magnetic Resonance 6
Boron-11 6
Proton 8
Spin Coupling 9
2-4. Vibrational Spectroscopy 9
2-5. Electronic Transitions 11
2-6. Mass Spectrometry 13
2-7. Molecular Orbital Calculations 15
2-8. Thermodynamic Properties 16
2-9. Miscellaneous Properties 18

Chapter 3 **Three-Coordinate Organoboranes**

3-1. Boron–Carbon Bond Construction 19
Transmetallation 19

Redistribution and Exchange Reactions 24
Hydroboration 26
Haloboration 35
Modified Friedel–Crafts Reaction 36
Cyclization by Loss of H_2 38
Diboration with Diboron(4) Compounds 38
Formation of Boron–Carbon Bonds by the Action of Metals (or Metal Boron Compounds) on Halides 40
Miscellaneous Preparative Reactions 41
3-2. Boron–Carbon Bond Cleavage and Other Reactions 91
Protodeboronation and Hydrolysis of the Carbon–Boron Bond 91
Autoxidation 97
Action of Peroxides and Related Compounds 100
Halogens 102
Reactions of Organoboranes with *N*-Chloroamines 106
Other Oxidative Reactions 107
Radical Reactions 108
Transmetallation 109
Dehydroboration and Isomerization Reactions 111
Other Elimination Reactions 113
Reactions with Unsaturated Systems 115
Migration of Halogen from Carbon to Boron 124
Migration of Boron-Attached Organic Group to an Electropositive Center: Organoboranes as Alkylating (or Arylating) Agents 125
Hydrogenolysis 131
Reactions of Organoboranes with Alkali Metals 131
Miscellaneous Reactions 133

Chapter 4 **Four-Coordinate Organoboranes**

4-1. Organoborohydride Ions 136
Synthesis 136
Reactions and Properties 137
4-2. Organoborane–Lewis Base Adducts, RBXYD (D = Donor Group) 147
Synthesis and Stability 147
Reactions 154
4-3. Organoboronium Ions 157

4-4. Carbon Monoxide–Borane; H_3BCO 159
Synthesis 159
Reactions 160
Physical Properties 162

Chapter 5 Organodiboranes

5-1. Synthesis 164
Exchange Reactions 164
Reductions with Metal Hydrides 167
Hydroboration 169
Reduction of Organoboron Esters and Halides with Diborane 172
Dehydroboration 173
Insertion 173
Miscellaneous Preparative Reactions 173
5-2. Reactions 175
Hydrolysis, Alcoholysis, and Related Reactions 175
Oxidation and Oxidative Hydrolysis 175
Disproportionation and Exchange Reactions 179
Hydroboration 181
Coordination Compounds with Lewis Bases 183
Diborane–Borane Equilibrium 186
Miscellaneous Reactions 188
5-3. Physical Properties 189
Structure 189
Nuclear Magnetic Resonance 190
Vibrational Spectroscopy 190
Mass Spectrometry 192

Chapter 6 Other Organopolyboranes

6-1. Carbon Monoxide–Triborane 193
6-2. Organotetraboranes 193
Alkyltetraborane(10) Compounds 193
Tetraborane(8)carbonyl 195
6-3. Organopentaborane(9) Compounds 196
Synthesis 196
Reactions 202
Physical Properties 203

6-4. Organopentaborane(11) Compounds 204
6-5. Organohexaborane(10) Compounds 206
6-6. Organodecaborane(14) Compounds 207
Synthesis 207
Reactions and Properties 209
6-7. Organic Derivatives of Large Closed-Cage Boron Hydrides 210

Chapter 7 **Supplementary Chemistry**

7-1. Cyclic Boron–Carbon Systems 214
7-2. Influence of the Organic Group on the Properties of the Attached Boron-Containing Moiety 218
7-3. Influence of Boron Substitution on the Properties of the Attached Organic Group 219
7-4. Analysis of Organoboron Compounds 222
7-5. Uses and Applications of Organoboron Compounds 223

Supplementary Sources of Information 225

Bibliography 230

Author Index 287

Subject Index 340

Preface

The major purpose of this book is to provide the organoboron specialist with a timely summary of the field and those working in peripheral subject areas the basic knowledge and organized introduction useful in the application of this chemistry.

The upsurge of interest in organoboron chemistry within the past two decades can largely be attributed to (a) the investment in boron research by the United States and to some extent by the USSR in attempts to obtain high-energy organoboron hydride fuels during the 1950's and to prepare high temperature-resistant polymers during the past decade and (b) the utility of organoboron compounds as intermediates in organic synthesis, a field initially developed and expanded by the group led by H. C. Brown. In addition, the potential for using organoboron compounds in biomedical research adds to the contention that organoborane chemistry will continue to expand in future years.

No attempt is made to cover in detail every aspect of organoboron chemistry as this would require a volume many times the size of this one. Only those compounds which contain a C–B bond are included. Brief mention is made of carboranes; this subject has received attention in an earlier volume in this series (R. N. Grimes, *Carboranes*, Academic Press, New York, 1970). A few sections of the review are classified in terms of reaction mechanism types; however, the lack of definitive mechanistic studies in most instances prevents this approach from being used throughout. Because of the space limitations an attempt has been made to reference many studies in tabular form in lieu of a detailed treatment within the descriptive text.

Considerable credit and thanks are due my wife, Sharon, for some of the more difficult aspects of the literature search, in particular the seeking out of foreign journal articles. I also wish to give special acknowledgment to The

Chemical Society Library, Burlington House, London, for lightening the burden of the literature review.

Thomas Onak

Organoborane Chemistry

Chapter 1

Introduction and Nomenclature

A relationship between organoboron compounds and organic species commonly found in textbooks can be appreciated when it is realized that R_3B and R_3C^+ are isoelectronic, as also are the following pairs of compounds: H_3BCO and CO_2, $RB(OH)_2$ and RCO_2H, $(MeBNH)_3$ and mesitylene; and within each pair there are found similar structural and chemical properties. The theoretical implications are manifold when considering possible aromaticity in cyclic compounds such as 1,2-borazobenzene and $C_5BH_6^-$, as well as comparative resonance interactions in the isoelectronic structural unit pair of B—C=C and C^+—C=C.

Space limitations do not permit a full examination of all permutations of multistep synthetic or degradative reactions involving organoboron compounds. Some are mentioned briefly, such as the dehydroboration–hydroboration rearrangement, hydroboration–oxidation, haloorganoborane synthesis–halogen migration, hydroboration–boron ligand exchange, e.g.,

$$\text{>C=C<} \xrightarrow{\text{HB<}} \text{HC—C—B<} \xrightarrow{\text{BCl}_3} \text{HC—C—BCl}_2$$

but mostly the reader will have to master the individual reaction types and piece together for himself possible combinations that may be useful.

Some attempt has been made to adopt prevailing IUPAC nomenclature rules (*18a*, *19*, *20*, *437*, *1453a*). Occasionally, there may be a lapse into historically useful names that have since been abandoned by the rules committee [e.g., tetraphenylboron for tetraphenylborate(1 –); methylboronic acid for dihydroxymethylborane; borazole for borazine; borane carbonyl for carbon monoxide(C-B)borane], but side by side usage of both the approved and common nomenclature should prove helpful during the present transition period.

Chapter 2

Structure and Physical Properties of Organoboranes

2-1. Boron–Carbon Bond Lengths

All trivalent boron compounds are coplaner with respect to the three atoms directly bonded to boron, whereas the four groups of the tetravalent compounds assume tetrahedral or nearly tetrahedral geometry. From the measured C—B bond distances (Table 2-1) one can observe that the C—B bond tends to lengthen (a) with decreasing *s*-orbital-character in the C—B bond, which, in effect, increases the radii of the sp^x hybrid atomic orbitals; thus alkyl-B > aryl-B > alkynyl-B, and tetravalent BR_4^- > trivalent BR_3; (b) when atoms (other than the organo group in question) bonded to trivalent boron contain an unshared pair of electrons that are available for potential back bonding; e.g., $B—\ddot{X} \leftrightarrow B^-{=}X^+$.

A factor that may foreshorten the C—B bond in arylboron compounds, in addition to orbital hybrid consideration in rule (a), is a resonance interaction between the boron and the aryl group:

$$C_6H_5—B\langle \longleftrightarrow {}^{+}C_6H_5{=}\bar{B}\langle \longleftrightarrow {}^{+}C_6H_5{=}\bar{B}\langle \longleftrightarrow {}^{+}C_6H_5{=}\bar{B}\langle$$

In Me_3B the p-π orbital of boron is unoccupied except by methyl group delocalization, a situation that invites hyperconjugative shortening of the B—C bond. For those compounds with potential back bonding to the boron from other atoms, rule (b), the p-π boron orbital is filled, or partially filled, and therefore the hyperconjugative effect from the organo groups is not encouraged.

TABLE 2-1

C—B Bond Length[a]

Compound	C–B bond length[a] (Å)	Reference
BPh_4^-	1.69 (XR)	*114a, 1569a*
	1.67 (XR)	*138*
	1.643 (XR)	*550*
	1.631–1.646 (XR)	*1709a*
$Rh[P(OMe)_3]_2[\pi\text{-}Ph—BPh_3]$	1.66 (XR)	*1453*
Me_3BNMe_3	1.65 (MW)	*647, 1116, 1117*
Me_2BNMe_2	1.65 ± 0.02 (XR)	*341*
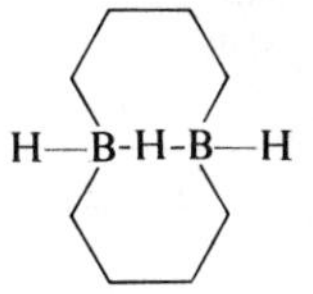 H—B-H-B—H	1.596–1.610 (XR)	*416a*
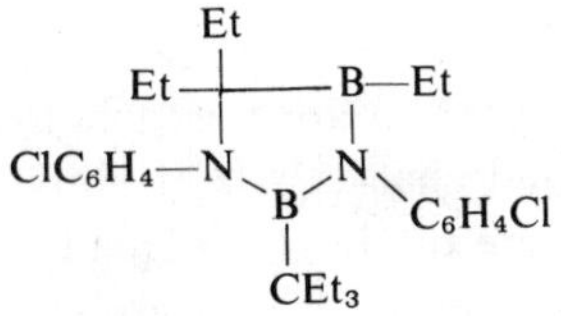 Et; Et—C—B—Et; ClC_6H_4—N, N—C_6H_4Cl; B; CEt_3	B(2)–CEt_3 1.600 ± 0.010 (XR) B(4)–C(5) 1.576 ± 0.011 B(4)–Et not accurately known	*1810a*
$\pi\text{-}(Et_6B_3N_3)Cr(Co)_3$	1.60 ± 0.03 (XR)	*850a*
$MeBF_2$	1.60 ± 0.02 (ED)	*87*
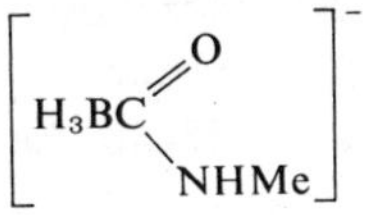 $[H_3BC(=O)NHMe]^-$	1.600	*1505*
$Me_4B_2H_2$	1.59 (ED)	*377*
$\pi\text{-}(MeBC_5H_5)_2Co$	1.594 (B—Me) (XR) 1.505, 1.522 (ring)	*850b*
$1\text{-}EtB_{10}H_{13}$	1.59 ± 0.01 (XR)	*1563*
$(mesityl)_2BN{=}CPh_2$	1.58 (XR)	*342*
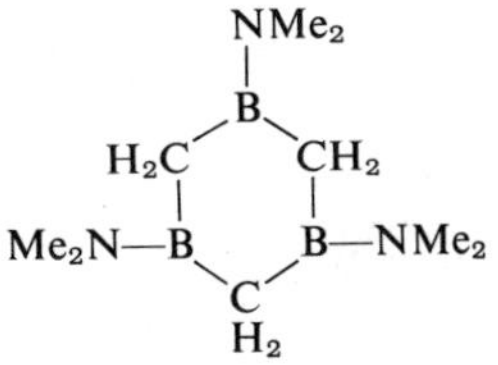 NMe_2; B; H_2C, CH_2; Me_2N—B, B—NMe_2; C; H_2	1.58 (XR)	*797*
$Cl_2BCH_2CH_2BCl_2$	1.58 ± 0.05 (XR)	*1389a*
$Me_3B_3O_3$	1.57 ± 0.03 (ED)	*86*

(continued)

TABLE 2-1 (*continued*)

Compound	C–B bond length[a] (Å)	Reference
$(t\text{-BuB})_4(NH)_8$	1.57 (XR)	*1769a*
Me_3B	1.5783 ± 0.0011 (ED)	*75*
	1.56 ± 0.02 (ED)	*1113*
$2,3\text{-}Me_2B_5H_7$	1.55 (XR)	*625, 1501*
Me_2BF	1.55 ± 0.02 (ED)	*87*
H_3BCO	1.540 (MW)	*671–673*
$p\text{-}BrC_6H_4B(OH)_2$	1.54 (XR)	*1961a, 1961b*
$Me_3B_3N_3H_3$	1.52–1.53 (XR)	*34b*
$PhBCl_2$	1.52 ± 0.07 (ED)	*425*
$\pi\text{-}(MeOBC_5H_5)_2Co$	1.526, 1.522 (ring) (XR)	*850b*
$HC\equiv CBF_2$	1.513 ± 0.005 (MW)	*1024, 1025*

[a] Method used: XR, X-ray diffraction; MW, microwave spectroscopy; ED, electron diffraction.

2-2. Boron–Carbon Bonding Characteristics

It is reasonably safe to state that in every known organoborane the carbon-to-boron link has largely "single" sigma C—B bond character. Certainly, there is little doubt about such bonds in BR_4^-, where the boron is quadrivalent; and because compounds such as $R_2B^-{=}CR_2$ and $RB{=}CR_2$ are unknown, one must turn to trivalent vinyl- or phenylboron systems to find at least partial boron–carbon double bond character. A considerable body of evidence, which includes bond distance comparisons (Section 2-1), vibrational spectroscopy (Section 2-4), and boron-11 nmr (Section 2-3), points to the suggestion of a hybrid containing some minor contribution of (**II**) to the

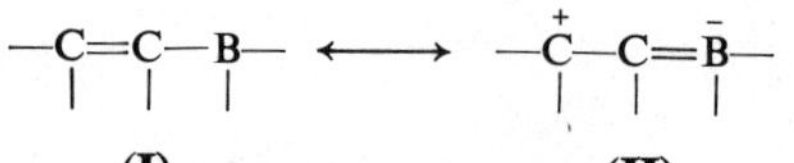

$$\underset{\textbf{(I)}}{-\overset{}{\underset{|}{C}}{=}\underset{|}{C}-\underset{|}{B}-} \longleftrightarrow \underset{\textbf{(II)}}{-\overset{+}{\underset{|}{C}}-\underset{|}{C}{=}\overset{-}{\underset{|}{B}}-}$$

vinylboron hybrid, and comparable dipolar contributions to phenylboron systems:

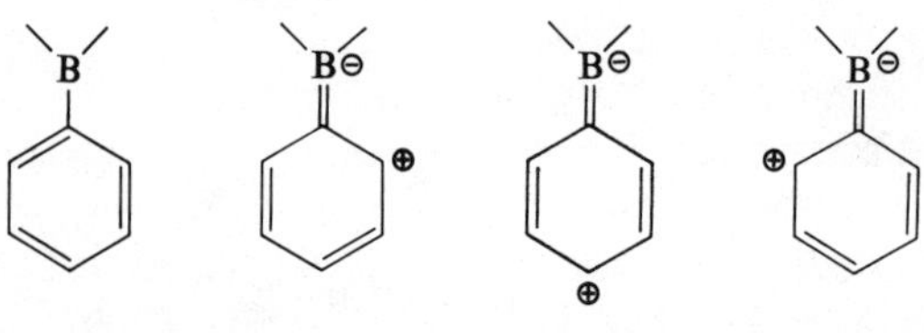

B B⊖ ⊕ B⊖ ⊕ ⊕ B⊖

Additional supporting evidence for partial double bond character in related systems includes: (a) the transmittal of resonance effects through the aromatic ring in a study of acid strengths of *o*-, *m*-, and *p*-$XC_6H_4B(OH)_2$ derivatives (*119*, *151*, *417*); (b) an LCAO-MO treatment for $CH_2{=}CHBMe_2$, which indicates a B—C bond order of 1.35 (*670*), and a modified F.E.M.O. approach to the π-electronic structure of vinylboranes lending support to the assignment of appreciable π-charge on the boron atom (*39*); (c) lower C=C stretching frequency in $CH_2{=}CHBX_2$ as compared to $CH_2{=}CHCH_2BF_2$ (*448*); (d) tetrahedral vinylboron compounds exhibit a C=C stretching bond at higher frequencies than trivalent vinylboron analogs (*448*) (however, this may be due in part to the additional formal negative charge on the tetrahedral boron); (e) a significant increase in electric moment when phenyl is substituted for *n*-amyl in $RBCl_2$, suggesting resonance contributions involving $^{+}C_6H_5{=}B^{-}Cl_2$ structures (*465*); (f) ^{19}F nmr evidence supports some π C—B bonding in vinylBF_2 (*448*, *449*) in that $J(^{11}B{-}^{19}F)$ decreases and the ^{19}F chemical shift increases in the series CH_3BF_2, *n*-$PrBF_2$, $C_2H_5BF_2$, $CH_2{=}CHBF_2$, with a substantial change in the last compound; this trend is in part attributed to less B—F π-bonding when there is more π-bonding from the organic groups; (g) carbon–boron π-bonding may be responsible for the decrease in Lewis acidity of $CH_2{=}CHBF_2$ relative to alkylBF_2 compounds (*449*) and the reduction of B—N bond order in *B*-aryl aminoboranes (*881*). The resonance energy of the $CH_2{=}CH{-}B\langle$ group is estimated at ca. 7.5 kcal mole^{-1} (*1529*).

There are a number of indications that the empty boron *p* orbital of trialkylboron compounds invite hyperconjugation with neighboring C—H bonds:

(a) A comparison of C—B bond lengths of a number organoboron compounds (see Section 2-1).

(b) The effect of a methyl substitution on the bridge bonds of diborane (*377*), i.e., the B—B distance is 0.065 ± 0.01 Å longer in $Me_4B_2H_2$ than in B_2H_6. Some CH_3—B π delocalization in $Me_4B_2H_2$ dissipates the boron *p* orbital conjugatively, thus reducing the strength of the BHB bridge bond proportionately. Steric effects, however, could also account for bridge lengthening.

(c) Although the barrier to internal rotation in $MeBF_2$ is about that expected for a potential function with sixfold symmetry (ca. 12–14 cal) (*403*, *1415*, *1918*) a barrier of 1720 cal was calculated for methyl rotation in Me_3B (*622*). In the latter compound, this high barrier might be explained by involving substantial contribution of $H^{+}CH_2{=}B^{-}Me_2$, whereas in the $MeBF_2$, back bonding from the fluorine would dissipate the boron *p* orbital toward hyperconjugation. However, much lower estimates for methyl rotational barriers in BMe_3 have since been observed (Section 2-4).

(d) Although both B_2H_6 and Al_2Me_6 are both the stable forms of BH_3 and $AlMe_3$, the occurrence of substantial hyperconjugation in BMe_3 may be partly responsible for the nonexistence of the dimer B_2Me_6 (*1396*). Trigonal conjugation should be much weaker for aluminum–carbon bonds, since only the first row elements form strong double bonds; therefore, the $AlMe_3$ monomer would not be stabilized to the same extent that BMe_3 is stabilized in a monomer–dimer equilibrium (*1396*). Other factors must also be playing some part in this, for both trialkylgallium, trialkylindium, and bulky trialkylaluminum compounds are monomeric (*1055*, *1571*).

(e) Dipole moment and quadrupole coupling constants (Section 2-9).

Although alkylboron ligand exchange and isomerization is known to occur (Sections 3-1, 3-2), RR′C*HB$\langle$ is capable of maintaining asymmetry without significant racemization over periods of several hours (*320*).

By thermal analysis, there is no evidence for association between the π-electrons of ethylene and the vacant orbital in trimethylborane (*402*); however, there is some indication that olefin–BF_3 complexes may exist (*1406*).

The threefold barrier to methyl rotation in tetramethyldiborane is ca. 1 kcal mole^{-1} (*76*). From potential energy calculations and proton nmr data, the geometries of a number of triarylboron compounds with ring methyl groups are found to involve large twist angles (ca. 45°) of the aryl group about the C—B bonds (*1865*); and in trimesitylborane and tri(2-methoxynaphth-1-yl)borane, the rings must be perpendicular to the C_3B plane (*1038*). The existence of two isomers of tri-α-naphthylborane may be attributed to the hindered rotation of such large groups about the C—B bond (*299*). The plane of aromatic rings in certain *B*-triarylborazines may well be perpendicular to the plane of the borazine ring (*103*); however, insertion of a C≡C group between the two rings releases the steric hindrance to allow a coplanar structure (*1859*). For boranes of the type $(Ar)_2BOR$ with ortho substituents (*t*-Bu on ring A and Me or H on ring B) potential barriers to B—C rotation were found to be in the region $\Delta G^{\ddagger}$ 11.5–12.5 kcal mole^{-1}. Hindrance to rotation is caused essentially by steric effects, and π–B–C_{Ar} bonding is only of minor importance (*1215a*).

2-3. Nuclear Magnetic Resonance

Boron-11

From the chemical shift data given in Table 2-2 for several organoboron compounds, it is seen that electron density at, or very near, the boron is a leading contributor to these shifts. Thus, quadrivalent boron resonances occur at considerably higher field than do the trivalent counterparts. Upfield shifts are also to be associated with groups capable of π interaction with the empty *p* orbital of a trivalent boron (e.g., vinyl, phenyl).

TABLE 2-2

Boron-11 Chemical Shifts

Compound	$\delta(BF_3\cdot Et_2O{=}0)$	Reference
BMe_3	−86.4	*1569*
	−86.0	*1488*
BEt_3	−85	*1503*
	−84.8	*1569*
	−86.5	*1488*
$ViBMe_2$	−74.5	*670*
Vi_2BMe	−64.4	*670*
BPh_3	−60	*1032*
Vi_3B	−55.2	*670*
B–Ph	−48	*50*
B^-–Ph	−26	*50*
BPh_4^-	+8.2	*1503*
	−2.1	*1569*
	+6.8	*1771*
	+6.3	*1488*
BVi_4^-	+16.1	*1771*
BEt_4^-	+16.6	*1488*
BMe_4^-	+20.5	*1771*
	+21.1	*1488*
$B(C{\equiv}CPh)_4^-$	+31.2	*1569*

The ^{11}B chemical shifts of substituted phenylboronic acids give a linear correlation with corresponding Hammett σ constants for the substituents (*90*). A correlation, $\delta = 480 - 201.5E_X + 1.79\delta_\pi$, where the first two terms represent the δ_σ contribution (E_X=electronegativity) and the last term represents the paramagnetic contribution, has been developed for the ^{11}B nmr chemical shifts of BR_4^- (R = alkyl, vinyl, phenyl, alkynyl) (*1771*). A similar correlation, $\delta = 156 - 54$ (group electronegativity), for BR_4^- seems to work satisfactorily because of the small π–π interaction in this type of compound (*1488*). Nuclear magnetic resonance studies on heteroaromatic organoboron compounds (*510*, *531*) show a ^{11}B chemical shift dependency on number of neighboring unshared electron pairs as well as on electronegativity and π bonding.

Substitution of H at boron by alkyl, vinyl, or phenyl groups results in a downfield displacement of the ^{11}B chemical shift (*515*, *1066*, *1119*). ^{11}B nmr shift data indicate that the order of conjugative ability toward boron is vinyl > cyclopropyl > alkyl (*442*). A correlation between ^{11}B and ^{13}C chemical shifts which involves pairwise interactions and assigned constitutive bond parameters (*1582*) has been successfully developed (*1581*, *1582*). A much more complete listing of ^{11}B nmr data is to be found in the monograph by Eaton and Lipscomb (see Supplementary Sources of Information).

Proton

Electronegativity values of 2.2 for the tricoordinated boron in BEt_3 and 1.9 for the tetracoordinated boron atoms in both Me_3NBEt_3 and BR_4^- have been estimated from the proton nmr chemical shift difference between methyl and methylene resonances of the *B*-ethyl groups (*828*, *1168*, *1491*, *1494*). The effect of substituting boron for carbon in an alkyl chain is to shift an α-CH proton upfield, usually ca. 0.4–0.6 ppm (*442*, *467*, *1068*, *1119*, *1384*); typical values are given in Table 2-3.

Proton nuclear magnetic resonance has been used to provide information about rotational barriers about the B—N bond in a number of aminoorganoboranes (*61*, *71*, *72*, *165*, *545*, *1627*, *1663*). The barriers are similar for both

TABLE 2-3

Proton Chemical Shifts

Compound	τ(TMS = 0)	Reference
BMe_3	9.21	*670*
	9.24	*1491*
BEt_3	α, 8.21; β, 9.05	*1491, 1494*
Me_2BVi	Me, 9.18; Vi, 3.65	*670*
$MeBVi_2$	Me, 9.14; Vi, 3.52	*670*
BVi_3	3.45	*670*
MeB_2H_5	Me, 9.47; bridge, 10.09	*1068*
$Me_4B_2H_2$	Me, 9.66; bridge, 8.90	*1068*
1-MeB_5H_8	Me, 9.96	*1815*
2-MeB_5H_8	Me, 9.66	*1815*
H_m H_o H_p—BF_2	2.45 (*m,p*); 2.78 (*o*)	*1405*
BMe_4^-	10.50	*1491*
BEt_4^-	α, 10.24; β, 9.36	*1491*

(p-$MeOC_6H_4$)B(Cl)NMe_2 and PhB(Cl)NMe_2, which probably indicates that the methoxy substituent has little effect on π bonding between the aryl group and boron or between nitrogen and boron (*72*). The pmr nonequivalence of the *N*-methyl protons in CH_2=CH—B(Me)NMe_2 over the temperature range of $-90°$ to $+70°$ suggests a simultaneous restriction of rotation about both the N—B and B—C_{vinyl} bonds. This is associated with conjugation of the entire four-atom system of the N—B—C—C grouping in analogy with the butadiene system, suggesting that some boron–carbon double bonding is probable (*1663*). Proton chemical shift data has been used to support the hypothesis that the phenyl rings in $Ph_3B_3N_3R_3$ are perpendicular to the borazine ring (*1384*).

The ^1H shifts of *B*-methyl protons of Lewis base adducts of trimethylboron are in the order $Me_3BNEt_3 < Me_3BPMe_3 < Me_3BNH_XMe_{3-X}$ ($X = 1, 2$) (*776*). With the assumption that a stronger coordinating bond shifts the B–Me proton to higher field, these results correlate well with enthalpies of dissociation (*184*). A separate study, however, indicates that proton shift data is not uniformly successful in predicting the stabilities of addition compounds (*450*).

The difference between $J(2,3)$ and $J(2',3')$ proton spin couplings in tricyclopropylborane may be attributed to conjugation between the extraannular orbitals on the cyclopropane moiety and the formally vacant $B(2p_Z)$ orbital (*443*).

Spin Coupling

For a number of Lewis base adducts of BMe_3, the $J(^{13}C—H)$ for the B-methyl group is found to be consistently in the region of 109 Hz (*776*), which indicates about 22% *s* character for the orbital contributed by the carbon to the C—H bond, and consequently ca. 34% *s* character contributed toward the C—B bond.

The absence of two-bond coupling, $^{11}B—C—H$, in Me_2BX, $MeBX_2$, and $(MeBO)_3$ is attributed to the high quadrupole moment of boron (*518*). Such coupling has been observed, however, in quaternary boron compounds such as $LiBMe_4$ (*1168a*) and several other compounds given in Table 2-4.

2-4. Vibrational Spectroscopy

As might be expected, the metal–carbon $\nu_3(a_1')$ stretching frequency and force constant for the series Me_3M (M = B, Al, Ga, In) is highest for Me_3B (*1493*). The force constant, k(BC), for CH_2=$CHBBr_2$ is substantially greater than that of $MeBBr_2$; also k(CC) of CH_2=$CHBBr_2$ is clearly smaller than the

TABLE 2-4

Spin Coupling

Compound	Reference
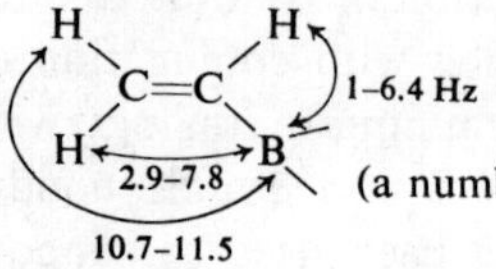 H(cis)–B 1–6.4 Hz; H(gem)–B 2.9–7.8; H(trans)–B 10.7–11.5 in H₂C=CH–B (a number of vinyl compounds)	*141, 141a, 141b*
$CH_3CH_2BEt_2$ (7)	*1168, 1494*
$CH_3CH_2BEt_3^-$ (3.0; 0.8, 4.0)	*1168*
1-H_3C—B_5H_8 (^{13}C–$^{11}B(1)$, 73 Hz; H,B(1), 6–7 Hz)	*580, 1067*
$CH_3BMe_3^-$ (3.9)	*1168a*
$Me_3CNCBMe_3$ (40)	*58a, 382*
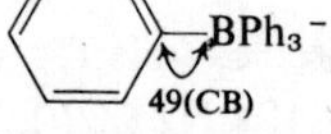 $C_6H_5BPh_3^-$ (49(CB))	*58a, 1862a, 1862b*

corresponding constant of ethylene or the double bond in propylene. Both observations point to the existence of electron delocalization in vinyldibromoborane (*1446*). The B—C force constant for H_3BCN^- (4.56 mdynes Å) is considerably larger than that for H_3BCO (2.78) (*114*, *440*, *1750*, *1764*, *1765*), whereas k(B—H) for H_3BCN^- (3.00) is smaller than that of H_3BCO (3.31) (*114*). This is taken to indicate that the adducts with strong donors force greater p character into the B—H bonds than do adducts of the weaker donor group (*114*). Other B—C force constants are given in Table 2-5.

The *B*-aryl stretching vibration of $(p\text{-}XC_6H_4)_4B^-$ ions (force constant of 1.75×10^5 dynes cm^{-1}) is primarily dependent upon the mass of the para-substituted phenyl ring and varies between 506 (X = Br) and 565 cm^{-1} (X = OCH_3) (*1832*). From infrared and Raman spectral data, the barrier to rotation about the boron–carbon link of $PhBCl_2$ is calculated as ca. 45 kcal

TABLE 2-5

Force Constants for B—C Bonds

Compound	k (mdyne Å^{-1})	Reference
Me_3B	4.09	*1493*
	3.64	*95, 675, 1699*
Me_2BCl	3.64	*95*
$MeBCl_2$	3.10	*95*
Me_2BNH_2	3.73	*105*
$ViBBr_2$	4.23	*1446*
$MeBBr_2$	3.60	*1643*

mole^{-1}, signifying considerable π character in the bond (*1129*). Internal methyl group rotation in Me_3B is essentially unhindered, being less than 30 cal/°C (*1739, 1924*).

Stretching frequencies for the B—C bond of a number of organoboranes are given in Table 2-6; a number of caveats appear certain but no attempt is made to give a confidence factor to each assignment.

2-5. Electronic Transitions

An absorption at 1770 Å in the uv spectrum of Me_3B is assigned to the hyperconjugative $n_B \leftarrow \pi_{CH}$ transition in which an electron is promoted from a C—H σ bond to the empty p_Z orbital of boron; and weak bonds found between 2250 and 1950 Å are assigned to the $n_B \leftarrow \sigma_{B-C}$ transitions (*1586*). The latter is also probably responsible for the end absorption observed for a number of trialkylboranes (ca. 215 nm, ϵ = 215–240) (*484, 1585, 1586, 1611, 1612*), in which the transition energy is expected to be proportional to the ionization of the hydrocarbon (*1585, 1586*).

Although the saturated tri-*n*-butylborane shows a uv maximum at 205 nm only, the unsaturated tri-3-butenyl and tri-4-pentenylborane give significant absorption at ca. 240 nm. Because a similar effect can be observed when increasing the concentration of $(n\text{-}Bu)_3B$ in an alcoholic solution of 1,1-phenylethylene, an interaction between the boron atom and the ethylenic π electrons is proposed (*1143*). An alternate explanation attributes the longer wavelength absorption to an ROB$\langle$ impurity (*484, 1611*). 2-Methyl-1-propene-1-boronic acid shows an absorption maximum at 223 nm (ϵ = 1000) in cyclohexane. As neither alkenes nor alkaneboronic acids absorb in this region, or at longer wavelengths, this property suggests resonance interaction involving the π electrons of the carbon–carbon double bond and the vacant orbital of the boron atom (*1105*).

TABLE 2-6

Boron–Carbon Stretching Frequencies

Compound	Asym (cm^{-1})	Sym (cm^{-1})	Reference
Me_3B	1116–1189	675–680	*371, 675, 1080, 1084, 1493, 1699, 1924*
	1298	620	*134*
	1295–1309	675	*1739*
Et_3B	1116–1150	620–675	*1081, 1084*
Me_2BF	1152	715	*98*
Me_2BNH_2	1187	705	*99, 105*
Me_2BNMe_2	1163, 1169	645	*99, 102*
$Me_2BNMePh$	1120	665	*102*
$MeBF_2$	775–777		*98, 869a, 1600*
	910–926		—
$MeBCl_2$	1010		*95*
Alkyldiboranes	1100–1150	516–844	*1078, 1371, 1602*
Alkylboroxines	1160–1265		*56*
Ph_2BX	1166–1375	571–671	*107, 1405*
PhMeBNMePh	875, 1215	680	*102*
Ph_2BNMe_2	1195	680	*102*
$PhB(OR)_2$	1130–1432		*344, 362*
$PhBX_2$	1208–1340	585–646	*37, 361, 695, 1130, 1405*
$(PhBO)_3$	1087		*56*
$(CF_2{=}CF)_3B$	1363	—	*1734, 1738*
$(CF_2{=}CF)_2BCl$	1361	—	*1734*
$CF_2{=}CFBCl_2$	1350		*1734*
H_3BCO	691–692		*116, 440, 1750*
$H_2BCO_2^-$	855		*1152*
H_3BCN^-	875–890		*113, 114*
$(Aryl)_4B^-$	506–565		*1832*
$ViBF_3^-$	1107, 1132		*1733*

The electronic spectra of several vinylboranes (Me_2BVi, 195, 228 nm; MeClBVi, 204, 228 nm; $MeBVi_2$, 201, 221, 253 nm) (*44*, *670*) are used with SCF-MO calculations to estimate B—C bond orders and charges on individual atoms (Section 2-7). In all of the compounds, considerable π-electron density is found to reside on the boron (*44*).

The ultraviolet spectra of tribenzylboranes of the type $(p\text{-}XC_6H_4CH_2)_3B$ (X = H, F, Me, —OMe), in addition to the benzenoid ^1L_b transition maxima, exhibit a medium intensity absorption in the region 240–285 nm, which is assigned to an intramolecular charge transfer from an aryl group to the boron vacant $2p$ orbital (*1588*, *1589*).

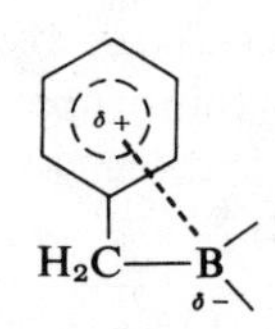

The light absorption by triorganoboron compounds containing two aromatic groups surpasses the additive absorption of light by two noninteracting aromatic groups (*1450*); and it is also found that trivalent boron atoms interact more strongly with the electronic transitions of attached phenyl rings than with an attached vinyl group (*1449*).

	λ_{max} (nm)	ϵ_{max}
Ph_3B	240	10,850
	270	2,230
Ph_3BNH_3	250–265	1,600

Intramolecular charge-transfer transitions have been identified in the ultraviolet spectra of several triarylboranes (*1585*, *1587*, *1590*, *1591*), and a polar excited state species is also proposed to account for the fluorescent properties of para-substituted phenyl(dimesityl)boranes (*561*) as well as for the colors of diphenylbipyridylylboronium salts (*69*). The twisting of B—Ph bonds up to 60% does not appreciably affect π-electronic transitions for Ph_2BCl and Ph_2BOH (*38*).

2-6. Mass Spectrometry

Molecular ion, Me_3M^+, abundances for the series of Me_3M compounds follows the order M = Al > B > Ga > In > Tl (*669*). The most prominent peak for Me_3B (Table 2-7) corresponds to Me_2B^+; for Et_3B; two prominent peaks are observed, one corresponding to Et_2B^+ and the other at mass 41, which arises from either $HBEt^+$ or from Me_2B^+ formed in some secondary process (*779*, *1065*, *1082*, *1683*). The stability of R_2B^+ is also manifested in the form of boronium salts (Section 4-3), which have been isolated and characterized as the bis-solvated species of this ion.

The mass spectra of both trimethylboroxine (*1089*) and trimethylborane have formal similarity in that the principal peaks in both represent the loss of one methyl group.

With phenylboronic acid, significant C—B bond cleavage occurs giving both $C_6H_6^+$ (by rearrangement) and $(HO)_2B^+$ ions (*460*). Tropilium ion has been identified as a mass spectral fragment of a number of phenylboronic esters and is obviously a result of carbon transfer from the B—O—C portion of the molecule to the phenyl ring (*147*, *430*, *458*). Metastable ions in the mass spectrum of phenyldioxaborolane, $Ph\overline{BOCH_2CH_2O}$, indicate that the $C_7H_7^+$ is formed directly from the parent ion and indirectly via the species $PhBOCH_2^+$ (*457*, *458*). This electron impact-induced rearrangement to form tropilium

TABLE 2-7

Mass Spectra of BMe_3 and BEt_3: Natural Isotopic Abundance

BMe_3 m/e	BMe_3 Relative abundance	BEt_3 m/e	BEt_3 Relative abundance
56	1.4	98	3.7
50	1.7	70	2.2
49	1.6	69	68.2
42	2.2	68	17.2
41	100.0	67	1.1
40	25.5	54	1.4
39	10.7	53	5.6
38	3.9	52	1.9
37	8.5	51	2.7
36	3.1	50	2.5
27	7.4	49	1.3
26	4.8	42	2.3
25	1.6	41	100.0
15	3.0	40	31.4
13	2.4	39	21.2
11	3.5	38	6.5
		37	11.6
		36	3.2
		29	2.6
		28	2.1
		27	22.7
		26	7.8
		15	2.2
		13	2.8
		11	1.2

ion from phenylborolanes containing oxygen and/or sulfur as the heteroatoms can be rationalized in terms of the relative bond energies between the atoms in the heterocyclic ring (*452*). Acyclic and cyclic derivatives of organoboranes containing at least one boron–nitrogen bond; e.g., $PhBOCH_2CH_2NH$ give peaks in their mass spectra assignable to boracyclopentadienyl ions (**III**) as well as boratropylium ions (**IV**) (*430, 459*).

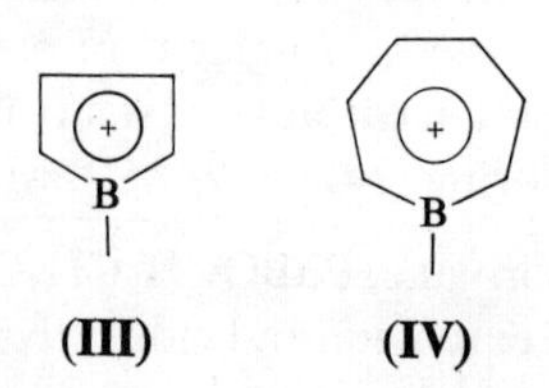

(**III**) (**IV**)

TABLE 2-8

Electron Impact Studies on RBF_2 Compounds

Compound	$D(R—BF_2)$ (kcal $mole^{-1}$)
$Vi—BF_2$	111
$Me—BF_2$	95–100
$Et—BF_2$	101
i-$Pr—BF_2$	96

N-Methyl substitution has a greater effect than *B*-methyl substitution on the ionization potential of substituted borazines (*1023*), an observation which is also predicted from MO calculations (*182*).

The molecular beam mass spectrum of H_3BCO is shown to have fragments of the type BH_xC^+, BH_yCO^+, and BH_x^+ ($x = 0, 1, 2$; $y = 0, 1, 2, 3$) (*788*). Evidence for HBC_2 was found from a mass spectrometric investigation of the high temperature reaction of hydrogen with boron carbide (*1736*). From electron impact studies on RBF_2 compounds (Table 2-8), the increase in B—C bond strength for the vinyl derivative is in part attributed to the change in hybridization of the bonded carbon atom but also is probably related to vinyl–boron π interaction (*1737*) as indicated by other work (*448*) (Section 2-2).

2-7. Molecular Orbital Calculations

Ab initio SCF-MO calculations suggest that C—B π bonding in Me_3B contributes less than 10% of the total bonding (*722*); and from an extended Hückel treatment, the π-bond character of Me_3B is about nine times that of Me_3Al, which helps to explain the stability of the trialkylboron monomers (cf. Me_6Al_2) (*1494a*). Also, by the latter method it has been demonstrated that positive charge on the central boron atom decreases as hydrogen replaces alkyl in the R_3B series (Table 2-9) (*1534*) and as alkyl replaces chlorine in the series $BCl_3 > RBCl_2 > R_2BCl > R_3B > R_2BSPh$ (*1533*).

TABLE 2-9

Electron Densities, Extended Hückel Method

Compound	Positive charge on B	Negative charge on H
BH_3	0.495	0.165
RBH_2	0.65	0.184
R_2BH	0.81	0.197
R_3B	0.98	—

It has been estimated that a number of vinylboranes—e.g., Vi_3B, $MeBVi_2$, Me_2BVi, Vi_2BX, $ViBX_2$ ($X =$ halogen)—have a modest π charge residing on the boron atom; and in the first three compounds of this series, this is a combined result of π interaction of the vinyl group and hyperconjugative contributions from the methyl groups, whereas in the halogenated molecules the halogens contribute considerably more π-electron density to the boron atom than the vinyl group (*39*, *43*, *44*, *670*).

Electronic spectral changes, as well as MO calculations, on both phenylboronic acid and triphenylboroxine support an extended conjugation throughout the boroxine (*148*, *1637*); but the phenyl group probably supplies little π-electron density to the boron as compared to the oxygen atoms (*41*).

Calculations on substituted aminoboranes (*392*, *881*, *1562*) correlate well with infrared stretching frequencies and ^{11}B nmr chemical shifts (*392*) and indicate that the B—N π-bond order for aminodimethylboranes is less than that for the hypothetical aminoborane (*392*), and that the B—N bond order is also slightly reduced by *B*-aryl substitution (*881*).

The aromatic character, π-electron density, and overall stability has been assessed for a considerable number of carbon–boron–nitrogen heterocycles and related compounds (*509*, *528*, *535*, *538*, *544*, *546*, *812*, *813*, *882*, *1147*) and several attempts have been made to correlate this information with known chemical properties (see Section 7-1).

2-8. Thermodynamic Properties

Considerable difficulty has been experienced in obtaining reliable heats of formation and other thermodynamic data for organoboranes (Table 2-10) (*604*, *639*, *640*, *640a*, *745*, *864*, *1135*, *1162*, *1574*, *1614*, *1710*); and as might be expected, this is largely attributed to the problems associated in knowing the exact nature of the obtained products using chemical techniques normally employed in gathering basic thermodynamic data. Not infrequently with organoboron compounds, there are suspicions of an incomplete reaction in the combustion approach. Even the more reliable heats of hydroboration of olefins in ethers are subject to the criticism that the products are not solely the result of boron addition to the least hindered carbon (Section 3-1).

Derived mean B—C bond dissociation energies for the simple trialkylboranes lie between 80 and 90 kcal mole^{-1}; the values for a given compound vary rather widely depending upon the source of thermodynamic data used for the calculations, but perhaps the bond energies shown by the following tabulation are to be accepted for the present (*603*, *1709d*).

Molecule	$\bar{D}$(B—C) (kcal mole^{-1})
BMe_3	86.8 ± 2.7
BEt_3	81.7 ± 2.7
B(*n*-Bu)$_3$	82.1 ± 2.7
$(C_6H_{11})_3B$	88

Bond strengths of phenylboron compounds (see tabulation below) have been analyzed in terms of resonance stabilized contributions, and it is concluded that the π contribution diminishes with increasing number of phenyl groups about the central boron (*1603*).

	$PhBCl_2$	Ph_2BCl	$PhBBr_2$	Ph_2BBr	Ph_3B
E(B—C)	116	112	115	110	106
E(B—C)$_\pi$	28	24	27	22	18

The secondary H—C bond energy of BEt_3 has been estimated at 80 ± 3 kcal mole^{-1}. This is rather low when compared to the C(2)—H of propane (ca. 95 kcal mole^{-1}) and might be explained in terms of stabilization of the unpaired electron by the adjacent vacant boron orbital (*712*).

TABLE 2-10

Heats of Formation of Organoboranes (Gaseous State)[a]

Molecule	H_f° (kcal mole^{-1}, 25°C)
BMe_3	−29.3 ± 5.5
BEt_3	−36.5 ± 2.5
B(*n*-Bu)$_3$	−67.7 ± 2.6
B(*i*-Bu)$_3$	−66.9 ± 1.3
B(*s*-Bu)$_3$	−58.7 ± 6
B(*n*-C_8H_{17})$_3$	−124.5 ± 2
n-Bu_2BCl	−86.7 ± 3
n-Bu_2BBr	−71.3 ± 2.7
n-Bu_2BI	−53.3 ± 2.6
n-Bu_2BOH	−130.6 ± 4
$MeBF_2$	−199 ± 3
$EtBF_2$	−209 ± 8
i-$PrBF_2$	−212 ± 8
$ViBF_2$	−171 ± 8

[a] From Ref. *1709d.*

2-9. Miscellaneous Properties

Trimethylborane does not exhibit a dipole moment, a result which is expected from the trigonal planar geometry (*1747*). Dipole measurements and calculated molecular moments of several methyl boron derivatives suggest that hyperconjugative resonance contributions of the type

$$\overset{-}{\mathrm{B}}{=}\mathrm{CH_2}\ \mathrm{H^+}$$

are not to be ignored (*519*).

From nqr measurements on trialkylborons the ^{10}B and ^{11}B quadrupole moments are $+0.111$ and $+0.053 \times 10^{-24}\ cm^2$, respectively (*516, 517*). The ^{11}B quadrupole coupling constants of the simple aliphatic symmetrical trialkylboranes appear to be a function of the number of α-C—H bonds in these compounds, for the average value of e^2Qq_{zz} in BEt_3, $B(n\text{-}Pr)_3$ and $B(n\text{-}Bu)_3$ is 125 ± 4 kHz sec^{-1} higher than that in BMe_3 (4.876 ± 0.004 MHz sec^{-1}), and the quadrupole coupling constant in $B(i\text{-}Pr)_3$ is increased by an additional 249 ± 4 kHz sec^{-1} above the values of the primary alkyl boron compounds. The α-secondary hydrogen isotope effect in BMe_3 increases e^2Qq_{zz} by 28 ± 4 kHz sec^{-1}. These results are in seemingly good agreement with classic carbon–hydrogen bond delocalization theory in which α-C—H orbital electron density overlaps the vacant p_z orbital on boron (*1140*). ^{35}Cl nqr measurements on a number of *B*-halogenated phenylboron compounds (*1715, 1716*) show a regular drop in the chlorine nqr frequency in the series BCl_xPh_{3-x} (*1715*). Such a trend is expected from changes in boron electronegativity as seen by chlorine as the latter are replaced by phenyl groups (*1715*).

Comparison of the photoelectron spectra of trivinylboron and triethylboron suggests there is only limited conjugation in the former. If the boron is sp^2 hybridized, it is suggested that the vinyl groups are twisted out of the plane leading to reduced conjugation. Also, Dreiding models indicate there is some nonbonding interaction between hydrogens on adjacent vinyl groups in a completely planar structure (*824*).

Evidence of a fluorescent enhancement of the BO_2 band at 5450 Å was found during the reaction of triethylboron with atomic oxygen (*1688*). Bond and molar refractivities as well as molar volume relationships have been developed (*408, 409, 1332, 1638, 1862*); and boiling points calculated from Kinney boiling point constants agree reasonably well with those observed for a number of organoboron compounds (*33*).

Chapter 3

Three-Coordinate Organoboranes

By far, the largest number of known organoboron compounds contain trivalent boron. The geometry of the groups about the central boron atom are arranged in a trigonal planar fashion with the potential of back bonding to the empty p boron orbital from adjacent groups containing an unshared electron pair or a conjugated π-bond

3-1. Boron–Carbon Bond Construction

Transmetallation

Transmetallation refers to the transfer of an organic group (alkyl, alkenyl, alkynyl, aryl) from a metal to boron by the reaction of an organometallic reagent with an appropriate boron compound, e.g.:

$$3\ RMgX + BCl_3 \longrightarrow R_3B + 3\ MgXCl \tag{1}$$

Besides the Grignard reagent common organometallics include RLi, R_3Al, R_3Al, R_2Zn, and R_4Sn. Most often, the boron compound contains at least one halogen atom or alkoxy group. With those having a B—Cl, B—Br or B—I bond, care must be taken to avoid the presence of C—O linkages elsewhere (such as an ether solvent), for such groups can react rapidly with these boron substrates. The alkylating power of organic aluminum derivatives decreases in the series $BF_3 > BCl_3 > BBr_3 > BI_3$, and the tendency to form BR_3 rather than organohaloboranes R_2BX or RBX_2 decreases in the order $BF_3 > BBr_3 > BI_3 > BCl_3$ (*1487*). Transmetallation of $(n\text{-BuO})_3B$ with aryl metallic reagents appears to be most facile with derivatives of the most electropositive metals, ArMgBr > $CdAr_2$ > $ZnAr_2$ > $HgAr_2$ (*661*). With mixed organotin compounds, it has been found that a trans-cinnamyl group is transferred more rapidly to BBr_3 than either ethyl or phenyl groups (*1763*). Often the equilibrium in a transmetallation reaction can be favorably shifted by selecting the substituent so that, for instance, a low-boiling R_3B can be pumped off as the reaction progresses (*1939*). A few examples of transmetallation reactions are shown in the following equations, and a more complete listing of reagent combinations that lead to transmetallation is given in Table 3-1.

$$Me_4Sn + 2\,BF_3 \xrightarrow[(358)]{} MeBF_2 + Me_3SnBF_4 \quad (2)$$

$$n\,Me_3Al + B(OMe)_3 \xrightarrow[(381)]{} Me_3B + [Al(OMe)_3]_n \quad (3)$$

$$Ar_2Hg + BX_3\ (X = Cl, Br) \xrightarrow[(661,\ 1236)]{} ArBX_2 + ArHgX \quad (4)$$

$$Al_2Me_6 + 4\,B_2H_6 \xrightarrow[(1653)]{} 2\,BMe_3 + 2\,Al(BH_4)_3 \quad (5)$$

$$Vi_2Zn + 2\,BF_3 \xrightarrow[(169)]{} 2\,ViBF_2 + ZnF_2 \quad (6)$$

$$3\,CpK + Cl_3B_3N_3Me_3 \xrightarrow[(1768)]{} Cp_3B_3N_3Me_3 + 3\,KCl \quad (7)$$

$$6\,ArMgX + (MeOBO)_3 \xrightarrow[(1575)]{} 3\,Ar_2BOMgBr + 3\,MeOMgBr \quad (8)$$

$$(n\text{-}BuBO)_3 \xrightarrow[(1564)]{Me_3Al\ \text{or}\ MeMgX} n\text{-}BuBMe_2 \quad (9)$$

$$\text{2,3,4,5-tetraphenyl-1,1-dimethylstannole (Ph, Ph, Ph, Ph; Sn; Me, Me)} \xrightarrow[(579)]{PhBCl_2} \text{1,2,3,4,5-pentaphenylborole (Ph, Ph, Ph, Ph; B–Ph)} \quad (10)$$

$$X_2B{-}\overset{|}{\underset{|}{C}}{-}\overset{|}{\underset{|}{C}}{-}BX_2 \xrightarrow[(390)]{Me_2Zn} Me_2B{-}\overset{|}{\underset{|}{C}}{-}\overset{|}{\underset{|}{C}}{-}BMe_2 \quad (11)$$

$$4\,H_3BNR_3 + 3\,MR \xrightarrow[(926)]{} 3\,MBH_4 + 4\,NR_3 + BR_3\ (M = Li, Na, K) \quad (12)$$

$$3\,Me_4Pb + BCl_3 \xrightarrow[(817)]{} 3\,Me_3PbCl + Me_3B \quad (13)$$

$$HC{\equiv}CMgBr \xrightarrow[(1184)]{B(OR)_3} HC{\equiv}CB(OR)_2 \quad (14)$$

$$2\,RMgX + Cl_2BNR'_2 \xrightarrow[(1434)]{} R_2BNR'_2 + MgX_2 + MgCl_2 \quad (15)$$

$$BrCH_2CH_2B(OR)_2 + HC{\equiv}CMgBr \xrightarrow[(1186)]{} HC{\equiv}CB(OR)_2 + MgBr_2 + C_2H_4 \quad (16)$$

With a number of transmetallation reactions, e.g., Eqs. 17–19, it is possible to obtain intermediate mono- and diorganoboron products. And as anticipated, these products are favored over the triorganoboron product when low

TABLE 3-1

Transmetallation Reagent Combinations

Group X to be replaced by R in —B—X	Organometallic reagent[a]	Reference
H	RLi	*926, 1567, 1576*
R′	RLi	*367a, 663, 1635*
OR	RLi	*171, 392, 478, 662, 667, 1065a, 1076, 1094, 1096, 1103, 1105, 1108, 1110, 1147, 1148, 1241, 1242, 1244, 1250, 1323, 1328, 1330, 1354, 1423, 1447, 1635, 1668, 1729, 1730, 1794, 1833*
F	RLi	*959, 1232, 1680, 1789, 1798, 1906*
SR′	RLi	*1297*
Cl	RLi	*115, 154, 406, 427, 700, 889, 946, 947, 1046, 1164, 1166, 1167, 1223, 1249a, 1383, 1466, 1468, 1470*
Br	RLi	*406, 579, 1529*
R′	R_2Be	*418*
H	RNa	*926*
F	RNa	*49, 962*
Cl	RNa	*725, 962, 999, 1451*
Br	RNa	*1527, 1529*
H	R_2Mg	*85, 926*
H	RMgX	*100, 161, 163, 512, 1066, 1231, 1379, 1713, 1884*
R′	RMgX	*22, 1186, 1207, 1939*
OR	RMgX	*93, 422, 438, 441, 568, 617, 620, 664, 667, 754, 784, 866, 883, 920, 1040, 1043, 1091, 1096, 1102, 1104, 1106, 1107, 1171, 1181, 1184, 1185, 1187, 1239, 1276, 1292, 1325, 1328, 1337, 1339, 1352a, 1353, 1356, 1357, 1359, 1427, 1455, 1456, 1557, 1564, 1575, 1619, 1645, 1664, 1712, 1719, 1722, 1729, 1730, 1769, 1785, 1789–1791, 1799, 1850, 1851, 1921, 1923, 1931, 1944, 1945*
F	R_2Mg	*1426*
F	RMgX	*183, 203, 385, 489, 505, 650, 653, 781–784, 880, 992–994, 996, 997, 1001, 1143, 1389, 1421, 1422, 1426, 1450, 1579, 1614, 1616, 1640, 1670, 1781, 1784, 1789, 1798, 1831, 1851, 1906*
Cl	RMgX	*145, 416, 543, 586, 667, 707, 740, 781, 782, 784, 888, 1166, 1217, 1218, 1220, 1223,*

(*continued*)

TABLE 3-1 (*continued*)

Group X to be replaced by R in —B(—)—X	Organometallic reagent[a]	Reference
		1230, 1301, 1383 1434, 1438, 1565, 1629, 1729, 1745, 1778, 1781, 1859, 1883, 1923, 1943
Br	RMgX	*1177*
I	RMgX	*742*
H	R_3Al	*1653*
R′	R_3Al	*129, 936, 948, 951, 1939, 1941*
OR′	R_3Al	*48, 381, 858, 932, 933, 935, 980, 1564, 1937*
OR′	$RAlCl_2$, R_2AlCl	*1944*
OR′	$RAlBr_2$, R_2AlBr	*1937*
F	R_3Al	*928, 932, 1487, 1938*
F	$RAlCl_2$, R_2AlCl	*1090, 1944*
Cl	R_3Al	*1487, 1922, 1940*
Cl	$RAlCl_2$, R_2AlCl	*74, 858*
Br	R_3Al	*1487*
I	R_3Al	*1487*
NR_2	R_4Si	*660*
Cl	R_4Si	*115, 588*
H	RK	*926*
OR′	RK	*1768*
R′	R_4Ti	*1946*
F	R_2Zn	*169, 338, 1789*
R′	R_2Zn	*1939*
OR′	R_2Zn	*616–620, 1937*
Cl	R_2Zn	*78, 390, 421, 670, 1632, 1740, 1827*
I	R_2Se	*1702*
Cl	RAg	*115, 159, 391, 588, 724*
F	R_4Sn	*169, 358, 396, 397, 1606, 1735*
Cl	R_4Sn	*58, 68, 168, 169, 346, 396, 397, 579, 603, 874, 877, 1111, 1436, 1490, 1735, 1802*
Br	R_4Sn	*50, 626, 655, 874, 1446, 1490, 1763*
I	R_4Sn	*1473, 1490, 1660*
H	R_2Hg	*164*
H	RHgCl	*164*
H	RHgBr	*164*
F	R_2Hg	*77, 169*
Cl	R_2Hg	*78, 396, 661, 1234, 1236, 1239*
Cl	RHgCl	*649, 984*
Br	RHgCl	*649*
Br	RHgI	*1177*
H	R_4Pb	*818*
Cl	R_4Pb	*817*

[a] R can be, variously, alkyl, alkenyl, or aryl in many, if not most, instances.

ratios of organometallic reagent to boron substrate are used. In some instances, it is convenient to add a Lewis base to trap a partially transmetallated

$$BCl_3 \xrightarrow[(1735)]{(CF_2{=}CF)_2SnMe_2} CF_2{=}CFBCl_2 + (CF_2{=}CF)_2BCl + (CF_2{=}CF)_3B \quad (17)$$

$$3\,BCl_3 + R_3Al \xrightarrow[(1940)]{} 3\,RBCl_2 + AlCl_3 \quad (18)$$

$$(RO)_3B \xrightarrow[(1065a)]{RLi} (RO)_2BR \quad (19)$$

product as a complex (Eq. 20) (*1102*, *1107*, *1109*, *1171*). In certain cases (e.g., Eq. 21) it has been found impossible to transfer more than one organo

$$(i\text{-}BuO)_3B \xrightarrow{PhMgX} \xrightarrow{NH_3} Ph_2BOBu \cdot NH_3 \quad (20)$$

$$BF_3 \xrightarrow[(169)]{(CH_2{=}CH)_2Hg} CH_2{=}CHBF_2 \quad (21)$$

group from the metal to boron (*169*, *1735*). This appears to be a problem primarily when preparing vinyl boron compounds from vinylmetallics and BF_3 and can be rectified by using BCl_3 instead. The reluctance of BF_3 to be di- and trivinylated is attributable to two factors, (a) the stability of $ViBF_2$ toward disproportionation and (b) an unfavorable rate, or equilibrium, for the reaction of $ViBF_2$ with the vinylmetallic reagent.

The relative reactivities of bond types within boron substrates in transmetallation reactions follow the general order B—Cl > B—OR > B—NR_2, and B—C > or < B—OR, depending on structural features (*422*, *586*, *601*, *1043*, *1186*, *1187*, *1301*, *1433*, *1434*, *1440*, *1465*, *1466*, *1468*, *1945*). Virtually no definitive work has been conducted on the mechanism of the transfer of organic groups from metals to boron. For present purposes, however, it is helpful to mention at least two possible general routes, Eqs. 22 and 23.

$$BX_3 + R_nM\ (M = \text{metal}) \longrightarrow \left[X_2B(\mu\text{-}R)(\mu\text{-}X)MR_{n-1}\right] \longrightarrow RBX_2 + R_{n-1}MX \quad (22)$$

$$BX_3 + R_nM \longrightarrow [RBX_3^-][R_{n-1}M^+] \longrightarrow RBX_2 + R_{n-1}MX \quad (23)$$

Obviously, Eq. 22 should be favored with less polar solvents (or absence of solvent), and Eq. 23 is favored when the metal can easily assume a full positive charge (*399*, *400*, *865*, *1244*, *1323*, *1354*, *1850*, *1851*). It should be noted that replacement of magnesium with boron in certain steroidal Grignard reagents proceeds with retention of configuration (*161*), a fact that must be taken into account for any proposed mechanism. Isomerism of a bulky alkyl group may also take place in the course of transmetallation (*784*)

$$3\,t\text{-}BuMgCl + BCl_3 \longrightarrow i\text{-}Bu_3B + 3\,MgCl_2 \quad (24)$$

but this may be in part attributed to the ease with which organoboron compounds can undergo a dehydroboration–hydroboration sequence (Section 3-2).

Diphenylmercury does not transmetallate B_2F_4 in a normal manner; instead, free mercury and phenylboron difluoride are formed (*825*).

$$B_2F_4 + Ph_2Hg \longrightarrow 2\,PhBF_2 + Hg \tag{25}$$

Redistribution and Exchange Reactions

This section is confined to those ligand–ligand exchange reactions between boron containing molecules in which B—C bonds are formed, e.g.,

$$BX_3 + BR_3 \rightleftharpoons R_2BX + RBX_2 \tag{26}$$

$$R_2BX \rightleftharpoons R_3B + RBX_2 \tag{27}$$

Obviously, in the process of creating a B—C bond by this type of redistribution reaction, another B—C bond is severed, and thus this section also covers some of the subject matter that falls under Section 3-2.

The nature of R in the general exchange reaction may be alkyl, alkenyl or aryl, and X may be any one of the groups listed in Table 3-2, H, R, NR_2, OR, F, SR, Cl, Br, and I. Factors that either slow or prevent this type of exchange reaction from occurring include:

(a) The presence of bulky groups, e.g., *t*-butyl attached to boron (*782, 784, 1207*).

TABLE 3-2

X—B< Bond Exchange with R—B< Compounds

X	Reference
—H	*263, 264, 304, 977, 1137, 1213, 1277, 1279, 1343, 1431, 1650, 1652, 1654, 1830, 1843, 1876*
—alkyl or aryl	*130, 343, 581, 782, 783, 938, 958, 1207, 1333d, 1344, 1346, 1527*
—NR_2	*421, 814, 943, 963, 1433*
—OR	*4, 48, 331, 525, 682, 780, 814, 848, 935, 936, 938, 1203, 1279, 1318, 1320, 1343–1348, 1352, 1891, 1944*
—F	*98, 343, 414, 938, 1329, 1812*
SR	*353, 1309, 1332a, 1344, 1890*
Cl	*6, 343, 414, 665, 666, 723, 822, 938, 945, 947, 985, 1044a, 1059, 1205, 1220, 1307, 1390, 1431, 1432, 1444, 1883, 1931–1936*
Br	*6, 938, 1390, 1643*
I	*938, 1706*

(b) Formation of an internal Lewis base–boron coordinate bond, surrounding the boron with four pairs of electrons (*153, 1102, 1107, 1109, 1542*), e.g.:

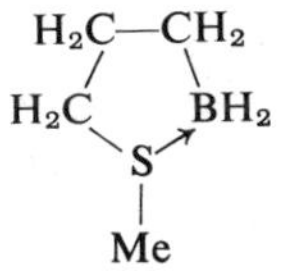

$$\begin{array}{c} H_2C\text{—}CH_2 \\ H_2C \quad\quad BH_2 \\ \searrow S \nearrow \\ | \\ Me \end{array}$$

(c) The presence of alkenyl groups (*169, 448, 1527, 1529*) or phenyl groups (*766*) in which a double bond is adjacent to the boron.

(d) The presence of B–F bonds (*79, 348, 1204, 1811*) (e.g., RBF_2, R_2BF) in which back-bonding of the type $B^-{=}F^+$ is probably important. Back bonding probably accounts for the observed relative rate of exchange $R_2BOR' > RB(OR')_2 > B(OR')_3$ with BR_3 (*1348*).

Factors that accelerate the ligand exchange are

(a) The addition of catalytic amounts of a B–H-containing compound (which is not necessary when X = H, of course) (*130, 208, 211, 938, 956, 958, 959, 979, 985, 1309, 1344, 1347, 1348*).

(b) When X is a group known to participate in bridge bonds (i.e., NR_2) (*421*).

All of the above suggests a bridged transition state, Eq. 28 (*1205, 1527*).

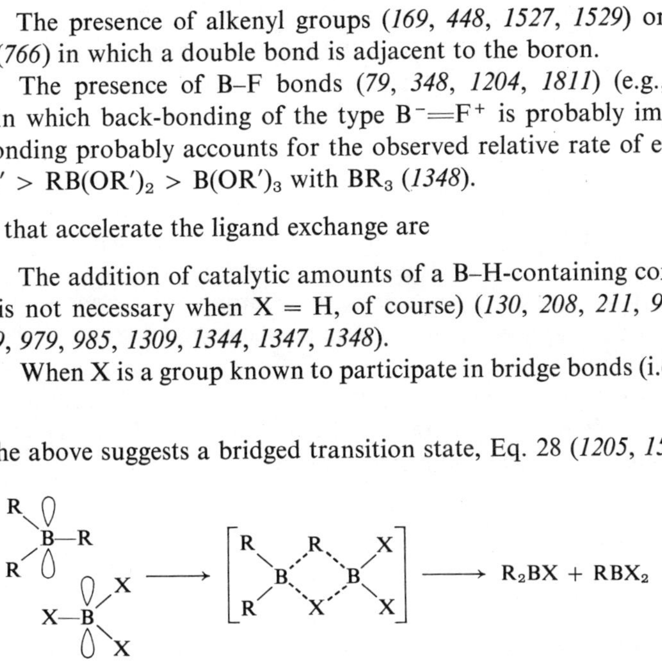

$$BR_3 + BX_3 \longrightarrow \left[\begin{array}{c} R \quad R \quad X \\ B \cdots B \\ R \quad X \quad X \end{array} \right] \longrightarrow R_2BX + RBX_2 \qquad (28)$$

Those factors, which either prevent close approach of the two molecules (steric) or which inhibit the acceptor power of the vacant *p* orbital on the boron atom (i.e., by coordination, or by enhanced $B\text{—}\ddot{X} \leftrightarrow B^-{=}X^+$, or $B\text{—}C{=}C \leftrightarrow B^-{=}C\text{—}C^+$ resonance interaction with the p–π electrons of an appropriate element(s) in the same periodic row as boron), will serve to raise the energy of the transition state. The resonance effect will tend to restrain the boron from assuming tetrahedral configuration in the bridge-bonded dimer that is expected for the activated complex. This may be overridden when $\text{—}\ddot{X}$ is a good bridging group.

The catalytic effect of B–H containing compounds (*1780*) (e.g., $R_4B_2H_2$) (see Chapter 5) is easily rationalized in terms of the above factors: (a) hydrogen is a good bridging atom, (b) hydrogen has a low steric requirement, and

(c) no possibility of π back bonding to boron. A plausible general sequence of reactions illustrating the use of a B—H compound as a catalyst is:

$$BX_3 + \tfrac{1}{2}(R'_2BH)_2 \rightleftharpoons [X_2BH] + R'_2BX \tag{29}$$

$$[X_2BH] + R_3B \rightleftharpoons X_2BR + \tfrac{1}{2}(R_2BH)_2 \tag{30}$$

There are a few redistribution reactions in which one or two trisubstituted boron compounds undergoing ligand exchange is complexed with a Lewis base (*977*, *1329*) and several other reports (*525*, *678*) in which strong bases (RO^-, NH_3) appear to catalyze the exchange. The mechanisms for these reactions must certainly differ from that proposed above, although insufficient data makes it difficult to suggest reasonable alternatives.

Hydroboration

The hydroboration reaction serves not only as an important preparative method for organoboron compounds but also as the first step in several reaction sequences of wide application in synthetic organic chemistry. For an extensive discussion of the role of the hydroboration reaction in organic syntheses one is advised to consult two excellent monographs on the subject by the leading contributor to this field, H. C. Brown (*184a*, *184b*). The hydroboration section in this book will be mainly confined to a brief summary of the B—C bond-making process via the addition of a B–H unit to an appropriate unsaturated compound, e.g., alkene, alkyne.

$$\rangle B{-}H + \rangle C{=}C\langle \longrightarrow -\overset{H}{\underset{|}{\overset{|}{C}}}-\overset{B\langle}{\underset{|}{\overset{|}{C}}}- \tag{31}$$

The first successful attempt to add diborane to olefins required elevated temperatures, 100°, and resulted in a low yield of the trialkylborane (*848*).

$$CH_2{=}CH_2 \xrightarrow{B_2H_6} BEt_3 \tag{32}$$

When it was realized that ethers and similar weak bases catalyze the reaction to the extent that the addition occurs quantitatively at room temperature, this manner of preparing organoboron compounds blossomed to its existing importance (*184a*, *184b*, *293*, *294*, *296*, *297*).

Although a number of B—H sources have been used (Table 3-3), the two most convenient procedures for adding B—H across the unsaturated bond of an olefin or alkyne are: (a) adding tetrahydrofuran–borane (in excess THF) to the unsaturated compound, and (b) addition of boron trifluoride etherate to a suspension of sodium borohydride in a solution of the unsaturated compound in ether (diglyme, ethyl ether, or THF). In the majority of instances the organoborane is not isolated, but immediately subjected to

$$BH_3 \cdot THF + 3\,RCH{=}CH_2 \xrightarrow[0^\circ]{THF} (RCH_2CH_2)_3B + THF \quad (33)$$

$$4\,BF_3 \cdot OEt_2 + 3\,NaBH_4 + 12\,RCH{=}CH_2 \xrightarrow[0\text{–}25^\circ]{ether} 4\,(RCH_2CH_2)_3B + 3\,NaBF_4 + 4\,Et_2O \quad (34)$$

oxidation and/or hydrolysis (Section 3-2) to a desired organic moiety containing no boron. The nature of the organic end product usually permits reasonable conclusions to be reached about the details of the initial hydroboration step (see below). In the few cases, where the "intermediate" organoborane is analyzed by nonchemical means, physical support of these conclusions is generally observed.

Hydroboration of the great majority of olefins is usually allowed to proceed to the corresponding trialkylborane. For olefins with a large degree of steric hindrance, the reaction may stop at the dialkylborane (dimer) or monoalkylborane (dimer) stage (*226*, *243*, *297*, *1968*, *1969*).

$$4\,Me_2C{=}CHMe + 2\,BH_3 \cdot THF \longrightarrow \left[(H{-}\overset{Me}{\underset{Me}{C}}{-}\overset{Me}{\underset{H}{C}}{-})_2BH \right]_2 \quad (35)$$

$$2\,Me_2C{=}CMe_2 + 2\,BH_3 \cdot THF \longrightarrow \left[H{-}\overset{Me}{\underset{Me}{C}}{-}\overset{Me}{\underset{Me}{C}}{-}BH_2 \right]_2 \quad (36)$$

Bulky alkylboranes can be used for the selective largely sterically controlled hydroboration reactions (*224*, *225*, *242*, *244*, *290*, *306*, *317*, *318*, *321*, *592*, *893*, *1966*, *1968*). For model examples, with respect to Sia_2BH (dimer), reactivities decrease in the following order: 1-hexene > 3-methyl-1-butene > 2-methyl-1-butene > 3,3-dimethyl-1-butene > *cis*-2-hexene > 2-methyl-2-butene > 2,3-dimethyl-2-butene.

Hydroboration will stop at the monoalkylboron stage, rather than proceeding to the trialkylboron, when the monoalkylboron product is internally stabilized (**V**, **VI**) as in the hydroboration of ureido-substituted olefins (*363*) or of *N*,*N*-dimethylallylamine (*21*).

H
N
H_2C C=O
H_2C NH_2
B
H_2

(V)

$\overset{+}{N}{-}\overset{-}{B}H_2$
Me_2

(VI)

TABLE 3-3

Reagents Used for Formation of B—C Bonds via Hydroboration

Reagent	Reference
BH_3	*594*
$PhSBH_2$	*1538*
Cl_2BH	*1144*
$ClBH_2$	*1692*
B_2H_6 (no solvent)	*79, 414, 689, 937, 1118, 1568, 1741, 1954*
B_2H_6 (solvent—usually an ether)	*226, 229, 240, 244, 260, 294, 315, 331a, 367, 414, 570, 593a, 771, 796, 1133, 1254, 1316, 1320, 1327, 1333d, 1935*
Organodiboranes (organoboranes)	*129, 130, 132, 185–187, 195, 224, 242, 244, 260–262, 269, 276, 290, 306, 317, 318, 320–322, 331, 592, 794, 893, 894, 929, 937, 958, 959, 975, 1133, 1207, 1278, 1284–1286, 1312, 1318, 1321, 1390, 1418, 1561, 1631, 1965–1969, 1971*
R(or Ar)$BH_2 \cdot$LB (LB = alkyl or arylamine)	*268, 364–366, 693, 695, 759–761, 766, 768, 1874*
$RBH_2 \cdot SR_2$	*153*
$ClBH_2 \cdot$etherate	*275, 275a, 275b, 1533, 1542, 1692, 1888, 1962*
$ClBH_2 \cdot$amine	*63, 1888*
$Cl_2BH \cdot$etherate	*275c, 1533, 1542, 1962*
LiH, $BF_3 \cdot Et_2O$	*248*
NaH, $BF_3 \cdot Et_2O$	*248*
$LiBH_4$, BF_3, ether	*204, 248*
$LiBH_4$, H_2SO_4, ether	*248*
$LiBH_4$, $B(OMe)_3$, $AlCl_3$, ether	*248*
$LiAlH_4$, BF_3, ether	*248, 1492, 1728, 1917, 1936*
$LiAlH_4$, $AlCl_3$, $B(OMe)_3$	*248, 314*
$LiAlH_4$, BCl_3	*1931*
$Na(HBEt_3)$	*833*
$Al(BH_4)_3$	*176, 177, 296*
KBH_4, BF_3, ether	*248*
KBH_4, BCl_3, ether	*248*

Reagent	References	Reagent	References
$(Et_2NBH_2)_2$	*1280*	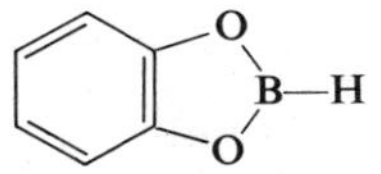	*209, 214*
R_2NBH_2	*943*		*608a, 608b, 1923a*
$(n\text{-}BuSBH_2)_x$	*1330, 1331, 1333, 1538, 1691*	$NaBH_4$, HCl	*314*
$(EtS)_3B_2H_3$	*1689*	$NaBH_4$, BF_3, ether	*186, 187, 193, 243, 245, 248, 271, 288, 294, 297, 317, 319, 321, 323, 325, 1687, 1969*
$[(n\text{-}BuS)_2BH]_2$	*1331*	$NaBH_4$, BCl_3, ether	*248, 294, 297*
$R_3N \cdot BH_3$ or pyridine $\cdot BH_3$	*21, 48, 63, 512, 528, 554, 692, 693, 755, 929, 936, 937, 979, 1285, 1617, 1618, 1669*	$NaBH_4$, $AlCl_3$, ether	*248, 292, 293, 296*
$R_3N \cdot BH_3$ or pyridine $\cdot BH_3$, BF_3	*248*		
$THF \cdot BH_3$	*186, 200, 223, 224, 227, 243, 253, 257, 260, 264–266, 271, 294, 297, 304, 306, 330, 746, 891, 892, 1189, 1248, 1373, 1418, 1543*		
$Me_2S \cdot BH_3$	*152, 153*		

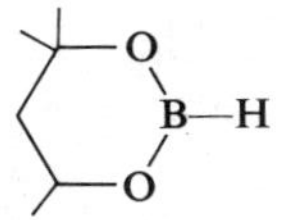

The stereochemistry of hydroboration probably involves cis addition of the B—H bond to the π bond of the olefin, entailing a four-center transition state (*312, 319, 322, 1836, 1861, 1917, 1969*). The direction of addition favors boron attachment to the least substituted carbon. It is thought that this is controlled both by the preferred polarization (or polarizability) of the B—H and C=C bonds and by steric factors (*319, 891*).

$$R{-}CH{=}CH_2 \xrightarrow{\;\rangle B{-}H\;} R{-}\overset{\delta+}{C}H{-}\overset{\delta-}{C}H_2 \cdots \overset{\delta-}{H}\text{-----}\overset{\delta+}{B}{<} \tag{37}$$

It is sometimes believed that borane (BH_3), or a substituted borane unit, is probably an important unstable intermediate in the hydroboration process, whether carried out in the gas phase with diborane, or conducted in ether solution. A rate study on the reaction of free borane with ethylene has given a molecular rate expression: $\log k_1$ (liters mole^{-1} sec^{-1}) $= 10.2 - (2000/4.575T)$; and the reaction of ethylborane with ethylene in the gas phase is about 20 times slower than the reaction of borane with ethylene (*594*). In another study, the kinetics of the gas phase reaction between $Al(BH_4)_3$ and simple olefins is found to be first order with respect to $Al(BH_4)_3$ and independent of the olefin concentration, with an activation energy of 30 kcal mole^{-1}. Prior dissociation of the $Al(BH_4)_3$ to borane before a reaction with the olefin takes place is implied (*176*).

Several questions have been raised about the validity of the "monoborane" mechanism which includes a need to consider a borane dimer, rather than the monomer, as part of an initial transition state (*181*). In accounting for the second-order kinetics (first order both in alkene and dimeric organoborane) in the reaction of cyclopentene with bis-3-methyl-2-butylborane (dimer) using tetrahydrofuran solvent two modes of hydroboration have been suggested. This involves dimeric borane in a slow step, Eq. 38, followed by a fast step, Eq. 39, which utilizes the remaining B—H-containing monomer (*242*). Also,

$$R_2BH_2BR_2 + C_5H_8 \rightleftharpoons R_2B(\mu\text{-}H)_2BR_2\cdot C_5H_8 \longrightarrow R_2BC_5H_9 + [R_2BH] \tag{38}$$

$$[R_2BH] + C_5H_8 \xrightarrow{\text{fast}} R_2BC_5H_9 \tag{39}$$

to account for the high stereospecificity upon hydroboration of *cis*-1-butene-1-*d* with diisopinocampheylborane (dimer), an alternate transition state involving a small perturbation from a triangular bridged complex is proposed (*1746*):

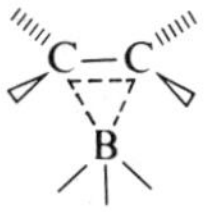

This is supported by orbital symmetry considerations indicating that the four-center transition state usually proposed for hydroboration may have significant symmetry barriers (*869*). On the other hand, if the hydroboration reaction is highly exothermic with low activation energy, then orbital symmetry control may not be important in an early transition state (*1545*).

The rate of formation of 2,3-dimethyl-2-butylborane from tetramethylethylene and borane in THF is first order in both borane and TME. The activation energy and entropy for the reaction are 9.2 ± 0.4 kcal mole^{-1} and -27 ± 1 eu, respectively. The reaction is considered to involve the direct reaction between a molecule of the $BH_3 \cdot THF$ complex and alkene in a very early transition state (*1545*). In the case of terminal alkenes, the dialkylboranes are formed via consecutive addition reactions in which $k_2 > k_1$ and not via the disproportionation of the alkylboranes. The greater reactivity of the monoalkylborane relative to borane is attributed to a weaker complexing interaction in the monoalkylborane–tetrahydrofuran complex than in the borane–tetrahydrofuran complex, thus resulting in a lowering of the activation energy for nucleophilic attack by an alkene at the boron atom (*1544*).

The presence of reasonably inert functional groups in the olefin molecule such as ethers and halogens (except for fluorine) (*200, 224, 288, 306, 771, 1254, 1327*), does not normally cause problems in the ether-catalyzed hydroboration reaction. Esters (*225, 306*) have not posed much of a problem either; however, the high reactivity of the aldehyde and free carboxylic acid groupings make it difficult to avoid simultaneous reduction of these during the hydroboration of a double bond in the same molecule. The presence of substituents may markedly influence the mode of B—H addition to the double bond (*200, 224, 225, 227, 288, 289, 306, 319, 891, 1190, 1540, 1548, 1568, 1669, 1681*). For example, an alkoxy substituent in the 1-position of 1-butene directs the boron to the 2-position, whereas a chlorine substituent directs the boron to the 1-position (*289, 1548*). To enhance the attack at the 1-position of some alkene derivatives, a bulky hydroborating agent has been used (*306*). Hydroboration of β-styreneboronic acid ester gives exclusively $PhCH_2CH_2[B(OR)_2]_2$, whereas α-styreneboronic acid ester gives a 95:5 mixture of $PhCHB(OR)_2$-$CH_2B(OR)_2$ and $PhC[B(OR)_2]_2CH_3$ (*1535, 1536*).

The α-haloorganoboranes from the hydroboration of vinyl halides do not give rise to free carbenes by α-elimination, but undergo a transfer reaction replacing the halogen by hydrogen (*1548*). A similar rearrangement occurs with the hydroboration of vinylthioesters in which the RS group migrates to the boron subsequent to B—H addition to the olefin (*1546*, *1547*). The β-haloorganoboranes generally undergo a rapid elimination of boron halide to give olefins in tetrahydrofuran solution (*1548*).

$$\mathrm{MeCH{=}CMeX} \xrightarrow{-\mathrm{BH}_2} \mathrm{MeCH_2C(Me)(X)\text{—}B{<}(H)} \longrightarrow \mathrm{MeCH_2\text{—}C(Me)(H)\text{—}B{<}(X)}$$

$$+$$

$$\mathrm{{>}B\text{—}CH(Me)\text{—}CH(Me)\text{—}X} \longrightarrow \mathrm{MeCH{=}CHMe + XB{<}} \qquad (40)$$

The usual terminal mode of addition is reversed upon hydroboration of $CF_3CH{=}CH_2$ (*132*, *1568*). But in this molecule, the effect of the electron-withdrawing CF_3 group may be to shift the π electrons of the double bond toward the central carbon atom, thus making this atom more susceptible for bonding to the vacant orbital of the attacking boron. The products often isolated are the result of fluorine migration from carbon to boron subsequent to the hydroboration step (*1568*, *1742*).

Asymmetric hydroboration of suitable olefins with optically active diisopinocampheylborane (dimer) has been found useful in preparing organic derivatives of high optical purity (*185–187*, *1746*, *1965*, *1967*). This work also establishes that a boron atom bonded to an asymmetric center RR′HC*—B is capable of maintaining optical activity without significant racemization over reasonable periods of time.

As mentioned earlier, hydroboration with diborane in the absence of catalytic ether solvents has not proved convenient for the preparation of organoboron compounds (*848*, *1742*); however, under carefully controlled conditions, simple alkenes and diborane (in excess) will react in the gas phase to give alkyldiboranes in respectable yield (*1118*) (see Chapter 5).

It appears that the preferred terminal mode of B—H addition to olefins is thermodynamically controlled; for, a ΔH of ca. -136 kcal mole^{-1} is observed for the addition of B_2H_6 to 1-alkenes (where it is known that the boron becomes attached to the terminal carbon) and ΔH of -116 kcal mole^{-1} is found for the addition to 2-alkenes (*109*) (where boron can only be attached to an internally situated carbon).

Hydroboration of allene and 1,1-dimethylallene results in terminal attachment of boron (*436*, *519b*, *608a*, *1118*, *1333d*), whereas central attack is dominant in 1,3-dimethylallene, tetramethylallene, 2,3-nonadiene, and cyclic allenes (*519b*, *608a*). Butylallene and phenylallene show both terminal and central attack (436). Dienes in which the double bonds are separated by at least one single C—C bond react with or without (at elevated temperatures) ether solvent to give a product in which the boron(s) is (are) predominantly bonded to the least substituted carbon(s) (*315*, *321*, *936*, *937*, *1866*, *1973*, *1974*) with evidence for small amounts of internally substituted products.

Acetylenes are generally more reactive than olefins toward hydroboration (*244*). Hydroboration of internal triple bonds can be controlled to produce the trivinylborane (*323*) with "borane" as the hydroborating agent. Terminal acetylenes, under similar conditions, give primarily the 1,1-dibora derivatives. There is some evidence, however, for vincinal rather than geminal diboro products (*414*, *746*, *1531*). In order to stop the reaction at the vinylboron stage bulky hydroborating agents such as thexylborane(dimer) (*1968*) or Sia_2BH may be employed (*323*). 1,3,2-Benzodioxaborole (**VII**) is a useful reagent in that it can monohydroborate alkynes to give alkeneboronic esters with the boron being attached to the less hindered carbon (*214*).

(VII)

As with olefins, cis addition of the B—H unit to the triple bond occurs. Steric control is exhibited in the hydroboration of diynes whereby the less sterically hindered triple bond is attacked first (*1970*). Also, preferential hydroboration of the alkyne group over the olefinic bond occurs in enynes (*1970*).

A number of hydroborating agents in addition to diborane, borane, and alkyldiboranes have been used for the preparation of organoboranes. These reagents and references to these studies are listed in Table 3-3, and only a few brief comments are given here. The reactivity of borane adducts ($LB \cdot BH_3$) appears to be inversely a function of the strength of the coordination link, i.e.,

$R_2O{:}BH_3 >$ pyridine$:BH_3 > R_3N{:}BH_3$ (*48*, *752*, *929*)

$THF{:}BH_3 > THF \cdot H_2BCl$ and $THF \cdot HBCl$ (*1533*, *1962*)

$R_2S \cdot RBH_2 > R_3N \cdot RBH_2$ (*21*, *153*)

$Et_2O \cdot H_2BCl > THF \cdot H_2BCl$ (*275*)

$RBH_2 \cdot NEt_3 > RBH_2 \cdot NMe_3$ (*268*)

The air-stable 9-borabicyclo[3.3.1]nonane, 9-BBN (*892*), may be conveniently used to prepare unsymmetrical trialkylboranes in which further chemistry may be carried out on a unique group (*893*, *894*), e.g., Eq. 41.

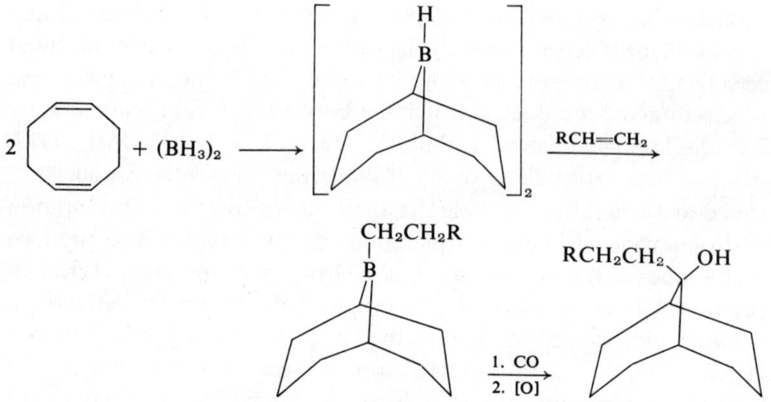

$$2\,C_8H_{12} + (BH_3)_2 \longrightarrow [\text{9-BBN-H}]_2 \xrightarrow{RCH{=}CH_2} \text{9-}(CH_2CH_2R)\text{-9-BBN} \xrightarrow[2.\ [O]]{1.\ CO} \text{9-}(RCH_2CH_2)\text{-9-OH-bicyclononane} \tag{41}$$

Hydroboration of olefins with trimethylamine-*t*-butylborane provides an alternative route to unsymmetrical trialkylboranes. An interesting application of this reaction is the synthesis of 1-bora-1-*t*-butyl-4,4-dimethyl-4-silacyclohexane (**VIII**) (*760*, *766*). The rate of addition of trimethylamine *t*-butylborane

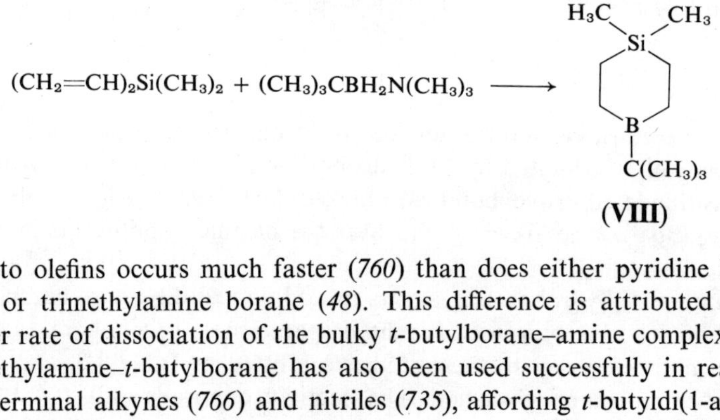

$$(CH_2{=}CH)_2Si(CH_3)_2 + (CH_3)_3CBH_2N(CH_3)_3 \longrightarrow \text{(VIII)} \tag{42}$$

(**VIII**)

(*759*) to olefins occurs much faster (*760*) than does either pyridine borane (*755*) or trimethylamine borane (*48*). This difference is attributed to the greater rate of dissociation of the bulky *t*-butylborane–amine complex (*760*). Trimethylamine–*t*-butylborane has also been used successfully in reactions with terminal alkynes (*766*) and nitriles (*735*), affording *t*-butyldi(1-alkenyl) boranes and dimeric alkylideneamine-*t*-butylboranes (**IX**), respectively. Other

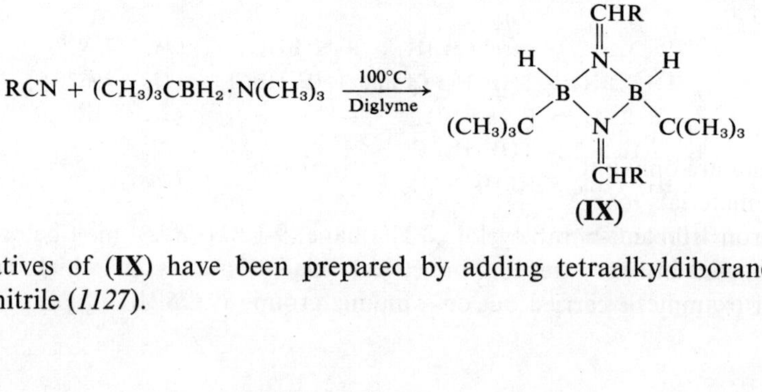

$$RCN + (CH_3)_3CBH_2\cdot N(CH_3)_3 \xrightarrow[\text{Diglyme}]{100°C} \text{(IX)} \tag{43}$$

(**IX**)

derivatives of (**IX**) have been prepared by adding tetraalkyldiboranes to acetonitrile (*1127*).

The reduction of allyl alkylamines with trialkylaminephenylborane results in two products, a stable solid and an air-sensitive liquid, to which structures (**X**) and (**XI**) have been assigned, respectively (*365*, *366*, *1874*). When allylamine is added to an ether solution of a tetraalkyldiborane, the predominant product is the inner complex (**XII**) (*1284*).

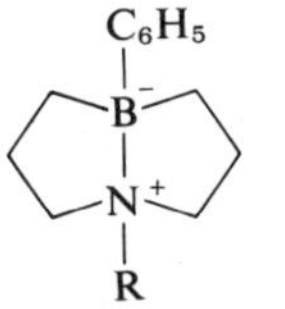

(X) 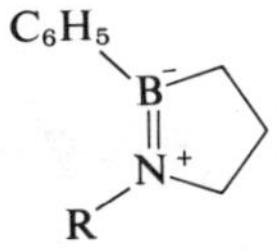 (XI) 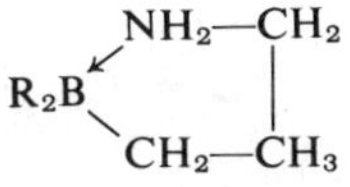 (XII)

A P → N ring system analogous to **X** results from the hydroboration of diallylphenylphosphine with triethylamine–phenylborane (*364*). Kinetic data for the hydroboration of olefins with H_2BCl in THF support a four-centered transition state in which very little charge density is developed at the olefin carbon atom (*1542*). A catalyst such as BF_3 reverses the direction of B—H addition to 4-*t*-butylethoxycyclohexene (*1537*).

Haloboration

Haloboranes such BBr_3 and $PhBCl_2$ react with acetylenes to afford 2-halo- or 2-phenylalkenyl boranes, depending on the nature of the acetylene. The steric course of reactions is cis for the substituted acetylenes but trans for acetylene itself (*1046*). Both 2-chlorovinyldichloroborane and bis-2-chlorovinylchloroborane are the products from the haloboration of acetylene in the

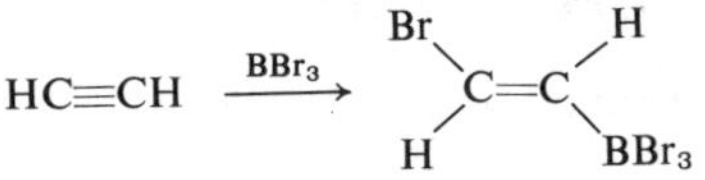

$$HC{\equiv}CH \xrightarrow{BBr_3} (Br)(H)C{=}C(H)(BBr_3) \quad (44)$$

$$BuC{\equiv}CH \xrightarrow{\frac{1}{2}\,PhBCl_2} [(Bu)(Cl)C{=}C(H)]B(Cl)[C(H){=}C(Bu)(Ph)] \quad (45)$$

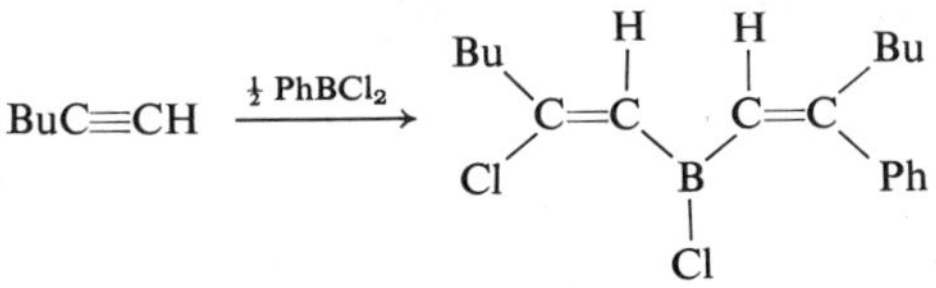

presence of an activated charcoal catalyst (*665*, *666*). Tris(2-chlorovinyl)-boranes are only obtained on disproportionation of the mono- and dichlorovinyl materials (*666*).

Boron tribromide is more reactive than BCl_3 toward olefins (*871*, *1315*) but may lead to products which are the result of secondary reactions (Eq. 46). Other examples of reactions involving haloboration are given in Eqs. 47–51.

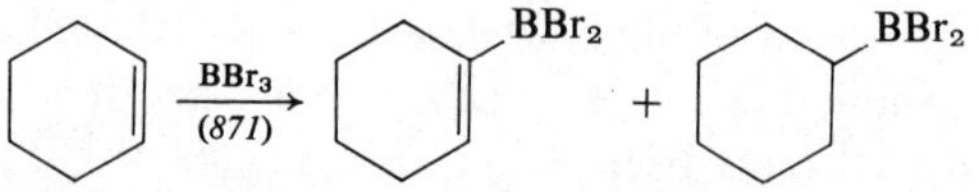

$$\xrightarrow[(871)]{BBr_3} \quad BBr_2 \quad + \quad BBr_2 \qquad (46)$$

$$\text{Alkenes} \xrightarrow[(1315)]{BBr_3} RCHBrCH_2BBr_2 \qquad (47)$$

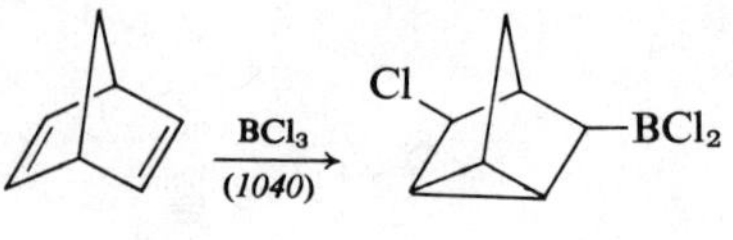

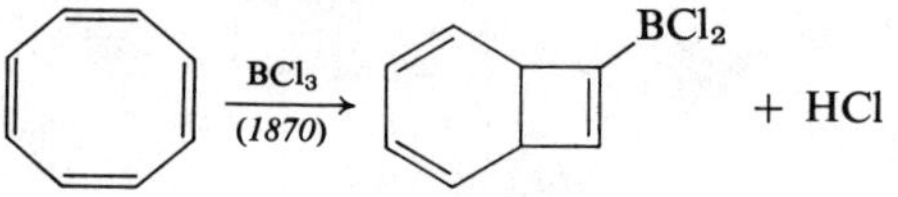

$$\xrightarrow[(1040)]{BCl_3} \quad Cl \quad BCl_2 \qquad (48)$$

$$\xrightarrow[(1870)]{BCl_3} \quad BCl_2 \quad + HCl \qquad (49)$$

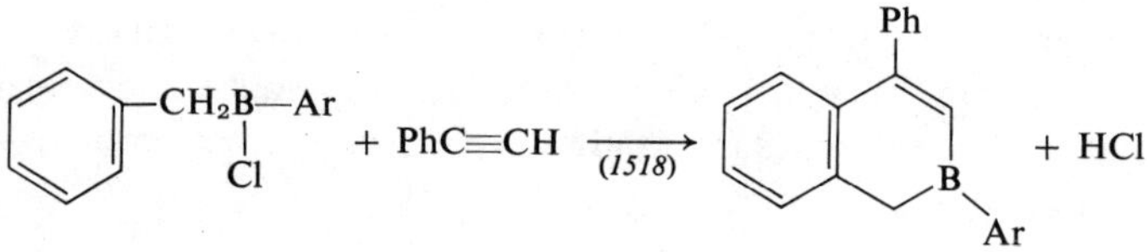

$$CH_2B\text{—}Ar,\ Cl \quad + PhC{\equiv}CH \xrightarrow[(1518)]{} \quad Ph,\ B,\ Ar \quad + HCl \qquad (50)$$

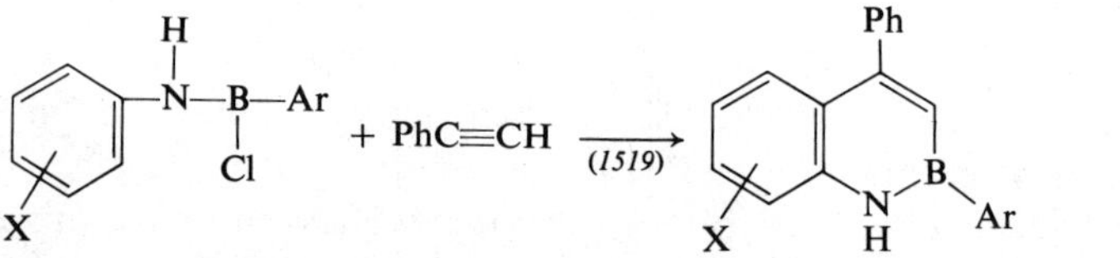

$$H,\ N\text{—}B\text{—}Ar,\ Cl,\ X \quad + PhC{\equiv}CH \xrightarrow[(1519)]{} \quad Ph,\ B,\ Ar,\ N,\ H,\ X \qquad (51)$$

Modified Friedel–Crafts Reaction

Introducing a boron-containing group onto an aromatic hydrocarbon is effected by the use of a boron halide, with Friedel–Crafts-type catalysts used occasionally (Eq. 52) (*340*, *1076*, *1100*, *1393*, *1394*), but care must be taken

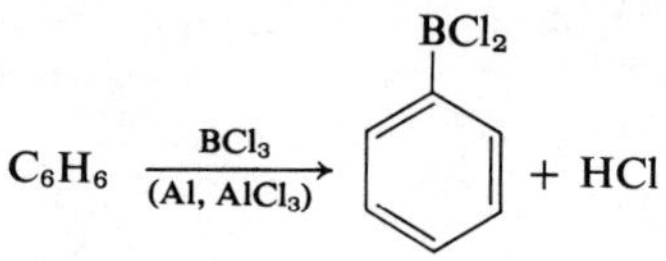

$$C_6H_6 \xrightarrow[(Al,\ AlCl_3)]{BCl_3} \quad BCl_2 \quad + HCl \qquad (52)$$

to avoid the reversible deboronation reaction in the presence of hydrogen halide (*340*, *1395*).

Other examples of related reactions include the syntheses of a substantial variety of cyclic heteroaromatic boron compounds for which only a few examples are cited, Eqs. 53–59.

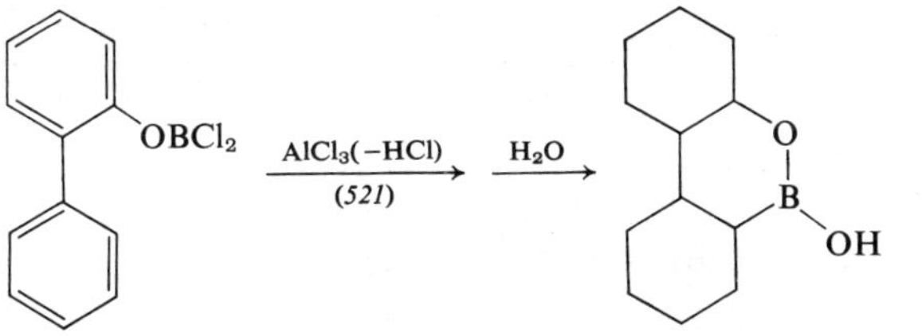

$\xrightarrow[(521)]{AlCl_3(-HCl)}$ $\xrightarrow{H_2O}$ (53)

OBCl$_2$; O, B, OH

$$(ClBNH)_3 \xrightarrow[AlCl_3 \quad (1435)]{\text{benzene/chlorobenzene}} (PhBNH)_3 \quad (54)$$

24%

$\xrightarrow[AlCl_3 \quad (543)]{BCl_3}$ (55)

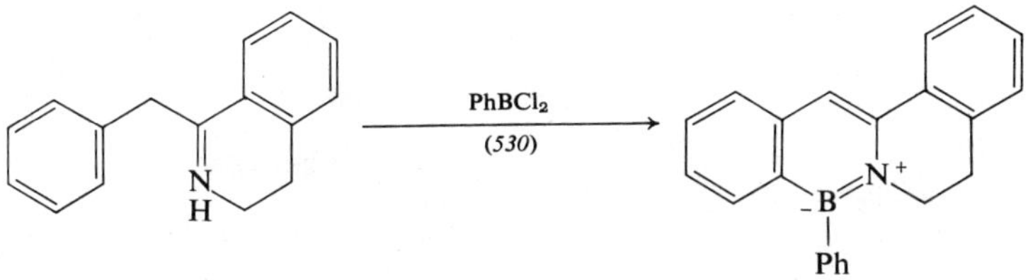

NH$_2$; B, N$^+$, H, Cl

$\xrightarrow[(530)]{PhBCl_2}$ (56)

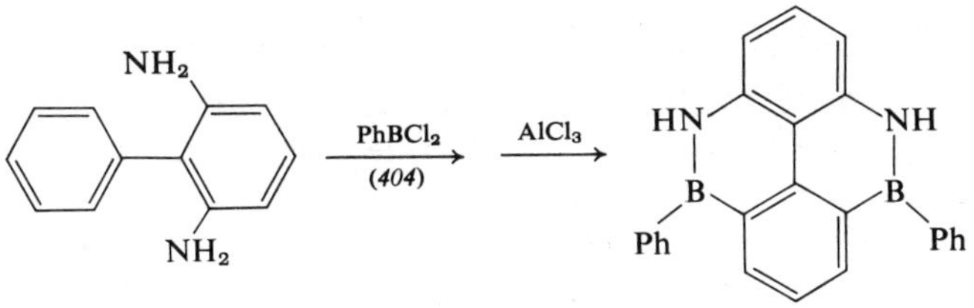

N, H ; B, N$^+$, Ph

$\xrightarrow[(404)]{PhBCl_2}$ $\xrightarrow{AlCl_3}$ (57)

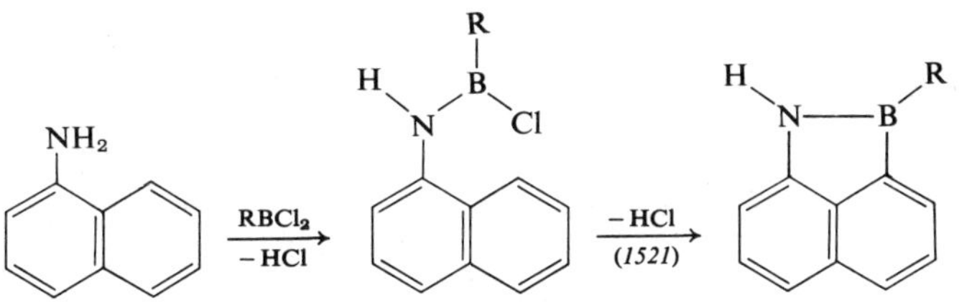

NH$_2$, NH$_2$; HN, NH, B, B, Ph, Ph

$\xrightarrow[-HCl]{RBCl_2}$ $\xrightarrow[(1521)]{-HCl}$ (58)

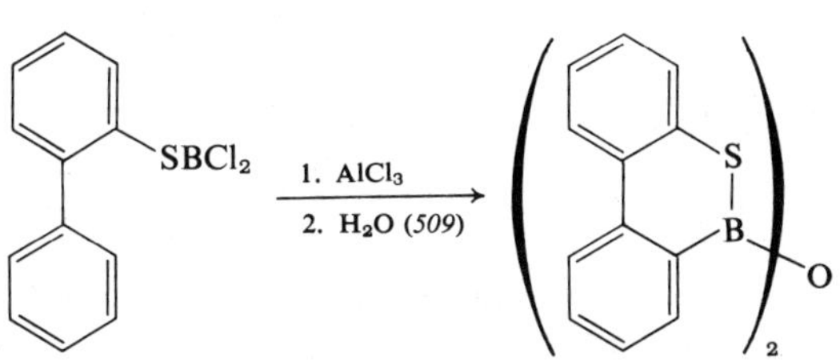

NH$_2$; R, H, B, N, Cl ; H, R, N, B

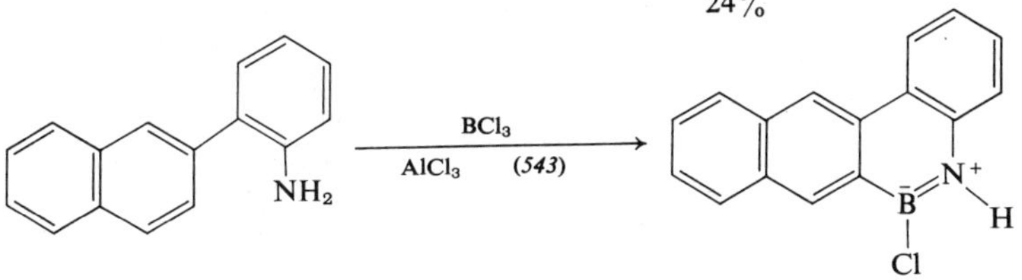

$\xrightarrow[2.\ H_2O\ (509)]{1.\ AlCl_3}$ (59)

SBCl$_2$; S, B, O, 2

Cyclization by Loss of H_2

A number of pyrolysis reactions of organoboron compounds afford cyclic products in which new B—C bonds are formed by loss of hydrogen from both carbon and boron (e.g., Eq. 60) (*960, 964, 965, 977, 978, 1835a, 1901*). Many of these reactions also appear to be accompanied by a loss of a boron

$$PhCH_2CH_2NMeBH_2 \xrightarrow[-H_2\ (964)]{200^\circ} \text{(fused benzo ring with B(H)–NMe)} \quad (60)$$

attached alkyl group reminiscent of dehydroboration reactions (Section 3-2) (*947, 965, 970, 974, 978, 1901*). A four-center transition state (Eq. 61) has

$$n\text{-}Bu_3B \xrightarrow{300^\circ} n\text{-}Bu\text{—}B\text{(cyclopentane ring)} \quad (61)$$

$$\xrightarrow{-C_4H_8} \left[n\text{-}Bu\text{—}B(H)\text{—}CH_2\text{—}CH_2\text{—}CH_2\text{—}CH_2\text{—}H \right] \longrightarrow \left[n\text{-}Bu\text{—}B(\cdots H\cdots H\cdots)CH_2\text{—}CH_2\text{—}CH_2\text{—}CH_2 \right] \xrightarrow{-H_2}$$

been proposed for the final step in which the new B—C is formed (*965*). To facilitate the overall reaction a trialkylborane can be equilibrated initially with diborane (*247*) prior to cyclization. This relatively low-energy exchange procedure (Chapter 5) builds up the necessary intermediate R_2BH species without requiring a sluggish dehydroboration step.

Diboration with Diboron(4) Compounds

Under relatively mild conditions, both boron atoms of diboron tetrahalides, B_2X_4 (X = F, Cl), add to the unsaturated bonds of alkenes (*390, 445, 446, 821, 823, 1622, 1827, 1948, 1951*) and alkynes (*390, 394, 395, 1870*) to give 1,2-addition products.

$$Cl_2B\text{—}BCl_2 + \ >C{=}C< \ \longrightarrow Cl_2B\text{—}\overset{|}{\underset{|}{C}}\text{—}\overset{|}{\underset{|}{C}}\text{—}BCl_2 \quad (62)$$

$$Cl_2B\text{—}BCl_2 + \text{—}C{\equiv}C\text{—} \longrightarrow Cl_2B\overset{|}{C}{=}\overset{|}{C}BCl_2 \quad (63)$$

Stereochemical investigations indicate a cis mode of addition to both double and triple bonds (*446, 1622, 1948*). As the reactants approach the proposed

4-center transition state, the B_2Cl_4 assumes a near planar configuration in order to allow for maximum overlap between the "vacant" p orbitals of the boron atoms and the p–π orbital of the hydrocarbon (Eq. 64) (*1622, 1948*).

$$\ce{>B-B< + >C=C< -> [>B---B<, >C=C<] -> -B(-C-)-B(-C-)} \qquad (64)$$

Alkyl, halogen and dichloroboryl substituents on the olefinic carbon reduce the reactivity of the unsaturated compound toward B_2Cl_4 (*390, 1827, 1870*).

Tetrachlorodiborane(4) also adds to the olefinic bonds of a number of vinyl-metal compounds (*445, 822, 823*), including $(CH_2{=}CH)_3B$ (*823*). The reaction of B_2Cl_4 with vinyl chloride in a 2:1 ratio yields 1,1,2-tris(dichloroboryl)ethane. A plausible pathway involves elimination of boron trichloride from a diboryl intermediate, Eq. 65 (*1607*). Cis addition products are obtained from the reaction of B_2Cl_4 with cyclopropene (Eq. 66) and cyclobutene (*1610*) and is the presumed mode of addition to cyclohexene (*122*), but there appears to be some disagreement as to whether cis or trans addition predominates in the case of cyclopentene (*1610, 1632*).

$$CH_2{=}CHCl + B_2Cl_4 \longrightarrow (Cl_2B)CH_2CHCl(BCl_2) \xrightarrow{-BCl_3} CH_2{=}CHBCl_2 \xrightarrow{B_2Cl_4} Cl_2BCH_2CH(BCl_2)_2 \qquad (65)$$

$$\text{cyclopropene} + B_2Cl_4 \longrightarrow \text{cis-1,2-bis(dichloroboryl)cyclopropane} \qquad (66)$$

Diboration of nonconjugated cyclic dienes can be effected in a stepwise fashion (*1632*):

$$\text{1,4-cyclohexadiene} \xrightarrow{B_2Cl_4} \text{4,5-bis(dichloroboryl)cyclohexene} \xrightarrow{B_2Cl_4} \text{1,2,4,5-tetrakis(dichloroboryl)cyclohexane} \qquad (67)$$

The 1,2,3,4-tetrakis(dichloroboryl) derivative of cyclohexane (**XIII**) can be obtained from the reaction of 2 moles of B_2Cl_4 to 1,3-cyclohexadiene

(*1951*). Although benzene reacts with B_2Cl_4 to give only a monoboryl derivative, naphthalene adds 2 moles of the diboron tetrachloride to give 1,2,3,4-tetrakis(dichloroboryl)-1,2,3,4-tetrahydronaphthalene (**XIV**) (*614*).

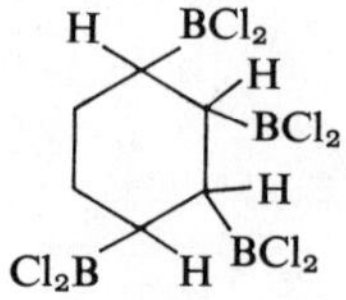

(XIII) 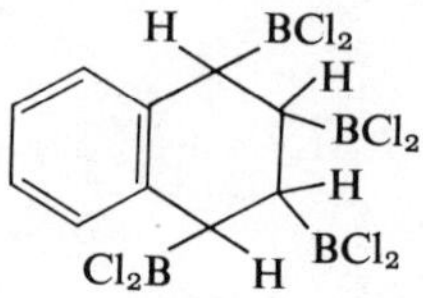 **(XIV)**

Cyclopropane is cleaved by B_2Cl_4 to give $Cl_2BCH_2CH_2CH_2BCl_2$ (*1949*).

Formation of Boron–Carbon Bonds by the Action of Metals (or Metal Boron Compounds) on Halides

Tetrakis(dimethoxyboryl)methane can be prepared by the reaction of dimethoxyboron chloride and carbon tetrachloride with lithium metal (*386*):

$$4\,(MeO)_2BCl + CCl_4 + 8\,Li \xrightarrow{THF} [(MeO)_2B]_4C + 8\,LiCl \qquad (68)$$

Methane triboronic esters and methane diboronic esters are obtained by the same method, using $CHCl_3$ and CH_2Cl_2, respectively, instead of CCl_4. It is likely that this reaction proceeds through an R_2BM intermediate as is proposed for the methylation of a Bu_2BCl, sodium/potassium alloy mixture (*57*):

$$Bu_2BCl + 2\,M \longrightarrow Bu_2BM + MCl \qquad (69)$$
$$Bu_2BM + MeI \longrightarrow Bu_2BMe + MI \qquad (70)$$

In a similar fashion, keto, thioketo, and carboxyl derivatives are obtained (Eq. 71) (*1656*); and a trifluoromethyl boron compound is prepared from an

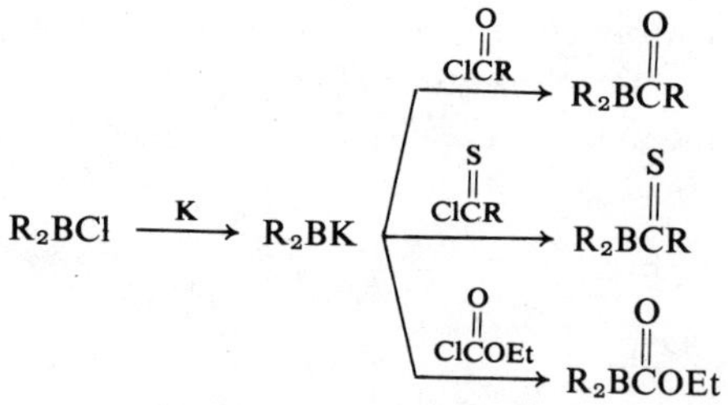

$$R_2BCl \xrightarrow{K} R_2BK \begin{cases} \xrightarrow{ClC(=O)R} R_2BC(=O)R \\ \xrightarrow{ClC(=S)R} R_2BC(=S)R \\ \xrightarrow{ClC(=O)OEt} R_2BC(=O)OEt \end{cases} \qquad (71)$$

amine adduct of the metal boryl compound (*1526*, *1528*):

$$(n\text{-}Bu)_2BK\cdot NEt_3 + CF_3I \longrightarrow n\text{-}Bu_2BCF_3\cdot NEt_3 + KI \qquad (72)$$
$$n\text{-}Bu_2BCF_3\cdot NEt_3 + HCl \longrightarrow n\text{-}Bu_2BCF_3 + Et_3NHCl \qquad (73)$$

A metal boryl compound is probably an intermediate in the production of (**XV**), along with carborane side products, from the following reaction (*946*):

$$6\,Et_2BCl + 6\,Li + 3\,C_2H_4 \longrightarrow \text{(XV)} + 3\,BEt_3 + 6\,LiCl \qquad (74)$$

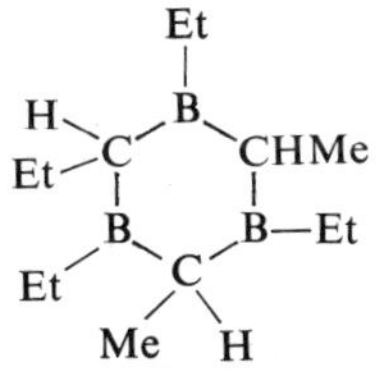

(**XV**): six-membered ring B(Et)–CHMe–B(Et)–C(H)(Me)–B(Et)–C(H)(Et)

Using aluminum metal as a reactant, the class reactions, Eqs. 75 and 76 have been observed (*1394*):

$$RCH{=}CHR \xrightarrow{Al,\ BX_3} H(RCHCHR)_yBX_2\ (y = 1, 2, \ldots) \qquad (75)$$

$$3\,RX + 2\,Al + B(OR')_3 \longrightarrow R_3B + AlX_3 + Al(OR')_3 \qquad (76)$$

Miscellaneous Preparative Reactions

Under photolytic conditions, the diboron(4) compound (**XVI**) reacts with carbon tetrachloride to give two monoboron products, one of which contains a boron-attached carbon with three chlorine atoms (*733*) (Eq. 77). Ferrocene

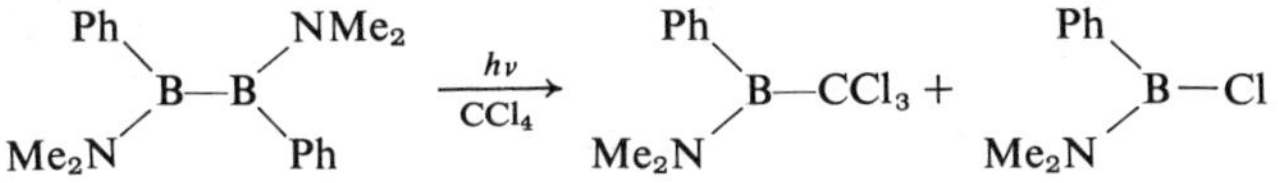

$$\underset{\textbf{(XVI)}}{(Ph)(Me_2N)B{-}B(NMe_2)(Ph)} \xrightarrow[CCl_4]{h\nu} (Ph)(Me_2N)B{-}CCl_3 + (Ph)(Me_2N)B{-}Cl \qquad (77)$$

is attacked by diboron tetrachloride to give a dichloroboryl substitution on one of the cyclopentadienyl rings (*983*, *984*) (Eq. 78).

$$(\pi\text{-}C_5H_5)_2Fe \xrightarrow{B_2Cl_4} (\pi\text{-}C_5H_5)Fe(\pi\text{-}C_5H_4BCl_2) \qquad (78)$$

Fluorocarbon groups can be attached to boron by the reaction of diborane with CF_3SCl (*1526*).

$$CF_3SCl \xrightarrow{B_2H_6} CF_3BF_2 \qquad (79)$$

Carbon vapor reacts with boron halides to give a variety of dichlorobora-substituted methanes and at least one two-carbon species (*552*):

$$BCl_3 \xrightarrow{C} Cl(Cl_2B)C{=}C(BCl_2)Cl + (Cl_2B)_3CCl + (Cl_2B)_2CCl_2 \qquad (80)$$

$$B_2Cl_4 \xrightarrow{C} C(BCl_2)_4 + (Cl_2B)_3CCl + (Cl_2B)_2CCl_2 \qquad (81)$$

Interaction of carbon vapor with Me_2NBH_2 yields Me_2NBHMe as the main product (*746a*). The cocondensation of boron monofluoride and acetylene gives the novel compound $(F_2BCH{=}CH)_2BF$ (*1774*).

The boronium ion, $(Me_3N)_2BH_2^+$, reacts with butyl lithium in what appears to involve the formation of a carbanion followed by internal rearrangement (*1366*, *1368*, *1369*).

$$(Me_3N)_2BH_2^+ \xrightarrow{BuLi} Me_3NBH_2CH_2NMe_2 \xrightarrow{slow} Me_3N + \tfrac{1}{2}[BH_2CH_2NMe_2]_2 \quad (82)$$

Methyl chloroboranes are obtained from the action of BCl_3 on aluminum carbide in the presence of HCl at elevated temperatures (*74*). The reaction probably proceeds according to the sequence:

$$Al_4C_3 + 9\,HCl \longrightarrow 3\,MeAlCl_2 + AlCl_3 \quad (83a)$$

$$MeAlCl_2 + BCl_3 \longrightarrow MeBCl_2 + AlCl_3 \quad (83b)$$

Diborane reacts with the silyl-substituted sulfonium ylide, $Me_3SiCHSMe_2$, to form the substituted diborane, $(Me_3SiCH_2BH_2)_2$ (*1213*). This probably occurs via a hydride shift of a simple adduct with the concomitant liberation of dimethyl sulfide.

$$Me_3Si\overset{\overset{\ominus BH_2\text{—}H}{|}}{\underset{\underset{\overset{\oplus}{S}Me_2}{|}}{C}}H \longrightarrow \underset{(dimer)}{Me_3SiCH_2BH_2} + Me_2S \quad (84)$$

With trimethylboron, the product $Me_3SiCHMeBMe_2$, is obtained. This procedure, when coupled with an oxidation step, is used to obtain homologated organic derivatives from functionally substituted ylides of sulfur and of arsenic (*1818*).

$$R_3B + Me_2\overset{+}{S}\overset{-}{C}HCO_2Et \longrightarrow [R_2BCHRCO_2Et] + Me_2S \xrightarrow{alkali} RCH_2CO_2Et \quad (85)$$

Cleavage of cyclopropanes occurs with B_2H_6 and B_4H_{10} to give, ostensibly, organoboron products; however, these were not isolated but were oxidatively converted to alcohols (*1604*).

Isocyanides react with diborane (*156*, *176*) to give cyclic N—C—B compounds. This probably proceeds through an unstable intermediate adduct, $RNC:BH_3$. Tetrabis(dimethylamino)diboron reacts with hydrogen cyanide to give a tetracyano derivative:

$$B_2[N(CH_3)_2]_4 + 4\,HCN \longrightarrow (CH_3)_2NH \rightarrow B(CN)_2{-}B(CN)_2 \leftarrow HN(CH_3)_2 + 2\,(CH_3)_2NH \quad (86)$$

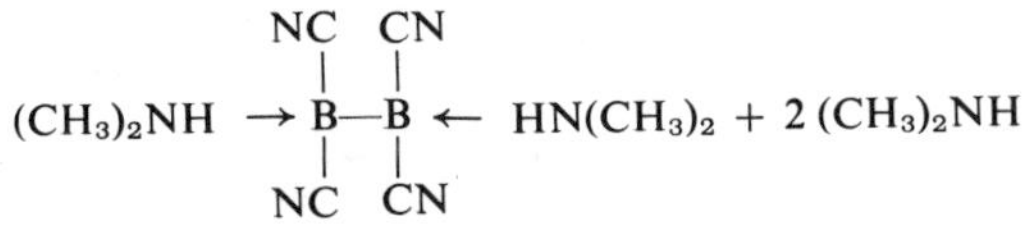

The same product is obtained from $B_2Cl_4 \cdot 2HN(CH_3)_2$ and AgCN (*1149*).

Alkyl and aryl groups may be introduced onto a cyclic dialkylboronhydride by a two-step sequence which entails (a) the preparation of a quaternary salt and (b) removal of the hydridic hydrogen by an appropriate acid (*279*).

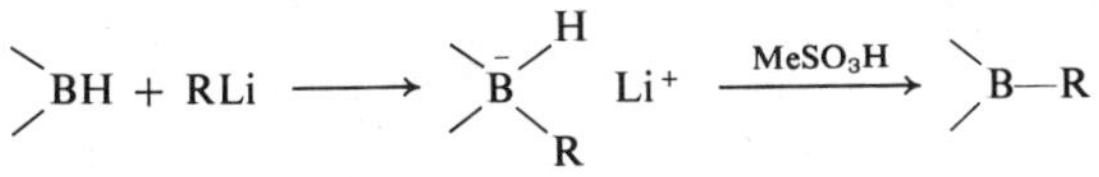

$$\rangle BH + RLi \longrightarrow \rangle \overset{-}{B}(H)(R)\ Li^+ \xrightarrow{MeSO_3H} \rangle B{-}R \quad (87)$$

B-Trichloroborazine undergoes an insertion reaction with diazomethane to give $(ClCH_2BNH)_3$.

Phenylboron dibromide can be prepared by the direct interaction of benzene with elemental boron and bromine in the presence of a nickel catalyst (*623*):

$$C_6H_6 + B + 1.5\,Br_2 \longrightarrow PhBBr_2 + HBr \quad (88)$$

Aryl iodides react with boron triiodide in the absence of a catalyst to give good yields of $ArBI_2$ (*1660*, *1703*, *1705*) by a mechanism which probably entails electrophilic attack at the iodine-attached carbon (*1703*).

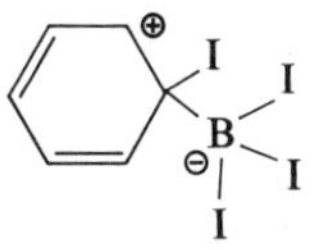

I I B ⊖ I I ⊕

Photolysis of a benzene/BX_3 (X = Br, I) mixture gives the dihaloborylbenzene. This reaction probably proceeds by photoactivation of an intermediate complex to give an addition product (**XVII**), which subsequently eliminates hydrogen halide (*150*).

Phenyldichloroborane appears to behave as an organometallic reagent, Eq. 89a, when reacted with certain unsaturated organic groups (*870*, *871*, *1046*).

$$C_6H_6 \xrightarrow[h\nu]{BX_3} C_6H_5BX_2; \quad C_6H_6 \xrightarrow{BX_3} \text{complex} \xrightarrow{h\nu} \text{(XVII)} \xrightarrow{-HX} C_6H_5BX_2 \tag{89}$$

(XVII)

$$\text{norbornadiene} \xrightarrow{PhBCl_2} \text{Ph-nortricyclyl-}BCl_2 \tag{89a}$$

Carbon monoxide reacts with diborane in the presence of catalytic amounts of borohydride ion to give trimethylboroxine (*1593*) (Eq. 90) Analogous

$$3\,CO + 1.5\,B_2H_6 \xrightarrow{BH_4^-} (MeBO)_3 \tag{90}$$

carbonylation reactions have been carried out using trialkylboron compounds (Section 3-2).

Thioboration of acetylenes give rise to β-thioalkenylboranes (*1330a*).

$$B(SEt)_3 + 3\,CH{\equiv}COEt \longrightarrow B\left(CH{=}C\begin{matrix}OEt\\SEt\end{matrix}\right)_3 \tag{91}$$

Several approaches to the preparation of polyboron substituted organic compounds, in addition to the addition of B_2X_4 to unsaturated bonds (see above), are worth special mention. Reminiscent of the Wurtz reaction, active metals can be used to piece together a B—B bond from an alkylboron halide (*179*, *1467*) (Eq. 92), or to piece together a B—C bond from a mixture of a boryl halide and organic halide (*386*) (Eq. 93). In an attempt to prepare a

$$2\,Me_2N\overset{R}{\overset{|}{B}}Cl + 2\,Na \longrightarrow Me_2N\overset{R}{\overset{|}{B}}-\overset{R}{\overset{|}{B}}NMe_2 + 2\,NaCl \qquad (R = Et, Ph) \tag{92}$$

$$4\,(MeO)_2\,BCl + CCl_4 + 8\,Li \xrightarrow{THF} [(MeO)_2B]_4C + 8\,LiCl \tag{93}$$

tetraalkyldiboron compound from an analogous reaction, only disproportionation products are obtained (*1883*) (Eq. 94). Hydroboration of vinyl-

$$BCl_3 + 2\,BMe_3 \longrightarrow 3\,[Me_2BCl] \xrightarrow[(-3\,NaCl)]{Na} B + 2\,Me_3B \tag{94}$$

boronic esters leads to both 1,1- and 1,2-diboryl products (*1189, 1190, 1247*) (Eq. 95). Acetylene is hydroborated to give both *gem*-diboryl (*130, 323, 367a, 1962a, 1963*) and 1,2-addition products (*414*) (Eq. 96). A common way to

$$CH_2{=}CH{-}B(OBu)_2 \xrightarrow[2.\ BuOH\ (1189)]{1.\ BH_3\cdot THF} CH_3\underset{B(OBu)_2}{\underset{|}{\overset{B(OBu)_2}{\overset{|}{C}}}}H + (BuO)_2BCH_2CH_2B(OBu)_2 \tag{95}$$

$$C_2H_2 + B_2H_6 \xrightarrow{MeOCH_2CH_2OMe} [H_2BCH_2CH_2BH_2] \xrightarrow{H_2O} (HO)_2BCH_2CH_2B(OH)_2$$

$$[H_2BCH_2CH_2BH_2] \xrightarrow{2\,C_2H_2} [B_2(C_2H_4)_3]_n \tag{96}$$

$$F_2BCH_2CH_2BF_2 \xleftarrow{BF_3} [B_2(C_2H_4)_3]_n \xrightarrow{BCl_3} Cl_2BCH_2CH_2BCl_3$$

produce α,ω-diboron compounds takes advantage of the terminal mode of B—H addition to olefins as well as the ease with which B,B exchange can occur (*1931–1935*) (Eqs. 97–98).

$$CH_2{=}CH(CH_2)_nCH{=}CH_2 \xrightarrow{hydroboration} \xrightarrow{BCl_3} Cl_2B(CH_2)_{n+4}BCl_2 \tag{97}$$

$$HC{\equiv}CH + B_2H_6 \xrightarrow{THF} \underset{polymer}{(C_2H_4B_2)_x} \xrightarrow[180^\circ\text{–}200^\circ C]{BCl_3} Cl_2BCH_2CH_2BCl_2 + Cl_2B\underset{Me}{\underset{|}{C}}HBCl_2 \tag{98}$$

Metal–boron exchange has been useful in preparing diborylmethanes such as $CH_2[B(OH)_2]_2$ (*1177*).

Nearly complete listings of the various types of tricoordinate boron compounds are given in Tables 3-4 through 3-11.

TABLE 3-4

Borane Derivatives (BRR′R″)

Compound	Reference	Compound	Reference
$H_2BCH_2CH_2CH_2NMe_2$	*21*	Et_2B(1-bromoethyl)	*309*
$H_2BCH_2CH_2CH_2SMe_2$	*153*	$Et_2BCEt{=}CMe(SiMe_3)$	*131a*
$H_2BCH(t\text{-Bu})CH_2(t\text{-Bu})$	*1133*	$EtB(CH_2CH_2OR)_2$	*1254*
$H_2B[o\text{-}(R_2NCH_2)C_6H_4]$	*388*	$EtB(CH{=}CHCl)_2$	*665, 666*
$H_2B[o\text{-}(R_2NCH_2CH_2)C_6H_4]$	*388*	$B(CH_2CH_2OR)_3$	*1254, 1327*
$HB(CH_2CH_2CH_2NR_2)(n\text{-Bu})$	*1390*	$B(CH_2CH_2SiR_3)_3$	*1253, 1669, 1670*
$HB(CMe_2CHMe_2)_2$	*1419*	$B[CH(CH_3)SiMe_3]_3$	*1669, 1670*
$HB(Me)(NMe_2)$	*351, 746a*	BVi_3	*670, 1529, 1781*
$HB(Et)(NR_2)$	*942, 943*	$B(CH{=}CHOR)_3$	*331, 331a*
$HB(n\text{-Bu})(NR_2)$	*1468, 1624*	$B(CF{=}CF_2)_3$	*447, 1735, 1738*
$HB(i\text{-Bu})(NR_2)$	*767*	$B(CH{=}CHCl)_3$	*666*
$HB(s\text{-Bu})(NR_2)$	*767, 1624*	$B[CH{=}C(OEt)(SEt)]_3$	*1330a*
$HB(t\text{-Bu})(NR_2)$	*767, 1624*	$Et_2B(n\text{-Pr})$	*951, 959, 972*
$HB(Ph)(NR_2)$	*1280*	$Et_2B(i\text{-Pr})$	*951*
$HB(Me)(PR_2)$	*359*	Et_2B(cyclopropyl)	*132*
BMe_3	*91, 106a, 183, 190, 338, 381, 617, 618, 680, 817, 858, 972, 1080, 1084, 1490, 1653, 1683, 1740, 1939, 1941, 1946*	$Et_2BC{\equiv}CR$	*131*
$Me_2BCH_2NH_2$	*1644*	$Et_2B(n\text{-Bu})$	*416, 932*
$Me_2BCH_2N_3$	*1644*	$Et_2B(i\text{-Bu})$	*951*
Me_2BCH_2OH	*1644*	$Et_2B(t\text{-Bu})$	*760, 766*
$Me_2BCH_2PMe_2$	*1592*	Et_2BCp	*719, 719a*
$Me_2BCH_2SMe_2$	*1592*	$Et_2B(MeCp)$	*719*
Me_2BCH_2Cl	*1644*	$Et_2BC_{12}H_{25}$	*932*
$Me_2BCH_2AsMe_2$	*1592*	$Et_2B(o\text{-}PhC_6H_4)$	*947*
Me_2BCH_2I	*1592*	$EtB(n\text{-Pr})_2$	*951, 959*
$B(CN)_3$	*391, 729*	$EtB(n\text{-Pr})(i\text{-Pr})$	*951*
$B(CH_2SiR_3)_3$	*1022, 1213, 1670*	$H_2NCH_2CH_2B(n\text{-Pr})_2$	*1284*
		$ROCH{=}CHBPr_2$	*331, 331a*
		$EtB(n\text{-Bu})_2$	*1323*
		$EtB(i\text{-Bu})_2$	*951*
		$H_2NCH_2CH_2B(n\text{-Bu})_2$	*1284*
		$ROCH{=}CHBBu_2$	*331, 331a*

Compound	References
Me_2BEt	*132*
$Me_2BCH_2CH_2SiMe_3$	*1213*
Me_2BVi	*670, 1527, 1529*
$Me_2B(n\text{-}Pr)$	*132*
Me_2B(cyclopropyl)	*442*
$Me_2BCH_2CH_2CF_3$	*132*
$Me_2BCH(CF_3)CH_3$	*132*
$Me_2BCH_2CH{=}CH_2$	*1338*
$Me_2BCH{=}CHCH_3$	*1529*
$Me_2B(n\text{-}Bu)$	*1551, 1564*
$Me_2B(i\text{-}Bu)$	*966*
Me_2B(2-thiophene)	*1700*
$Me_2BCMe{=}CMe(SiMe_3)$	*131a*
$MeBEt_2$	*132*
$MeBVi_2$	*670, 1529*
$MeB(n\text{-}Pr)_2$	*132*
$MeB[CH(CF_3)CH_3]_2$	*132*
$MeB(CH{=}CHCH_3)_2$	*1527, 1529*
$MeB(n\text{-}Pr)(i\text{-}Bu)$	*966*
$MeB(n\text{-}Bu)_2$	*57*
$MeB(i\text{-}Bu)_2$	*966*
$(CO_2Et)B(n\text{-}Bu)_2$	*1626*
$NCB(n\text{-}Bu)_2$	*588, 1256*
$NCB(i\text{-}Bu)_2$	*1256*
$NCB(i\text{-}C_5H_{11})_2$	*1256*
$(CF_3)B(n\text{-}Bu)_2$	*1528*
BEt_3	*26, 48, 183, 338, 616–619, 932, 952, 980, 1487, 1490, 1614, 1683, 1740, 1751, 1922, 1937–1939, 1941, 1956*
	1059, 1134, 1138, 1614, 1937, 1944, 1954
$(n\text{-}Pr)_2B(i\text{-}Pr)$	*951*
$B(i\text{-}Pr)_3$	*994, 1359*
B(cyclopropyl)$_3$	*130, 442*
$(n\text{-}Pr)_2BCH_2CH_2CH_2NH_2$	*1390*
$B[CH_2CH_2CH_2B(OMe)_2]_3$	*1320*
$B(CH_2CH_2CH_2OR)_3$	*1143, 1327*
$B(CH_2CH_2CF_3)_3$	*132*
$B[(CH_2)_3SiMe_3]_3$	*1327, 1681*
$B(CH_2CH_2CH_2Cl)_3$	*771*
$B(CH_2CHBrCH_2Br)_3$	*1785*
$B(CH_2CH{=}CH_2)_3$	*1337, 1781, 1783, 1785, 1944*
$(n\text{-}Pr)_2B(i\text{-}Bu)$	*966*
$(n\text{-}Pr)_2B(t\text{-}Bu)$	*766, 1207*
$(CH_2{=}CHCH_2)_2B(n\text{-}Bu)$	*1337*
$(CH_2{=}CHCH_2)_2B[CH_2{=}C(R)CH_2CH{=}CH_2)$	*333, 334, 1273, 1275*
$(CH_2{=}CHCH_2)_2B[CH_2C(Me){=}CHCH_2CH{=}CH_2]$	*336*
$(CH_2{=}CHCH_2)_2B(\overline{CHCH_2C}MeCH_2CH{=}CH_2]$	*336*
$(n\text{-}Pr)B(n\text{-}Bu)_2$	*865, 1323*
$(n\text{-}Pr)B(i\text{-}Bu)_2$	*966*
$(H_2NCH_2CH_2CH_2)B(n\text{-}Bu)_2$	*1390*
$(CH_2{=}CHCH_2)B(n\text{-}Bu)_2$	*1337*
$(CH_2{=}CHCH_2)B[CH_2{=}C(OR)CH_2CH{=}CH_2]_2$	*1273*
$B(n\text{-}Bu)_3$	*14, 338, 505, 781, 951, 1059, 1323, 1359,*

(*continued*)

TABLE 3-4—(*continued*)

Compound	Reference
B(*n*-Bu)$_3$ (*continued*)	*1614, 1937, 1942, 1943, 1954*
(*n*-Bu)$_2$B(*t*-Bu)	*760, 766*
(*n*-Bu)B(*i*-Bu)$_2$	*212*
(*n*-Bu)B(*s*-Bu)$_2$	*212*
(*n*-Bu)B(*t*-Bu)$_2$	*783, 1207*
B(*i*-Bu)$_3$	*48, 212, 487, 489, 565, 781, 784, 918, 932, 951, 992, 1937, 1938, 1954*
(*i*-Bu)$_2$B(*s*-Bu)	*212*
(*i*-Bu)$_2$B(*t*-Bu)	*487, 489, 490, 760, 766, 784*
B(*s*-Bu)$_3$	*483, 505, 781*
(*s*-Bu)$_2$B(*i*-Bu)	*212*
B(*t*-Bu)$_3$	*994*
(*t*-Bu)B(1-butenyl)$_2$	*766*
B[(CH$_2$)$_4$OR]$_3$	*1143*
B[(CH$_2$)$_3$CH$_2$Cl]$_3$	*959*
B(CH$_2$CHMeCH$_2$Cl)$_3$	*959*
B(CH$_2$CH$_2$CH=CH$_2$)$_3$	*1143*
B(CH$_2$CMe=CH$_2$)$_3$	*140a*
(*t*-Bu)B(*i*-Bu)(*n*-amyl)	*787, 1207*
(*t*-Bu)$_2$B(*n*-amyl)	*787, 1207*
(*n*-Bu)$_2$B(C≡CCMe=CH$_2$)	*1565*
(*n*-Bu)$_2$B(cyclopentyl)	*1561*
(*i*-Bu)$_2$B(cyclopentyl)	*212*
(*n*-Bu)$_2$B(cyclohexyl)	*1561*
(*i*-Bu)$_2$B(CH$_2$Ph)	*883*
(*n*-Bu)$_2$B[C(O)Ph]	*1656*
(*n*-Bu)$_2$B[C(S)Ph]	*1656*
B(2-Et-hexyl)$_3$	*1891*
B[CH$_2$CHBr(CH$_2$)$_5$CH$_3$]$_3$	*1315*
B(CH$_2$CH$_2$CH—CH$_2$CH$_2$CH=CH—CH$_2$)$_3$ (ring: CH and CH$_2$ joined)	*959*
B(cyclooctyl)$_3$	*1891*
B(CH$_2$CH$_2$Ph)$_3$	*752, 1741*
B(CH=CHPh)$_3$	*1784*
B(C≡CPh)$_3$	*999–1001*
[PhC(Cl)=CH]$_2$BPh	*1046*
(PhC≡C)$_2$BPh	*1046*
(PhC≡C)B(Ph)$_2$	*1046*
B(CH$_2$CHPhCH$_3$)$_3$	*913, 1891*
B(*n*-decyl)$_3$	*298, 932*
B(undecyl)$_3$	*1891*
B(C$_{12}$H$_{25}$)$_3$	*880*
B(cyclododecyl)$_3$	*1891*
B(C$_{18}$H$_{37}$)$_3$	*880*
B(*cis*-myrtanyl)$_3$	*959*
B(Ph)$_3$	*603, 951, 952, 971, 982, 990, 993, 1278, 1279, 1426, 1448, 1449, 1450, 1451, 1908, 1910*
(Ph)$_2$B(*o*-tolyl)	*1301*
(Ph)$_2$B(*p*-tolyl)	*1355*
PhB(*o*-tolyl)$_2$	*1789*
PhB(*o*-tolyl)(*p*-tolyl)	*1301*
PhB(*o*-tolyl)(*p*-ClC$_6$H$_4$)	*1301*
PhB(α-naphthyl)$_2$	*1355*
(*o*-Tolyl)B(α-naphthyl)$_2$	*1355*

Compound	References	Compound	References
$(n\text{-Bu})_2B(n\text{-}C_8H_{17})$	*1561*	$B(m\text{-tolyl})_3$	*1906*
$(i\text{-Bu})_2BPh$	*883*	$B(p\text{-tolyl})_3$	*990, 996, 1585, 1906*
$(i\text{-Bu})_2B(o\text{-MeC}_6H_4)$	*883*	$B(o\text{-PhC}_6H_4)_3$	*1906*
$(t\text{-Bu})B(n\text{-amyl})_2$	*782*	$B(p\text{-R}_2NC_6H_4)_3$	*1906*
$(t\text{-Bu})B(1\text{-pentenyl})_2$	*766*	$B(p\text{-MeOC}_6H_4)_3$	*994, 1426*
$(t\text{-Bu})B(1\text{-hexenyl})_2$	*766*	$B(ClC_6H_4)_3$	*1074*
$(s\text{-Bu})B(\text{isopinocampheyl})_2$	*26, 186, 320*	$(p\text{-XC}_6H_4)B(\text{mesityl})_2$	*561*
$B(n\text{-}C_5H_{11})_3$	*880, 1059*	$B(p\text{-xylyl})_3$	*997*
$B(i\text{-}C_5H_{11})_3$	*992, 1359, 1937*	$B(\text{mesityl})_3$	*203, 748, 1038, 1585*
$B(CH_2CHMeCH_2CH_3)_3$	*243, 752*	$B(2,4,6\text{-Me}_3\text{-}3,5\text{-NO}_2C_6)_3$	*748*
$B(CH_2CMe_3)_3$	*1616*	$B(C_6F_5)_3$	*1164, 1165, 1167*
$B(\text{cyclopentyl})_3$	*752*	$B(\alpha\text{-naphthyl}_3$	*203, 299, 990, 997, 1585*
$B[(CH_2)_5OR]_3$	*1143*	$B(2\text{-MeO-naphth-1-yl})_3$	*1038*
$B(CH_2CH_2CH_2CH{=}CH_2)_3$	*1143*	B[2-(5-methylthienyl)]$_3$ (structure: B[thiophene ring with S and Me]$_3$)	*1700*
$B[CH_2{=}CHCH_2C(R){=}CH_2]$	*1273, 1275*	$Me_2B(NR_2)$	*8, 61, 94, 96, 97, 99, 102, 105, 139, 165, 350, 351, 421, 586, 587, 681, 686, 746a, 829, 862, 1225, 1226, 1403, 1433, 1434, 1440, 1443, 1459, 1460, 1462, 1489, 1554, 1651, 1856, 1878–1880*
$B(Cp)_3$	*1579*	$(NC)_2B(NR_2)$	*115*
$B(1\text{-}C_6H_{13})_3$	*48, 294, 298, 755, 959, 1059*	$MeB(Et)(NR_2)$	*1438, 1592*
$B[CH_2CH(Me)(n\text{-Pr})]_3$	*485, 755*	$MeB(Et)(N_3)$	*1592*
$B[(CH_2)_4CH{=}CH_2]_3$	*1141*	$MeB(Vi)(NMe_2)$	*1440*
$B[CH_2CHBr(CH_2)_3CH_3]_3$	*1315*	$MeB(n\text{-Bu})(NR_2)$	*1468*
$B[CH(Et)CH(Br)(Et)]_3$	*1315*		
$B(\text{cyclohexyl})_3$	*48, 603, 755, 931, 932, 990, 996, 1891*		
$(\text{Cyclohexyl})_2B(n\text{-}C_8H_{17})$	*1561*		
$B(\text{exonorbornyl})_3$	*233*		
$B(CH_2Ph)_3$	*335a, 951, 972, 997, 1589, 1891*		
$B[CH_2(p\text{-XC}_6H_4)]$ X = Me, F, MeO	*1589*		
$(n\text{-}C_7H_{15})_2B(n\text{-}C_8H_{17})$	*1561*		
$(n\text{-}C_7H_{15})B(n\text{-}C_8H_{17})_2$	*1561*		
$B(1\text{-octyl})_3$	*48, 298, 755, 880, 932, 959, 1891*		

(continued)

TABLE 3-4—(*continued*)

Compound	Reference	Compound	Reference
$MeB[C(Br)Ph_2](NR_2)$	*1225*	$MeB(Ph)(N_3)$	*1514*
$MeB(Ph)(NR_2)$	*61, 62, 71, 101, 165, 1436*	Bu_2BN_3	*1514*
$(Cl_3C)B(Ph)(NMe_2)$	*733*	$BuB(Ph)(N_3)$	*1514*
$(Et)_2B(NR_2)$	*862, 931, 942–944, 952, 1264, 1403, 1433, 1434, 1436, 1438, 1443, 1460, 1475, 1489, 1491, 1856, 1857*	$(Cyclohexyl)B(Ph)(N_3)$	*1514*
$EtB(Ph)(NR_2)$	*1436*	Ph_2BN_3	*1510*
$ViB(Ph)(NMe_2)$	*1440*	$PhB(o\text{-tolyl})(N_3)$	*1514*
$(n\text{-}Pr)_2B(NR_2)$	*556b, 558, 587, 942, 943, 1058–1060, 1258–1261, 1264, 1284, 1361, 1460, 1461, 1472, 1475, 1624*	$PhB(p\text{-tolyl})(N_3)$	*1514*
$(i\text{-}Pr)_2B(NR_2)$	*1259*	$PhB(p\text{-}ClC_6H_4)(N_3)$	*1514*
$(CH_2{=}CHCH_2)_2B(NR_2)$	*1337, 1341*	$PhB(m\text{-}CF_3C_6H_4)(N_3)$	*1514*
$(n\text{-}Bu)_2B(NR_2)_2$	*144, 427, 492, 555, 556, 648, 653, 815, 853, 942, 943, 964, 1058–1060, 1240, 1257, 1258, 1260, 1261, 1264, 1265, 1280, 1284, 1345, 1346, 1361, 1403, 1443, 1459–1461, 1469, 1472, 1475, 1624, 1690, 1856*	$PhB(naphthyl)(N_3)$	*1514*
		$(o\text{-}Tolyl)_2B(N_3)$	*1514*
		$(p\text{-}Tolyl)_2B(N_3)$	*1514*
		$(p\text{-}MeOC_6H_4)_2B(N_3)$	*1514*
		$(p\text{-}ClC_6H_4)_2B(N_3)$	*1514*
		Me_2BOH	*680, 1654, 1821, 1822*
		$MeB(Et)(OH)$	*1592*
		$MeB(Ph)(OH)$	*1245*
		Et_2BOH	*1215*
		$EtB(s\text{-}Bu)(OH)$	*309, 311*
		$EtB(Ph)(OH)$	*563, 1245*
		$(R_3SiCH_2CH_2)_2BOH$	*1253*
		$(n\text{-}Pr)_2BOH$	*742, 1328*
		$(i\text{-}Pr)_2BOH$	*742*
		$(n\text{-}Pr)B(n\text{-}Bu)(OH)$	*1328*
		$(n\text{-}Pr)B(Ph)(OH)$	*1245*
		$(i\text{-}Pr)B(Ph)(OH)$	*563*
		$(n\text{-}Bu)_2BOH$	*650, 742, 1106, 1328*
		$(i\text{-}Bu)_2BOH$	*742, 775*
		$(n\text{-}Bu)B(Ph)(OH)$	*1245*
		$(t\text{-}Bu)B(Ph)(OH)$	*563*
		$(i\text{-}Pentyl)_2BOH$	*1106*

Compound	References
$(C_5H_{11})_2B(NR_2)$	861, 1060, 1201, 1264, 1433, 1444
$(n\text{-}C_5H_{11})_2B[N(Ph)CONEt_2]$	120
$(n\text{-}C_6H_{13})_2B(NR_2)$	943, 1060
$n\text{-}C_6H_{11}B[(CH_2)_4CHClCH_3]NEt_2$	1836c
(*n*-Hexyl)B(Ph)(NR_2)	1433
$(PhCH_2)B(Ph)(NR_2)$	165
$(n\text{-}C_8H_{17})_2B(NEt_2)$	1280
$(n\text{-}C_{10}H_{21})_2B(NR_2)$	943
$[Ph_2C(Br)]B(Ph)(NR_2)$	1225
$(Ph)_2B(NR_2)$	35, 61, 101, 102, 420, 427, 455, 466, 648, 653, 815, 860, 862, 1043, 1048, 1049, 1225, 1243, 1280–1282, 1289, 1291, 1293, 1295, 1311, 1432–1434, 1436, 1445, 1450, 1460, 1462, 1475, 1476, 1479, 1509, 1624, 1633, 1748, 1749, 1824, 1846
PhB(*p*-BrC_6H_4)(NR_2)	1433
(*o*-Tolyl)$_2$B(NR_2)	1438, 1449
(*p*-Tolyl)$_2$B(NR_2)	420, 422, 1434, 1438
(*p*-FC_6H_4)$_2$B(NR_2)	1438
(*p*-BrC_6H_4)$_2$B(NR_2)	420, 1433
(Mesityl)$_2$B(NR_2)	422, 862, 1073, 1748, 1749
(α-Naphthyl)$_2$BNHR	1281, 1282, 1293, 1295, 1311, 1450
Me_2BN_3	1516
	703b, 920, 1234, 1237, 1352a, 1575, 1850
PhB(*o*-tolyl)(OH)	1301
PhB(*p*-tolyl)(OH)	1292
PhB(*p*-ClC_6H_4)(OH)	1301
$PhB(CH_2Ph)(OH)$	1448
PhB(2-biphenyl)(OH)	478
PhB(1-naphthyl)(OH)	1104, 1292
(*o*-Tolyl)$_2$BOH	1357, 1575
(*m*-Tolyl)$_2$BOH	1575
(*p*-Tolyl)$_2$BOH	1357, 1450, 1575
(*p*-Anisyl)$_2$BOH	920, 1575
(*p*-Chlorophenyl)$_2$BOH	1575
(*p*-BrC_6H_4)$_2$BOH	920
(Mesityl)$_2$BOH	748
$(C_6F_5)_2BOH$	396
(2-Biphenyl)$_2$BOH	478
(*p*-Biphenyl)$_2$BOH	1575
(1-Naphthyl)$_2$BOH	1356, 1450, 1575
(β-Naphthyl)$_2$BOH	920
Me_2BOR	348, 359, 655, 684, 1530, 1555, 1822, 1881
Me_2BOOMe	1566
MeB(Et)(OR)	1592
MeB(Ph)(OR)	1244, 1245, 1791
Et_2BOR	597, 976, 1214, 1215, 1360, 1491, 1696a, 1807
$EtB(CH_2CH_2OR)(OR)$	1254
EtB(*n*-Pr)(OR)	1328
EtB[CH(*o*-Bu)CH_2CCl_3](OR)	1181

(*continued*)

TABLE 3-4—(*continued*)

Compound	Reference
EtB[CH(Br)CH_2CCl_3](OR)	*1181*
EtB(*n*-Bu)(OR)	*1328*
EtB(*s*-Bu)(OR)	*309*
EtB[CEt_2CHMe($SiMe_3$)]OR	*131a*
EtB(Ph)(OR)	*563, 588a, 1244, 1245, 1791*
($R_3SiCH_2CH_2$)$_2$BOR	*1022, 1253*
($ROCH_2CH_2$)$_2$BOR	*1254*
(ROCH=CH)$_2$BOR	*331*
ViB(Ph)(OR)	*1181*
ViB(*o*-tolyl)(OR)	*1181*
ViB(2,5-$Me_2C_6H_3$)(OR)	*1181*
ViB(mesityl)(OR)	*1181*
HC≡CB(Ph)(OR)	*1187*
HC≡CB(2,5-$Me_2C_6H_3$)(OR)	*1187*
(*n*-Pr)$_2$BOR	*976, 1059, 1240, 1258, 1261, 1263, 1282a, 1323, 1345–1348, 1459, 1550, 1696a*
(*i*-Pr)$_2$BOR	*1359, 1361*
(*n*-Pr)B[(CH_2)$_3$OH]OR	*976*
(*n*-Pr)B(*n*-Bu)(OR)	*1322*
(*n*-Pr)B(Ph)(OR)	*1241, 1245*
(*i*-Pr)B(Ph)(OR)	*563*
$R_2NCH_2CH_2CH_2$B(*n*-Bu)(OR)	*1390*
$HSCH_2CH_2CH_2$B(*n*-Bu)(OR)	*1286*
$Cl_3CCH_2CHBrB[CH_2CH_2CH(CH_3)_2]$(OR)	*1181*
Cl_3CCH_2CHBrB(Ph)(OR)	*1181*
Cl_3CCH_2CHBrB(2,5-$Me_2C_6H_3$)(OR)	*1181, 1187*
Cl_3CCH_2CHBrB(mesityl)(OR)	*1181*
Ph_2BOR	*2–6, 9–11, 407, 422, 563, 572, 576a, 651, 652, 758, 1049, 1107, 1109, 1242, 1277, 1352a, 1356, 1357, 1382a, 1427, 1449, 1450, 1481, 1609, 1711, 1729, 1798, 1824, 1825, 1959*
PhB(*o*-tolyl)(OR)	*667, 1301, 1729, 1790, 1798*
PhB(*m*-tolyl)(OR)	*667, 1729*
PhB(*p*-tolyl)(OR)	*667, 1292, 1729*
PhB(*p*-$MeOC_6H_4$)(OR)	*1729*
PhB(*p*-BrC_6H_4)(OR)	*1292*
PhB(2-biphenyl)(OR)	*1798*
PhB(2,4,6-*t*-$Bu_3C_6H_2$)OR	*1732b*
PhB(1-naphthyl)(OR)	*1104, 1109, 1292, 1450, 1798*
PhB(2-thiophene)(OR)	*1619*
(*o*-Tolyl)$_2$BOR	*422, 1356, 1357, 1449*
(*o*-Tolyl)B(*m*-tolyl)(OR)	*1790*
(*o*-Tolyl)B(*p*-tolyl)(OR)	*1790*
(*o*-Tolyl)B(*p*-ClC_6H_4)(OR)	*1790*
(*o*-Tolyl)B(*p*-BrC_6H_4)(OR)	*1301*
(*m*-Tolyl)$_2$BOR	*1798*
(*m*-Tolyl)B(*p*-ClC_6H_4)(OR)	*1798*
(*p*-Tolyl)$_2$BOR	*422, 758, 1356, 1357, 1427, 1449, 1450, 1798*
(*p*-Tolyl)B(*p*-$MeOC_6H_4$)(OR)	*667*

Compound	References
$(CH_2{=}CHCH_2)_2BOR$	1267, 1621, 1782, 1783, 1785
$(n\text{-}Bu)_2BOR$	211, 275, 572, 609, 650–652, 853, 865, 867, 913, 976, 1059, 1106, 1162, 1240, 1263, 1264, 1322, 1347–1349, 1359, 1361, 1395a, 1696a
$(s\text{-}Bu)_2BOR$	211, 275
$(i\text{-}Bu)_2BOR$	211, 275, 488, 489, 1361
$(i\text{-}Bu)B(t\text{-}Bu)(O_2CH_2CHMe_2)$	488
$n\text{-}BuB(CH_2CH_2CH_2CH_2Br)OR$	1836a
$n\text{-}BuB(CH_2CH_2CH_2CH_2NR_2)OR'$	1836b
$[Br(CH_2)_4]_2BOR$	1836a
$(n\text{-}Bu)B(Ph)(OR)$	1102, 1241, 1278, 1449
$(i\text{-}Bu)B(Ph)(OR)$	563
$(n\text{-}Pentyl)_2BOR$	304, 305, 1059, 1444
$(i\text{-}Pentyl)_2BOR$	1106, 1347, 1348, 1359, 1361
$(CH_2CH_2CHMeCH_2)_2BOMe$	243
$(Me_2CHCHMe)_2BOMe$	226
$(Cyclopentyl)_2BOR$	211, 275, 304, 305
$(n\text{-}C_6H_{13})_2BOR$	1059, 1348
$n\text{-}C_6H_{13}B[(CH_2)_4CHXMe]OMe$ X = Cl, Br	1836c
$(Cyclohexyl)_2BOR$	211, 1928
$(n\text{-}C_7H_{15})_2BOR$	1313
$(Norbornyl)_2BOR$	275
$(Isopinocampheyl)_2BOMe$	226
$PhCH_2B(Ph)(OR)$	563, 1102, 1448
$(PhCH_2)_2BOR$	951, 976
$(PhC{\equiv}C)_2BOR$	422
$(o\text{-}MeOC_6H_4)_2BOR$	422
$(3,4\text{-}Me_2C_6H_3)_2BOR$	422
$(p\text{-}ClC_6H_4)_2BOR$	422, 758, 1277, 1353, 1450
$(p\text{-}BrC_6H_4)_2BOR$	422, 758, 1107, 1353
$(p\text{-}CH_3OC_6H_4)_2BOR$	758, 1798
$(p\text{-}Vinylbenzene)_2BOR$	1098
$(Mesityl)_2BOR$	748
$(2\text{-}Biphenyl)_2BOR$	478
$(4\text{-}Biphenyl)_2BOR$	422
$(1\text{-}Naphthyl)_2BOR$	422, 976, 1104, 1109, 1353, 1358
$(2\text{-}Thiophene)_2BOR$	1619
$(Me)_2BF$	96, 98, 132, 348, 350, 358, 956, 1490
$MeB(Et)(F)$	1592
Et_2BF	132, 956, 1064, 1490, 1742
Vi_2BF	169
$(n\text{-}Pr)_2BF$	956, 1064
$(i\text{-}Pr)_2BF$	956
$(n\text{-}Bu)_2BF$	343, 547, 956, 1064
$(i\text{-}Bu)_2BF$	784
$(Cyclohexyl)_2BF$	956
$(PhCH_2)_2BF$	956
$(n\text{-}C_8H_{17})_2BF$	956
$(n\text{-}C_{10}H_{21})_2BF$	956
Ph_2BF	956, 1490
$(Mesityl)_2BF$	203, 422
Me_2BPH_2	359
Me_2BPMe_2	359, 829
$(n\text{-}Pr)_2BPR_2$	1484
Ph_2BPR_2	420
$(p\text{-}Tolyl)_2BPR_2$	420

(continued)

TABLE 3-4—(*continued*)

Compound	Reference
(*p*-BrC$_6$H$_4$)$_2$BPR$_2$	*420*
(*n*-Pr)$_2$BSH	*1256*
(*n*-Bu)$_2$BSH	*1256*
(*i*-C$_5$H$_{11}$)$_2$BSH	*1256*
Me$_2$BSR	*353, 360, 1701, 1828, 1829*
Et$_2$BSR	*1263, 1701*
(R$_3$SiCH$_2$CH$_2$)$_2$BSR	*1253*
(*n*-Pr)$_2$BSR	*1059, 1061, 1240, 1257, 1258, 1263, 1331, 1349, 1361, 1689*
(*i*-Pr)$_2$BSR	*1701*
R$_2$NCH$_2$CH$_2$CH$_2$B(*n*-Bu)(SR)	*1390*
(CH$_2$=CHCH$_2$)$_2$BSR	*1340, 1342*
(*n*-Bu)$_2$BSR	*760, 1059, 1061, 1240, 1257, 1258, 1263, 1349, 1361, 1701, 1704*
(*i*-Bu)$_2$BSR	*1331, 1361*
(*n*-Bu)B(Ph)(*n*-BuS)	*1278*
(*n*-Amyl)$_2$BSR	*760, 1207*
(*i*-C$_5$H$_{11}$)$_2$BSR	*1361*
(*n*-C$_5$H$_{13}$)$_2$BSR	*1061*
(*n*-C$_8$H$_{17}$)$_2$BSR	*1689*
[CH$_3$(CH$_2$)$_5$CH(*s*-Bu)CH$_2$]$_2$BSR	*1315*
(C$_6$H$_5$)$_2$BSR	*1311, 1701, 1704, 1706, 1829*
(α-Naphthyl)BSR	*1311, 1450*
Me$_2$BCl	*74, 95, 1490*
MeB(Et)(Cl)	*956, 1592, 1883*
MeB(Vi)(Cl)	*670*
MeB(Ph)(Cl)	*1244, 1514*
PhB(*o*-tolyl)(Cl)	*1301, 1514*
PhB(*p*-tolyl)(Cl)	*1292, 1514*
PhB(*p*-ClC$_6$H$_4$)(Cl)	*1301, 1514*
PhB(*m*-CF$_3$C$_6$H$_4$)(Cl)	*1514*
PhB(α-naphthyl)(Cl)	*1292, 1450*
(*o*-Tolyl)$_2$BCl	*1514*
(*p*-Tolyl)$_2$BCl	*1514*
(*p*-ClC$_6$H$_4$)$_2$BCl	*1514*
(*p*-MeOC$_6$H$_4$)$_2$BCl	*1514*
(C$_6$F$_5$)$_2$BCl	*396*
(α-Naphthyl)$_2$BCl	*1288, 1450*
Ph$_2$BAsPh$_2$	*420*
(*p*-Tolyl)$_2$BAsPh$_2$	*420*
(*p*-BrC$_6$H$_4$)$_2$BAsPh$_2$	*420*
Me$_2$BSeR	*1702*
Ph$_2$BSeR	*1702*
Me$_2$BBr	*655, 956, 1490, 1643*
Et$_2$BBr	*956, 1487, 1490*
EtB(*s*-Bu)(Br)	*309*
(*n*-Pr)$_2$BBr	*956, 1138*
(*i*-Pr)$_2$BBr	*956*
(*n*-Bu)$_2$BBr	*650, 655, 865, 956, 1162*
[CH$_3$(CH$_2$)$_3$CHBrCH$_2$]$_2$BBr	*1315*
(EtCHBrCHEt)$_2$BBr	*1315*
(Cyclohexyl)$_2$BBr	*956*
(*n*-C$_7$H$_{15}$)$_2$BBr	*1313*
(PhCH$_2$)$_2$BBr	*956*
(*n*-C$_8$H$_{17}$)$_2$BBr	*956*
[CH$_3$(CH$_2$)$_5$CHBrCH$_2$]$_2$BBr	*1315*
(*n*-C$_{10}$H$_{21}$)$_2$BBr	*956*
Ph$_2$BBr	*5, 603, 956, 1237, 1490*

Et_2BCl	*343, 616, 617, 956, 1044a, 1138, 1487, 1490*
$EtB(CH{=}CHCl)(Cl)$	*665, 666*
$EtB(Ph)(Cl)$	*1244*
Vi_2BCl	*169, 823*
$(ClCH{=}CH)_2BCl$	*666*
$(CF_2{=}CF)_2BCl$	*1735*
$(n\text{-Pr})_2BCl$	*553, 815, 956, 1138*
$(n\text{-Pr})B(n\text{-Bu})(Cl)$	*1324*
$(n\text{-Pr})B(Ph)(Cl)$	*1244*
$(i\text{-Pr})_2BCl$	*956*
$[Cl(CH_2)_3]_2BCl$	*771*
$(n\text{-Bu})_2BCl$	*144, 343, 650, 655, 815, 956, 1059, 1162, 1205, 1304*
$(n\text{-Bu})B(Ph)(Cl)$	*1244, 1514*
$(i\text{-Bu})_2BCl$	*1205*
$(s\text{-Bu})_2BCl$	*1205*
$(n\text{-Bu})B(Ph)(Cl)$	*2*
$(n\text{-Amyl})_2BCl$	*1059, 1063, 1444*
$n\text{-}C_6H_{13}B[(CH_2)_4CHClMe]Cl$	*1836c*
$(\text{Cyclopentyl})_2BCl$	*275*
$(n\text{-Hexyl})_2BCl$	*275, 1059, 1063*
$(\text{Cyclohexyl})_2BCl$	*742, 956*
$(\text{Cyclohexyl})B(Ph)(Cl)$	*1514*
$[BuC(Cl){=}CH]B[CH{=}C(Bu)(Ph)](Cl)$	*1046*
$(PhCH_2)_2BCl$	*956*
$(n\text{-}C_8H_{17})_2BCl$	*956*
$[PhC(Cl){=}CH]_2BCl$	*1046*
$(n\text{-}C_{10}H_{21})_2BCl$	*956*
$[PhC(Bu){=}CH]B(Ph)(Cl)$	*1046*
Ph_2BCl	*2, 5, 422, 603, 815, 956, 1234, 1288, 1432, 1490, 1802*
Me_2BI	*956, 1490*
Et_2BI	*956, 1487, 1490*
$(n\text{-Pr})_2BI$	*553, 956, 1134*
$(i\text{-Pr})_2BI$	*742, 956*
$(n\text{-Bu})_2BI$	*742, 956, 1162*
$(i\text{-Bu})_2BI$	*742*
$(\text{Cyclohexyl})_2BI$	*742, 956*
$(PhCH_2)_2BI$	*956*
$(n\text{-}C_8H_{17})_2BI$	*956*
$(n\text{-}C_{10}H_{21})_2BI$	*956*
Ph_2BI	*956, 1473, 1490, 1706*
$MeB(NR_2)_2$	*139, 454, 585, 646, 660, 686, 803, 1066, 1441, 1452, 1459, 1463, 1466, 1477, 1478, 1651, 1822, 1860*
$NCB(NR_2)_2$	*115*
$EtB(NR_2)_2$	*1066, 1441, 1466, 1491*
$ViB(NR_2)_2$	*626, 1066, 1170, 1441*
$(n\text{-Pr})B(NR_2)_2$	*759, 763, 1060, 1305, 1311, 1466*
$(i\text{-Pr})B(NR_2)_2$	*759, 763*
$(C_3F_7)B(NR_2)_2$	*406*
$(CH_2{=}CHCH_2)B(NR_2)_2$	*1342*
$(n\text{-Bu})B(NR_2)_2$	*529, 759, 763, 767, 1058, 1060, 1097, 1305, 1345, 1346, 1441, 1466, 1624, 1860*
$(i\text{-Bu})B(NR_2)_2$	*763*
$(s\text{-Bu})B(NR_2)_2$	*759, 763, 1624*
$(t\text{-Bu})B(NR_2)_2$	*763, 798, 1624*
$(n\text{-}C_5H_{11})B(NR_2)_2$	*763*
$(i\text{-}C_5H_{11})B(NR_2)_2$	*1058, 1302, 1305, 1311*
$(n\text{-}C_6H_{13})B(NR_2)_2$	*763, 1349*

(continued)

TABLE 3-4—*(continued)*

Compound	Reference
$CH_2CH(Me)C(Me)_2B(NR_2)_2$	*564*
(Cyclohexyl)$B(NR_2)_2$	*763*
$Et_3CB(NR_2)_2$	*380*
$PhCH_2B(NR_2)_2$	*763*
$PhCH_2CH_2B(NR_2)_2$	*1926a*
$(n\text{-}Bu)_3CB(NR_2)_2$	*798*
$(PhCH_2)_3CB(NR_2)_2$	*798*
$PhB(NR_2)_2$	*1, 70, 71, 120, 347, 383, 405, 432, 451, 454–456, 473, 474, 529, 537, 560, 608, 778, 803, 806, 860–862, 922, 1044, 1045, 1048, 1066, 1095, 1097, 1281, 1282, 1311, 1371, 1429, 1432, 1438, 1441, 1442, 1452, 1462, 1474–1478, 1486, 1510, 1512, 1520, 1624, 1749, 1860, 1926a, 1927*
$[p\text{-}(R_2N)_2BC_6H_4]B(NR_2)_2$	*68*
(*p*-Tolyl)$B(NR_2)_2$	*1045*
$(p\text{-}MeOC_6H_4)B(NR_2)_2$	*71, 1097*
$(p\text{-}BrC_6H_4)B(NR_2)_2$	*1097*
$PhCH_2B(NR_2)_2$	*1927*
(Mesityl)$B(NR_2)_2$	*748*
(α-Naphthyl)$B(NR_2)_2$	*1097, 1281, 1282*
$MeB(NR_2)(OH)$	*1879, 1880*
$MeB(NR_2)(OR)$	*61, 1491, 1822, 1879*
$CH_3CH(Me)C(Me)_2B(NR_2)(OR)$	*564*
$MeB(OH)_2$	*348, 680, 883, 1201, 1880*
$HOCH_2B(OH)_2$	*1151, 1177*
Naphthyl$CH_2B(OH)_2$	*1065a*
$EtB(OH)_2$	*616, 617, 1201, 1564*
$ROCH_2CH_2B(OH)_2$	*1254*
$H_2CONHCH_2CH_2B(OH)_2$	*363*
$RSCH_2CH_2B(OH)_2$	*1191*
$ViB(OH)_2$	*1455*
$AcSCH{=}CHB(OH)_2$	*1185*
$(n\text{-}Pr)B(OH)_2$	*883, 992, 1719, 1848*
$(i\text{-}Pr)B(OH)_2$	*1065a, 1201*
$Me_2C(OH)B(OH)_2$	*1188*
$MeCH{=}CHB(OH)_2$	*1455*
$CH_2{=}CMeB(OH)_2$	*1455*
$(n\text{-}Bu)B(OH)_2$	*1065a, 1719*
$(i\text{-}Bu)B(OH)_2$	*489, 883, 992, 1105, 1719*
$(s\text{-}Bu)B(OH)_2$	*1201*
$(t\text{-}Bu)B(OH)_2$	*489, 866*
$Me_2CHCH(Cl)B(OH)_2$	*1541*
$Me_2C{=}CHB(OH)_2$	*1105, 1455*
(*n*-Pentyl)$B(OH)_2$	*1507, 1719*
(*i*-Pentyl)$B(OH)_2$	*883, 992*
$Cl(CH_2)_3CH{=}CHB(OH)_2$	*214*
(2-Furan)$B(OH)_2$	*866*
(2-Thiophene)$B(OH)_2$	*866*
(5-Br-2-thiophene)$B(OH)_2$	*1065b*
(*n*-Hexyl)$B(OH)_2$	*1962*
$CH_3CH_2CH_2CHMeCH_2B(OH)_2$	*485*
(Cyclohexyl)$B(OH)_2$	*742, 1201, 1254*
(*n*-Heptyl)$B(OH)_2$	*1799*

Compound	References
[illegible]	[illegible], 906, 1043, 1099, 1474, 1520, 1927
(Mesityl)B(NR_2)(OR)	1927
MeB(NR_2)(F)	1491, 1882
PhB(NR_2)(F)	71
(*n*-Bu)B(NMe_2)(SiH_3)	32
(*n*-Bu)B(NR_2)(PR_2)	1485
MeB(NR_2)(SR)	1226
EtB(NR_2)(SR)	1306
(*n*-Pr)B(NR_2)(SR)	1305, 1306
$EtSCH_2CH_2CH_2B(NR_2)(SR)$	1342
$CH_2{=}CHCH_2B(NR_2)(SR)$	1342
(*n*-Bu)B(NR_2)(SR)	1305, 1306, 1467
(*i*-Bu)B(NR_2)(SR)	1305
(*i*-C_5H_{11})B(NR_2)(SR)	1305, 1306
PhB(NR_2)(SR)	537, 1226, 1282, 1439
(*p*-Tolyl)B(NR_2)(SR)	71, 1282
MeB(NR_2)(Cl)	723, 1466, 1467, 1879
EtB(NR_2)(Cl)	179, 1467, 1468
$BrCH_2CHBrB(NR_2)Cl$	426b
(*n*-Pr)B(NR_2)(Cl)	1466, 1468
BuB(NR_2)(Cl)	1465, 1468
ViB(NR_2)Cl	426b
PhB(NR_2)(Cl)	70, 71, 133, 179, 733, 862, 1042, 1433, 1436, 1470, 1474, 1874, 1888
(*p*-$MeOC_6H_4$)B(NR_2)(Cl)	71, 1465
MeB(NR_2)(Br)	1226, 1466, 1468
ViB(NR_2)(Br)	626
PhB(NR_2)(Br)	71, 1226
(*n*-Bu)B(NR_2)(Br)	1466, 1485
MeB(NR_2)(I)	1491
(*n*-Bu)B(NR_2)(I)	1466
[illegible]	[illegible], 1926
(*n*-C_8H_{17})B$(OH)_2$	665, 932, 1799
$PhCH_2CH_2B(OH)_2$	1535, 1926
(1-Phenylethyl)B$(OH)_2$	497
$PhCH{=}CHB(OH)_2$	214, 386, 871, 1640
$PhC{\equiv}CB(OH)_2$	1096
(*o*-$NH_3^+C_6H_4$)$CH_2CH_2B(OH)_2$	522
o[(2-C_5H_4N)C≡C—]$C_6H_4B(OH)_2$	1110
(Cyclohexyl)$CH{=}CHB(OH)_2$	214
(*n*-$C_{10}H_{21}$)B$(OH)_2$	932, 1799
(Lauryl)B$(OH)_2$	1799
(Myristyl)B$(OH)_2$	1799
(Cetyl)B$(OH)_2$	1799
(Stearyl)B$(OH)_2$	68, 119, 600, 661, 883, 993, 1799
PhB$(OH)_2$	93, 1011, 1235, 1450, 1507, 1664, 1745, 1789, 1850, 1851, 1876, 1925, 1926
[*p*-$(HO)_3BC_6H_4$]B$(OH)_2$	441, 1399
(*o*-Tolyl)B$(OH)_2$	883, 920, 1234, 1801, 1926
(*m*-Tolyl)B$(OH)_2$	920, 1011, 1926
(*p*-Tolyl)B$(OH)_2$	93, 920, 1011, 1926
[*o*-$(HOCH_2)C_6H_4$]B$(OH)_2$	1793
[*p*-$(R_2NCH_2)C_6H_4$]B$(OH)_2$	1793
o-$BrCH_2C_6H_4B(OH)_2$	1720, 1793
m-$BrCH_2C_6H_4B(OH)_2$	1793
p-$BrCH_2C_6H_4B(OH)_2$	1720, 1793
p-$(Br_2CH)C_6H_4B(OH)_2$	1793
o-$(Br_2CH)C_6H_4B(OH)_2$	1793
m-$(Br_2CH)C_6H_4B(OH)_2$	1793
o-$CF_3C_6H_4B(OH)_2$	1636
o-$CHOC_6H_4B(OH)_2$	1720, 1793
m-$CHOC_6H_4B(OH)_2$	1793

(continued)

TABLE 3-4—*(continued)*

Compound	Reference	Compound	Reference
p-$CHOC_6H_4B(OH)_2$	*1720, 1793*	2-NH_2-4-CO_2H-$C_6H_3B(OH)_2$	*1792*
m-$(CO_2H)C_6H_4B(OH)_2$	*118, 119, 920, 1801*	4-CO_2H-3-NH_2-$C_6H_3B(OH)_2$	*1801*
p-$(CO_2H)C_6H_4B(OH)_2$	*90, 118, 119, 1237*	3-CO_2H-5-$NO_2C_6H_3B(OH)_2$	*1801*
p-$(CO_2R)C_6H_4B(OH)_2$	*90, 1668, 1792*	2-NO_2-4-CO_2H-$C_6H_3B(OH)_2$	*1725, 1792*
p-$(COCl)C_6H_4B(OH)_2$	*1792*	4-CO_2H-3-$(NO_2)C_6H_3B(OH)_2$	*1801*
p-$(O_2NCH{=}CH)C_6H_4B(OH)_2$	*1720*	3-CO_2R-5-$NO_2C_6H_3B(OH)_2$	*1801*
p-$(AcOCH{=}CH)C_6H_4B(OH)_2$	*1720*	2-NO_2-4-$(CO_2R)C_6H_3B(OH)_2$	*1792*
p-$[HO_2CCH(NH_2)CH_2]C_6H_4B(OH)_2$	*1720*	4-CO_2R-3-$(NO_2)C_6H_3B(OH)_2$	*1801*
p-Ethylbenzene$B(OH)_2$	*1091*	3-COCl-5-$NO_2C_6H_3B(OH)_2$	*1801*
o-Vinylbenzene$B(OH)_2$	*1096*	4-COCl-3-$NO_2C_6H_3B(OH)_2$	*1801*
p-Vinylbenzene$B(OH)_2$	*1091, 1098, 1455, 1557*	3,5-$(NH_2)_2C_6H_3B(OH)_2$	*1801*
o-$NCCH_2C_6H_4B(OH)_2$	*387*	3-NH_2-5-$(NO_2)C_6H_3B(OH)_2$	*1794*
p-$NCCH_2C_6H_4B(OH)_2$	*1668, 1712*	4-NH_2-2-$(NO_2)C_6H_3B(OH)_2$	*1792*
p-Isopropylbenzene$B(OH)_2$	*568*	4-NH_2-3-$NO_2C_6H_3B(OH)_2$	*1801*
p-Isopropenylbenzene$B(OH)_2$	*1098*	3-NO_2-4-$BrC_6H_3B(OH)_2$	*93*
o-$(PhCONH)C_6H_4B(OH)_2$	*417*	2,6-$(CH_3O)_2C_6H_3B(OH)_2$	*1015*
m-$(PhCONH)C_6H_4B(OH)_2$	*417*	2-HO-6-$MeOC_6H_3B(OH)_2$	*1635*
p-$(RCONH)C_6H_4B(OH)_2$	*417, 1792*	2-HO-5-$ClC_6H_3B(OH)_2$	*1667, 1668*
o-$NH_2C_6H_4B(OH)_2$	*90, 93, 417*	2-HO-5-$BrC_6H_3B(OH)_2$	*1635, 1636, 1667, 1668*
m-$NH_2C_6H_4B(OH)_2$	*93, 417, 1664*	2-MeO-5-$ClC_6H_3B(OH)_2$	*1636, 1667, 1668*
p-$NH_2C_6H_4B(OH)_2$	*417, 1792*	2-RO-5-$BrC_6H_3B(OH)_2$	*1635, 1636, 1667, 1668*
o-$R_2NC_6H_4B(OH)_2$	*1664, 1712*	3-MeO-4-$ClC_6H_3B(OH)_2$	*1009*
m-$R_2NC_6H_4B(OH)_2$	*1664, 1712, 1721, 1727*	2-Cl-5-$MeOC_6H_3B(OH)_2$	*1009*
p-$R_2NC_6H_4B(OH)_2$	*90, 1732b*	2-Br-5-$MeOC_6H_3B(OH)_2$	*1009*
m-$(H_2NCONH)C_6H_4B(OH)_2$	*1725*	2-I-5-$MeOC_6H_3B(OH)_2$	*1009*
m-$NOC_6H_4B(OH)_2$	*1011*	(Mesityl)$B(OH)_2$	*748, 1732b*
o-$NO_2C_6H_4B(OH)_2$	*119, 743, 1664*	3,5-Me_2-4-$XC_6H_2B(OH)_2$ X = Me_2N, MeO	*1732b*
m-$NO_2C_6H_4B(OH)_2$	*119, 743, 1664*		
p-$NO_2C_6H_4B(OH)_2$	*119, 743, 1664, 1732b, 1792*	2-Me-3,5-$(NH_2)_2C_6H_2B(OH)_2$	*1794*
		2-Me-3,5-$(NO_2)_2C_6H_2B(OH)_2$	*1794*
m-$HOC_6H_4B(OH)_2$	*93, 417*	2,4-Br_2-5-$MeOC_6H_2B(OH)_2$	*1009*
[illegible]	[illegible]	4-CO_2H-3-NH_2-2,6-$Br_2C_6HB(OH)_2$	*1801*

Compound	References	Compound	References
m-ROC$_6$H$_4$B(OH)$_2$	*920*	(o-Phenetyl)B(OH)$_2$	*151*
p-ROC$_6$H$_4$B(OH)$_2$	*93, 577, 923, 1011, 1234, 1732b*	(m-Phenetyl)B(OH)$_2$	*151*
		(p-Phenetyl)B(OH)$_2$	*151*
o-PhenoxyphenylB(OH)$_2$	*119, 478*	(α-Naphthyl)B(OH)$_2$	*328, 920, 1234, 1609, 1926*
o-(HO)$_2$BOC$_6$H$_4$B(OH)$_2$	*577*		
m-EtCO$_2$C$_6$H$_4$B(OH)$_2$	*1011*	(Substituted α-naphthyl)B(OH)$_2$	*1065c, 1722*
p-EtCO$_2$C$_6$H$_4$B(OH)$_2$	*1011*	(β-Naphthyl)B(OH)$_2$	*328, 920, 1234, 1926*
m-FC$_6$H$_4$B(OH)$_2$	*119*	(Substituted-β-naphthyl)B(OH)$_2$	*1065c*
p-FC$_6$H$_4$B(OH)$_2$	*119*	(9-Phenanthryl)B(OH)$_2$	*1769*
p-MeSC$_6$H$_4$B(OH)$_2$	*1635*	MeB(OR)$_2$	*56, 684, 1696, 1822, 1881*
o-ClC$_6$H$_4$B(OH)$_2$	*151, 920*		
m-ClC$_6$H$_4$B(OH)$_2$	*119, 151, 1011, 1019*	ClCH$_2$B(OR)$_2$	*1174*
p-ClC$_6$H$_4$B(OH)$_2$	*119, 151, 1011*	ICH$_2$B(OR)$_2$	*1177*
m-BrC$_6$H$_4$B(OH)$_2$	*119, 1447*	EtB(OR)$_2$	*616, 617, 619, 932, 1057, 1325, 1491, 1923*
p-BrC$_6$H$_4$B(OH)$_2$	*93, 119, 920, 1011*		
m-IC$_6$H$_4$B(OH)$_2$	*1008*		
p-IC$_6$H$_4$B(OH)$_2$	*1008*	H$_2$NCONHCH$_2$CH$_2$B(OCH$_3$)$_2$	*363*
(o-Diphenyl)B(OH)$_2$	*1926*	ROCH$_2$CH$_2$B(OR)$_2$	*1188, 1254*
(m-Diphenyl)B(OH)$_2$	*1926*	R$_3$SiCH$_2$CH$_2$B(OR)$_2$	*1253, 1681*
(p-Diphenyl)B(OH)$_2$	*1926*	R$_3$SiCH(CH$_3$)B(OR)$_2$	*1681*
m-(CO$_2$R)C$_6$H$_4$C$_6$H$_4$B(OH)$_2$	*1801*	RSCH$_2$CH$_2$B(OR)$_2$	*1170, 1920*
o-(N-Morpholinomethyl)benzeneB(OH)$_2$	*749*	CH$_3$CH(SPh)B(OR)$_2$	*1188*
2,6-Me$_2$C$_6$H$_3$B(OH)$_2$	*748*	BrCH$_2$CH$_2$B(OR)$_2$	*1183*
3,5-Me$_2$C$_6$H$_3$B(OH)$_2$	*1732b*	CH$_3$CHBrB(OR)$_2$	*1188*
3-H$_2$NCONH-4-MeC$_6$H$_3$B(OH)$_2$	*1725*	BrCH$_2$CHBrB(OR)$_2$	*1184, 1185, 1246, 1920*
2-Me-5-NO$_2$C$_6$H$_3$B(OH)$_2$	*1794*	ICH$_2$CH$_2$B(OR)$_2$	*1179, 1188*
4-Me-3-NH$_2$C$_6$H$_3$B(OH)$_2$	*93*	CH$_3$CHIB(OR)$_2$	*1188*
4-Me-3-(NO$_2$)C$_6$H$_3$B(OH)$_2$	*93, 1801*	ViB(OR)$_2$	*1170–1172, 1339, 1449, 1455, 1920, 1921, 1923*
4-Me-3-HOC$_6$H$_3$B(OH)$_2$	*93*		
3-CONH$_2$-5-NO$_2$C$_6$H$_3$B(OH)$_2$	*1801*		
2-NO$_2$-4-(CONH$_2$)C$_6$H$_3$B(OH)$_2$	*1792*	ROCH=CHB(OR)$_2$	*331, 331a, 1185*
4-CONH$_2$-3-(NO$_2$)C$_6$H$_3$B(OH)$_2$	*1801*	C$_6$H$_{13}$SCH=CHB(OR)$_2$	*1184, 1185*
3-CO$_2$H-5-NH$_2$C$_6$H$_3$B(OH)$_2$	*1801*	PhSCH=CHB(OR)$_2$	*1185*

(*continued*)

TABLE 3-4—*(continued)*

Compound	Reference
$BrCH{=}CH_2B(OR)_2$	*1046, 1185*
$BrCH{=}CHBrB(OR)_2$	*1184, 1185*
$HC{\equiv}CB(OR)_2$	*1184–1186, 1923*
$ROC{\equiv}CB(OR)_2$	*1186*
$(n\text{-}Pr)B(OR)_2$	*655, 1057, 1324, 1325, 1345–1349, 1467*
$(i\text{-}Pr)B(OR)_2$	*1325, 1717*
$R_2NCH_2CH_2CH_2B(OR)_2$	*1390*
$CCl_3CH_2CH(OBu)B(OR)_2$	*1183*
$Me_2C(OBu)B(OR)_2$	*1178*
$C_3F_7B(OR)_2$	*406*
$RSCH_2CH_2CH_2B(OR)_2$	*1342*
$Cl_3CCH_2CH(SBu)B(OR)_2$	*1183*
$(CH_3)_2C[SC(NH_2)_2]^+{-}B(OR)_2$	*1188*
$CCl_3CH_2CHClB(OR)_2$	*1170*
$CCl_3CH_2CHBrB(OR)_2$	*1170*
$CCl_3CH_2CHIB(OR)_2$	*1183*
$CH_3CHBrCH_2B(OR)_2$	*426b, 1178*
$Me_2CBrB(OR)_2$	*426b, 1178, 1188*
$CH_2BrCHBrCH_2B(OR)_2$	*1246*
$(MeCHBrCHBr)B(OR)_2$	*426b*
[structure: cyclopropane ring bearing Cl, Cl and $B(OR)_2$]	*1920*
$CH_2{=}CHCH_2B(OR)_2$	*1267, 1320, 1337, 1782, 1783, 1785, 1944*
$CH_3CH{=}CHB(OR)_2$	*426b, 1170, 1455*
$CH_2{=}C(CH_3)B(OR)_2$	*426b, 1170, 1455*
$ClCH_2CH{=}CHB(OR)_2$	*214*
$(Cyclopentyl)B(OR)_2$	*208, 209, 268, 275c, 305*
$(1\text{-Penten-1-yl})B(OR)_2$	*214*
$(1,3\text{-Pentadien-1-yl})B(OR)_2$	*1249a*
$(n\text{-}PrCR{=}CH)B(OR)_2$	*1275*
$Cl(CH_2)_3CH{=}CHB(OR)_2$	*214*
$CH_2{=}CHCH_2CR{=}CHB(OR)_2$	*1275*
$(NC)_2CHCH_2CHBrB(OR)_2$	*1188*
$(RO_2C)_2CHCH_2CHBrB(OR)_2$	*1188*
$CpB(OR)_2$	*719*
[structure: cyclopropane ring bearing NC, CN and $B(OR)_2$]	*1188*
$(n\text{-}C_6H_{13})B(OR)_2$	*1057, 1347–1349, 1812*
$(3\text{-Hexyl})B(OR)_2$	*268*
$CH_3(CH_2)_2CH(CH_3)CH_2B(OR)_2$	*268*
$Me_2CHCMe_2B(OMe)_2$	*226*
$n\text{-}BuC(Br)(Me)B(OR)_2$	*1545a*
$(Cyclohexyl)B(OR)_2$	*208, 226, 268, 1256*
$(t\text{-}Bu\text{-}CH{=}CH)B(OR)_2$	*214*
$(3\text{-Hexen-3-yl})B(OR)_2$	*214*
$CH_3{=}CHCH_2C(CH_3){=}CHB(OMe)_2$	*628*
$(n\text{-}Bu\text{-}C{\equiv}C)B(OR)_2$	*1184, 1185*
$Et_3CB(OR)_2$	*808*
[structure: norbornyl bearing $B(OR)_2$]	*208, 209, 1194, 1920*

Compound	References	Compound	References
$Cl_3CCH{=}CHB(OR)_2$	*1182*	B(OR)$_2$	*1193, 1196, 1197, 1920, 1923*
$Cl_3CCH{=}CBrB(OR)_2$	*1184, 1185*	B(OR)$_2$	*1184*
$MeC{\equiv}CB(OR)_2$	*1945*	$PhCH_2B(OR)_2$	*588a*
$(n\text{-}Bu)B(OR)_2$	*14, 171, 208, 209, 576, 601, 602, 655, 867, 1057, 1162, 1345, 1347, 1348, 1360, 1467*	$4\text{-}ClC_6H_4CH_2B(OR)_2$	*588a*
$(s\text{-}Bu)B(OR)_2$	*208, 209, 490, 1106*	B(OR)$_2$	*1193*
$(s\text{-}Bu)B(O_2R)_2$	*483*	(1-Octyl)B(OR)$_2$	*655*
$(i\text{-}Bu)B(OR)_2$	*208, 209, 488–490, 1360*	(2,4,4-Trimethyl-1-pentyl)B(OR)$_2$	*210*
$(i\text{-}Bu)B(O_2R)_2$	*487–490*	$[Cl_2C{=}CHCH(i\text{-amyl})]B(OR)_2$	*1182*
$(t\text{-}Bu)B(OR)_2$	*489*	$(CH_2{=}CHCH_2)(CH_2{=}CRCH_2)CHCH_2B(OR)_2$	*1271*
$(t\text{-}Bu)B(O_2CH_2CHMe_2)_2$	*488*	(Cyclohexylethen-1-yl)B(OR)$_2$	*214*
$Cl_3CH_2CH(Et)B(OR)_2$	*1183*	(Cyclooctyl)B(OR)$_2$	*871*
$Br(CH_2)_4B(OR)_2$	*1836a*	(1-Phenylethyl)B(OR)$_2$	*497, 1176*
$Me_2CBrCH_2B(OR)_2$	*426b*	$PhCH_2CH_2B(OR)_2$	*1535, 1536*
$Me_2CBrCHBrB(OR)_2$	*426b*	$(PhCHBrCH_2)B(OR)_2$	*1536*
$CH_3CH_2CBr(Me)B(OR)_2$	*1178*	$(PhCH{=}CH)B(OR)_2$	*214, 871, 1535, 1536*
$CH_3CHBrCBr(Me)B(OR)_2$	*1180*	$PhC(Cl){=}CHB(OR)_2$	*1046*
(2-Butenyl)-2-B(OR)$_2$	*1180*	(α-Styryl)B(OR)$_2$	*1176, 1535, 1536*
$Me_2C{=}CHB(OR)_2$	*1455*	$Cl_3CCH_2CH(Ph)B(OR)_2$	*1183*
cis-$CH_3CH_2CH{=}CHB(n\text{-}BuO)_2$	*1645*	$(CyclohexylCH_2CH{=}CH)B(OR)_2$	*214*
trans-$CH_3CH_2CH{=}CHB(OR)_2$	*1645*	$Cl_2C{=}CHCHPhB(OR)_2$	*1182*
$MeCH{=}CMeB(OR)_2$	*426b*	$CH_2{=}CHCH_2C(n\text{-}Bu){=}CHB(OR)_2$	*332*
$CH_3CHBrCH{=}CHB(O\text{-}t\text{-}Bu)_2$	*1645*		
$CH_3CH{=}CHCHBrB(O\text{-}t\text{-}Bu)_2$	*1645*		
$CH_2{=}CHC{\equiv}CB(OR)_2$	*1945*		
$(NC)_2CHCH_2B(OR)_2$	*1177*		
(1-Pentyl)B(OR)$_2$	*208, 209, 304, 305, 1057*		
$(i\text{-}C_5H_{11})B(OR)_2$	*1302*		
$(i\text{-}Pr)(Me)CHB(OR)_2$	*268*		

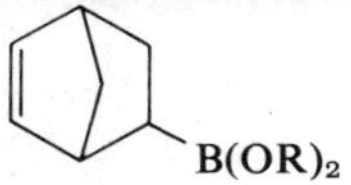

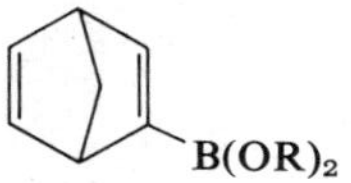

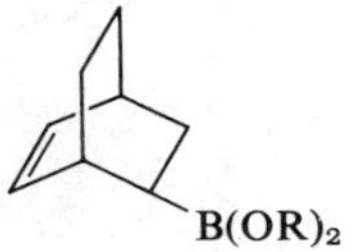

(*continued*)

TABLE 3-4—*(continued)*

Compound	Reference
$CH_2{=}C(Me)CH_2CH(CH_2CH{=}CH_2)CH_2B(OMe)_2$	*628*
(1-Decyl)$B(OR)_2$	*209*
$CH_2{=}C(n\text{-}Bu)CH_2CH(CH_2CH{=}CH_2)CH_2B(OR)_2$	*332*
$(n\text{-}Bu)_3CB(OR)_2$	*808*
$(i\text{-}Bu)_3CB(OR)_2$	*212*
$(i\text{-}Bu)_2(n\text{-}Bu)CB(OR)_2$	*212*
$(s\text{-}Bu)_2(n\text{-}Bu)CB(OR)_2$	*212*
$(t\text{-}Bu)(s\text{-}Bu)(n\text{-}Bu)CB(OR)_2$	*1417*
$(n\text{-}Bu)_2(n\text{-}pentyl)CB(OR)_2$	*210*
$(i\text{-}Bu)_2(n\text{-}pentyl)CB(OR)_2$	*210*
$(s\text{-}Bu)_2(n\text{-}pentyl)CB(OR)_2$	*210*
$(i\text{-}Bu)_2(cyclopentyl)CB(OR)_2$	*212*
$(s\text{-}Bu)_2(cyclopentyl)CB(OR)_2$	*212*
$(t\text{-}Bu)(cyclopentyl)_2CB(OR)_2$	*1417*
(Cyclopentyl)$(n\text{-}pentyl)_2CB(OR)_2$	*268*
$(n\text{-}hexyl)_3CB(OR)_2$	*808*
$X(CH_2)_yB(OR)_2$ (X = Cl, Br, I, COR, NR_2, OH, CN, CO_2R; y = 3,4,5,6)	*1239*
$PhB(OR)_2$	*6, 9, 11, 64, 68, 73, 171, 173, 344, 469, 473, 537, 572, 601, 602, 652, 757, 778, 804, 919, 976, 1012, 1049, 1099, 1107, 1211, 1235, 1242, 1290, 1381, 1388, 1449, 1450, 1520, 1572, 1682, 1717, 1789, 1796, 1850, 1930, 1959*
$(n\text{-}Bu)B(OR)(Br)$	*1162*
$PhB(OR)(Br)$	*6*
$(n\text{-}Bu)B(OR)(I)$	*1162*
$MeBF_2$	*132, 168, 348, 403, 598, 869a, 956, 1490, 1600*
FCH_2BF_2	*685*
CF_3BF_2	*98, 1520, 1528*
$EtBF_2$	*79, 132, 598, 690, 955, 956, 1490, 1742*
$ViBF_2$	*77, 169, 598, 825*
$CF_2{=}CFBF_2$	*447, 1735*
$HC{\equiv}CBF_2$	*1606*
$(n\text{-}Pr)BF_2$	*169, 598, 956, 1568*
$CF_3CH_2CH_2BF_2$	*1568*
$(i\text{-}Pr)BF_2$	*956*
(Cyclopropyl)BF_2	*442*
$CH_2{=}CHCH_2BF_2$	*169*
$(n\text{-}Bu)BF_2$	*172, 343, 956, 1064, 1204, 1811*
$(n\text{-}C_5H_{11})BF_2$	*1064, 1204, 1811*
$(i\text{-}C_5H_{11})BF_2$	*1329*
$(2\text{-}s\text{-}C_5H_{11})BF_2$	*1204*
$(t\text{-}C_5H_{11})BF_2$	*1204*
$CpBF_2$	*1579*
$(n\text{-}C_6H_{13})BF_2$	*1064, 1204, 1329, 1811*
(Cyclohexyl)BF_2	*956, 1204*
$PhCH_2BF_2$	*956*
$(n\text{-}C_8H_{17})BF_2$	*956*
$(m\text{-}C_{10}H_{21})BF_2$	*956*
$PhBF_2$	*825, 956, 1206, 1490, 1789*

Compound	References	Compound	References
$PhB(OSiEt_3)_2$	*10*	$(C_6F_5)BF_2$	*396, 397*
$(o\text{-}MeC_6H_4)B(OR)_2$	*388, 588a, 778*	$(p\text{-}MeC_6H_4)BF_2$	*1206*
$(m\text{-}Tolyl)B(OR)_2$	*588a*	$BuB(F)(Cl)$	*810*
$(p\text{-}Tolyl)B(OR)_2$	*588a, 1012, 1572*	$PhB(PR_2)_2$	*419*
$p\text{-}(CH_2{=}CH)C_6H_4B(OR)_2$	*1455*	$PhB(PHPh)(Cl)$	*419*
$4\text{-}Me_2NC_6H_4B(OR)_2$	*1732b*	$MeB(SR)_2$	*353, 1696, 1829, 1886, 1890*
$4\text{-}NO_2C_6H_4B(OR)_2$	*1732b*	$Et_2MeSiCH_2CH_2B(SR)_2$	*1331*
$p\text{-}MeOC_6H_4B(OR)_2$	*1107, 1572, 1732b*	$(n\text{-}Pr)B(SR)_2$	*765, 1061, 1306, 1311, 1332a, 1349*
$p\text{-}ClC_6H_4B(OR)_2$	*1012, 1290*	$(i\text{-}Pr)B(SR)_2$	*1306, 1332a*
$p\text{-}BrC_6H_5B(OR)_2$	*1012, 1107*	$EtSCH_2CH_2CH_2B(SR)_2$	*1342*
$p\text{-}(RO)_2BC_6H_4B(OR)_2$	*68*	$CH_2{=}CHCH_2B(SR)_2$	*1340, 1342*
$3,5\text{-}Me_2C_6H_3B(OR)_2$	*1732b*	$(n\text{-}Bu)B(SR)_2$	*765, 1061, 1305, 1306, 1332a, 1344, 1349, 1657*
$(\alpha\text{-}Naphthyl)B(OR)_2$	*1107, 1572*	$i\text{-}BuB(SR)_2$	*765, 1305, 1332a*
$(\beta\text{-}Naphthyl)B(OR)_2$	*1234*	$(s\text{-}Bu)B(SR)_2$	*765*
$(n\text{-}Bu)B(OR)(F)$	*172, 173, 343*	$(t\text{-}Bu)B(SR)_2$	*765*
$PhB(OR)(F)$	*172, 173*	$(i\text{-}C_5H_{11})B(SR)_2$	*1061, 1311*
$(n\text{-}Pr)B(OR)(SR)$	*1266*	$(n\text{-}C_6H_{13})B(SR)_2$	*765, 1061, 1331, 1349*
$(n\text{-}Bu)B(OR)(SR)$	*1266*	$(Cyclohexyl)B(SR)_2$	*1331*
$PhB(OR)(SR)$	*451a*	$(n\text{-}C_8H_{17})B(SR)_2$	*1331, 1689*
$EtB(OR)(Cl)$	*1063, 1325*	$PhCH_2CH_2B(SR)_2$	*1331*
$(n\text{-}Pr)B(OR)(Cl)$	*1324*	$PhB(SR)_2$	*73, 451, 456, 537, 607, 1047, 1311, 1448, 1657, 1705, 1829*
$ViB(OR)(Cl)$	*1339*	$(p\text{-}Tolyl)B(SR)_2$	*1705*
$(n\text{-}Bu)B(OR)(Cl)$	*172, 1063, 1162, 1324*	$(3,5\text{-}Me_2C_6H_3)B(SR)_2$	*1705*
$(n\text{-}C_6H_{13})B(OR)(Cl)$	*1063*	$(2,5\text{-}Me_2C_6H_3)B(SR)_2$	*1705*
$PhB(OR)(Cl)$	*172, 173, 468, 469, 1299*	$(2,4,6\text{-}Me_3C_6H_2)B(SR)_2$	*1705*
$(o\text{-}Tolyl)B(OR)(Cl)$	*1299*	$(3\text{-}ClC_6H_4)B(SR)_2$	*1705*
$(p\text{-}CH_3C_6H_4)B(OR)(Cl)$	*1300*	$(4\text{-}ClC_6H_4)B(SR)_2$	*1705*
$(Cyclohexyl)B(OR)(Cl)$	*1250*		
N B(OR)(Cl)	*1303*		

(*continued*)

TABLE 3-4—*(continued)*

Compound	Reference	Compound	Reference
$(4\text{-}IC_6H_4)B(SR)_2$	*1705*	$(m\text{-}BrC_6H_4)BCl_2$	*1437*
$(n\text{-}Pr)B(SR)(Cl)$	*1306*	$(2,5\text{-}Me_2C_6H_3)BCl_2$	*1394*
$(i\text{-}Pr)B(SR)(Cl)$	*1306*	$(3,4\text{-}Me_2C_6H_3)BCl_2$	*1394*
$(n\text{-}Bu)B(SR)(Cl)$	*1306*	$[3,5\text{-}(CH_3)_2C_6H_3]BCl_2$	*1237, 1394, 1395*
$(i\text{-}Pr)B(SR)(Br)$	*1306*	$C_6F_5BCl_2$	*396, 397*
$(i\text{-}C_5H_{11})B(SR)(Br)$	*1306*	$(\alpha\text{-Naphthyl})BCl_2$	*1234, 1394*
PhB(SR)RBr)	*1306*	$(\beta\text{-Naphthyl})BCl_2$	*1234*
$MeBCl_2$	*74, 95, 397, 956, 1090, 1490, 1605, 1883*	$(Ferrocenyl)BCl_2$	*983, 984*
$EtBCl_2$	*343, 655, 956, 1044a, 1144, 1202, 1325, 1487, 1490, 1940*	$PhC(Cl){=}CHBCl_2$	*1046*
$ViBCl_2$	*78, 168, 169*	$(n\text{-}Bu)B(Cl)(Br)$	*1162*
$ClCH{=}CHBCl_2$	*666, 1607*	PhB(Cl)(Br)	*606*
$CF_2{=}CFBCl_2$	*447, 877, 1735*	$(n\text{-}Bu)B(Cl)(I)$	*1162*
$HC{\equiv}CBCl_2$	*1606*	$MeB(SeR)_2$	*1702*
$(n\text{-}Pr)BCl_2$	*655, 956, 1202, 1325*	$(n\text{-}Bu)B(SeR)_2$	*1326*
$(MeCH{=}CH)BCl_2$	*426b*	$PhB(SeR)_2$	*1702*
$(i\text{-}Pr)BCl_2$	*956, 1144, 1202, 1325*	PhB(SeR)(I)	*1702*
$(Cyclopropyl)BCl_2$	*442*	$MeBBr_2$	*655, 956, 1490, 1643*
$(n\text{-}Bu)BCl_2$	*144, 234, 343, 655, 956, 1162*	$EtBBr_2$	*956, 1487, 1490*
$(i\text{-}Bu)BCl_2$	*1202, 1940*	$ViBBr_2$	*626, 1446*
$(s\text{-}Bu)BCl_2$	*1063, 1202*	$BrCH{=}CHBBr_2$	*1046*
$(t\text{-}Bu)BCl_2$	*1144, 1202*	$(n\text{-}Pr)BBr_2$	*956*
$(n\text{-}C_5H_{11})BCl_2$	*1063, 1202*	$(i\text{-}Pr)BBr_2$	*956, 1304*
$(i\text{-}C_5H_{11})BCl_2$	*1302*	$(n\text{-}Bu)BBr_2$	*655, 865, 956, 1162*
		$(i\text{-}Bu)BBr_2$	*1305*
		$(Cyclohexyl)BBr_2$	*871, 956*
		$(1\text{-Cyclohexenyl})BBr_2$	*871*
		$PhCH_2BBr_2$	*956*

Compound	References	Compound	References
(Cyclopentyl)BCl_2	*234*	(n-C_7H_{15})BBr_2	*1313*
$CpBCl_2$	*1131a*	(n-C_8H_{17})BBr_2	*956, 1313*
(n-C_6H_{13})BCl_2	*234, 1063, 1202*	(Cyclooctyl)BBr_2	*871*
(2-Hexyl)BCl_2	*234*	(n-$C_{10}H_{21}$)BBr_2	*956*
(3-Hexyl)BCl_2	*234*	PhC(Br)=C(Ph)BBr_2	*1046*
$CH_3CH_2CH_2CHMeCH_2BCl_2$	*234*	$PhBBr_2$	*6, 603, 623, 649, 956, 1237, 1490, 1763*
(Cyclohexyl)BCl_2	*234, 956, 1144, 1202, 1250*	(p-Tolyl)BBr_2	*649, 1237*
(Norbornyl)BCl_2	*234*	(n-Bu)B(Br)(I)	*1162*
$PhCH_2BCl_2$	*871, 956*	$MeBI_2$	*956, 1490*
(n-C_8H_{17})BCl_2	*655, 956*	$EtBI_2$	*956, 1487, 1490*
(n-$C_{10}H_{21}$)BCl_2	*956*	(n-Pr)BI_2	*956*
$PhBCl_2$	*2, 346, 468, 603, 614, 649, 956, 1090, 1206, 1236, 1393, 1394, 1447, 1490, 1729, 1802*	(i-Pr)BI_2	*956*
		(n-Bu)BI_2	*956, 1162*
		(Cyclohexyl)BI_2	*742, 956*
		$PhCH_2BI_2$	*956*
		(n-C_8H_{17})BI_2	*956*
(o-Tolyl)BCl_2	*649, 1729*	(n-$C_{10}H_{21}$)BI_2	*956*
(m-Tolyl)BCl_2	*649, 1394*	$PhBI_2$	*956, 1473, 1490, 1660, 1703*
(p-Tolyl)BCl_2	*649, 1206, 1235, 1393, 1394, 1729*	(o-Tolyl)BI_2	*1703*
(m-EtC_6H_4)BCl_2	*1394*	(m-Tolyl)BI_2	*1703*
(p-EtC_6H_4)BCl_2	*1394*	(p-Tolyl)BI_2	*1703*
$PhC_6H_4BCl_2$	*1394*	4-$FC_6H_5BI_2$	*1705*
(o-ROC_6H_4)BCl_2	*1234*	3-$ClC_6H_5BI_2$	*1705*
(p-ROC_6H_4)BCl_2	*1234, 1729*	4-$ClC_6H_5BI_2$	*1705*
$PhOC_6H_4BCl_2$	*1394*	3,5-$Me_2C_6H_3BI_2$	*1705*
[o-(Cl_2BO)C_6H_4]BCl_2	*577*	3,6-$Me_2C_6H_3BI_2$	*1705*
(p-ClC_6H_4)BCl_2	*649*	2,4,6-$Me_3C_6H_2BI_2$	*1705*
(p-BrC_6H_4)BCl_2	*649*	(2-Thiophene)BI_2	*1700*

TABLE 3-5

Cyclic (CBC) Organoboron Compounds

Compound	Reference	Compound	Reference
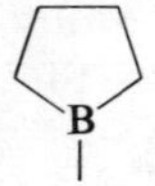	259, 262, 264–267, 416, 605, 695, 760, 766, 936, 943, 945, 949, 956, 959, 963, 965, 976, 977, 979, 985, 1252, 1273a, 1276, 1278, 1309, 1313, 1330, 1333, 1333a, 1350, 1352, 1418, 1631, 1692, 1697, 1789, 1901, 1932, 1934	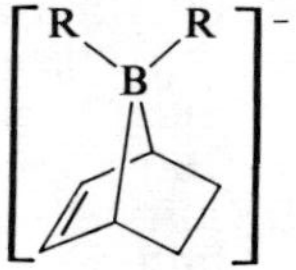	700
		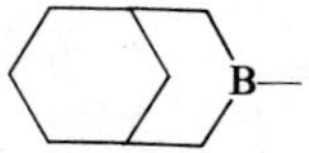	1333b, 1333c
		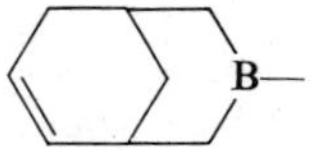	333, 334, 334a, 1249, 1255a, 1268, 1270, 1271a, 1274, 1276a, 1276c
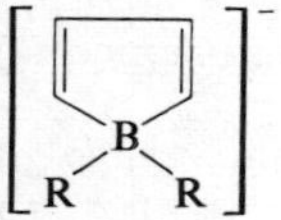	700	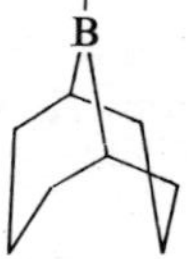	262, 265, 267, 269, 360a, 892, 937, 943, 956, 958, 959, 965, 976, 978, 1035, 1687
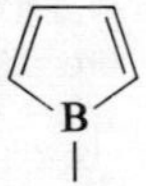	65, 154, 579	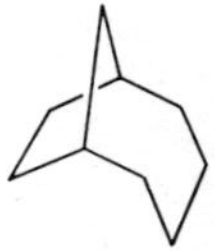	262, 265

Structure	References	Structure	References
	247, 260–263, 265, 267, 332, 360a, 416, 766, 959, 965, 976, 977, 979, 985, 1330, 1418, 1631, 1691, 1692, 1789, 1901, 1931		1333b
	332–334, 628, 1249, 1255a, 1271, 1276b, 1276c		965, 974
	50		697, 956, 970, 976
	50		257, 965, 974
	260, 262, 265, 269, 360a, 959, 965, 979, 985, 1307, 1333, 1352, 1418, 1691, 1692, 1931		956, 970
	579		1518
			58, 1111

(continued)

TABLE 3-5—(*continued*)

Compound	Reference	Compound	Reference
	956		*1798*
	947, 956, 1483		
	874, 874a, 1834, 1835		*1798*
	1276b		*963*
	1108		*128*
	1833		*963, 975, 1252*

Structure	References	Structure	References
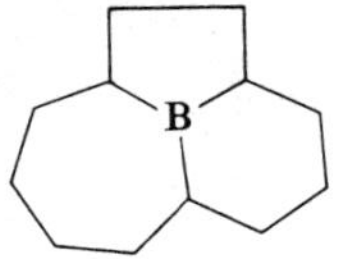	*1617, 1618*	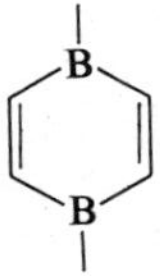	*871, 1774*
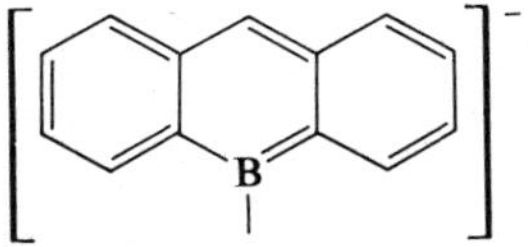	*1835*	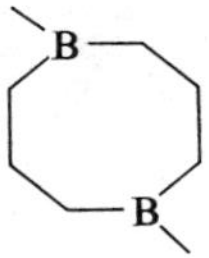	*985, 1252, 1316, 1320, 1338*
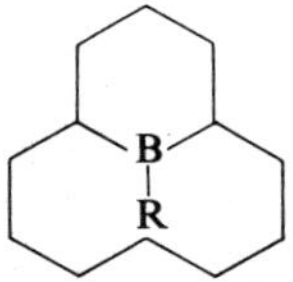	*976, 1618*	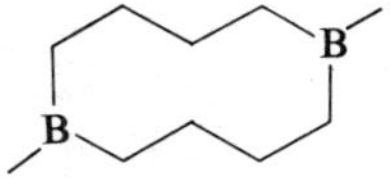	*976, 985, 1252, 1352*
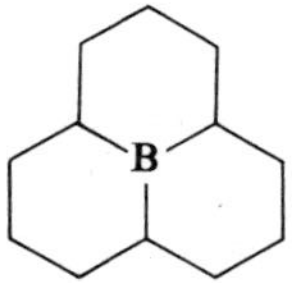	*202, 691, 692, 929, 943, 958, 959, 965, 978, 1617, 1618*	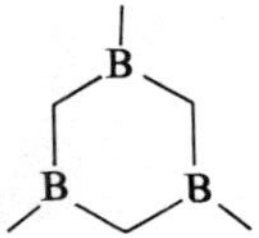	*797, 946, 998a*
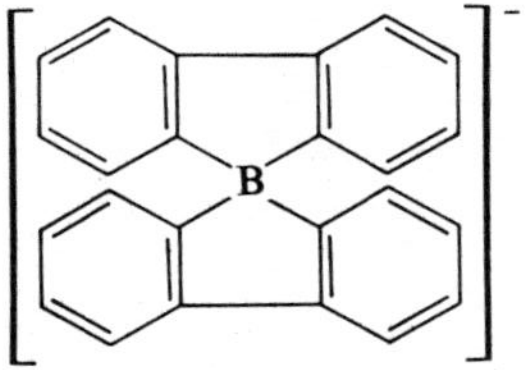	*947, 1789, 1906*	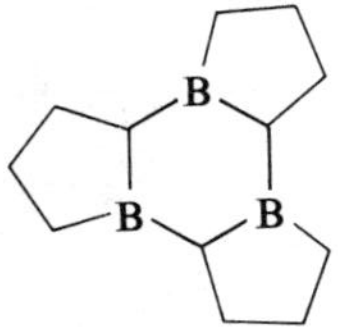	*945, 949*

(*continued*)

TABLE 3-5—*(continued)*

Compound	Reference	Compound	Reference
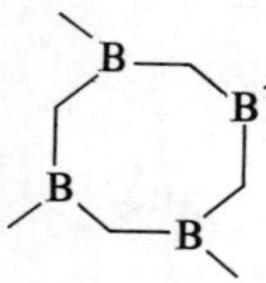	*676, 677*	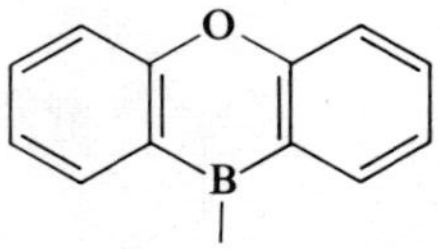	*478*
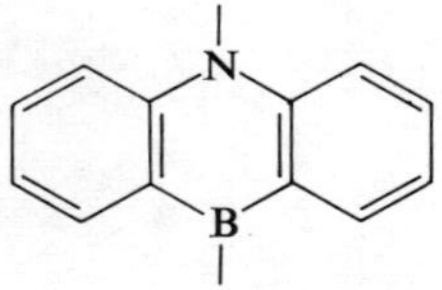	*1147*	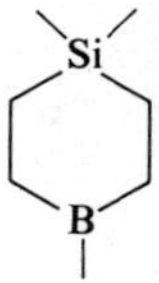	*760, 766*
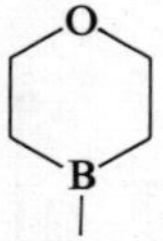	*766*	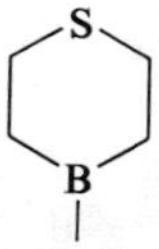	*1171*

TABLE 3-6

Heterocyclic Organoboron Compounds (Parent Ring System)

Compound	Reference	Compound	Reference
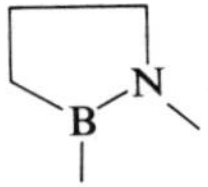	*365, 366, 554, 554b, 557, 1283, 1285, 1286a, 1874, 1927*	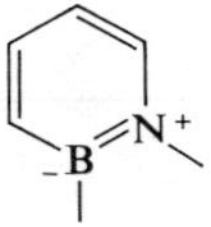	*463, 512, 542, 1874*
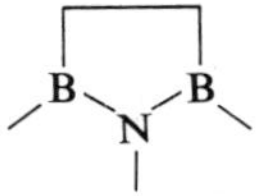	*963*	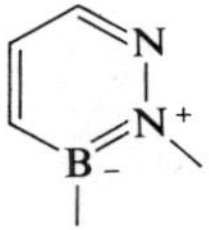	*703a, 1413*
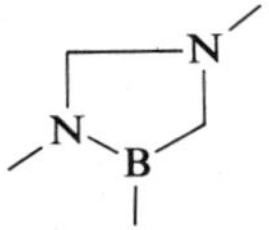	*802*	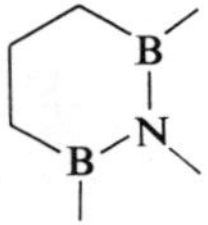	*1319*
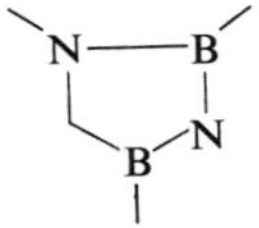	*379, 1810, 1902*	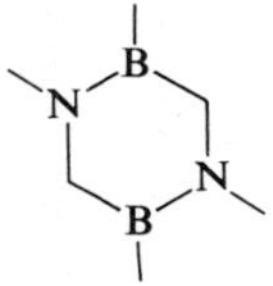	*379, 380, 661a, 799–802, 1762*
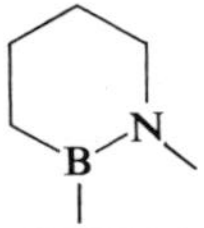	*512, 1351, 1836b, 1874, 1888*	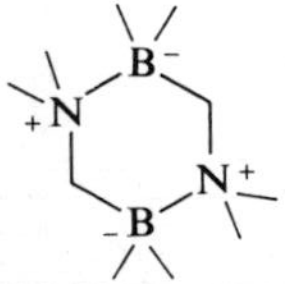	*160, 1366–1369*

(*continued*)

TABLE 3-6—*(continued)*

Compound	Reference	Compound	Reference
	156, 160, 799, 800, 1902a		*1147*
	1350		*404*
	139, 365		*463, 524, 530, 532–535, 537, 540, 541, 699, 701, 845*
	527, 749, 963, 964, 1720		*532*
	749		*532*
	528		

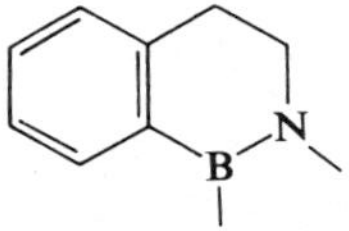	387, 388, 964	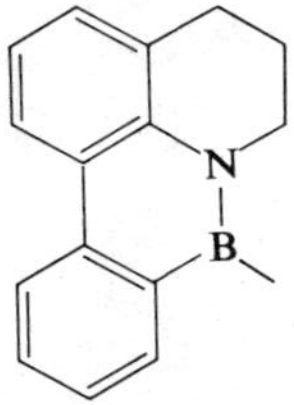	532
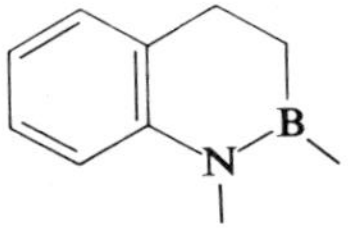	522	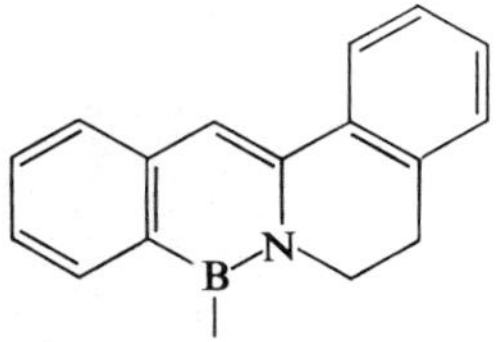	532
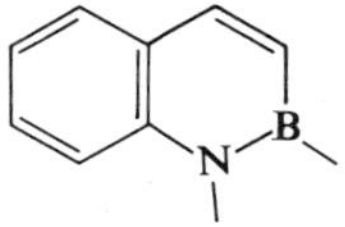	522–524, 526, 532	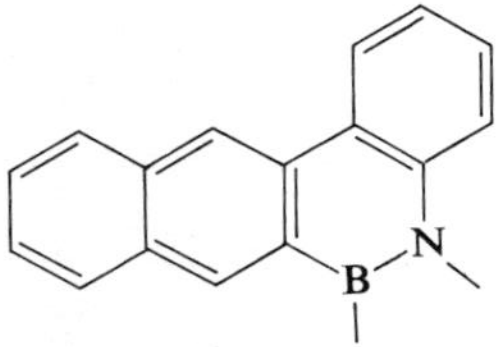	543
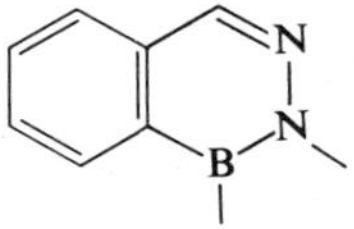	526, 546	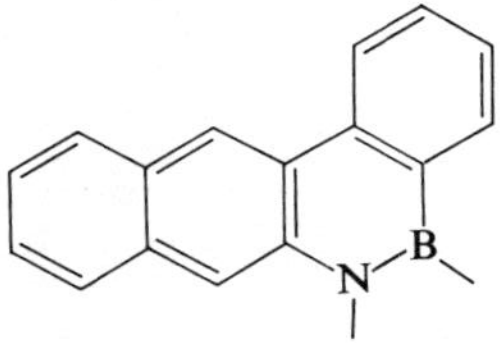	543
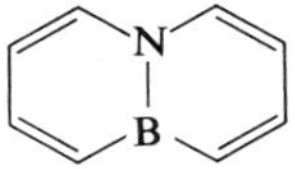	528		
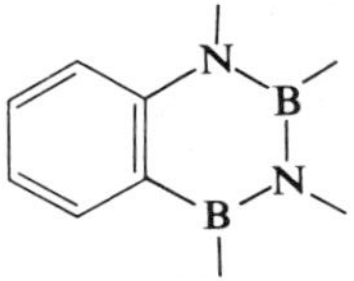	963		

(continued)

TABLE 3-6—(*continued*)

Compound	Reference	Compound	Reference
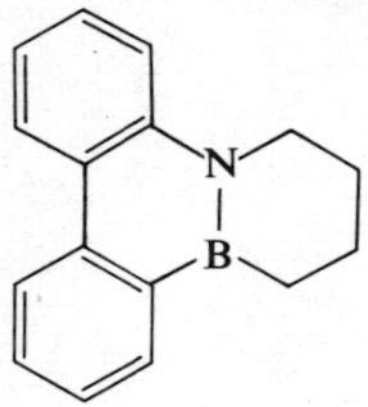	*463*	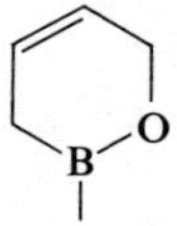	*1752b*
		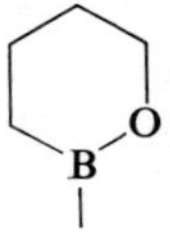	*949, 1836a*
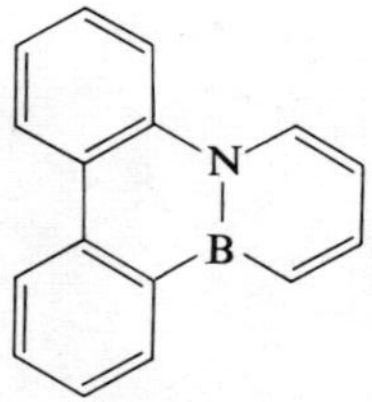	*463*	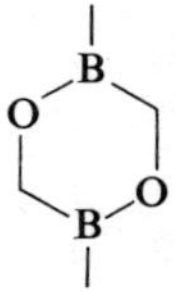	*807, 1159*
		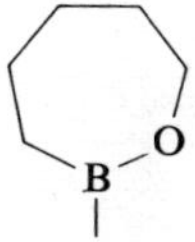	*1249, 1836c*
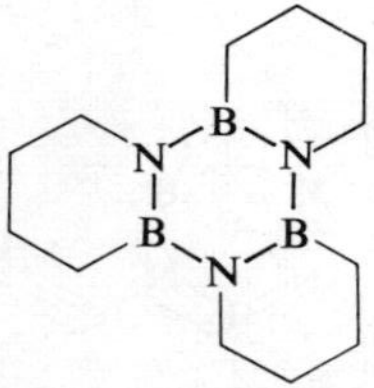	*463*	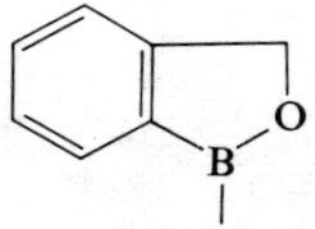	*387, 388, 748, 1092, 1720*

	463		387, 1103
			521, 539, 701
	404		543
	979		364
	127		554a, 1286
	809		509, 701

(continued)

TABLE 3-6—*(continued)*

Compound	Reference	Compound	Reference
	158		*703*
	1651		*704, 705*
	526, 1720		*1191*
	1609a		*554a*
	703–705, 1412		*703*

TABLE 3-7

Symmetrical Borazines ($R_3B_3N_3R'_3$)

R	Reference
Me	*22, 104, 180, 345, 401, 591, 611, 654, 674, 683, 750, 857, 1125, 1217, 1233, 1401, 1474, 1495, 1576, 1599, 1649, 1651, 1652, 1713, 1842, 1854, 1855, 1858, 1875, 1878–1880, 1916*
NC—	*159, 724*
R_3SiCH_2—	*1671, 1673, 1674*
$ClCH_2$—	*1820*
$CHCl_2$	*1227*
CCl_3	*1227*
$R_2NC(O)$—	*925*
Et	*180, 750, 931, 952, 1576, 1693, 1853–1855, 1858, 1953*
$R_3SiCH_2CH_2$—	*1674, 1677, 1681*
$Ph_2PCH_2CH_2$—	*1676*
Vi	*626, 1556*
ClCH=CH—	*1625, 1626*
F_2C=CF—	*889*
n-Pr	*655, 759, 764, 952, 1304*
i-Pr	*759, 764, 1304*
CX_2CH_2CHX— (X = Cl, Br)	*1674, 1678*
CH_2=$CHCH_2$—	*1342, 1556, 1629*
MeCH=CH—	*1859*
MeC≡C—	*1859*
n-Bu	*376, 571, 655, 759, 764, 922a, 1304, 1625, 1629, 1713, 1858*
i-Bu	*764*
s-Bu	*759, 764*
t-Bu	*764*
n-Pentyl	*764*
i-C_5H_{11}	*1265, 1302, 1311*
Cp	*725, 1768*
n-Hexyl	*764*
MeCp	*1768*
$PhCH_2$—	*764*
$C_6F_5CH_2$—	*1231*
PhC≡C—	*1859*
Ph	*1, 22, 103, 104, 180, 347, 376, 429, 466, 508, 656, 659, 707, 750, 922, 922a, 1217, 1230, 1243, 1311, 1401, 1435, 1450, 1459, 1474, 1486, 1495, 1509, 1511, 1512, 1625, 1666, 1698, 1713, 1858, 1888*
p-MeC_6H_4	*1300*
p-ClC_6H_4	*1450*
2,6-Xylyl	*1402*
3-Acetyl-2,6-xylyl	*1402*
C_6F_5	*1166, 1230*
$C_6H_5C_6H_4$	*1383*
p-$C_6H_5C_6H_4C_6H_4$	*1383*
α-$C_{10}H_7$	*1300, 1383*
(π-Cp)Fe(Cp—)	*981*
Other (complex)	*627, 960, 964, 1779*

TABLE 3-8

Unsymmetrical Borazines ($RR'R''B_3N_3R_3'''$)

R	R′	R″	Reference
H	H	Me	*55a, 92, 1567, 1576, 1651, 1652, 1713, 1842, 1875*
H	H	Et	*1567, 1575*
H	H	Ph	*1379, 1698, 1713*
H	Me	Me	*1567, 1576, 1651, 1652, 1713, 1842, 1875*
H	Me	Et	*1567, 1576, 1713*
H	Et	Et	*1567, 1576*
H	Ph	Ph	*1379, 1698, 1713*
Me	Me	Et	*1217, 1567, 1576*
Me	Me	Vi	*888*
Me	Me	$F_2C{=}CF$—	*889*
Me	Me	$MeC{\equiv}C$—	*1220*
Me	Me	*n*-Bu	*1713*
Me	Me	$PhC{\equiv}C$—	*1220*
Me	Me	Ph	*22, 1217*
Me	Et	Et	*1576*
Me	Et	*n*-Pr	*1713*
Me	Ph	Ph	*22, 1217*
Me	$MeC{\equiv}C$—	$MeC{\equiv}C$—	*1220*
Me	$PhC{\equiv}C$—	$PhC{\equiv}C$—	*1220*
NC—	NC—	*n*-Bu	*724*
NC—	*n*-Bu	*n*-Bu	*724*
$CHCl_2$	CCl_3	CCl_3	*1227*
Bu	Bu	Cp	*725*
Bu	Cp	Cp	*725*
n-Bu	Ph	Ph	*922*
Bu	Bu	NR_2	*726, 727*
Me	Me	F	*1232*
Bu	Bu	F	*1232*
Me	Me	Cl	*1217, 1220, 1431, 1778, 1842*
Me	Ph	Cl	*1217*
Me_3SiCH_2—	Me_3SiCH_2—	Cl	*1671*
n-Bu	*n*-Bu	Cl	*1218, 1629*
Cp	Cp	Cl	*725*
Ph	Ph	Cl	*1217*
Me	Me	Br	*1220, 1842*
Et	R_2N	R_2N	*1297*
n-Bu	R_2N	R_2N	*727, 1297, 1298*
Ph	EtNH	*n*-Bu	*1043*
Bu	RO	RO	*1297, 1298*
Bu	F	F	*1232*
Et	RS	RS	*1297*
n-Bu	RS	RS	*1297*
Me	Cl	Cl	*55a, 1217, 1431, 1778, 1842*
n-Bu	Cl	Cl	*1218, 1629*
Ph	Cl	Cl	*55a, 1217*
Me	Br	Br	*55a, 1220, 1842*
Ph	Br	Br	*55a*

TABLE 3-9

Boroxine Derivatives [$(RBO)_3$]

R	Reference	R	Reference
Me	*56, 348, 682, 683, 1089, 1201, 1593, 1600, 1822, 1879, 1880*	p-MeC_6H_4	*1729*
Et	*48, 617, 780, 1201*	o-$CF_3C_6H_4$	*1636*
Vi	*1172*	m-$CF_3C_6H_4$	*1404*
n-Pr	*780, 1201, 1203*	o-$(BrCH_2)C_6H_4$	*749*
i-Pr	*1201*	p-$(BrCH_2)C_6H_4$	*1718*
$EtSCH_2CH_2CH_2$—	*1342*	o-$(Me_2NCH_2)C_6H_4$	*749*
$CH_2{=}CHCH_2$	*1337*	p-$(Br_2CH)C_6H_4$	*1718*
n-Bu	*780, 1201, 1203, 1564, 1719*	o-$(PhCH{=}CHCH{=}CH)C_6H_4$	*1148*
i-Bu	*489, 780, 1201, 1203, 1564*	o-(o-$HSC_6H_4)C_6H_4$	*509*
s-Bu	*28, 1203*	o-(o-$EtO_2CNHC_6H_4)C_6H_4$	*509*
t-Bu	*487, 489, 1201*	o-HOC_6H_4	*662*
$EtCH{=}CH$—	*1564, 1645*	m-HOC_6H_4	*662*
n-C_5H_{11}	*1201, 1203, 1444*	p-HOC_6H_4	*662*
n-C_6H_{13}	*1201*	m-ROC_6H_4	*920*
Cyclohexyl	*780, 990, 1250*	p-$MeOC_6H_4$	*1729*
Et_3C—	*807*	o-ClC_6H_4	*882a*
$CH_2{=}C(R)CH_2CH(CH_2CH{=}CH_2)CH_2$—	*332, 334, 628*	m-BrC_6H_4	*1437*
$(n$-$Bu)_3C$—	*807*	p-BrC_6H_4	*920*
$(n$-$C_6H_{13})_3C$—	*807*	2,6-$Me_2C_6H_3$	*748*
$(n$-$C_8H_{17})_3C$—	*807*	3,5-$Me_2C_6H_3$	*1237*
Ph	*2, 5, 6, 441, 453, 656, 780, 882a, 1070, 1235, 1292, 1515, 1609, 1729, 1852*	3-HO-5-ClC_6H_3	*1636*
o-Tolyl	*163a, 749, 882a, 1234, 1729*	2-HO-3-BrC_6H_3	*1636*
		2-HO-5-BrC_6H_3	*1636*
		2-RO-5-BrC_6H_3	*1636*
		α-Naphthyl	*920, 1234*
		β-Naphthyl	*920, 1234*

TABLE 3-10

Polyboron Compounds

Compound	Reference	Compound	Reference
$(Me)(R_2N)B—B(NR_2)(Me)$	*1467*	$(RS)_2B(CH_2)_3B(SR)_2$	*1318, 1319, 1321*
$(Et)(Me_2N)B—B(NMe_2)(Et)$	*179, 1467*	$Cl_2B(CH_2)_3BCl_2$	*390, 1936, 1949*
$(Pr)(R_2N)B—B(NR_2)(Pr)$	*1467*	$Cl_2BCH_2CH_2CHMeBCl_2$	*1309*
$(Bu)(Me_2N)B—B(NMe_2)(Bu)$	*1465, 1467*	$(n\text{-}Bu)_2B(CH_2)_4B(n\text{-}Bu)_2$	*416*
$(Ph)(Me_2N)B—B(NMe_2)(Ph)$	*71, 179, 1465*	$(RO)_2BCHPhCHBrCH_2B(OR)_2$	*1763*
$Bu(R_2N)B—B(OR)(Bu)$	*1467*	$n\text{-}Bu(MeO)B(CH_2)_4B(OMe)_2$	*263, 264*
$Pr(RO)B—B(OR)Pr$	*1467*	$(RO)PhB(CH_2)_4BPh(OR)$	*438*
$(Me)(Cl)B—B(Me)(Cl)$	*1773*	$(R_2N)_2B(CH_2)_4B(NR_2)_2$	*68, 1350*
$(Me)(Cl)B—BCl_2$	*1826*	$(RO)(R_2N)B(CH_2)_4B(SR)(NR_2)$	*1350*
$R_2BCH_2BR_2$	*325a, 998a*	$(HO)_2B(CH_2)_4B(OH)_2$	*68, 1933*
$(HO)_2BCH_2B(OH)_2$	*1177*	$(RO)_2B(CH_2)_4B(OR)_2$	*68, 1352*
$(MeO)_2BCH_2B(OMe)_2$	*386, 588a*	$(RS)_2B(CH_2)_4B(SR)_2$	*1309, 1350*
$Cl_2BCH_2BCl_2$	*1183a*	$Cl_2B(CH_2)_4BCl_2$	*985, 1309, 1933*
$Br_2BCH_2BBr_2$	*1177*	$(HO)_2BCH_2CH_2CHMeCH_2B(OH)_2$	*1934*
$Et_2BCHMeBEt_2$	*963*	$(RO)_2BCH_2CH_2CHMeCH_2B(OR)_2$	*1936*
$(HO)_2BCHMeB(OH)_2$	*1190, 1247*	$Cl_2BCH_2CH_2CHMeCH_2BCl_2$	*1932, 1934*
$(BuO)_2BCHMeB(OBu)_2$	*1189, 1190, 1247*	$(HO)_2B(CH_2)_5B(OH)_2$	*68*
$(R_2N)BCHMeB(NR_2)_2$	*1247*	$(HO)_2B(CH_2)_5BCl_2$	*1931*
$Cl_2BCHMeBCl_2$	*1935*	$(RO)_2B(CH_2)_5B(OR)_2$	*68, 1936*
$(MeO)_2BCHPhB(OMe)_2$	*386*	$Cl_2B(CH_2)_5BCl_2$	*985, 1931*
$(RO)_2BC(Me)(Ph)B(OMe)_2$	*1535, 1536*	$(HO)_2B(CH_2)_6B(OH)_2$	*68*
$(RO)_2BCH(CH_2Ph)B(OR)_2$	*1535, 1536*	$(RO)_2B(CH_2)_6B(OR)_2$	*68, 1352, 1936*
$(Cl_2B)_2CCl_2$	*552*	$(HO)_2B(CH_2)_6BCl_2$	*1931*
$Me_2BCH_2CH_2BMe_2$	*390, 1827*	$Cl_2B(CH_2)_6BCl_2$	*985, 1931*
$Et_2BCH_2CH_2BEt_2$	*963*	$Ph(RO)B(CH_2)_{10}B(OR)Ph$	*438*
$(R_2N)_2BCH_2CH_2B(NR_2)_2$	*819, 1247*	$(HO)_2B(CH_2)_{10}B(OH)_2$	*68*
$(HO)_2BCH_2CH_2B(OH)_2$	*414, 819, 1190, 1247*	$(RO)_2B(CH_2)_{10}B(OR)_2$	*68*
$(RO)_2BCH_2CH_2B(OR)_2$	*1189, 1190, 1247, 1827*	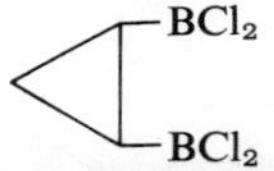	
$F_2BCH_2CH_2BF_2$	*390, 414, 445*	BCl_2 / BCl_2	*1610*
$Cl_2BCH_2CH_2BCl_2$	*414, 820, 1827, 1935*		

Compound	References
$R_3N \cdot Cl_2BCH_2CH_2BCl_2 \cdot NR_3$	*819*
$Me_2BCHMeCHMeBMe_2$	*390*
$(RO)_2BCHPhCH_2B(OR)_2$	*1535, 1536*
$F_2BCHMeCHMeBF_2$	*390*
$F_2BCH_2CH(BF_2)CMe_3$	*445*
$F_2BCH_2CH(BF_2)SiMe_3$	*445*
$Cl_2BCH_2CHMeBCl_2$	*390*
$Cl_2BCH_2CHEtBCl_2$	*1309*
$Cl_2BCHMeCHMeBCl_2$	*390*
$Cl_2BCH_2CH(BCl_2)CMe_3$	*445*
$Cl_2BCH_2CH(BCl_2)SiMe_3$	*445*
$Cl_2BCH_2CH(BCl_2)SiCl_3$	*445*
$Cl_2BCH_2CH(BCl_2)SnCl_3$	*445, 1949*
$Cl_2BCH_2CH(BCl_2)GeCl_3$	*445*
$Me_2BCH{=}CHBMe_2$	*1527*
$Et_2BCMe{=}CPr(BPr_2)$	*131*
$Et_2BCMe{=}CEt(BEt_2)$	*126, 131*
$Et_2BCEt{=}CEt(BEt_2)$	*126, 131*
$Et_2BCEt{=}CPh(BEt_2)$	*131*
$Et_2BCPh{=}CPh(BEt_2)$	*131*
$F_2BCH{=}CHBF_2$	*390*
cis-$Cl_2BCH{=}CHBCl_2$	*390, 395, 446, 823, 1622*
trans-$Cl_2BCH{=}CHBCl_2$	*446*
cis-$(X_2B)(Ph)C{=}C(Ph)(BX_2)$ X = Cl, OR, NR_2	*1870*
$Cl(Cl_2B)C{=}C(BCl_2)Cl$	*552*
$(n\text{-}Pr)_2BC{\equiv}CB(n\text{-}Pr)_2$	*742*
$(i\text{-}Pr)_2BC{\equiv}CB(i\text{-}Pr)_2$	*742*
$(n\text{-}Bu)_2BC{\equiv}CB(n\text{-}Bu)_2$	*742*
$(i\text{-}Bu)_2BC{\equiv}CB(i\text{-}Bu)_2$	*742*
$R_3BC{\equiv}CBR_3^{2-}$	*1906, 1915*
$(HO)_2B(CH_2)_3B(OH)_2$	*1316, 1936*
$(RO)_2B(CH_2)_3B(OR)_2$	*1316, 1318, 1321, 1936*
$(RS)_2B(CH_2)_3B(SR)(NR_2)$	*1319*
$(RS)(R_2N)B(CH_2)_3B(SR)(NR_2)$	*1319*
(cyclobutane ring) BCl_2, BCl_2	*1610, 1949*
(cyclopentane ring) BCl_2, BCl_2	*1610, 1632*
(cyclohexane ring) BX_2, BX_2; X = H, F, Cl, Br	*122*
(cyclohexene ring) BCl_2, BCl_2	*1632*
(cyclohexene ring) BMe_2, BMe_2	*1632*
(benzene ring) $(HO)_2B$, $B(OH)_2$	*1447*
(benzene ring) Cl_2B, BCl_2	*1447*
(benzene ring) $(HO)_2B$, $B(OH)_2$, X; X = H, NO_2	*68, 1399, 1447, 1726*

(continued)

TABLE 3-10—(*continued*)

Compound	Reference	Compound	Reference
R, R′O, B, B, R, OR′ (structure)	*438*	$(Cl_2B)_4C$	*552*
 R = Ph, *p*-$MeOC_5H_4$, Bu, cyclohexyl		$[(HO)_2B]_2CHCH[B(OH)_2]_2$	*394*
Cl_2B, BCl_2 (structure)	*1447*	$(Cl_2B)_2CHCH(BCl_2)_2$	*394, 1607*
Ph, RO, B, B, Ph, OR (structure)	*438*	$[(BuO)_2BCH_2CH_2]_mB[CHMeB(OBu)_2)_n$ $m + n = 3$	*1190*
$(Et_2B)_3CMe$	*946*	BCl_2, BCl_2, BCl_2, BCl_2 (structure)	*1949, 1951*
$(Et_2B)_2CMe(BHEt)$	*946*	Cl_2B, Cl_2B, BCl_2, BCl_2 (structure)	*1632*
$[(MeO)_2B]_3C—X$ X = H, Me, Ph	*386, 588a*	H, BH_2, H, BH_2, H, BH_2, H, BH_2 (structure)	*1950*
$(Cl_2B)_3CH$	*1183a*	H, BCl_2, H, BCl_2, H, BCl_2, BCl_2, H (structure)	*614, 1947, 1951*
$(Cl_2B)_3CCl$	*552*	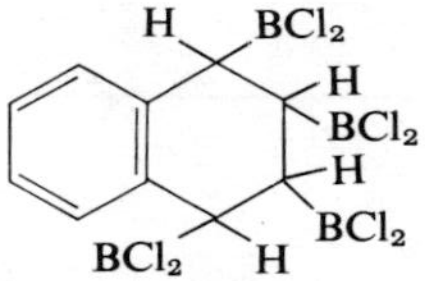	
$(F_2B)_2CHCH_2BF_2$	*1607*		
$[(HO)_2B]_2CHCH_2B(OH)_2$	*823*		
$(Cl_2B)(Me_2B)CHCH(BMe_2)_2$	*817*		
$(Vi_2B)(F_2B)CHCH_2(BF_2)$	*823*		
$(Vi_2B)(Cl_2B)CHCH_2(BCl_2)$	*822, 823*		
$(Cl_2B)_2CHCH_2BCl_2$	*445, 822, 823, 1607, 1949*		
$(F_2BCH{=}CH)_2BF$	*1774*		
$[(RO)_2B(CH_2)_3]_2B(OR)$	*1317, 1320*		
$[(RO)_2B(CH_2)_4]_2BOR$	*1352*		
$[(MeO)_2B]_4C$	*386, 588a*		

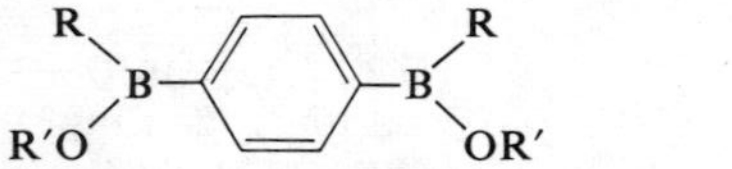

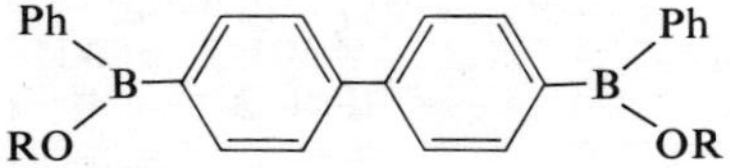

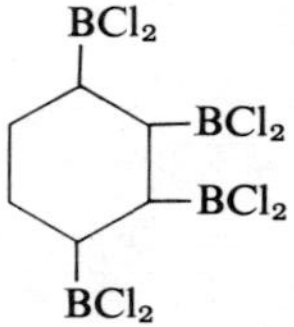

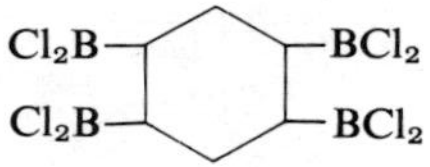

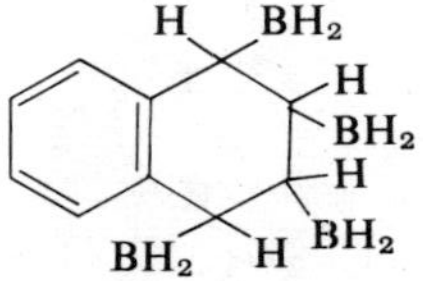

TABLE 3-11

Miscellaneous Organoboron Compounds

Compound	Reference
$n\text{-}Pr_3BHBH_3^-$	*46*
$n\text{-}Bu_3BHBH_3^-$	*46, 140*
$[H_3BCNBH_3]^-$	*23, 1840*
$[B_3H_7CN]^-$	*23* 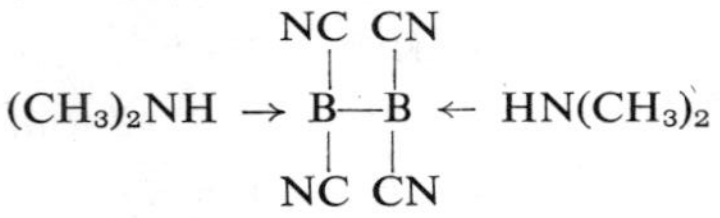
$H_3BCNBH_2NH_3$	*23*
$(PhB)_n$ (n = 9–12)	*1003*
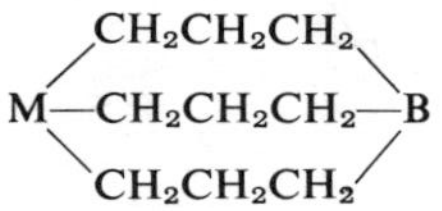 $M(CH_2CH_2CH_2)_3B$ (cage: M—$CH_2CH_2CH_2$—B, three bridges) M = N, B	*693*
Ph—B—N—R (bicyclic, two five-membered rings)	*366*
$Me_3BN(Et)_2CH_2CH_2N(Et)_2BMe_3$	*329*
$[t\text{-}BuBNHNH]_4$	*1364, 1365*
$R_2BNHNHBR_2$ R = Et, Pr, Bu	*1475*
PhB(NH—NH)$_2$BPh (six-membered ring)	*1308, 1475*
$C_6F_5B{=}NAr$	*1517*
$(BH_2CN)_n$ n = 4, 5, 6, 7, 8, 9	*1732*
$(R_2BCN)_4$	*1256*
$(CH_3)_2NH \rightarrow B(CN)_2{-}B(CN)_2 \leftarrow HN(CH_3)_2$	*1149*
$(RCH{=}NBHt\text{-}Bu)_2$	*753*
$(R_2C{=}NBR'_2)_2$ R′ = Me, Et, Pr, Bu, allyl, Ph	*330, 335, 1127, 1222, 1223, 1228, 1229, 1287*
$(R_2C{=}NBXR')_2$ R′ = Me, Bu, Ph, C_6F_5; X = halogen	*1221–1224, 1228*
$R_2B^{\ominus}(NR')_2$ ring fused to cyclohexene ring ($\oplus$) R = n-Pr, n-Bu	*559*
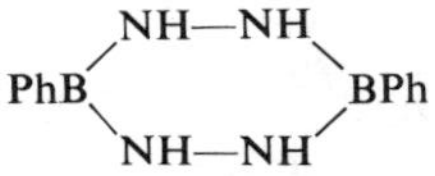 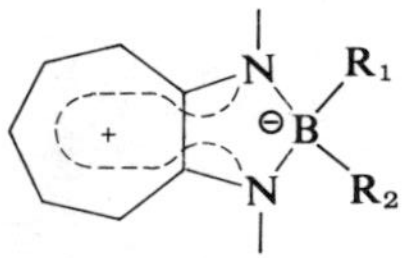 $R_1R_2B^{\ominus}$ chelated by two N (each N–CH₃) of a cycloheptatrienyl ring (+) R_1, R_2 = alkyl, aryl	*830*

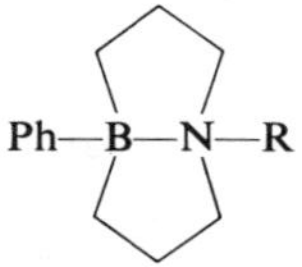

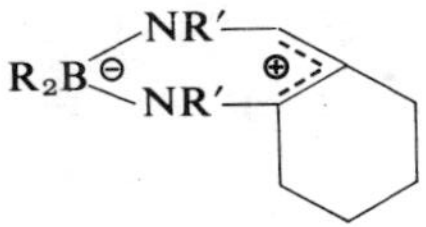

(*continued*)

TABLE 3-11—(*continued*)

Compound	Reference	Compound	Reference
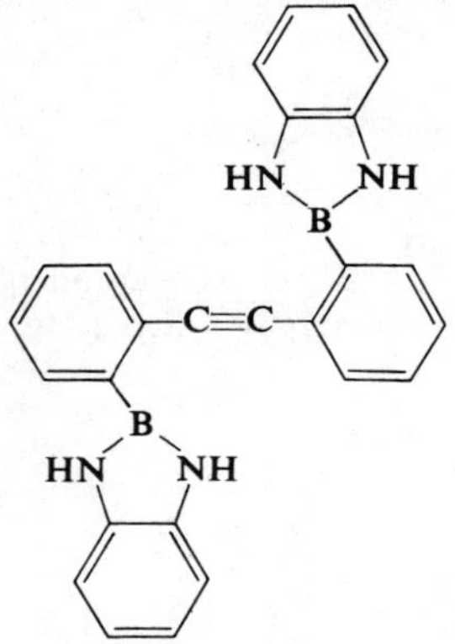	*1103*	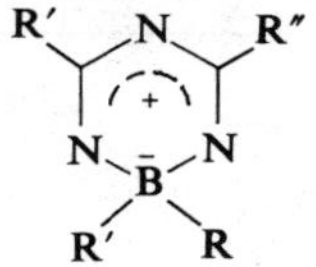	*559a, 1286b*
		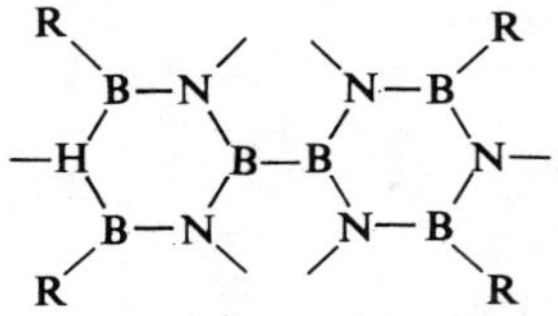 R = Me R = Bu	 *1220* *728*
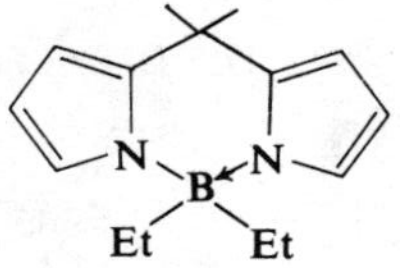	*108*	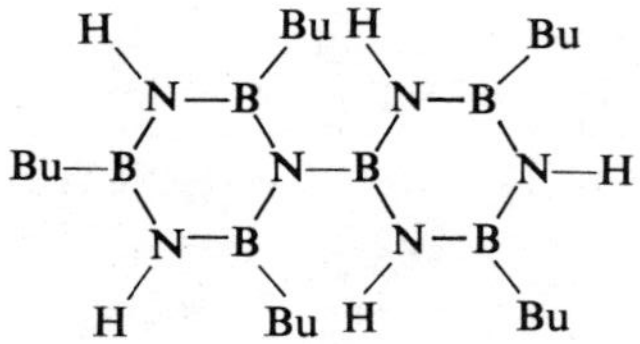 R = Me R = Bu	 *1219* *740*
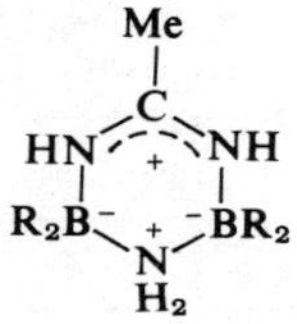	*559a*		
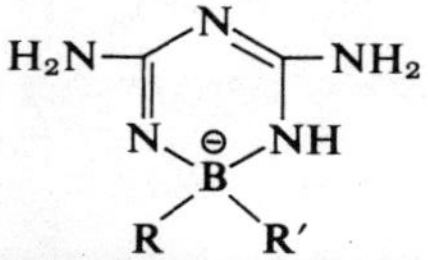	*1371*	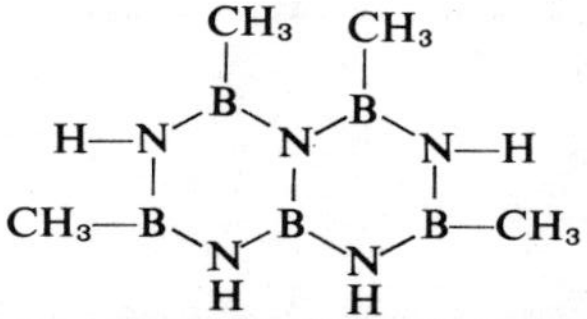	*145*

R	R′
C_6H_5	C_6H_5
C_6H_5	OH
C_6H_5	OEt
C_4H_9	OH
[illegible]	OH

Compound	Ref.	Compound	Ref.
	1807		1671
R = Et, Ph	1806, 1809	Z = O, NR	1809
		R = Pr, Bu, cyclic	575, 1283
R′, R = alkyl, aryl, H	1803–1805, 1807		1572

(*continued*)

TABLE 3-11—(*continued*)

Compound	Reference	Compound	Reference
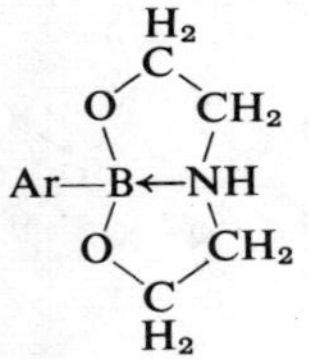	*1400*	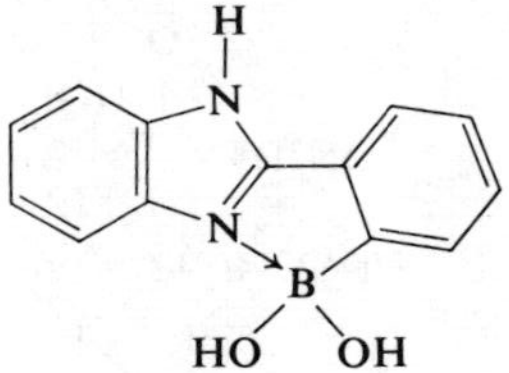	*1100*
R = Ph, *o*-tolyl	*525*	R = *n*-Bu	*1004*
		R = Ph, 2-thiophene	*1620*
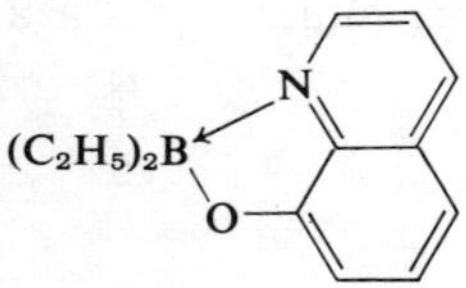	*1788*	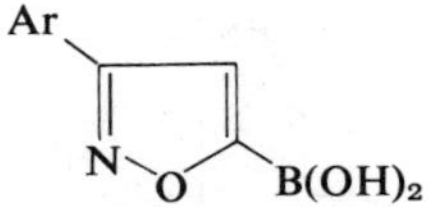	*121*
$[(C_2H_5)_2B(HNCOCH_3)]_2$	*1788*	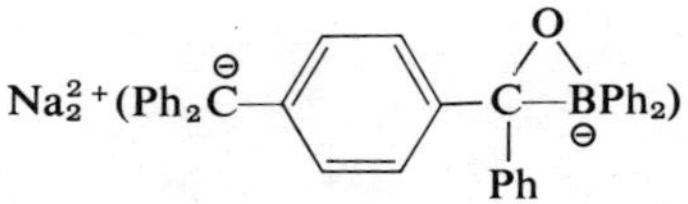	*1904*
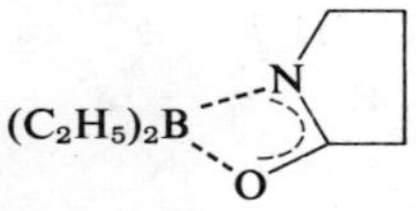	*1788*	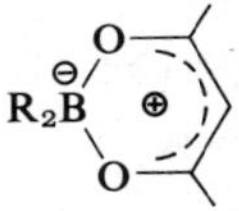 R = $C_6H_5C{\equiv}C$—, C_6H_5, C_2H_5, Pr, *n*-Bu *i*-Bu, 2-Bu, *n*-Hexyl, *cyclo*-C_6H_{11}, etc.	*64, 66, 556a, 774, 976, 1262, 1788, 1827a*
$[(C_2H_5)_2B(C_6H_4NO)]_2$	*1788*		

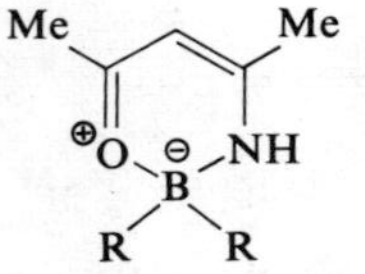

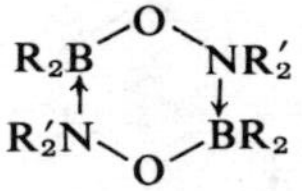

Compound	Ref.	Compound	Ref.
Me; Et_2—B; B—Et_2; H_2	*557*	$(C_2H_5)_2B$; C—CH_3	*1787, 1788*
R; R—C; C—R; NH; Et; Et	*953*	$B(OH)_2$; C≡C; $(HO)_2B$	*1103*
HO; C—X; $B(OH)_2$; X = CH_3, NH_2	*1726*	$B(OH)_2$	*1093, 1094*
HO; CH_3	*1726*	H; $B(OH)_2$	*1093*
		R; R; $B(OH)_2$	*1173*

(continued)

TABLE 3-11—*(continued)*

Compound	Reference	Compound	Reference
Et_2N, $B(OH)_2$, N=N, CO_2H (structure)	*1723*	H, Me_2, P, Me_2, P, H_2B, B, BH_2, Me_2P, P, PMe_2, B, Me, B, H_2, H_2 (structure)	*1844*
R, $B(OH)_2$, R, S (structure)	*703*	(RO)PhB, S, BPh(OR) (structure)	*438*
Cl, Cl, Cl, Cl, Cl, Cl, $B(OR)_2$ (structure)	*1920*	n-$Bu_2BSSBBu_2$	*1659*
R, B, O, O (structure)	*870*	$(MeBS)_3$	*1887, 1889*
		$(PhBS)_3$	*456, 1657b, 1885, 1887, 1890*
		$(n$-$BuBS)_3$	*1657b*
		S, S, B, B, S, R, R (structure); R = Ph, Me	*1658, 1660*
		BCl_2 (structure)	*870*
		$Na_2HB(CH_3)_2$	*352, 370*
		$Na_2HB(CH_3)_2 \cdot B(CH_3)_3$	*352, 370*
		$CrB(Ph)_4$	*1598*
		π-$(Me_3B_3N_3Me_3)Cr(CO)_3$	*1578, 1872*

Compound	Reference
[structure: benzocyclobutenyl-B(O₂C₆H₄)]	*870*
$Li^{+}\,^{-}C(BO_2C_3H_6)_3$	*1177a*
$[RO(BF_2CH_2)_2]^-$, R = Ph_3C, Me, H	*1697*
[structure: Y-nortricyclyl-BX_2], X = Cl, OH; Y = Ph, Cl	*870, 871*
$[H(Ph)P—B(Ph)Cl]_3$	*1767*
$[Me_3BPMe_2BMe_3]^-$	*1770*
$(Me_2B^{-}PMe_2^{+})_3$ (cyclic)	*1844*
π-$(Me_3B_3N_3Et_3)Cr(CO)_3$	*513b*
π-$(Et_3B_3N_3Me_3)Cr(CO)_3$	*513b*
$(HO)_2BC_5H_4Mn(CO)_3$	*1419b, 1426a*
$(Ph_2BMn)_2$	*1482*
$Bu_2BMn(CO)_4PPh_3$	*1482, 1483*
$Ph_2BMn(CO)_4PPh_3$	*1482, 1483*
$Ph(Me_2N)BMn(CO)_5$	*1482*
$MeB(NMe_2)_2Fe(CO)_3$	*1655a*
$(Me_2NBBrCH{=}CH_2)Fe(CO)_3$	*1655a*
RO—B(C₄H₄)Fe(CO)₃	*787*
$Ph_2BFe(C_5H_5)(CO)_2$	*1482*
$C_5H_5FeC_5H_4B(OH)_2$	*1157, 1420a, 1423–1425*
$(HO)_2BC_5H_4FeC_5H_4B(OH)_2$	*1423–1425*
$X_2BC_5H_4FeC_5H_4B(OH)_2$, X = Cl, Br	*1424*
$(TPA)_2CoBCl(Ph)$	*1655*
$(C_5H_5)Co^{+}(C_5H_5B—Ph)$	*784a, 785*

(continued)

TABLE 3-11—(*continued*)

Compound	Reference
[cyclopentadienyl]–Co–[borabenzene]B—X X = Ph, Me, Br, OH; paramagnetic	*784a, 786*
[borabenzene]B—R / Co / R—B[borabenzene]	*850b*
$(CoL_3)(BPh_4)_3$	*879d*
$(Me_2BNMe_2)Ni(\pi\text{-}C_3H_5)$	*1655a*
$(Ph_3P)_2Ni(\mu\text{-}BPh_2)_2Ni(PPh_3)_2$ (Ph₃P, Ph₃P / Ni—Ni / PPh₃, PPh₃; bridging B Ph₂)	*464*
$(MeSe—BMeI)_3$	*1702*
R—B(Se—Se)(Se)B—R ring, R = Bu, Ph	*1326, 1659, 1660, 1702a*
$Me_3B_3N_3Et_3Mo(CO)_3$	*5136*
$Et_3B_3N_3Me_3Mo(CO)_3$	*5136*
$\pi\text{-}C_5H_5RuB(C_6H_5)_4$	*731*
$Rh[P(OMe)_3]_2BPh_4$	*1453*
$(CH_3)_3SnB(C_6H_5)_2$	*1480, 1481*
$(CH_3)_2Sn[B(C_6H_5)_2]_2$	*1480, 1481*
$CH_3Sn[B(C_6H_5)_6]_3$	*1480, 1481*
$Sn[B(C_6H_5)_2]_4$	*1480, 1481*
$(HO)_2BC_5H_4Re(CO)_3$	*4119b*

3-2. Boron–Carbon Bond Cleavage and Other Reactions

Protodeboronation and Hydrolysis of the Carbon–Boron Bond

Cleavage of the C—B bond with protic agents, HZ (Table 3-12), takes place according to the general scheme outlined in Eq. 99. A major synthetic

$$R{-}B\langle + HZ \longrightarrow Z{-}B\langle + RH \qquad (99)$$

application includes the use of this reaction as a second step following hydroboration (Section 3-1) in a sequence that converts an olefin to a saturated paraffinic species (*184b*) or an internal acetylene to a cis olefin (*323*).

$$>C{=}C< \xrightarrow{H-B<} H{-}\overset{|}{\underset{|}{C}}{-}\overset{|}{\underset{|}{C}}{-}B\langle \xrightarrow{HZ} H{-}\overset{|}{\underset{|}{C}}{-}\overset{|}{\underset{|}{C}}{-}H + Z{-}B\langle \qquad (100)$$

$$R{-}C{\equiv}C{-}R \xrightarrow{H-B<} \begin{matrix} R & & R \\ & C{=}C & \\ H & & B{-} \end{matrix} \xrightarrow{HZ} \begin{matrix} R & & R \\ & C{=}C & \\ H & & H \end{matrix} + Z{-}B\langle \qquad (101)$$

Generally, carboxylic acids such as acetic acid (*245*, *246*, *323*, *461*, *677*, *684*) are more effective for this purpose than are common inorganic acids such as HCl (*144*, *461*) and HBr (*172*). This may be attributed to the potential for the organic acid to enter into a cyclic transition state (**XVIII**). The reaction

$$R_3B + RCO_2H \longrightarrow \begin{matrix} >B{\leftarrow}O \\ | \quad\quad \| \\ R \quad\quad C{-}R \\ H{-}O \end{matrix} \rightleftharpoons \begin{matrix} >B\cdots O \\ R \quad\quad C{-}R \\ H\cdots O \end{matrix} \;\textbf{(XVIII)} \longrightarrow RH + RCO_2BR_2 \qquad (102)$$

of triethylboron with carboxylic acids is first order in each reagent (*1787*), and the second-order rate constant is an inverse function of the pK_a of the acid. From several observations, including a primary isotope effect, it is suggested that a pre-rate-determining nucleophilic coordination of the organoborane by oxy functions of the acid occurs in a step which activates both

TABLE 3-12

Protodeboronation Reagents[a]

Reagent	References
H^+ (aq)	*4, 24, 327, 434, 866, 940, 999, 1014–1016, 1018, 1104, 1179, 1246, 1404, 1719, 1722, 1732*
H_2O	*24, 263, 266, 1000, 1104, 1135, 1173, 1184, 1185, 1234, 1251, 1273, 1293, 1337, 1361, 1444, 1564, 1735, 1822*
OH^- (aq)	*121, 203, 406, 1170, 1185, 1864*
ROH	*331, 334b, 586, 919, 1059, 1214, 1253, 1261, 1271, 1273, 1278, 1315, 1337, 1345, 1359, 1361, 1428, 1636, 1738, 1782, 1785, 1822, 1850*
RO^-	*386*
RCOOH	*125, 245, 246, 335a, 367, 395, 414, 461, 500, 650, 652, 684, 913, 1046, 1075, 1163, 1214, 1215, 1290, 1360, 1568, 1606, 1609, 1782, 1784, 1785, 1971*
Ketones	*597*
Acetylacetones	*64, 556a, 774, 976, 1262, 1618, 1787*
Oximes	*609, 1555*
NH_3, RNH_2, or R_2NH	*97, 466, 559b, 588, 686, 826, 952, 1059, 1261, 1264, 1265, 1283a, 1285, 1337, 1341, 1361, 1428, 1429, 1804, 1806, 1808, 1809, 1822, 1878–1880*
N_2H_4 or $RNHNH_2$	*1260, 1460, 1553*
$Ph_2C{=}NH$	*1554*
Aminoalcohols	*1040, 1104, 1609, 1822*
Amides	*556b, 1262, 1282b, 1285, 1958*
HF	*1392*
H_2S	*1822*
$Ph_3P{=}NH$	*35*
RSH	*760, 1059, 1240, 1253, 1259, 1261, 1263, 1266, 1278, 1315, 1337, 1340, 1342, 1361*
HCl	*54, 74, 144, 416, 616, 1245, 1842, 1875*
HBr	*6, 865, 1609, 1710*
8-Hydroxyquinoline	*976, 1428, 1428a*
Metal bisdimethylglyoxime	*1661, 1823*
RSO_3H (or $ArSO_3H$)	*1807*
H N (ring)	*108*
H_2O, $ZnCl_2$	*1424*

[a] For reagents with at least ten references, no attempt is made to be exhaustive.

the incipient carbanion and the electrophilic hydrogen of the acid (*1787*). As an alternative to the six-membered cyclic transition state in Eq. 102, a four-center transition state (**XIX**) can also account for the observed results (*336, 1787*). In the reaction of unsymmetrical trialkylboranes with propionic acid, there is little selectivity in the removal of primary or secondary alkyl

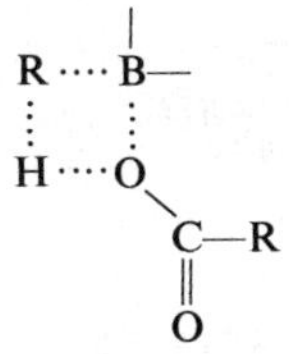

$$\begin{array}{l} R \cdots B - \\ \vdots \quad \vdots \\ H \cdots O \\ \quad\quad \searrow \\ \quad\quad\quad C-R \\ \quad\quad\quad \| \\ \quad\quad\quad O \end{array}$$

(XIX)

groups, but tertiary alkyl groups react much more slowly. The difference is attributed largely to steric rather than electronic effects (*125*) in a cyclic transition state, (**XVIII**) or (**XIX**). Steric factors have also been attributed to the observation that trimesitylborane is far less reactive toward water than either triphenyl or tri-α-naphthylborane (*203*).

The ease with which protodeboronation occurs in the sequence $R_3B >$ $R'CO_2BR_2 > (R'CO_2)_2BR$ using carboxylic acids is in agreement with "coordination" mechanism; for this reactivity sequence parallels the expected Lewis acidity and coordinative ability of the boron (*245*, *677*, *1214*). This principle has been applied in carrying out a controlled stepwise cleavage of triorganoboranes by adding only stoichiometric amounts of acid (*1340*, *1342*).

$$R_3B \xrightarrow{HZ} R_2BZ \xrightarrow{HZ} RBZ_2 \xrightarrow{HZ} BZ_3 \qquad (103)$$

Bond cleavage with carboxylic acids occurs with retention of configuration about the organic group (*246*, *500*, *1971*); and allylic rearrangement is encountered during protonolysis of allylorganoboranes (*1216*, *1251*).

$$>C=CHCH_2B< \xrightarrow{RCO_2H} >CH-CH=CH_2 \qquad (104)$$

Five-membered cycloboranes (**XX**) appear to be a good deal less stable toward ring opening protonolysis than are the six-membered cycloboranes (**XXI**). This is attributed to the strain in the borolane ring which is aggravated by the sp^2 hybridization of the boron atom (*263*, *266*, *1278*).

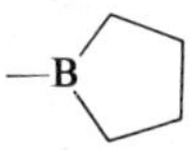

—B(five-membered ring) (**XX**) —B(six-membered ring) (**XXI**)

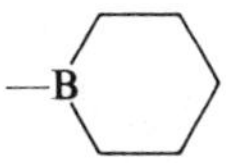

The kinetics of protodeboronation of phenylboronic acid and ring-substituted derivatives are consistent with an A-S_E2 or A-2 mechanism but not with an A-1 mechanism (*326*, *1014–1016*).

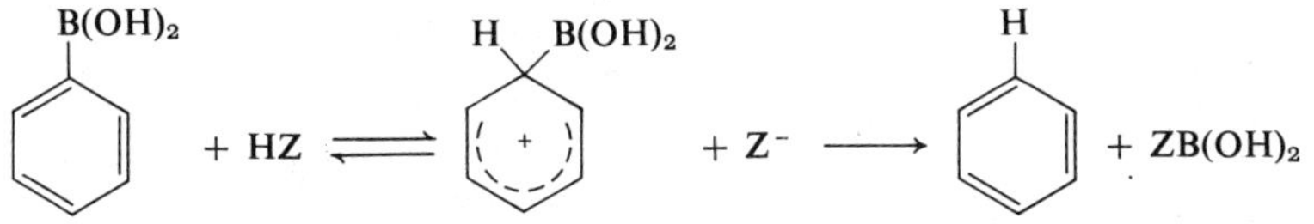

$$C_6H_5B(OH)_2 + HZ \rightleftharpoons [C_6H_5(H)B(OH)_2]^+ + Z^- \longrightarrow C_6H_5H + ZB(OH)_2 \qquad (105)$$

Alternatively, an anion (**XXII**) may be formed prior to internal electrophilic aromatic attack by H^+ (*1018*, *1404*). Protodeboronation of $ArB(OH)_2$ is

$$\mathrm{HO{-}\overset{\ominus}{B}(OH){-}OZ} \text{ (B bonded to phenyl)} \qquad Z = -SO_3H, -H$$

(XXII)

catalyzed by Cd^{2+}, Mg^{2+}, Zn^{2+}, Cu^{2+}, Pb^{2+}, Ni^{2+}, Co^{2+}, Ag^+, and Hg^{2+} (*866*, *1017*, *1719*), proceeding by a general two-step process, Eq. 106 (*24*) in which transmetallation (Section 3-1) is followed by hydrolysis.

$$ArB(OH)_2 + CdX_2 \xrightarrow[H_2O]{slow} ArCdX + B(OH)_3 + HX \tag{106a}$$

$$ArCdX + H_2O \xrightarrow{fast} ArH + Cd(X)OH \tag{106b}$$

The stability of several organoboron halides toward the formation of ethylenes upon the action of water follows the sequence $Vi_2BCl > ViBCl_2 > ViBF_2 > CF_2{=}CFBF_2$ (*169*).

Alkali fission of the C—B bond of boronic acids (Eq. 107) is less general than its acid-catalyzed counterpart and seems to depend on the ability of the

$$RB(OH)_2 \xrightarrow{OH^-} R{-}H + B(OH)_4^- \tag{107}$$

organic group to sustain a negative charge (*24*, *866*, *1184*, *1719*, *1818*, *1864*). This principle may explain the ease with which the ethynyl–boron linkage of $HC{\equiv}CB(OH)_2$ is cleaved by bases as mild as sodium bicarbonate solution (*1185*). Deuteriodeboronation of optically active 1-phenylethylboronic compounds with sodium deuteroxide gives 54% net inversion. The reaction probably proceeds by an S_E1 mechanism, with nucleophilic attack of the

$$R{-}B(OR')_2 + \bar{O}D \rightleftharpoons R{-}\bar{B}(OR')_2OD \xrightarrow{DOD} DO{-}D \; R{-}B(OR')_2{-}OD \longrightarrow DO^- + DR + B(OR')_2{-}OD \tag{108}$$

deuteroxide ion at boron followed by heterolysis of the C—B bond. The 1-phenylethyl anion is asymmetrically shielded by the departing borate directing electrophilic attack by deuterium preferentially to the opposite face of the carbanion (*500*).

In contrast, retention of configuration observed in the alkaline hydrolysis of the hydroboration products of *cis*- and *trans*-2,3-diphenylbut-2-enes is interpreted in terms of an S_Ei displacement at the carbon center (*1864*).

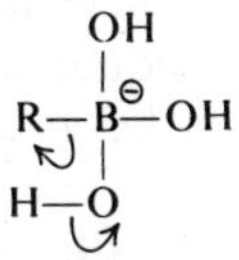

OH
R—B$^{\ominus}$—OH
H—O

Methoxide ion catalyzes the deboronation of tetrakis(dimethoxyboryl)-methane to bis(dimethoxyboryl)methane (*386*).

$$[(MeO)_2B]_4C + 2\ MeOH \xrightarrow{MeO^-} [(MeO)_2B]_2CH_2 + 2(MeO)_3B \qquad (109)$$

Basic hydrolysis of γ-chloropropylboranes provides a useful method for the synthesis of cyclopropane (*761*) (also see references in *Other Elimination Reactions* in Section 3-2).

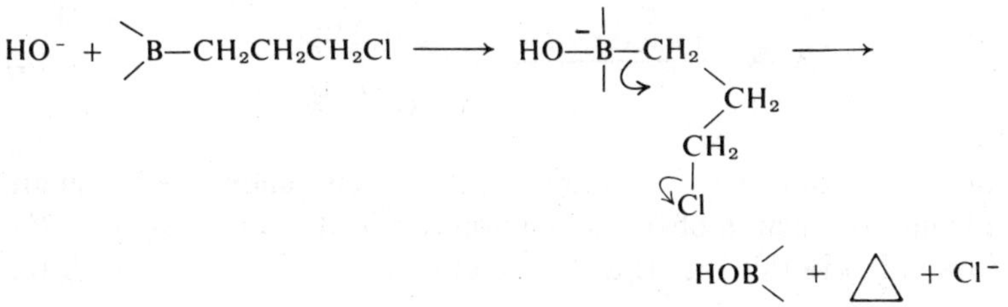

$$HO^- + \rangle B{-}CH_2CH_2CH_2Cl \longrightarrow HO{-}\overset{-}{B}{-}CH_2CH_2CH_2Cl \longrightarrow HOB\langle + \triangle + Cl^- \qquad (110)$$

Although alcohols react sluggishly with simple organoboranes (*1359*, *1822*), the presence of an ester, keto, or nitrile grouping on the α-carbon of the organoboron moiety promotes C—B bond cleavage (*250*, *251*, *283*) (Eq. 111).

$$Et_2BCHEtCO_2Et \xrightarrow{t\text{-BuOH}} EtCH_2CO_2Et + t\text{-BuOBR}_2 \qquad (111)$$

These appropriately situated functionalities serve to stabilize, usually by inductive effects, the incipient carbanion in the transition state. A similar argument is advanced to account for the ease with which aqueous ethanol cleaves the C—B bond of $C_6F_5B(OH)_2$ (*397*) as compared to other arylboronic acids. The cleavage is inhibited by acid, which supports the suggestion that the tetravalent boron anion derived from pentafluorophenylboronic acid readily loses pentafluorophenyl as a carbanion (*397*).

Alcoholysis of the ring compound (**XXIII**) preferentially cleaves the alkyl-boron ring rather than the phenyl—B bond (*1278*), but this result might be

$$\text{Ar—B}\langle\text{C}_4\text{H}_8\rangle \ (\textbf{XXIII}) + \text{ROH} \longrightarrow \text{ArB}(n\text{-Bu})\text{—OR} \qquad (112)$$

largely attributed to the instability of the five-membered ring (see above and Section 7-1). Boron (vinyl)carbon bonds in *trans*-2-alkoxyvinyl boranes are cleaved considerably more readily by alcohols than *B*-alkyl bonds (*334b*). Hydrolysis of phenyl(α-naphthyl)borinic acid in the presence of Me_2NCH_2-CH_2OH preferentially cleaves the naphthyl rather than the phenyl C—B bond (*1104*).

$$\text{Ph—B(OH)—C}_{10}\text{H}_7 \xrightarrow[\text{Me}_2\text{NCH}_2\text{CH}_2\text{OH}]{\text{H}_2\text{O}, \Delta} \text{PhB(OH)}_2 + \text{C}_{10}\text{H}_8 \qquad (113)$$

Alcoholysis of triarylborons with $HOCH_2CH_2NH_2$ stops after one C—B bond is cleaved (Eq. 114). In part, this can be attributed to chelation (*1609*).

$$\text{Ar}_3\text{B} \xrightarrow{\text{HOCH}_2\text{CH}_2\text{NH}_2} \text{Ar}_2\text{B}\langle^{\text{N}\to}_{\text{O}}\rangle(\text{CH}_2\text{—CH}_2) \qquad (114)$$

A series of chelate complexes with tetrahedral geometry about the boron are produced from the reaction of triorganoboranes with 1,3-diones (*774*, *1788*). Protodeboronation in this case (Eq. 115) undoubtedly proceeds through the

$$\text{R}_3\text{B} + \text{R}'\overset{\text{O}}{\overset{\|}{\text{C}}}\text{CH}_2\overset{\text{O}}{\overset{\|}{\text{C}}}\text{R}' \longrightarrow \text{RH} + \text{R}_2\overset{-}{\text{B}}\langle^{\text{O—C(R}')}_{\text{O—C(R}')}\rangle^{+}\text{C—H} \qquad (115)$$

enol form of the β-dicarbonyl compound in a fashion similar to the reaction of carboxylic acids with trialkylboranes (*245*).

The presence of catalytic amounts of thiols appears to catalyze the rupture of the B—C bond with alcohols (*1261*, *1263*, *1345*) or with amines (*1264*). Peroxides catalyze the cleavage of C—B bond with thiols (*1266*) (Eq. 116);

$$\text{R}_3\text{B} + \text{R}'\text{SH} \xrightarrow{\text{peroxide}} \text{R}_2\text{BSR}' + \text{RH} \qquad (116)$$

and protolysis of triorganoboranes can be substantially accelerated by catalytic amounts of pivolic acid (*940*).

Protodeboronation of trialkyl and trivinylboranes have also been accomplished with the use of ammoniacal silver hydroxide (Eq. 117), but a competing

$$R_3B \xrightarrow[NH_3]{AgOH} R{-}H + R{-}R \tag{117}$$

reaction involves the formation of R—R and metallic silver by what is probably a free-radical mechanism (*77, 866, 1527*). It is interesting to note that both ferrocene and biferrocene are formed by the interaction of ferrocenylboronic acid with ammoniacal silver oxide, and hydrolysis of phenylboronic acid under the influence of this same reagent yields benzene; alkenylboron compounds give alkenes, but alkylboric acids undergo coupling and disproportionation of the radicals (*281, 1105, 1423, 1424*) (see *Radical Reactions* in Section 3-2).

There are several examples in which apparent intramolecular protodeboronation occurs (Eq. 118). At temperatures approaching 200° both

$$[(HO)_2B]_2BCH_2CH_2[B(OH)_2] \xrightarrow[(394)]{\Delta} C_2H_6 + 2\,B_2O_3 + 2\,H_2O \tag{118}$$

diphenylboronous acid (in water) and phenylboronic acid are dephenylated to benzene and metaboric acid (*4*). The esters of these acids are more stable to dephenylation and prefer to disproportionate before dephenylation can occur (*4*); but those dialkylesters of phenylboronites that can produce intermediate phenylboronic acids by olefin elimination do undergo decomposition to give benzene (*7*).

Autoxidation

Generally, the autoxidation of organoboranes (*14, 67, 175, 203, 238, 259, 485, 488, 502, 616, 619, 736, 880, 1238, 1238a, 1530, 1753, 1754, 1763a*, and references within this section) proceeds only to the dialkoxyalkylborane (Eq. 119),

$$BR_3 + O_2 \longrightarrow RB(OR)_2 \tag{119}$$

but the presence of a solvent such as tetrahydrofuran (*239*) encourages the reaction to completion; (Eq. 120).

$$R_3B + 1.5\,O_2 \longrightarrow B(OR)_3 \xrightarrow{3\,H_2O} B(OH)_3 + 3\,ROH \tag{120}$$

When the alkoxyboron product is treated with water, this scheme provides a good source of alcohols from organoboron compounds, and when combined in sequence with hydroboration (Section 3-1) provides a convenient source of alcohols from alkenes.

A molecular mechanism proposed earlier for the autoxidation of C—B bonds has been discarded in favor of a radical chain process (*27*, *493*, *497*, *499*). The product-forming step involves attack of an alkylperoxy radical at the boron in a bimolecular homolytic substitution (S_H2) fashion (Eq. 120a).

$$ROO\cdot + BR_3 \longrightarrow ROOBR_2 + R\cdot \tag{120a}$$

$$R\cdot + O_2 \longrightarrow ROO\cdot \tag{120b}$$

The displaced alkyl radical then reacts with oxygen to regenerate the alkylperoxy radical (Eq. 120b), propagating the chain (*28*, *493*, *501*, *713*, *1002*). The radical mechanism is supported by the following observations:

(a) Reactive radical scavengers such as galvinoxyl (*27*, *28*, *493*, *497*, *499*, *501*, *502*, *921*) and iodine (*237*, *708*, *1238*) exhibit a measurable inhibiting effect.

(b) The reaction is initiated by radical generators such as *t*-butylhyponitrite (*502*, *854*).

(c) Stereoisomeric R groups lose their stereospecificity in forming the alkylperoxyboron compound (*28*, *497*, *499*, *501*).

The kinetic behavior is highly reminiscent of hydrocarbon autoxidation in that chain termination can occur by a combination of two alkoxy radicals (*854*). For the trialkylboranes, the rate constant of the S_H2 step is considerably (up to 10^7 times) faster than for the corresponding reaction at hydrogen in hydrocarbons.

The peroxide RO_2BR_2 or diperoxide $(RO_2)_2BR$ (*483*) formed in the initial steps may react with a second mole of oxygen in another S_H2 process or undergo an intermolecular redox reaction (*28*, *489*, *1375*, *1566*) to give an alkoxyboron product:

$$RO_2BR_2 + R_3B \longrightarrow 2\ ROBR_2 \tag{121}$$

The first boron–carbon bond of a trialkylborane oxidizes faster than the second and the second much faster than the third. This trend supports the view that the reactivity of the alkylboron compounds is diminished by attachment of atoms or groups that contain electron pairs capable of resonance interaction with an adjacent boron, thus reducing the acceptor ability of the boron. Any group that can donate electrons inter- or intramolecularly to the unfilled shell of the boron atom will, by doing so, render the compound less susceptible to oxidative attack. The following trends in reactivity are consistent with this (*493*, *865–867*, *921*, *1204*, *1374*, *1529*, *1719*, *1789*, *1790*):

$R_3B > R_2BOR > RB(OR)_2$, $R_3B > R_2BCl > RBCl_2$, $R_3B > R_2BOH > RB(OH)_2$, alkyl-$B\langle$ > vinyl-$B\langle$. Also, in a study carried out at 180°K, oxygen reacts more readily with Me_3B than with either 1,1-$Me_2B_2H_4$ or $Me_4B_2H_2$. This difference in reactivity demonstrates again that an unoccupied boron orbital is essential for the oxidation (*1530*). Both water (*866*, *867*, *1374*, *1719*) and amines (*14*, *483*, *487*, *489*, *490*, *493*, *617*, *618*, *620*, *1374*) inhibit the oxidation, and this inhibition can be attributed to the formation of a coordinate bond to the boron atom. Also, ethylacetoacetyl di-*n*-butylboronite (**XXIV**) is considerably more stable toward air than the simple $(n\text{-Bu})_2BOR$ esters and this is attributed to chelation (*651*).

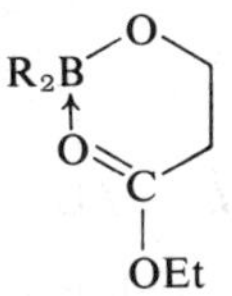

(XXIV)

The observed reactivity trends (*67*, *490*, *493*, *713*) *t*-alkyl-$B\langle$ > *s*-alkyl-$B\langle$ > *n*-alkyl-$B\langle$ > Me—$B\langle$ can be justified in terms of stabilizing effects on the incipient alkyl radical in both the transition state and the activated complex of the propagation step in which radical displacement on the boron atom occurs (Eq. 120a). Consistent with this is the high reactivity of tribenzylborane toward autoxidation (*921*), and the observation that perfluorotriphenylboron does not react with pure oxygen (*1165*). Steric retardation in the S_H2 step has also been observed (*493*).

The initiation step is still very much in doubt but may involve a transient O_2BR_3 complex (*1375*, *1961*), which rearranges followed by homolytic cleavage to give a supply of radicals. Alternatively, a displacement of the type $R_3B + O_2 \rightarrow R\cdot + [\cdot O_2BR_2]$ may occur (*170*). Studies on the nature of the initiation step show a large steric effect in that increased crowding around the boron atom increases the induction period (*237*), and that the rates of initiation, 1° > 2°, are in the opposite order to that observed in autoxidation, 3° > 2° > 1° (*718*). Also, trimesitylboron is far less reactive toward oxygen than either triphenyl or tri-α-naphthylboranes and this is attributed largely to steric factors (*203*). Another study suggests that steric control does not operate to the exclusion of electronic factors; for the primary initiation rates of borane autoxidations are tricyclohexylborane > tri-*s*-butylborane > triisobutylborane in an initiation reaction that is first order in borane and oxygen (*170*).

Mass spectral evidence for the intermediate cyclic 3,5-dimethyl-1,2,4-trioxa-3,5-diborolan (**XXV**), was obtained during the autoxidation of BMe_3

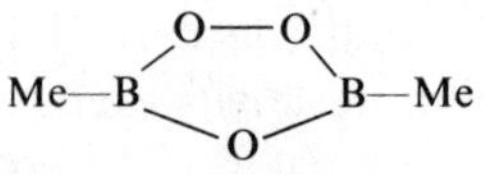

$$\mathrm{Me{-}B}\underset{\mathrm{O}}{\overset{\mathrm{O{-}O}}{\diagdown\!\!\diagup}}\mathrm{B{-}Me}$$

(XXV)

(*80*). The controlled low-temperature autoxidation of organoboranes gives a diperoxyborane which when treated with hydrogen peroxides yields alkyl hydroperoxides (*235*, *1038a*). Free radical autoxidation of dialkoxyalkylboranes may proceed through a C—H hydrogen abstraction rather than attack at boron (*921*).

Action of Peroxides and Related Compounds

The oxidation of C—B bonds with peroxides (*13*, *1006*, *1719*) is often a follow-up reaction to hydroboration (Section 3-1) in the conversion of olefins into alcohols (*293*, *294*). Peroxidation of phenylboron compounds results in the formation of phenols, and vinylboron compounds give aldehydes or ketones (*24*, *184a*, *184b*, *754*, *843*, *1105*, *1455*, *1698b*).

$$R_3B + 3\,H_2O_2 + OH^- \longrightarrow 3\,ROH + B(OH)_4^- \qquad (122)$$

Unlike autoxidation (see above), cleavage with acid- or base-catalyzed hydrogen peroxide is stereospecific and can be used in reactions where stereochemical properties are to be preserved. Normally, the peroxide reactions occur with retention of configuration about the R group (*320*, *322*, *1861*, *1969*).

Generally, the peroxide reaction, whether carried out with H_2O_2 or with other peroxides, such as $C_6H_5CO_3H$ (*867*), results in the quantitative cleavage of all C—B bonds in an organoborane. However, a reactivity trend $R_3B > R_2BX > RBX_2$ (X = halogen, OH, OR) for the boron moiety is noticed and is consistent with reduced acceptor ability of the boron when an electron pair of an adjacent group interacts with the vacant boron orbital ($R_2B\ddot{O}R \leftrightarrow R_2B^-{=}O^+R$) (*867*).

When using alkaline hydrogen peroxide, the mechanism probably does not involve free radicals as does autoxidation and also as do reactions with di-*t*-butyl peroxide (see below). From detailed studies on benzeneboronic acid, the following pathway is suggested (*495*, *1005*, *1007*, *1372*):

$$H_2O_2 + OH^- \rightleftharpoons HOO^- + H_2O \qquad (123a)$$

$$C_6H_5B(OH)_2 + HOO^- \longrightarrow \left[HO{-}\underset{OH}{\overset{Ph}{B}}{-}O{-}O{-}H \right]^- \qquad (123b)$$

$$\left[HO{-}\underset{OH}{\overset{Ph}{B}}{-}O{-}O{-}H \right]^- \longrightarrow C_5H_6OB(OH)_2 + OH^- \qquad (123c)$$

and it appears highly probable that an analogous mechanism is operating in the peroxidation of trialkylboranes. Hydroxydeboronation of alkylboronic acids with hydrogen peroxide is second order over the H_0 range of -1.5 to 6. Consistent with this, and with observed substituent effects, the proposed transition states for the base- and acid-catalyzed S_E2 reactions are, respectively (*1372*),

$$\left[HO{-}\underset{OH}{B}\overset{R}{\cdots}\overset{\delta-}{O}\cdots\overset{\delta-}{OH} \right]^{\ddagger} \quad \text{and} \quad \left[HO{-}\underset{OH}{B}\overset{R}{\cdots}\overset{\delta-}{O}(H)\cdots\overset{\delta+}{OH_2} \right]^{\ddagger}$$

Several variations in the transition state structure have been offered to account for the effect of different coreagents such as perchloric and phosphoric acid (*1006*).

Peroxidation of the dihydroboration product of acetylenes produces an alcohol predominantly, rather than the expected ketone. This result is attributed to the rapid hydrolysis of the dihydroboration intermediate (*323*).

Accumulation of *t*-butanol during the oxidation of benzeneboronic acid with *t*-butylhydroperoxide leads to a change in the reaction mechanism (*1148a*). The peroxidation of benzeneboronic acid is catalyzed by chelating agents such as pinacol and benzilic acid but retarded by malonate (*1020*).

By using peroxide reagents and conditions that favor the generation of RO· radicals (*496*, *1002*), an S_H2 displacement of a boron-attached organic group can occur. An esr examination of a reaction between trialkylboron and di-*t*-butyl peroxide under ultraviolet radiation shows no evidence for the formation of an $R_3\dot{B}O$-*t*-Bu intermediate but only the products which result

$$t\text{-BuO}\cdot + R_3B \longrightarrow R\cdot + R_2BO\text{-}t\text{-Bu} \qquad (124)$$

from the displacement depicted by Eq. 124 (*1002*). Alkoxy radicals also induce an S_H2 reaction at the boron atom of trialkylboroxines $(RBO)_3$; however, the rate is expectedly slower than for the corresponding trialkylboron, R_3B (*480*). Low activation energies, 0–5 kcal mole^{-1} are observed for these S_H2 reactions (*480*).

A radical-producing reaction is apparently operating in the formation of $C_{12}H_{26}$ from the vigorous reaction of $(n\text{-}C_6H_{13})_3B$ with neutral aqueous hydrogen peroxide in which two of the three B—C bonds are cleaved (*124*), for the yields of the dimeric product hydrocarbons are reduced considerably in the presence of radical scavengers (*123*). The reaction between tri-2-norbornylboranes and neutral 30% aqueous hydrogen peroxide may also proceed through an intermediate norbornyl radical (*507*). Bimolecular homolytic pathways are also suggested for the reactions of *t*-butylhypochlorite (*494*), Eq. 125, and of triplet ketones, Eq. 126 (*503*), with organoboranes.

$$t\text{-BuOCl} \longrightarrow t\text{-BuO}\cdot \xrightarrow{\text{RB}<} t\text{-BuOB}< + \text{R}\cdot \qquad (125)$$

$$\text{R}'_2\text{CO}^* + \text{BR}_3 \longrightarrow \text{R}'_2\text{COBR}_2 + \text{R}\cdot \longrightarrow \text{products} \qquad (126)$$

Trialkylboranes readily participate in radical chain reactions with organic disulfides; the reactions are initiated by oxygen or light and are inhibited by iodine (*236*).

$$\text{R}_3\text{B} + \text{O}_2 \longrightarrow \text{RBO}_2\cdot + \text{R}\cdot \qquad (127a)$$

$$\text{R}\cdot + \text{MeSSMe} \longrightarrow \text{RSMe} + \text{MeS}\cdot \qquad (127b)$$

$$\text{MeS}\cdot + \text{R}_3\text{B} \longrightarrow \text{MeSBR}_2 + \text{R}\cdot \qquad (127c)$$

$$\text{MeS}\cdot + \text{MeSBR}_2 \longrightarrow (\text{MeS})_2\text{BR} + \text{R}\cdot \quad \text{etc.} \qquad (127d)$$

$$\text{Overall: R}_3\text{B} + 2\ \text{MeSSMe} \longrightarrow (\text{MeS})_2\text{BR} + 2\ \text{RSMe} \qquad (128)$$

There is only one study indicating that basic solutions of hydrogen peroxide may produce radical reactions; this involves the use of Fenton's reagent (H_2O_2, Fe^{2+}) in which hydroperoxide radicals probably predominate (*1532*):

$$\text{PhCH=CH—B}\langle\text{O}_2\text{C}_2\text{H}_4\rangle \xrightarrow[\text{Fe}^{2+}]{\text{H}_2\text{O}_2} \text{PhCHO} + \text{PhCOMe} \qquad (129)$$

It is also proposed that radical intermediates may be formed during the oxidation of triphenylboron with organic hydroperoxides (*1698a*).

Halogens

Facile cleavage of B—C bonds with bromine or chlorine can occur with formation of an alkyl (or aryl) halide and a haloboron species (*6*, *24*, *866*, *867*, *1034*, *1836a*, *1836b*).

$$>\text{B—C}\!\!\!\!\!\!\!\quad\ \ + \text{X}_2 \longrightarrow\ >\text{B—X} + \text{X—C}\!\!\!\!\!\!\!\quad \qquad (130)$$

The cleavage mechanism for the reaction of alkylboranes that contain at least one α-C—H bond with bromine does not involve a simple rupture of the C—B bond; instead, the reaction can proceed through a fast α-bromination of the organoborane (*233a*, *309*, *311a*) followed by a subsequent cleavage of the intermediate with hydrogen bromide (*1034*, *1035*, *1836a*).

$$R_2BCH_2R' + Br_2 \longrightarrow R_2B\text{—}CHBrR' + HBr \quad (131a)$$

$$HBr + R_2BCHBrR' \longrightarrow R_2BBr + R'CH_2Br \quad (131b)$$

$$R_2BCHBrR' \longrightarrow (RR'CH)(R)BBr \quad (131c)$$

The first step, Eq. 131a, probably involves a free-radical chain mechanism, Eq. 132a,b (*309*, *1034*) which can be encouraged by uv irradiation (*233a*, *311a*, *1545a*), e.g. Eq. 133.

$$R_2B\overset{|}{\underset{H}{C}}\text{—} + Br\cdot \longrightarrow R_2B\text{—}\overset{|}{\dot{C}}\text{—} + HBr \quad (132a)$$

$$R_2B\text{—}\overset{|}{\dot{C}}\text{—} + Br_2 \longrightarrow R_2B\overset{|}{\underset{Br}{C}}\text{—} + Br\cdot \quad (132b)$$

$$n\text{-BuCHMe(B}\langle O\text{CH}_2\text{CH}_2O\rangle) \xrightarrow{Br,\ h\nu} n\text{-BuCBrMe(B}\langle O\text{CH}_2\text{CH}_2O\rangle) \quad (133)$$

If hydrogen bromide is removed or inactivated before protonolysis can occur, Eq. 131b, an alkyl rearrangement, Eq. 131c, can be induced by several catalysts, including water, methanol, and pyridine (*309*, *1036*). It appears that this rearrangement is preceded by coordination of the nucleophile to the boron atom (*309*):

$$R'\underset{Br}{\underset{|}{C}}H\text{—}\overset{R}{\overset{|}{B}}\text{—}R \;\; (\ddot{B}\leftarrow \text{Nucl.})$$

The relative rates for bromodeboronation decrease in the order: $R_3B > R_2BX > RBX_2$ (X = halogen, OH, OR) (*867*).

From a study of the reaction of bromine atoms with the α-hydrogens of triethylboron, an estimated 80 ± 3 kcal mole^{-1} was found for the α-C—H bond energy (*712*). This is rather low (cf. C(2)—H of propane) and might be explained in terms of stabilization of the unpaired electron of $>B—\dot{C}HCH_3$ by the adjacent vacant boron orbital (*712*).

The rate of brominolysis of several substituted benzene boronic acids, Eq. 134, (in 20% acetic acid) support an aromatic electrophilic displacement mechanism (*1008, 1011*).

$$ArB(OH)_2 + Br_2 \xrightarrow{H_2O} ArBr + B(OH)_3 + HBr \qquad (134)$$

The reaction is first order both in bromine and in boronic acid and shows a positive salt effect, catalysis by bases, and retardation by acids (*1010*), all of which suggest a quadricovalent boronate anion in the rate determining step (*1011, 1019*):

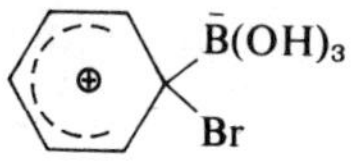

$\bar{B}(OH)_3$
$\oplus$
Br

A similar intermediate is proposed for the iodinolysis of phenyl (*327, 1021*) and of naphthalene boronic acids (*328*) as well as of alkenylboronic acids (*214a*).

In contrast to trialkylboranes the boron–carbon bonds in an alkylboroxine such as tri-*n*-butylboroxine are unaffected by chlorine (*1564*); instead, chlorine oxidizes the butyl to a butenyl group with concomitant formation of HCl (*1564*).

A reaction of iodine with vinylboranes provides for a convenient synthesis of cis–trans conjugated dienes (*1976*).

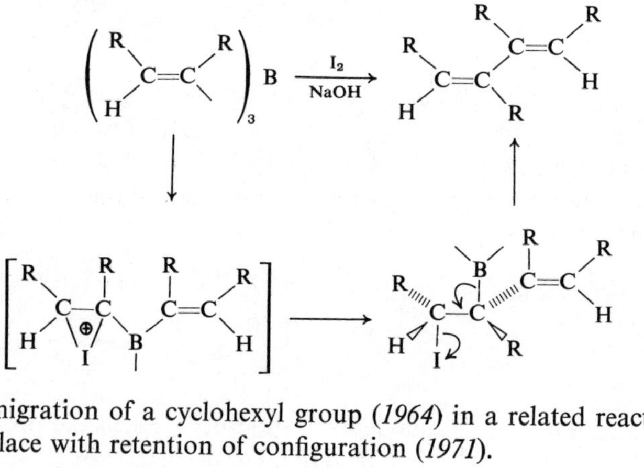

$$(RHC{=}CR)_3B \xrightarrow[NaOH]{I_2} RHC{=}CR{-}CR{=}CRH \qquad (135)$$

R R R R
C—C C=C
H ⊕ B H
I I

R B R R
C—C C=C
H I R H

A similar migration of a cyclohexyl group (*1964*) in a related reaction, Eq. 136 takes place with retention of configuration (*1971*).

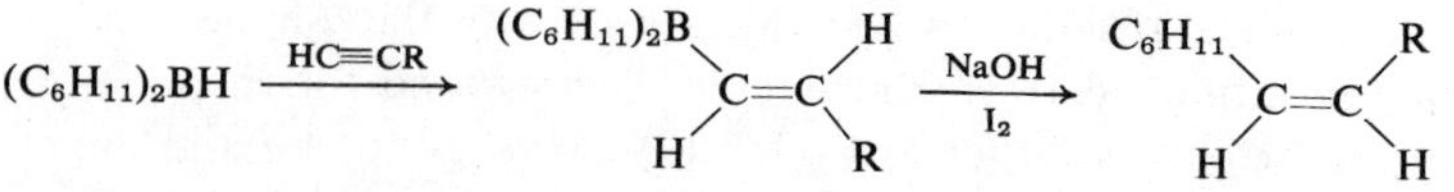

$$(C_6H_{11})_2BH \xrightarrow{HC\equiv CR} (C_6H_{11})_2B(H)C{=}C(H)R \xrightarrow[I_2]{NaOH} C_6H_{11}(H)C{=}C(H)R \qquad (136)$$

In the presence of sodium methoxide, organoboranes react rapidly with bromine (*232*) with inversion of configuration at the carbon (*233*):

$$[\text{norbornyl-H}]_3B \xrightarrow[\text{NaOMe, MeOH}]{\text{Br}_2\text{, THF, 0°C}} \text{norbornyl(H)Br} + R_2BOMe \qquad (137)$$

The acceleration in rate upon addition of base is presumably due to the formation of an intermediate quaternary boron. This increases the electron density on carbon and the ease of bond scission upon backside electrophilic attack by bromine (*233*):

$$\text{C—BR}_2 \underset{}{\overset{\text{NaOMe}}{\rightleftharpoons}} [\text{C—B(OMe)R}_2]^- Na^+ \xrightarrow{Br_2} [\text{Br---Br---C---B(OMe)R}_2] Na^+ \longrightarrow \text{Br—C} + R_2BOMe + NaBr \qquad (138)$$

"inversion"

An analogous cleavage of organoboranes with iodine in base yields the organic iodide (*274*). With primary alkyl groups the reaction stops after 2 mole equivalents of the C—B bonds are cleaved, Eq. 139; secondary alkyl groups react more sluggishly, and only one of the three bonds about boron is broken (*274*).

$$R_3B + 2\,I_2 + 2\,NaOH \longrightarrow 2\,RI + 2\,NaI + RB(OH)_2 \qquad (139)$$

Chlorination and bromination of *o*-hydroxy- and *o*-methoxyphenylboronic acids give the 5-halogenoderivatives, whereas the isomeric para acids undergo deboronization.

$$o\text{-}RO\text{-}C_6H_4B(OH)_2 \xrightarrow{Br_2} 5\text{-Br-}2\text{-}RO\text{-}C_6H_3B(OH)_2 \qquad (140)$$

$$p\text{-}RO\text{-}C_6H_4B(OH)_2 \xrightarrow{Br_2} p\text{-}RO\text{-}C_6H_4Br \qquad (141)$$

Deboronization occurs with both of the above isomers when treated with iodine (*1667*), and also when phenylboron dichloride reacts with chlorine (*1236*).

$$PhBCl_2 + Cl_2 \longrightarrow PhCl + BCl_3 \tag{142}$$

When Br_2 reacts with triallylboron there is rupture of the B—C bond as well as addition to the double bonds (*1944*).

$$(CH_2{=}CH{-}CH_2)_3B \xrightarrow{Br_2} BrCH_2CHBrCH_2Br \tag{143}$$

Sequential photolytic chlorination of *B,B,B*-trimethylborazine eventually results in deboronation of a *B*-perchloro intermediate (*1227*).

$$(MeBNH)_3 \xrightarrow[uv]{Cl_2} (Cl_2CHBNH)_3 \xrightarrow[uv]{Cl_2} (Cl_3CBNH)_3 \xrightarrow[uv]{Cl_2} (ClBNH)_3 \tag{144}$$

Reactions of Organoboranes with N-Chloroamines

Both possible modes of cleavage have been observed upon reacting *N*-chloroamines with trialkylboranes (*217, 491, 1686*).

$$\rangle B{-}R + Cl{-}N\langle \longrightarrow \rangle B{-}Cl + R{-}N\langle \tag{145a}$$

$$\rangle B{-}R + Cl{-}N\langle \longrightarrow \rangle B{-}N\langle + RCl \tag{145b}$$

Cleavage to give the alkyl amine, Eq. 145a, occurs in an alkaline tetrahydrofuran solution of the parent chloroamine with a stereochemical result that suggests a polar 1,2-rearrangement, Eq. 146 (*217*). *N*-Chlorodialkylamines react to give both the amine, Eq. 145a, and alkylchloride, Eq. 145b (*491, 1686*). When the reaction is carried out in chlorobenzene or in isopentane, the cleavage via Eq. 145b proceeds by a radical mechanism which involves the propagation steps Eqs. 147a and 147b (*491*).

$$R{-}B\langle + NH_2Cl \longrightarrow R{-}\overset{\ominus}{B}\langle{\leftarrow}\overset{\oplus}{N}H_2{-}Cl \longrightarrow Cl{-}B\langle{\leftarrow}H_2NR \tag{146}$$

$$Me_2N\cdot + BR_3 \longrightarrow Me_2NBR_2 + R\cdot \tag{147a}$$

$$R\cdot + ClNMe_2 \longrightarrow RCl + Me_2N\cdot \tag{147b}$$

Other Oxidative Reactions

Trimethylamine oxide reacts with trialkylboron compounds to form alkylborates (*967*, *968*). This reaction is also applicable to triarylboron and

$$BR_3 + 3\ Me_3NO \longrightarrow B(OR)_3 + 3\ Me_3N \tag{148}$$

heterocyclic boron compounds as well as to R_2BX (X = OR, halogen) (*968*), but not to carboranes nor a B—C bond of alkynylboron compounds (*968*). In a competitive reaction, trimethylamine oxide preferentially oxidizes the alkyl C—B bond rather than vinyl C—B bonds of $(RCH{=}CH)_2B$-thexyl (*1976*).

Nitrosocyclohexane appears to cleave tricyclohexylborane to give dicyclohexylborinic acid although the product analysis is not complete (*1928*). Nitric oxide reacts with n-Bu_3B at room temperature to yield three major products: R_2NOBR_2, $R_2BN(OR)NO$, and $R_2BNROBR_2$, with the liberation of some butene. n-Butyl nitrite also reacts with n-Bu_3B to yield, after hydrolysis, R_2BOH, ROH, RNHOH, 1-butene, and R_2NOH. The first step in this reaction appears to be

$$RONO + R_3B \longrightarrow R_2BOR + RNO \tag{149}$$

The nitroso compound then reacts with additional n-Bu_3B to yield R_2NOBR_2, $RNHOBR_2$, and butene (*853*). From the behavior of NO toward i-Bu_3B, it is proposed that an initial step involves bonding of the oxygen to boron and subsequent rearrangement of the complex (*15*).

The oxidation of alkylboronic acids by chromic acid, Eq. 150, (*1847*) is first

$$CrO_3 + RB(OH)_2 \xrightarrow{H_2O} ROH + H_3BO_3 + Cr^{3+} \tag{150}$$

order in each reagent over the wide pH range of −1 to 9. The pH rate profile indicates that alcohols are synthesized at pH ca. 5, whereas higher acidities (2 *N* acid) will produce ketones (*204*, *1847*). The chromic acid cleavage proceeds with the same stereochemistry as the H_2O_2 cleavage (see above) (*1847*).

Trialkylboranes are oxidized with sulfur at 145°C, with the formation of dialkylthioborinic esters, Eq. 151 (*1257*, *1258*), whereas Se reacts at 220–250°C

$$R_3B + S \longrightarrow R_2BSR \tag{151}$$

with the formation of cyclic boron selenium compounds of the type (*1326*):

```
    Se——Se
    |    |
R—B     B—R
    \Se/
```

The anodic oxidation of methaneboronate ions in alkaline solution produces ethane and methanol. In concentrated solutions ethane is the predominant product, the yield approaching that obtained by oxidation of acetate ion in acid solution (*846a*).

Radical Reactions

Two major courses of radical attack on organoboranes are known, Eqs. 152 and 153. Homolytic displacement at the boron site of the type given in Eq. 152

$$A\cdot + >B—R \longrightarrow A—B< + R\cdot \qquad (152)$$

$$A\cdot + >B—\overset{|}{\underset{H}{C}}— \longrightarrow >B—\dot{C}— + A—H \qquad (153)$$

have been demonstrated for A = $RO_2\cdot$ (see Section 3-2, *Autooxidation*), RO· (*482*, *1002*, *1174*) (see Section 3-2, *Peroxides*), RS· (*236*, *498*, *502*, *1263*, *1266*, *1342*), $R_2N\cdot$, Eq. 154 (*491*, *492*), $R_2C{=}CHO\cdot$ (*219*, *220*, *870*), and I· (*16*, *1122*, *1123*, *1134*),

$$Me_2N—N{=}N—NMe_2 \xrightarrow[(492)]{h\nu} 2\ Me_2N\cdot + N_2 \qquad (154a)$$

$$Me_2N\cdot + BBu_3 \longrightarrow Me_2NBBu_2 + Bu\cdot \qquad (154b)$$

Factors which control the rate of the homolysis reaction shown in Eq. 152 include: (a) the availability of an empty orbital on the boron atom in that it is observed that both complexation of the boron compound with a Lewis-base and the presence of $>B—\ddot{X}—$ (e.g. X = O, S) bonds retard the S_H2 reaction; (b) steric protection of the boron and of the attacking radical which also retards the reaction; and (c) the stability of the displaced alkyl radical which serves to increase the reaction rate with radical stability (*480*, *493*, *498*, *854*). A delicate balance between these factors appears to exist in many instances making it difficult to state the relative importance.

In contrast to I·, bromine atoms attack preferentially at the α-hydrogen of the alkyl groups, Eq. 153 (*233a*, *309*, *311a*, *712*, *1034*, *1536*, *1545a*). Methyl (*106a*, *709*, *710*, *714*), $CCl_3\cdot$ (*310*, *732*), and $CF_3\cdot$ radicals (*106a*) attack at both positions, Eqs. 152 and 153.

Trialkylboranes can quench acetone, acetophenone, and biacetyl triplets with the formation of ethyl and $R_2\dot{C}OBR_2$ radicals (*481*, *503*, *506*, *584*, *709–711*) by an S_H2 process. Excited-state homolysis of the boron–alkyl bond in *N*-methylanilinophenylalkylboranes occurs during the photocleavage of this compound in CCl_4 solvent (*732a*).

$$R_2C{=}O^* + BR'_3 \longrightarrow R_2COBR_2 + R\cdot \qquad (155)$$

The radicals $H_2\dot{C}—BMe_2$ and $H_2\dot{C}—BMe_2(NH_3)$ from γ-irradiated trimethylboron and its ammonia adduct have been identified by their esr spectra. The results suggest a π^1 structure with an estimated spin density of ca. 85% on carbon and ca. 15% on boron. It is concluded that these radicals are effectively planar at both carbon and boron centers (*1145*).

A reaction between $(n\text{-}Bu)_2BNa$ and RI is believed to occur by electron transfer from the former to the latter compound followed by decomposition of $RI\cdot^-$ to iodide ion and alkyl radical, for the major products are alkane and alkyl dimer with very little dibutylalkylborane being produced (*1551*). This work is in contrast to earlier work in which either triorganoboranes or carboranes were produced at some stage in the preparation of the R_2BM (M = alkali metal) or the subsequent reaction of this material with alkyl halides (*57*, *696a*, *946*, *949*).

Evidence has been obtained for the existence of $R_2\dot{B} \leftarrow NC_5H_4R$ and several other structurally related radicals that are obtained from dehalogenation of pyridine adducts of dialkylhaloboranes with an alkali metal (*941*, *944*, *950*, *1071*, *1869*).

Treatment of alkylboronic acids with ammoniacal silver oxide results in the coupling of alkyl groups to form higher alkanes (*866*, *1719*).

$$RB(OH)_2 \xrightarrow[NH_3]{Ag_2O} R—R \tag{156}$$

Trialkylboranes react similarly with alkaline silver nitrate (*215*, *291*, *308*) in what is presumed, from the products produced, to involve radical intermediates.

Transmetallation

As described in Section 3-1, transmetallation is an effective method for C—B bond construction. In many instances, transmetallation can also be used for C—B bond cleavage, Eq. 157. The metal may be Li^I (*541*), Be^{II} (*418*), B^{III} (*823*), Mg^{II} (*418*, *1939*), Al^{III} (*129*, *144*, *951*, *1051*, *1624*, *1939*, *1941*, *1955*), Si^{IV} (*1484*), Cu^{II} (*1424*), Zn^{II} (*1939*), Ag^I (*215*), Sn^{IV} (*648*), Sb^{III} (*547*),

$$BR_3 + {}^3/_x\, MZ_x \longrightarrow BZ_3 + {}^3/_x\, MR_x \tag{157}$$

Hg^{II} (*24*, *339*, *393*, *621*, *831*, *832*, *920*, *1052–1054*, *1175*, *1176*, *1190*, *1192*, *1193*, *1195–1197*, *1424*, *1640*, *1816*, *1944*), Tl (*393*, *1640*), or Pb^{II} (*832*).

Greatest attention has been focused on mercurideboronation, where it has been found that transfer organic groups can occur both by retention (*1051*, *1053*, *1176*) and by inversion of configuration (*1192*, *1193*, *1196*).

Organoboranes transmetallate mercuric acetate in THF to give dialkylmercury compounds. The first two alkyl groups rapidly at room temperature,

$$Et_3B \xrightarrow[(831)]{Hg(OAc)_2} Et_2Hg \tag{158}$$

$$(n\text{-}Pr)_3B \xrightarrow[(418)]{Et_2Be} (n\text{-}Pr)_2Be \tag{159}$$

$$Ph_3B \xrightarrow[(648)]{R_3SnNMe_2} PhSnR_3 \tag{160}$$

$$PhB(OH)_2 + HgBr_2 + H_2O \xrightarrow[(24)]{} PhHgBr + HBr + B(OH)_3 \tag{161}$$

while the third is more sluggish (*339*, *1053*). The reaction using trialkylboranes containing secondary alkyl groups proceeds much slower than the corresponding reaction of those containing primary alkyl groups. The reactivities of tri-*s*-alkylboranes toward mercuric benzoate decrease in the order: cyclopentyl > cyclohexyl > cyclooctyl > *s*-butyl > norbornyl (*1053*).

$$(C_5H_9)_3B + 2\,Hg(OAc)_2 \xrightarrow{THF} C_5H_9B(OAc)_2 + 2\,C_5H_9HgOAc \tag{162}$$

A reasonable mechanism for electrophilic transmetallation of $CH_2[B(OMe)_2]_2$ involves assistance by a neighboring boron (*1175*).

$$(MeO)_2BCH_2B(OMe)_2 + HgCl_2 + OAc^- + MeOH \longrightarrow \left[\begin{array}{c}(MeO)_2B\text{—}CH_2 \\ \text{Cl—Hg-----B(OMe)}_2 \\ \text{Cl} \quad \text{O} \\ \text{Me}\end{array}\right]^-$$

$$\longrightarrow (MeO)_2BCH_2HgCl + B(OMe)_3 + HOAc + Cl^- \tag{163}$$

Vinylboranes react with mercuric acetate to give the corresponding vinylmercury acetates, Eq. 164. This reaction is accompanied by a side reaction

$$\begin{array}{c}R\quad\;\; H\\ C{=}C\\ H\quad\;\; BR'_2\end{array} \xrightarrow{Hg(OAc)_2} \begin{array}{c}R\quad\;\; H\\ C{=}C\\ H\quad\;\; HgOAc\end{array} \tag{164}$$

leading to olefins and elemental mercury, the extent of which appears to be determined by the steric requirements of the vinylborane (*1054*). Mercurideboronation of benzeneboronic acid with phenylmercuric perchlorate in aqueous ethanol probably proceeds through the intermediate **(XXVI)** (*1013*).

PhHg $\bar{B}(OH)_3$ (cyclohexadienyl cation)

(XXVI)

Substituents increasing the electron density of the aromatic nucleus facilitate the mercury–boron exchange of benzeneboronic acids, whereas substituents withdrawing electrons from the nucleus have the opposite effect. This is in agreement with a mechanism in which mercuration takes place via electrophilic attack by positively charged mercuric ions (*1797*).

Dehydroboration and Isomerization Reactions

The reversibility of the hydroboration reaction (Section 3-1), Eq. 165a, has been directly observed in some instances (*48*, *823*, *932*, *978*, *1613*, *1901*) and

$$\underset{\mathrm{H}\quad\mathrm{B}-}{-\overset{|}{\underset{|}{\mathrm{C}}}-\overset{|}{\underset{|}{\mathrm{C}}}-} \rightleftharpoons \;\rangle\mathrm{C}{=}\mathrm{C}\langle\; + \underset{\text{(dimer)}}{\mathrm{H}-\mathrm{B}\langle} \tag{165a}$$

implied in a host of others involving alkyl isomerization (*191–193*, *243*, *295*, *298*, *324*, *325*, *892*, *930*, *1132*, *1418*, *1901*, *1968*).

The gas-phase unimolecular elimination of ethylene from triethylborane, $E = 33.7 \pm 1.2$ kcal mole^{-1} and $A = 4 \times 10^{11}$ sec^{-1}, is one of the slowest olefin eliminations observed from trialkylboranes (*17*). Dehydroboration has been involved as a primary step in alkylborane isomerization (*295*, *1209*), Eq. 165b, whereby the net effect is for the boron to move to the least sterically

$$(i\text{-Pr})_3\mathrm{B} \rightleftharpoons (i\text{-Pr})_2\mathrm{B}(n\text{-Pr}) \rightleftharpoons (i\text{-Pr})\mathrm{B}(n\text{-Pr})_2 \rightleftharpoons (n\text{-Pr})_3\mathrm{B} \tag{165b}$$

crowded carbon of an alkyl chain by an elimination–addition sequence (Eq. 165c) (*193*, *295*, *298*, *316*, *324*, *325*, *780*, *781*, *1209*).

$$\underset{\mathrm{B}\quad\mathrm{H}}{-\overset{|}{\mathrm{C}}-\overset{|}{\underset{|}{\mathrm{C}}}-\overset{|}{\underset{|}{\mathrm{C}}}-} \rightleftharpoons -\overset{|}{\underset{|}{\mathrm{C}}}-\overset{|}{\mathrm{C}}{=}\overset{|}{\mathrm{C}}- + \underset{\text{(dimer)}}{\mathrm{H}-\mathrm{B}\langle} \rightleftharpoons \underset{\mathrm{H}\quad\mathrm{B}}{-\overset{|}{\mathrm{C}}-\overset{|}{\underset{|}{\mathrm{C}}}-\overset{|}{\underset{|}{\mathrm{C}}}-} \tag{165c}$$

Ethers and the presence of B—H-containing compounds (*295*, *296*, *298*, *316*, *324*) catalyze the isomerization, but even then temperatures ca. 100–150° are usually required. From the equilibrium composition of alkylboron compounds, the stability of alkyl groups attached to boron decreases in the order: primary alkyl > secondary > tertiary (*298*, *581*, *1209*).

First-order kinetics are observed for the thermal isomerization of $(t\text{-Bu})\mathrm{B}(i\text{-Bu})_2$ to $(i\text{-Bu})_3\mathrm{B}$ and for the stepwise rearrangement of $(i\text{-Pr})_3\mathrm{B}$, $(s\text{-Bu})_3\mathrm{B}$, and $(i\text{-Bu})_3\mathrm{B}$ to the corresponding tri-n-alkylboranes (*424*, *1615*). Enthalpies and entropies of activation for these reactions are consistent with virtually complete disengagement of an alkyl group as olefin via a four-center transition state (*424*) in the case of (t-Bu)B(i-Bu)$_2$ (Eq. 166), but only partial disengagement in the case of secondary trialkylboranes (Eq. 167), the latter process perhaps involving a rotational inversion of a π complex (*1615*). An

$$R_2B{-}C(CH_3)_3 \rightleftharpoons [R_2B\cdots C(CH_3)_2{=}CH_2 \cdots H] \xrightarrow{\text{slow}} R_2BH + (CH_3)_2C{=}CH_2 \rightleftharpoons [R_2B\cdots CH_2{=}C(CH_3)_2\cdots H] \rightarrow R_2BCH_2CH(CH_3)_2 \qquad (166)$$

$$[R_2B\cdots CH(CH_3){=}CH_2\cdots H] \rightleftharpoons R_2BH \leftarrow CH(CH_3){=}CH_2 \rightleftharpoons R_2BH \leftarrow CH_2{=}CH{-}CH_3 \rightleftharpoons [R_2B\cdots CH_2{=}CH(CH_3)\cdots H] \qquad (167)$$

alternate general mechanism (*1897*) takes into account the above mentioned catalytic effect (*324*, *325*) of B—H-containing compounds on the alkyl isomerization and also the recognition that hydrogen tautomerization, known to occur in a member of boron hydrides, may be reasonably facile in these organoboranes (*1897*).

$$HB{<}C{-}C^{*}H \rightarrow [HB\cdots H\cdots C^{*}] \rightleftharpoons [H\cdots C{-}C^{*}{-}B(H)] \rightarrow H{-}C{-}C^{*}{-}BH \qquad (168)$$

In a reaction, Eq. 169, that may involve dehydroboration, carbon–carbon bond cleavage also occurs (*966*).

$$B(i\text{-}Bu)_3 \xrightarrow{300°} MeB(Pr)(i\text{-}Bu) + (CH_3)_2C{=}CH_2 \quad (169)$$

Other Elimination Reactions

Organoboron compounds containing a β-halogen substituent may undergo an elimination reaction involving transfer of the halogen to the boron atom.

$$-\overset{|}{\underset{B<}{C}}-\overset{|}{\underset{X}{C}}- \longrightarrow >C{=}C< + XB< \quad (170)$$

Dehaloboronation of this type is accomplished both by vacuum thermolysis (*1046*, *1536*) and by the use of a Lewis base such as pyridine, which encourages the abstraction of BX_3 (*1046*).

$$PhCH(Br){-}CH(Br){-}B\langle O{-}CH_2CH_2{-}O\rangle \xrightarrow{250°} PhCH{=}CHBr \ (\text{trans } 85\%, \text{ cis } 15\%) \quad (171)$$

$$H(Br)C{=}C(H)BBr_3 \xrightarrow{C_5H_5N} HC{\equiv}CH + Br_3B\cdot NC_5H_5 \quad (172)$$

Condensation of an organometallic reagent such as PhMgBr with a β-substituted organoboron compound can also result in β elimination (*1678*):

$$3\ PhMgBr + (BrCH_2CH_2BNPh)_3 \longrightarrow (PhBNPh)_3 + 3\ CH_2{=}CH_2 + 3\ MgBr_2 \quad (173)$$

β-Eliminations have been involved to account for the instability of the hydroboration products of vinyl chloride (*771*) and 3-chlorocyclohexene (*129*), and for the transfer of fluorine from carbon to boron in reactions of fluoroolefins with diborane (*79*, *1568*, *1742*), tetramethyldiborane (*132*), and with difluoroborane. Additional β-eliminations have been observed in still other instances (*29*, *445*, *1269a*, *1537*, *1548*, *1549*).

A displacement–elimination reaction requiring the presence of an appropriate nucleophile, such as hydroxide, alkoxide, or an amine, gives olefins with β-halogenated organoboranes (*771*, *1180*, *1246*).

$$ClCH_2CH_2B\langle \xrightarrow{OH^-} Cl{-}CH_2{-}CH_2{-}\overset{\ominus}{B}{-}OH \longrightarrow$$

$$Cl^- + CH_2{=}CH_2 + \rangle BOH \quad (174)$$

Organoboranes with a γ-halogen substituent, or other good leaving group, can undergo a similar displacement reaction in which cyclopropanes are produced (*129, 130, 276, 761, 771, 956, 1158*).

$$ClCH_2CH_2CH_2B\langle \xrightarrow{OH^-} Cl{-}CH_2{-}CH_2{-}CH_2{-}\overset{\ominus}{B}(OH) \xrightarrow{-Cl^-}$$

$$CH_2{-}CH_2{-}CH_2 \text{ (cyclopropane)} + HOB\langle \quad (175)$$

This approach has been used to synthesize *B*-cyclopropyl and *B*-cyclobutylboranes from open chain boron intermediates (*277*).

$$HC{\equiv}CCH_2CH_2OTs \xrightarrow{9\text{-}BBN} HC(B)(B)CH_2CH_2CH_2OTs \xrightarrow{CH_3Li} \text{cyclobutyl}{-}B \quad (176)$$

2-Alkoxyboron compounds are often found to be thermally unstable relative to the formation of alkenes by what may involve a cyclic four-membered transition state (*1254, 1269a, 1537, 1549*).

$$\overset{\ominus}{B}{-}CH_2{-}CH_2{-}\overset{\oplus}{O}(R) \text{ (four-membered ring)}$$

A related acid-catalyzed elimination probably proceeds by (**XXVII**) and a base-catalyzed elimination by (**XXVIII**) (*1549*).

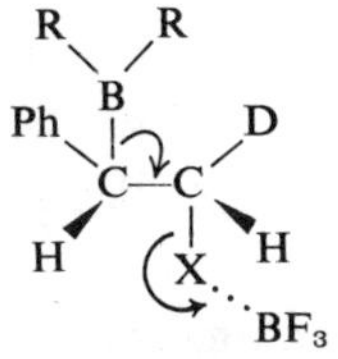

(XXVII) 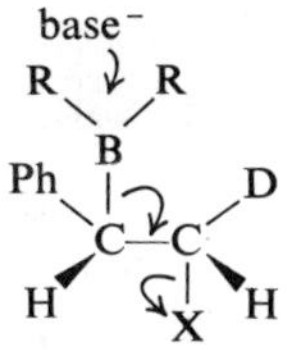 (XXVIII)

Two examples of deboronation involving the loss of α,β-situated borons are known (*819*, *1190*).

$$(HO)_2BCH_2CH_2B(OH)_2 \xrightarrow[(819)]{\Delta} H_2O + B_2O_3 + C_2H_4 \quad (177)$$

$$(HO)_2BCH_2CH_2B(OH)_2 \xrightarrow[OH^-\ (1190)]{HgCl} B(OH)_3 + C_2H_4 \quad (178)$$

Reactions with Unsaturated Systems

Organoboranes can react with multiple bonded groups to give addition (Eq. 179) or displacement (Eq. 180)products.

$$R_3B + Y{=}Z \longrightarrow R_2B{-}Y{-}Z{-}R \quad (179)$$

$$R_2B{-}\overset{|}{\underset{|}{C}}{-}\overset{|}{\underset{|}{C}}{-}H + Y{=}Z \longrightarrow R_2B{-}Y{-}Z{-}H + {>}C{=}C{<} \quad (180)$$

Because both pathways are potentially possible in any given interaction of an unsaturated compound with an organoborane, it is desirable to point out which factors are responsible for directing the reaction one way or the other. Unfortunately, the literature is nearly devoid of significant studies on this aspect of the subject, although there appear a substantial number of examples for each of the above two types of reaction, Eqs. 179 and 180. One obvious factor, however, is that arylboranes would prefer to add to a multiple bond, since the displacement type of reaction, Eq. 180, would be virtually energetically inaccessible.

Examples of the addition mode of reaction are found using alkenes (*336*, *870*, *932*, *1275b*, *1376*), alkynes (*332–334*, *628*, *1046*, *1246*, *1249*, *1268*, *1270*, *1271*, *1273–1275a*, *1276a*, *1753a*), vinyl ethers (*1269*), aldehydes (*590*, *1271a*, *1377*), α,β-unsaturated carbonyl compounds (*24a*, *219*, *220*, *222*, *238*, *273*, *282*, *773*, *775*, *870*, *875*, *1333d*, *1335*, *1639*, *1752*, *1755*, *1766b*), nitriles (*330*, *335*, *555*), isocyanates (*120*, *455*), and carbodiimide (*860*).

$$Et_3B + CH_2{=}CH{-}C_8H_{17} \xrightarrow{(932)} Et_2B{-}CH_2{-}CH(Et){-}C_8H_{17} \quad (181)$$

$$(CH_2{=}CHCH_2)_3B + 3\ RC(=O)H \xrightarrow{(1267)} B(OCH(R)CH_2CH{=}CH_2)_3 \quad (182)$$

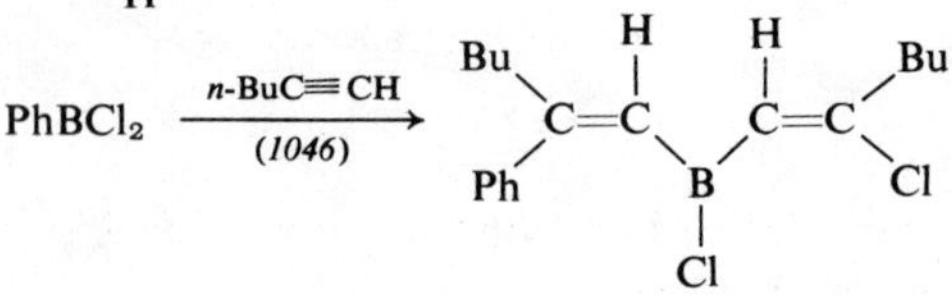

$$PhBCl_2 \xrightarrow[(1046)]{n\text{-}BuC{\equiv}CH} \begin{matrix} Bu \\ Ph \end{matrix}{>}C{=}C(H){-}B(Cl){-}C(H){=}C{<}\begin{matrix} Bu \\ Cl \end{matrix} \quad (183)$$

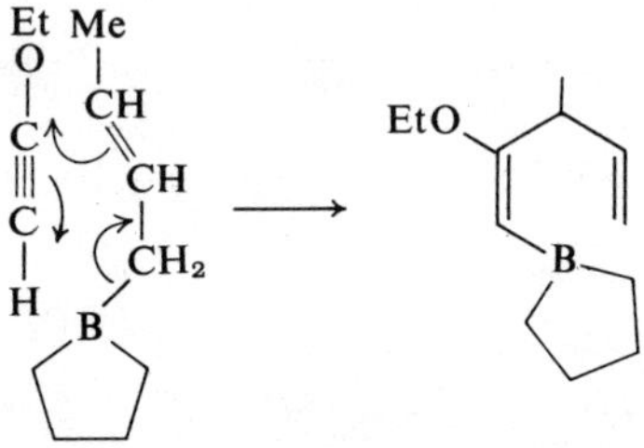

Et Me
O CH
C CH
C CH₂
H B

⟶ EtO B (184)

With the use of mixed trisubstituted boron compounds, the relative susceptibility of a phenyl group originally attached to the boron to migrate to an unsaturated carbon is generally greater than chlorine but less than $-NR_2$ (*120, 455, 860, 870, 1046*). Cis addition is observed with alkynes (*1046*). Also, allylboron compounds add cis to unsaturated compounds via a cyclic transition state (*332–334a, 628, 1249, 1268–1271, 1273–1275a, 1276a, 1333a, 1333c*).

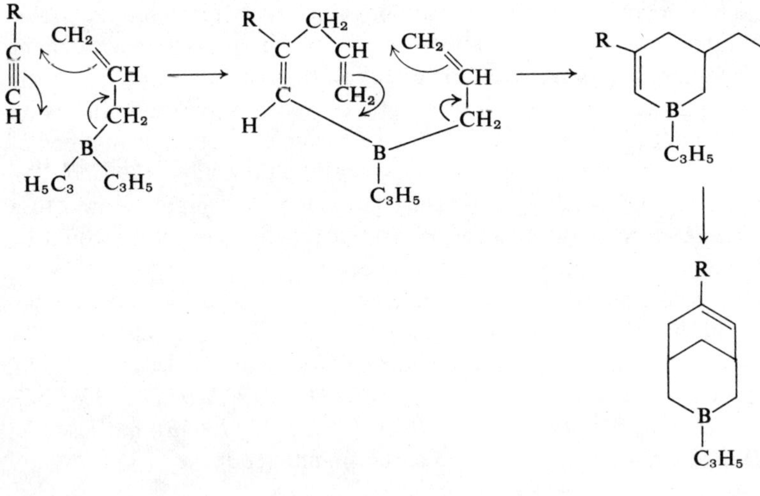

R
C CH₂
C CH
H CH₂
B
H_5C_3 C_3H_5

⟶

R CH_2
C CH CH_2
C CH_2 CH
H CH_2
B
C_3H_5

⟶

R
B
C_3H_5

↓ (185)

R
B
C_3H_5

The reaction of 1-methylcyclopropene with triallylborane occurs both with cis addition of allyl and allylboron fragments to the cyclopropene double bond, and by C(2)—C(3) ring cleavage (*336*).

$$(CH_2{=}CHCH_2)_3B + \text{1-methylcyclopropene} \xrightarrow{50\text{–}60\%} \text{1-Me-1-allyl-2-}[B(CH_2CH{=}CH_2)_2]\text{cyclopropane} \quad (H) \tag{186}$$

$$\xrightarrow{20\text{–}40\%} CH_2{=}CHCH_2CH{=}C(CH_3)CH_2B(CH_2CH{=}CH_2)_2$$

A photochemically induced reaction between triethylboron and cyclohexene gives indirect evidence of an addition across the double bond rather than an elimination–displacement (*1376*).

$$\text{cyclohexene} \xrightarrow[h\nu]{R_3B} \text{2-R-cyclohexyl-}BR_2 \xrightarrow{H_2O_2} \text{2-R-cyclohexanol} \tag{187}$$

Possibly all of the addition reactions of alkylboranes with carbonyl compounds proceed by way of a radical mechanism. Monomeric formaldehyde undergoes a facile reaction with $(n\text{-Bu})_2B$ to give displacement products, 1-butene and $n\text{-Bu}_2BOMe$; however, in the presence of oxygen, the addition product $n\text{-Bu}_2BOCH_2Bu$ is formed (*1377*). The latter reaction is sensitive to radical inhibiting agents and may well proceed by the mechanism shown in Eqs. 188a,b,c.

$$R_3B + O_2 \longrightarrow RBO_2\cdot + R\cdot \tag{188a}$$

$$R\cdot + CH_2O \longrightarrow RCH_2O\cdot \tag{188b}$$

$$RCH_2O\cdot + R_3B \longrightarrow R_2BOCH_2R + R\cdot \tag{188c}$$

An olefin intermediate is proposed for the alkylation of cyclic ketones via a Mannich Base (*273*), Eq. 189.

The reaction presumably proceeds through a 1,4 addition of the organoborane to the α,β-unsaturated carbonyl such as observed in other instances (*222*, *282*, *1752*), Eq. 190.

$$\text{(2-oxocyclopentyl)}CH_2N(CH_3)_2\cdot HCl \xrightarrow[K_2CO_3]{CH_3I} \left[\text{(2-oxocyclopentyl)}CH_2\overset{+}{N}(CH_3)_3I^-\right] \longrightarrow \left[\text{2-methylenecyclopentanone}\right] \xrightarrow{R_3B} \text{(2-oxocyclopentyl)}CH_2R \quad (189)$$

$$R_3B + CH_2{=}C(CH_3)CHO \xrightarrow{H_2O} RCH_2CH(CH_3)CHO \quad (190)$$

Both ring carbons and organo groups exo to the ring of boracyclanes are known to transfer from boron to the α,β-unsaturated carbonyl substrate. In the first instance, the intermediate boron product is used to prepare ω-hydroxyketones (*1756*); and in the latter case, it is known that transfer

$$\text{(boracyclane)}B{-} + CH_2{=}CH\overset{O}{\overset{\|}{C}}Me \xrightarrow{ROH} RO\overset{|}{B}(CH_2)_6\overset{O}{\overset{\|}{C}}Me \xrightarrow{(O)} HO(CH_2)_6\overset{O}{\overset{\|}{C}}Me \quad (191)$$

of the *B*-alkyl group takes place rather readily when the alkyl group is secondary or tertiary (*258*).

$$R{-}B\langle + \rangle C{=}\overset{|}{C}{-}\overset{O}{\overset{\|}{C}}{-} \xrightarrow{H_2O} R{-}\overset{|}{\underset{|}{C}}{-}\overset{|}{C}H{-}\overset{O}{\overset{\|}{C}}{-} + HOB\rangle \quad (192)$$

The 1,4-addition of trialkylboranes to α,β-unsaturated carbonyl compounds is inhibited by galvinoxyl (*876*) and "catalyzed" by a number of radical initiators—oxygen (*220*), diacyl peroxides, and photochemically (*219*).

The reaction of trialkylboranes with 1,4-benzoquinone, Eq. 193, produces the corresponding alkylhydroquinones in nearly quantitative yields, and alkylboronic acids may be obtained as by-products (*773*, *775*, *1334*, *1336*, *1766c*). The initial step may be a reductive alkylation (*775*), which produces a quinol ester (*1335*), and in turn rearranges in the presence of a Lewis acid (trialkylborane) to the corresponding hydroquinone (*775*). Alternatively, Eq. 194, a radical chain process has been proposed on the basis that oxygen catalyzes the reaction, and also that radical inhibitors such as iodine and galvinoxyl show a measurable effect (*875*).

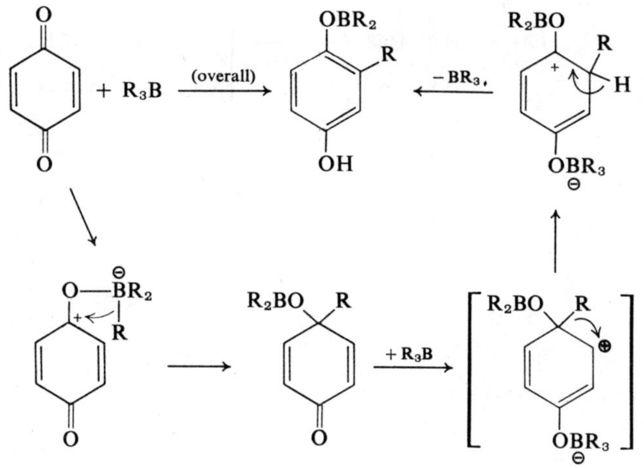

(193)

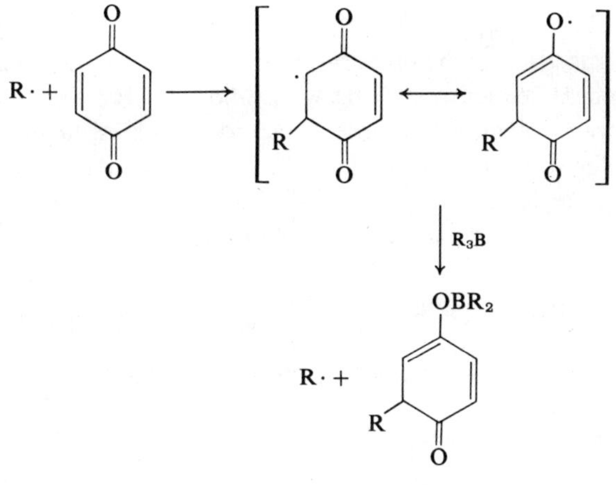

(194)

Displacement

The olefin–organoborane displacement reaction, Eq. 195, in the presence

$$B(C_nH_{2n+1})_3 + 3\ C_mH_{2m} \rightleftharpoons B(C_mH_{2m+1})_3 + 3\ C_nH_{2n} \qquad (195)$$

or absence of ether solvent, (*295*, *928*, *930*, *932*) has been used effectively for the contrathermodynamic conversion of internal to terminal olefins (*191–193*, *298*, *1775*), Eq. 196, for the preparation of mixed alkylboranes (*936*, *1037*), Eq. 197, for the syntheses of functional derivatives of trialkylboranes, Eq.

$$Et_2C{=}CHCH_3 \xrightarrow[\text{diglyme}]{(H{-}B\lessdot)_2} Et_2CH\underset{}{C}(B\lessdot)HCH_3 \xrightarrow[\text{diglyme}]{160^\circ} Et_2CHCH_2CH_2B\lessdot \xrightarrow{\text{1-decene, diglyme}} Et_2CHCH{=}CH_2 \quad (196)$$

$$Pr_3B + CH_2{=}C(Me){-}CH{=}CH_2 \xrightarrow{(936)} \text{PrB(cyclopentane)-Me} + 2\ C_3H_6 \quad (197)$$

$$i\text{-}Bu_3B + 3\ CH_2{=}CHCHOCF_2CHFCl \xrightarrow{(1327)} (ClFCHCF_2OCH_2CH_2CH_2)_3B + 3\ i\text{-}C_4H_8 \quad (198)$$

$$(CH_2{=}CHCH_2)_3B + (i\text{-}Bu)_3B \xrightarrow{(1338)} \xrightarrow{ROH} ROB\langle(CH_2)_3\rangle_2BOR \quad (199)$$

198 and 199, and also for the preparation of tritiated olefins via a hydroboration–displacement sequence starting with diborane-^{3}H (*1409*). Advantage of the trialkylborane–olefin displacement reaction has been taken in the equilibration of different positional isomers of macrocyclic alkadienes using triethylboron as a catalyst (*844*).
Alkyne–organoborane exchange is also known (*842*, *843*).

$$RC{\equiv}CR + BR'_3 \longrightarrow B(CR{=}CHR)_3 + \text{olefin} \quad (200)$$

The ease of olefin displacement depends much on the other two substituents

$$RCH_2CH_2BX_2 + R'CH{=}CH_2 \longrightarrow R'CH_2CH_2BX_2 + RCH{=}CH_2 \quad (201)$$

attached to boron (*1335*) with the following order for ease of reaction being observed (*1313*):

$$X = i\text{-Bu} > \text{(cyclopentyl)B} \geqslant Br > Cl > SR > OR,\ NR_2 \quad (202)$$

Two mechanisms appear plausible for the displacement reaction, the first involving an initial formation of olefin and dialkylborane through a dehydroboration step (see above), Eq. 203, followed by a reaction of the dialkylborane (monomer or dimer) with the unsaturated intermediate in a hydroboration step (Section 3-1), Eq. 204 (*295*, *298*). Alternatively, a cyclic six-membered

$$R_2BCH_2CH_2R' \longrightarrow R_2BH + CH_2{=}CHR' \quad (203)$$

$$R_2BH + CH_2{=}CHR'' \longrightarrow R_2BCH_2CH_2R'' \quad (204)$$

transition state (*1312*, *1314*) has been suggested in order to explain rate study observations (*1314*) as well as small differences between the isomer product distribution for this reaction and that obtained from a direct hydroboration of an appropriate olefin (*1312*).

$$RCH_2CH_2B\!\lt \overset{CH_2=CHR'}{\rightleftharpoons} [\text{R–C}\cdots H\cdots C\text{–R'};\ H_2C\cdots B\cdots CH_2] \rightleftharpoons RCH{=}CH_2 + R'CH_2CH_2B\!\lt \quad (205)$$

A cyclic transition state (*1272*) has also been suggested for the displacement reaction between aldehydes and organoboranes (*1214*, *1272*, *1275*, *1775*, *1782*, *1785*).

$$(RR'CHCH_2)_3B + PhCHO \xrightarrow{80-160^\circ} RR'C{=}CH_2 + (RR'CHCH_2)_2BOCH_2Ph \quad (206)$$

$$[\ {>}B\text{–}CH_2\text{–}C{<}\text{–H};\ B\leftarrow O{=}CH\text{–Ph}\ ]$$

In what probably involves similar transition states, 2-methyl-2-nitrosopropane, Eq. 207 (*609*), as well as *cis*-azobenzene, Eq. 208, undergo facile reactions with trialkylboranes to bring about stereospecific cis elimination of one *B*-alkyl group as alkene (*479*, *504*, *609*).

$$\text{H(Me)C=C(Me)Et} \xrightarrow[\text{2. } t\text{-BuNO}]{\text{1. Hydroboration}} [R_2B\text{–C–C–H};\ B\leftarrow O{=}N\text{–}t\text{-Bu}] \longrightarrow \text{H(Me)C=C(Me)Et} + t\text{-BuNHOBR}_2 \quad (207)$$

$$\text{PhN=NPh} + \text{H–C–C–BR}_2 \longrightarrow [\text{Ph–N}\cdots H\cdots C;\ \text{Ph–N}\cdots B\cdots C] \longrightarrow \text{Ph(H)N–N(Ph)BR}_2 + {>}C{=}C{<} \quad (208)$$

Carbonylation

Hexaalkyl-2,5-diboradioxanes and trialkylcarbinylboronic acids (or anhydrides) are obtained from reactions of trialkylboranes with carbon monoxide (*807*, *1582a*).

$$R_3B + CO \xrightarrow[50^\circ]{H_2O} \text{(}R_2C\text{–O–BR–CR}_2\text{–O–BR ring)} \quad (209a)$$

$$R_3B + CO \xrightarrow[2.\ \Delta,\ -H_2O]{1.\ H_2O,\ 150^\circ} (R_3CBO)_3 \quad (209b)$$

A probable mechanism involves the initial formation of R_3BCO followed by boron to carbon alkyl migration (*807*):

$$R_3B + CO \longrightarrow R_3\overset{-}{B}\text{—}\overset{+}{C}O \longrightarrow R_2B\text{—}\overset{\overset{O}{\|}}{C}R \longrightarrow \tfrac{1}{2}\ (R_2\overset{-}{B}\text{–O–}\overset{+}{C}R\text{–}\overset{-}{B}R_2\text{–O–}\overset{+}{C}R) \longrightarrow (RB\text{–O–CR}_2\text{–BR–O–CR}_2) \quad (210)$$

Oxidative hydrolysis of the boron-containing products gives the corresponding di- and trialkylcarbinols and dialkyl ketones in near theoretical yields (*253*, *270*, *271*, *807*). This has provided the basis of a reaction sequence (hydroboration–carbonylation–oxidation) of great utility in the preparation of monocyclic (*254*, *267*), unbridged polycyclic (*255–257*), and bridged polycyclic (*894*), as well as mixed acyclic, trialkylcarbinols (*267*, *1417*). The sequence tolerates other groups such as ester and cyano functionalities (*221*).

$$\text{BH (dimer)} \xrightarrow{RCH{=}CH_2} \text{B–CH}_2CH_2R \xrightarrow[2.\ [O]]{1.\ CO} RCH_2CH_2\text{–C–OH} \quad (211)$$

The hydroboration–carbonylation sequence has great synthetic utility in that the first step can serve to stitch carbon atoms together and then the second step replaces the boron by carbon; for example (*201*, *255*):

$$\xrightarrow{Et_3N:BH_3} \xrightarrow[(CH_2OH)_2]{CO,\ 150°} \xrightarrow{NaOH,\ H_2O_2} \quad (212)$$

Carbonylation of organoboranes can be controlled to achieve migration of all three organo groups from boron to carbon to provide the trialkylcarbinol (*270*), of two groups to provide the ketone or dialkylcarbinol (*271*), or of only one group to provide the corresponding methylol derivative (*1594*). Mixed trialkylboranes can lead to unsymmetrical ketones (*272*, *1416*). Both $NaBH_4$ and $LiBH_4$ catalyze the rate of CO absorption by trialkylboranes in ether solvent in a reaction that can be controlled to achieve the transfer of only one *B*-alkyl

$$R_3B + CO \xrightarrow{NaBH_4} \xrightarrow{KOH} RCH_2OH \quad (213)$$

group (*1594*) Eq. 213. Careful control of carbonylation reactions carried out in the presence of $LiAlH(OR)_3$ followed by oxidation, provides a convenient procedure for the preparation of aldehydes (*198*, *199*, *228*).

$$R_3B + CO \rightleftharpoons R_3\overset{-}{B}\overset{+}{C}O \rightleftharpoons R_2BC(=O)R \xrightarrow{LiAlH(OMe)_3} R_2BCH(OM)R \xrightarrow{[O]} RCHO + 2\,ROH \quad (214)$$

The carbonylation reaction is a highly stereospecific reaction in which migration of the organic group from boron to the carbon (of the CO) occurs with retention of configuration (*285*). The reaction of trialkylboranes with a mixture of carbon monoxide and aldehydes gives substituted 4-bora-1,3-

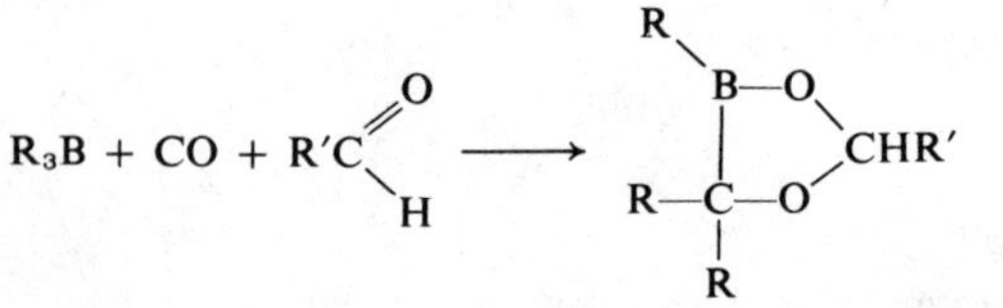

$$R_3B + CO + R'C(=O)H \longrightarrow \text{R(B—O—CHR'—O—C(R)(R)—)} \quad (215)$$

dioxolanes (*809*). Although the mechanism is not known, it probably involves the initial formation of R_3BCO (*809*). The carbonylation of trialkylboranes in the presence of glycols gives cyclic esters of trialkylcarbinylboronic acid (*808*).

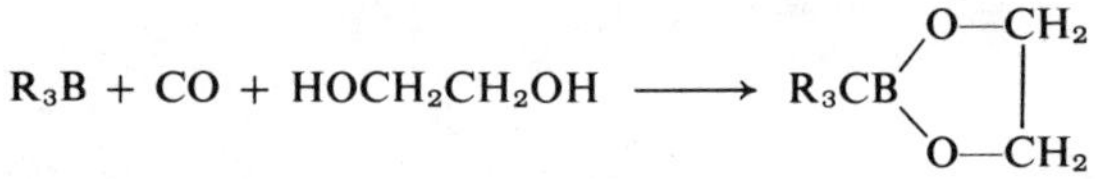

$$R_3B + CO + HOCH_2CH_2OH \longrightarrow R_3CB\langle(O\text{—}CH_2)(O\text{—}CH_2)\rangle \quad (216)$$

Migration of Halogen from Carbon to Boron

Many organoboron compounds having halogen atoms attached to the organic group readily rearrange to form a more stable boron halide (*79, 129, 132, 445, 771, 1026, 1046, 1114, 1536, 1539–1541, 1548, 1568, 1742, 1935*).

Halogens located on the β-carbon often lead to elimination of B—X and olefin (or alkyne) (see above), whereas α-haloorganoboranes may undergo a transfer reaction of the type of Eq. 217, in which inversion of configuration

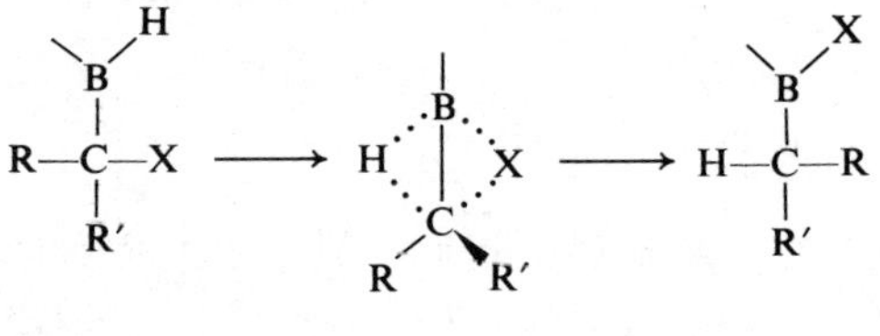

$$\text{(H)B—C(R)(R')X} \longrightarrow [\text{H}\cdots\text{B}\cdots\text{X bridged over C(R)(R')}] \longrightarrow \text{(X)B—C(H)(R)(R')} \quad (217)$$

at the carbon atom has been established in at least one instance (*1539, 1540*).

The rearrangement of (**XXIX**) to (**XXX**) is facilitated by the presence of

$$\underset{\textbf{(XXIX)}}{(Me_2CHCH(Cl)BH_2)_2} \xrightarrow{THF} \underset{\textbf{(XXX)}}{2\ Me_2CHCH_2BH(Cl)\cdot THF} \quad (218)$$

Lewis acids, and the observed first-order kinetics with respect to (**XXIX**) are interpreted in terms of an intramolecular hydride transfer (*1541*).

$$\overset{\delta-}{H}\cdots\underset{|}{\overset{\diagdown\,\diagup}{C}}\cdots X\cdots\overset{\delta-}{B}\lessgtr \qquad \cdots B^{\delta+}$$

The instability of R_FB compounds relative to F—B< (*79*, *877*, *1026*, *1526*, *1568*, *1742*) has sometimes made it difficult to prepare certain fluoroorganoboron compounds in satisfactory yields. Unlike the perfluoroalkyl (*1526*) or perfluorovinyl (*877*) compounds, the perfluorophenylboryl halides do not decompose by fluorine migration but will disproportionate upon heating (*396*).

$$2\,C_6F_5BX_2 \longrightarrow C_6F_5BX + BX_3\ (X = F, Cl) \tag{219}$$

Migration of Boron-Attached Organic Group to an Electropositive Center: Organoboranes as Alkylating (or Arylating) Agents

Migration of an organic group from boron to a neighboring atom containing a good leaving group is now a general reaction that has gained wide utility. Homologation of the organic group may be accomplished in this way when the organoborane is reacted with a variety of ylides (Eqs. 220–221) (*1817–1819*). Homologation is also achieved by reacting trialkyl (or triaryl)-

$$R_3B + \underset{CO_2Et}{\overset{-}{C}H\overset{+}{S}Me_2} \longrightarrow R_2\overset{R}{B^-}\!-\!\underset{CO_2Et}{CH}\!-\!\overset{+}{S}Me_2 \xrightarrow{-Me_2S} R_2B\!-\!\underset{CO_2Et}{CHR} \xrightarrow{alkali} RCH_2CO_2Et \tag{220}$$

$$R_3B + Ph\overset{-}{C}H\overset{+}{A}sPh_3 \longrightarrow \xrightarrow[alkali]{H_2O_2} PhCH_2R + PhCH_2CH(Ph)R + ROH \tag{221}$$

boranes with an appropriate halogenated organic compound in the presence of alkoxide, Eqs. 222–223. The reaction probably involves an intermediate

$$R_3B + BrCH_2Z + K^{+\,-}OR' \longrightarrow RCH_2Z + KBr + R'OBR_2 \tag{222}$$

$$2\,R_3B + Br_2CHCO_2Et + 2\,K^{+\,-}OR' \longrightarrow R_2CHCO_2Et + 2\,KBr + 2\,R'OBR_2 \tag{223}$$

α-halocarbanion that attacks the trialkylborane. This is followed by rapid rearrangement of the resulting quaternary boron compound. Protonolysis of the rearranged intermediate gives the desired homologated organic compound.

$$BrCH_2Z \xrightarrow{K^+{}^-OR'} {}^-CH(Br)Z \xrightarrow{R_3B} R_2\overset{-}{B}(R)\text{—}CH(Br)Z \xrightarrow{-Br^-} R_2B\text{—}CH(R)Z \xrightarrow{R'OH} RCH_2Z + R'OBR_2 \quad (224)$$

Group Z of the organic halides can be ester (—CO_2Et) (*252, 278–280, 283, 284, 1410*), vinylester (—CH═$CHCO_2Et$) (*249*), keto [—C(═O)R] (*250, 279–281*), nitrile (*251, 1410, 1411*), or halogen (*252, 278, 284*). Although *t*-butoxide ion will serve as the deprotonating agent in the first step, the preferred base is 2,6-di-*t*-butylphenoxide (*252*). The bulky substituents of the latter base are probably responsible for preventing two side reactions: the displacement of halide before the active methylene proton is abstracted, and the formation of a coordinated $ROBR_3^-$. The organo group of the borane may be alkyl or aryl; and the rearrangement step occurs with retention of configuration at the carbon atom undergoing migration (*285*).

Factors which appear to accelerate the rearrangement are (a) increased negative charge on the boron atom (Eq. 225) and (b) positive charge develop-

$$Me_2BCH_2Cl \xrightarrow[(1592)]{Me_2NH} \left[Me\text{—}\overset{-}{B}(Me)(\overset{+}{N}HMe_2)\text{—}CH_2\text{—}Cl \right] \xrightarrow{-HCl} Me\text{—}B(NMe_2)\text{—}CH_2Me \quad (225)$$

ment on the atom receiving the migrating organic group. The latter may be altered by the nature of the leaving group.

A competing reaction in the base-catalyzed rearrangement is the simple displacement of the leaving group at the α-carbon (*1592*). Use of an electrophilic catalyst may accelerate the loss of the leaving group and consequently facilitate the rearrangement (*311*).

$$Et\text{—}B(Et)\text{—}CH(Br)Me \xrightarrow{AlBr_3} Et\text{—}B(Et)\text{—}CH(Br\cdot\cdot AlBr_3)Me \longrightarrow Et\text{—}B(Br)\text{—}CH(Et)\text{—}Me \quad (226)$$

Phenyl sulfide anion appears to be an acceptable leaving group in the reaction of triorganoboranes with α-lithioorganosulfides (*1395b*). Organo group migration is also known to occur during bromination (light induced in some cases) of trialkylboranes and borinic acids in the presence of water (*233a*, *311a*, *311b*).

Trialkylboranes react with chlorodifluoromethane in the presence of lithium triethylcarboxide to give a good route to trialkylcarbinols (*197*). This

$$n\text{-}Bu_3B + HCClF_2 \xrightarrow[THF]{LiOCEt_3} \xrightarrow{[O]} n\text{-}Bu_3COH \tag{227}$$

and the reaction utilizing α,α-dichloromethyl methyl ether in place of $HCClF_2$ (*196a*) appear to be analogous to the preparation of diarylcarbinols from dichloromethyllithium and triarylboranes (*904*) or of trialkylcarbinols from trichloromethyl lithium and alkylboranes (*197*).

$$Ar_3B \xrightarrow{LiCHCl_2} \xrightarrow{[O]} Ar_2CHOH \tag{228}$$

Migration of an organic group from a quaternary boron atom to an attached electron-deficient carbon atom also occurs in the reactions of triorganoboranes with carbon monoxide (see above) (*807–809*), with organodiazo compounds, Eq. 229 (*486*, *685*, *835*, *836*, *837a*, *838–840*, *1072*), and with isonitriles Eq. 230 (*160*, *379*, *380*, *382*, *799–801*, *1902a*). With the

$$R_3B + N_2CHZ \longrightarrow \left[R_2\overset{R}{\underset{-}{B}}\text{—}\underset{N_2^+}{CHZ} \right] \xrightarrow{-N_2} R_2B\text{—}\overset{R}{C}HZ \xrightarrow{\text{hydrolysis}} RCH_2Z \tag{229}$$

R = organic group; Z = aldehyde (*840*), ketone (*834–838*), carboxyl ester (*834*, *835*, *838*), nitrile (*834*, *835*, *839*);

R = F, Z = H (*685*)

isonitrile reaction, the last step, (**XXXI**) → (**XXXII**), involving a Wagner–Meerwein type of shift can be catalyzed by $AlCl_3$(*1902*).

An alternate mechanism to that implied in Eq. 229 is suggested to account for the products when diazoketones are used, for $(n\text{-}Pr)_3B$ reacts with diazoacetophenone in THF to give both cis and trans (**XXXIII**) (*1550*). No β-ketoborane is observed though it may be an intermediate (*1550*).

Diazomethane reacts with trialkylboranes to give polymethylated products; by adding a nucleophilic reagent (amine, water, or alcohol) to the system, the reaction can be controlled to give monomethylated products, e.g., $RCH_2B\langle$ rather than $R(CH_2)_nB\langle$ (*486*). Both dialkylchloroboranes (*241*) and alkyldichloroboranes (*833a*) are more reactive than trialkylboranes as alkylating agents in obtaining alkylacetates from ethyl diazoacetate.

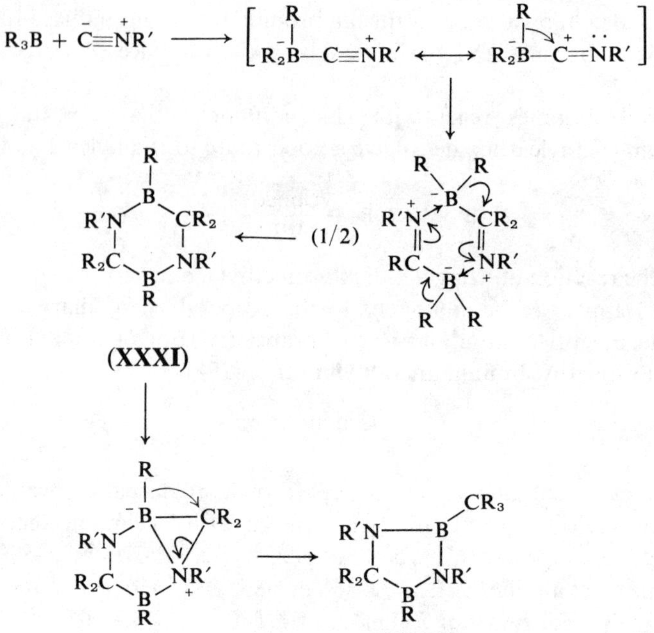

$$R_3B + \overset{-}{C}{\equiv}\overset{+}{N}R' \longrightarrow \left[R_2\overset{-}{B}(R){-}C{\equiv}\overset{+}{N}R' \longleftrightarrow R_2\overset{-}{B}(R){-}C{=}\overset{+}{N}R' \right]$$

(1/2) → (XXXI) → (XXXII) (230)

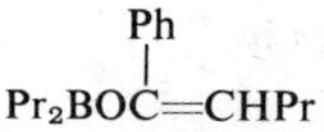

(XXXI)

(XXXII)

$Pr_2BOC(Ph){=}CHPr$

(XXXIII)

Similarly, dialkylchloroboranes (*240*) react with organic azides faster than do trialkylboranes (*1758*), in this case producing secondary amines, Eq. 231. The reaction in these instances, as well as when using alkyldichloroboranes (*241a*, *1112a*), is thought to proceed by a mechanism involving an initial coordination of the azide with the boron moiety. The increased reactivity of

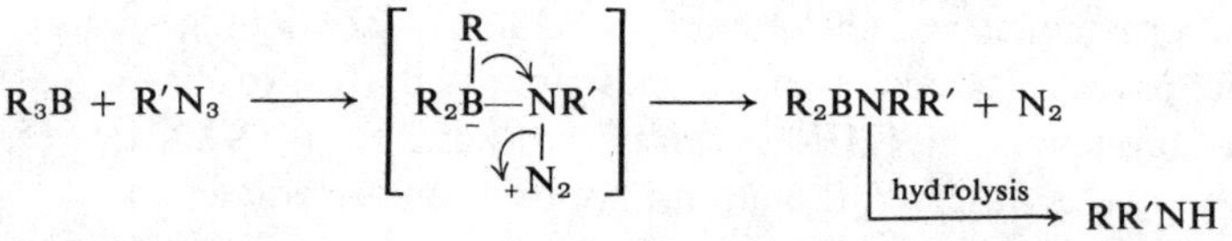

$$R_3B + R'N_3 \longrightarrow \left[R_2\overset{-}{B}(R){-}NR'({-}\overset{+}{N_2}) \right] \longrightarrow R_2BNRR' + N_2 \xrightarrow{\text{hydrolysis}} RR'NH \quad (231)$$

dialkylchloroboranes may be attributed to the greater Lewis acidity (*240*). Related to this type of rearrangement are: (a) the phenyl migration from boron to nitrogen in the thermal decomposition of Ph_2BN_3·pyridine to give $(PhBNPh)_3$ (*1509*, *1512*) (Eq. 232), (b) the decomposition of dimesitylboron

$$3\ Ph_2BN_3{\cdot}pyr \longrightarrow 3\ N_2 + 3\ pyr + (PhBNPh)_3 \quad (232)$$

azide in the presence of a Lewis acid to form, after hydrolysis, mesidine (*1073*), and (c) the alkylation of azide ion with chloromethyldimethylborane $ClCH_2BMe_2$ (*1644*). The thermal decomposition of dibutylboron azide, Bu_2BN_3, gives butyl-substituted borazenes, whereas diarylboron azides yield primarily tetraaryl-1,3,2,4-diazadiboretanes. The formation of diazadiboretanes and borazenes by the thermal decomposition of diorganoboron azides involves the migration of an organic group from the boron to the α-nitrogen in a Curtius-type rearrangement (*1515*). From the observed migratory aptitudes of organic groups, m-$CF_3C_6H_4$ > o-tolyl > naphthyl > Ph > p-ClC_6H_4 > Me > p-tolyl, as well as from other factors, it is concluded that the diorganoboron azides decompose in a synchronous process with anchimeric assistance by the organic ligands (*1513*).

Optically active 1-phenylethylboronic compounds undergo hydroxydeboronation with trimethylamine oxide with retention of configuration (*500*). This is interpreted in terms of a nucleophilic attack by the oxide at boron followed by a nucleophilic 1,2-rearrangement of the alkyl group from boron to oxygen (*500*).

$$BuO{-}B(R)(OBu) + ONMe_3 \rightleftharpoons BuO{-}\overset{-}{B}(R)(OBu){-}O{-}\overset{+}{N}Me_3 \longrightarrow (BuO)_2BOR + NMe_3 \tag{233}$$

Rearrangement following attack of an organoborane by a carbene provides another example of migration of a boron attached alkyl group to an electron-deficient center (*1675, 1757*).

$$Cl_2CHOMe \xrightarrow[(1757)]{LiMe} :CHOMe \xrightarrow{(RCH_2)_3B} \left[(RCH_2)_2\overset{-}{B}(CH_2R){-}\overset{+}{C}HOMe \right] \longrightarrow (RCH_2)_2B{-}CH(CH_2R)OMe \tag{234}$$

In Eq. 235, the α,α-dichloroalkylborane (**XXXIV**) can undergo further intramolecular rearrangement to (**XXXV**), which gives a dialkylcarbene by α-elimination; the latter undergoes subsequent rearrangement to the observed olefin (*1675*).

$$PhHgCCl_2 + (RCH_2)_3B \xrightarrow{(1675)} (RCH_2)_2\overset{-}{B}-\overset{+}{C}Cl_2(CH_2R) \longrightarrow RCH_2B(Cl)(CH_2R)-CCl(Cl)CH_2R \ \text{(XXXIV)} \longrightarrow RCH_2B(Cl)-CCl(CH_2R)_2 \ \text{(XXXV)} \longrightarrow RCH_2BCl_2 + (RCH_2)_2C: \longrightarrow RCH_2CH{=}CHR \quad (235)$$

Several additional rearrangements in which a 1,2-shift of a boron-attached organic group occurs are shown in Eqs. 236–237 (*903*, *904*, *1419a*) and Eq. 238, the latter resulting in C=C bond cleavage (*798*).

$$Ar_2C{=}C(Cl)\overset{-}{B}Ph_2(Ph) \xrightarrow[(904)]{-Cl} Ar_2C{=}C(Ph)BPh_2 \quad (236)$$

$$RCH{=}C(Br)B\text{(cyclohexyl ring)} \xrightarrow[(1970a)]{NaOH} RCH{=}C\text{(cyclic borepane, }B{-}OH) \xrightarrow{I_2} [RCH(I)C(\text{cyclohexyl})B<] \xrightarrow{-IB<} RCH{=}C_6H_{10} \quad (237)$$

$$\text{(1,1'-diphenyl-bis(imidazolidin-2-ylidene))} + 2\,BR_3 \xrightarrow{(778)} 2\ \text{(1,3-diphenyl-1,3,2-diazaborolidin-2-yl)}-CR_3 \quad (238)$$

The reactions between organoboranes and α-lithium furan give 1,6,6-trialkyl-1,2-boroxarocyclohex-4-enes in good yields (*1752b*).

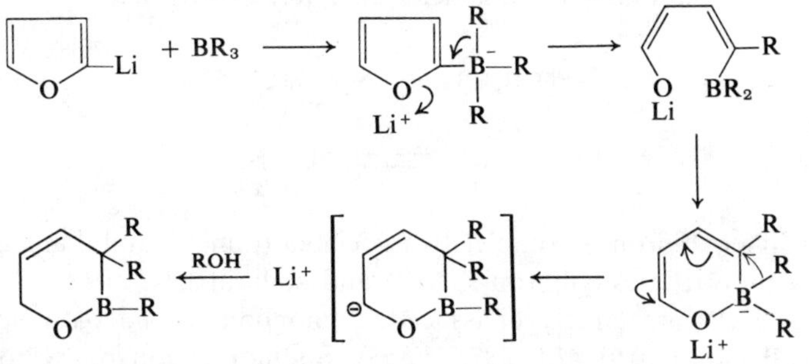

(239)

Evidence of carbon–carbon cleavage occurring within an organic moiety of an organoboron compound undergoing group transfer has been observed (*1114*).

Hydrogenolysis

In the presence of hydrogen and under rather forced conditions, trialkyl- and triarylboranes can be converted to diborane or to alkylboron hydrides (*890*, *934*, *952*). Tertiary amines and metals have occasionally been used as "catalysts."

$$R_3B + 3\ H_2 \xrightarrow[\text{(pres.) (890)}]{150\text{–}200°} \tfrac{1}{2}\ B_2H_6 + 3\ RH \qquad (240)$$

The ease with which trialkylamine–trialkylboron compounds undergo hydrogenolysis, as compared to other organometallic systems, is partially attributed to relief of B-strain in the transition state (*1573*).

$$Et_3B + Et_3N + 3\ H_2 \xrightarrow[(927)]{200°} H_3BNEt_3 + 3\ C_2H_6 \qquad (241)$$

$$2\ M + 2\ BR_3 + 7\ H_2 \xrightarrow[(928)]{} 2\ MBH_4 + 6\ RH \qquad (242)$$

M = alkali metal

Reactions of Organoboranes with Alkali Metals

One-electron reduction products are obtained from reversible (*111*, *413*) reactions between alkali metals and triarylboron compounds (*112*, *203*, *433*, *989*, *995*, *1869*), Eq. 243. A similar adduct is formed from tribenzylborane

$$Na + Ar_3B \rightleftharpoons NaBAr_3 \qquad (243)$$

(*951*), Eq. 244. The preparation of sodium tri-α-naphthylboron (TNB) is accompanied by the formation of a colored disodium compound (*110–112*), Eq. 245, but the second sodium is held much less firmly than the first (*112*).

$$Na + (PhCH_2)_3B \longrightarrow NaB(CH_2Ph)_3 \tag{244}$$

$$2\,Na + TNB \rightleftharpoons Na_2TNB \tag{245}$$

Sodium triphenylboron is completely associated (dimeric and diamagnetic), whereas sodium trimesitylboron (TMB) and sodium tri-β-methylnaphthylboron (TMeNB) are largely unassociated (monomeric and paramagnetic) in tetrahydrofuran (*410–413*, *1378*, *1868*). Sodium tri-α-naphthylboron is dimeric in ethyl ether but is paramagnetic and unassociated in THF (*1378*). These results, and observations on other sodium triarylboron systems, indicate that an equilibrium exists between monomeric and dimeric forms that is much dependent upon solvent polarity and steric effects (*410–413*, *1071*, *1074*).

$$2\,Ar_3B\cdot^- \rightleftharpoons (Ar_3B)_2^{2-} \tag{246}$$

The boron nuclear spin coupling constant (7.84G) of the triphenylboron anion radical, $Ph_3B\cdot^-$, is positive and coupling constants to the ring protons are: para, 2.73G; ortho, 1.99G; and meta, 0.670G (*1071*, *1074*). The esr data point to a nonplanarity of the hydrocarbon groups of the trimesitylboron negative ion in that an increase of twist angle can account for the Δg sign reversal in going from $Ph_3B\cdot^-$ to $Mes_3B\cdot^-$ (*696*). Several trialkylborons, e.g., Et_3B and Pr_3B, fail to give esr signals on treatment with Na—K, although metal is consumed (*1074*).

The difference in electron affinities between TMB and TMeNB in tetrahydrofuran solutions is demonstrated by the equilibium Eq. 247, that lies to the right (*413*).

$$NaTMB + TMeNB \rightleftharpoons NaTMeNB + TMB \tag{247}$$

Prolonged reaction of triphenylboron anion radical with alkali metals degrades it to biphenyl anion radical, both halves of which come from the same triphenylboron (1074).

Trimesitylboron–sodium adduct is probably involved in the use of trimesitylborane to mediate the transfer of electrons to organic systems in aprotic media (*471*); and the cleavage of ethers by Ph_3BNa has been investigated (*1909*).

Miscellaneous Reactions

In a manner similar to boron–boron and boron–metal exchange reactions (Sections 3-1 and 3-2, *Transmetallation*) both phosphorus and sulfur halides exchange groups with trialkylborons (*567*).

$$RSCl + R'_3B \longrightarrow RSR' + R'_2BCl \tag{248}$$

$$R_2PCl + R'_3B \longrightarrow R_2PR' + R'_2BCl \tag{249}$$

Cleavage of a C—B bond in trimethylboron with Me_2PNMe_2 probably occurs through an adduct intermediate (*829*).

$$Me_3B \xrightarrow{Me_2PNMe_2} Me_2PBMe_2 + Me_2NBMe_2 \tag{250}$$

Hydroxylamine-*o*-sulfonic acid reacts with primary or secondary alkylboron compounds to give amines (*217*, *307*, *1595*) with retention of configuration about the reacting carbon (*307*).

$$R{-}B\langle \xrightarrow[\text{diglyme}]{H_2NOSO_3H} R{-}NH_2$$

Benzeneboronic acid in water under relux reacts with cupric chloride or bromide to give chlorobenzene and bromobenzene, respectively, Eq. 251 (*24*). Other examples of similar reactions include Eqs. 252–255.

$$PhB(OH)_2 + 2\,CuX_2 + H_2O \xrightarrow{\Delta} PhX + Cu_2X_2 + HX + B(OH)_3 \tag{251}$$

$$\text{2-furyl}{-}B(OH)_2 \xrightarrow[H_2O\ (866)]{CuBr_2} \text{2-furyl}{-}Br \tag{252}$$

$$\text{Ferrocenyl—}B(OH)_2 \xrightarrow[(1424)]{CuCl_2} \text{ferrocenyl chloride} \tag{253}$$

$$(PhBO)_3 \xrightarrow[(882a)]{CuCl_2} PhCl \tag{254}$$

$$R_3B + 2\,CuBr_2 + H_2O \xrightarrow[(1033)]{THF} RBr + R_2BOH + Cu_2Br_2 + HBr \tag{255}$$

A two-step mechanism is proposed for the last reaction, Eq. 255, in which the first step is a direct electron transfer oxidation of a carbon–boron bond with the generation of an alkyl radical (*1033*).

Antimony pentachloride also serves to cleave the B—C bond of trialkylboranes (*941*).

$$BEt_3 \xrightarrow[-EtCl]{SbCl_5} ClBEt_2 \tag{256}$$

Photolysis of BPh_3 yields primarily phenol and phenylboronic acid (*1895*); and dienylboranes can undergo photocyclization (*415*).

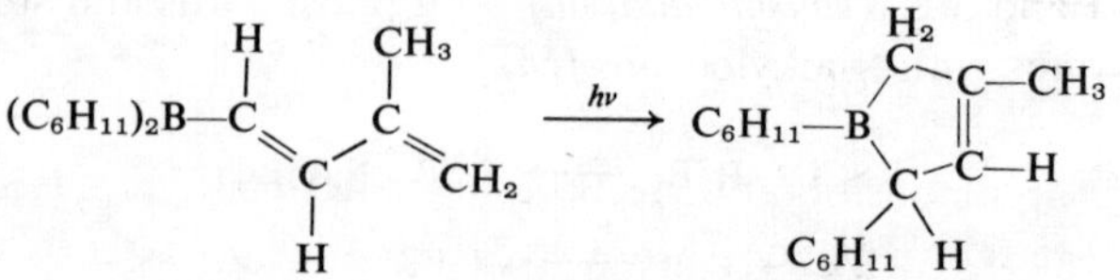

$$(C_6H_{11})_2B{-}CH{=}CH{-}C(CH_3){=}CH_2 \xrightarrow{h\nu} C_6H_{11}{-}\overset{}{B}\langle CH_2{-}C(CH_3){=}CH{-}CH(C_6H_{11})\rangle \tag{257}$$

Triallylboranes undergo allylic rearrangement (*140a*, *142*, *1255*) with an activation energy of ca. 11 kcal mole^{-1} found for the parent triallyl compound (*1255*).

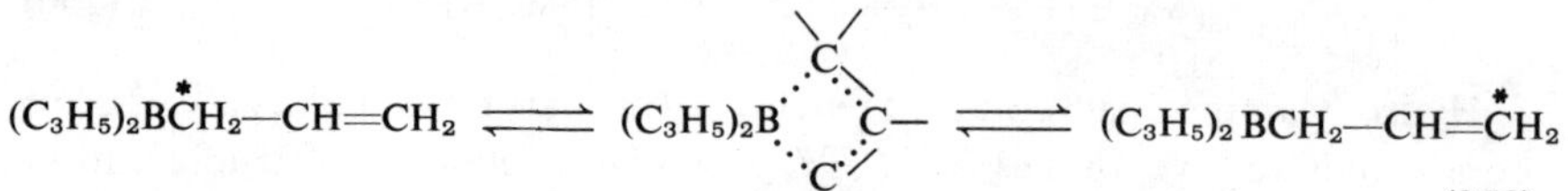

$$(C_3H_5)_2B\overset{*}{C}H_2{-}CH{=}CH_2 \rightleftharpoons (C_3H_5)_2B\cdots C{\cdots}C{\cdots}C \rightleftharpoons (C_3H_5)_2BCH_2{-}CH{=}\overset{*}{C}H_2 \tag{258}$$

Alkoxydivinylboranes react with $Fe_2(CO)_9$ to form (**XXXVI**), which contains a single complexed vinyl group. Irradiation of (**XXXVI**) produces the π-divinylborane complex (**XXXVII**) (*787*).

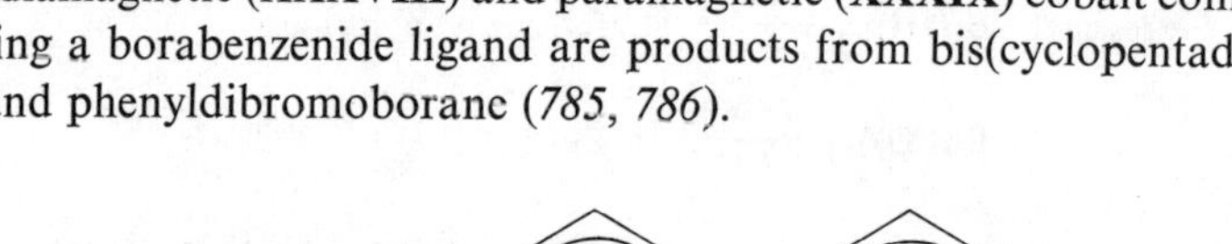

$$(CH_2{=}CH)_2B{-}OR \xrightarrow{Fe_2(CO)_9} RO{-}B(CH{=}CH_2)(CH{=}CH_2 \cdot Fe(CO)_4) \xrightarrow{h\nu} RO{-}B(CH{=}CH_2)_2 \cdot Fe(CO)_3 \tag{259}$$

(**XXXVI**) (**XXXVII**)

Both diamagnetic (**XXXVIII**) and paramagnetic (**XXXIX**) cobalt complexes containing a borabenzenide ligand are products from bis(cyclopentadienyl)-cobalt and phenyldibromoborane (*785*, *786*).

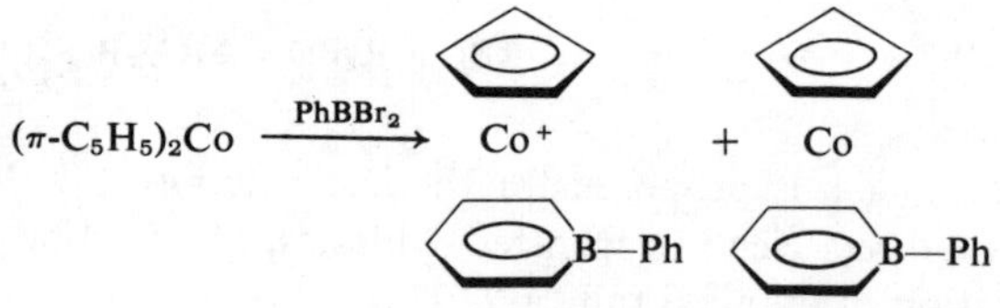

$$(\pi\text{-}C_5H_5)_2Co \xrightarrow{PhBBr_2} [(C_5H_5)Co(C_5H_5B{-}Ph)]^+ + (C_5H_5)Co(C_5H_5B{-}Ph) \tag{260}$$

(**XXXVIII**) (**XXXIX**)

Ring expansion of unsaturated cyclopropanes occurs with the oxygen-catalyzed addition of trialkylboranes in benzene–THF solution (*1414*).

Ph Ph O $\xrightarrow[(O_2 \text{ catalyst})]{R_3B}$ R Ph O (261)

The thermal decomposition of Me_3B between 400–600° gives tetra-β-methyl-*cyclo*-1,3,5,7-tetraboraoctane, $(MeBCH_2)_4$ (*337, 676, 677, 1075*), among other compounds that includes a mixture of carboranes (*325a*).

Preliminary results on the photolysis of trimethylboron indicate that methane is the major product (*337*). Boron–carbon homolysis is probably the primary step in the thermal decomposition of *B*-aryl and alkylborazoles (*1430*).

Chapter 4

Four-Coordinate Organoboranes

4-1. Organoborohydride Ions

Synthesis

Tetracoordinate anionic organoboron compounds (Table 4-1) are most often obtained from a direct interaction of a tricoordinate boron species with an anion (Eq. 262),

$$A^- + BXYZ \longrightarrow ABXYZ^- \tag{262}$$

with examples given in Eqs. 263–275.

$$\left.\begin{array}{r} 2\,LiH + (C_6H_5BH_2)_2 \\ 2\,LiC_6H_5 + B_2H_6 \end{array}\right\rangle \xrightarrow[(1877)]{} 2\,LiB(C_6H_5)H_3 \tag{263}$$

$$LiEt + BMe_3 \xrightarrow[(1646)]{} Li^+BMe_3Et^- \tag{264}$$

$$Ph_3P^+CH_2^- + \tfrac{1}{2}(BH_3)_2 \xrightarrow[(751,\ 762)]{} Ph_3P^+CH_2BH_3^- \tag{265}$$

$$MeN^+{\equiv}C^- + BMe_3 \xrightarrow[(382)]{} Me\overset{+}{N}{\equiv}C\overset{\ominus}{B}Me_3 \tag{266}$$

$$MgR_2 + B_2H_6 \xrightarrow[(84)]{} Mg(BRH_3)_2 \tag{267}$$

$$NaH + (C_2H_5)_3B \xrightarrow[(287,\ 833)]{} NaB(C_2H_5)_3H \tag{268}$$

$$BEt_3 + NOCl \xrightarrow[(1841)]{\text{liquid HCl}} NO^+[BEt_3Cl]^- \tag{269}$$

$$Me_3SnCF_3 + BF_3 \xrightarrow[(399,\ 400)]{} [Me_3Sn]^+[CF_3BF_3]^- \tag{270}$$

$$\text{PhC(Li)=C(Ph)-C(Ph)=C(Ph)Li} \xrightarrow[\substack{2.\ Me_4N^+Br^- \\ (700)}]{1.\ Ph_2BCl} \text{[Ph}_4\text{C}_4\text{BPh}_2]^{\ominus} \ (NMe_4)^+ \quad (271)$$

$$2\ Me_3B + KPMe_2 \xrightarrow[(1770)]{} K^+[Me_3BPMe_2BMe_3]^- \quad (272)$$

$$NaH\ (\text{or LiH or KH}) + (C_6H_5)_3B \xrightarrow[(1908)]{\text{ether or heat}} NaB(C_6H_5)_3H \quad (273)$$

$$NaCN + 2\ BH_3 \cdot OR_2 \xrightarrow[(23)]{R_2O} Na^+(H_3BCNBH_3)^- \cdot 2R_2O \quad (274)$$

$$BR_3 + BH_4^- \underset{(46)}{\rightleftharpoons} R_3BHBH_3^- \quad (275)$$

Alternatively, organosubstituted borohydrides can be obtained from the alcoholysis of sodium triphenylborane (Eq. 276) (*1908*), from the acid catalyzed rearrangement of an isocyanotrihydroborate (Eq. 277) (*1840*), and

$$2\ (C_6H_5)_3BNa + CH_3OH \longrightarrow NaB(C_6H_5)_3H + NaB(C_6H_5)_3OCH_3 \quad (276)$$

$$Na^+[H_3BNC] \xrightarrow[THF]{H^+} Na^+[H_3BCN]^- \quad (277)$$

by hydrogen elimination from a mixture containing hydridic and protonic compounds (Eq. 278–279) (*131*, *1840*, *1911*). Potassium and trimethylboron

$$LiBH_4 + HCN \xrightarrow[(1911)]{100°C} LiB(CN)H_3 + H_2 \quad (278)$$

$$Na^+[R_3BH]^- + HC{\equiv}CR \longrightarrow Na^+[R_3BC{\equiv}CR']^- + H_2 \quad (279)$$

in liquid ammonia react to form potassium aminotrimethylborate, $K^+[H_2NBMe_3]^-$ (*826*); and addition of tritylsodium and triphenylborane to butadiene gives the addition product **(XL)** (*1777*).

$$Ph_3C^-Na^+ + CH_2{=}CH{-}CH{=}CH_2 + Ph_3B \longrightarrow \text{Ph}_3\text{CCH}_2(\text{H})\text{C}{=}\text{C}(\text{H})\text{CH}_2\overset{\ominus}{\text{B}}\text{Ph}_3\ Na^+ \quad (280)$$

(XL)

Reactions and Properties

Reactions with Proton Sources

The kinetics of the reaction of H^+ with BPh_4^-, Eq. 281 (*1908*) shows no evidence for a molecular species $HBPh_4$ (*434*).

$$H^+ + BPh_4^- \longrightarrow C_6H_6 + BPh_3 \quad (281)$$

TABLE 4-1

Borohydride ($BWXYZ^-$) Derivatives

Compound	Reference	Compound	Reference
H_3BMe^-	*1577, 1647, 1849*	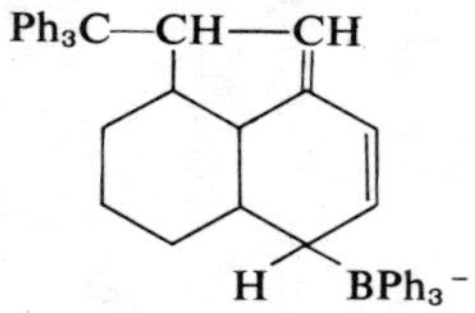 $Ph_3C—CH—CH$, H, BPh_3^-	*1912*
H_3BCN^-	*1120, 1840, 1910, 1911*		
$H_3BCO_2H^-$	*1152*		
$H_3BCO_2R^-$	*1150, 1152*		
$H_3BCONR_2^-$	*378, 811, 1525, 1952*		
H_3BEt^-	*84*		
H_3BPh^-	*1877*	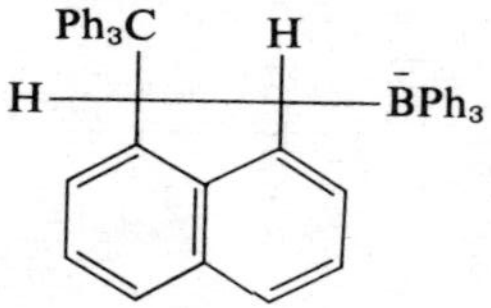	
$H_3B^-CHRP^+Ph_3$	*751, 762, 972, 1672*	Ph_3C, H, H, $\bar{B}Ph_3$	*1776*
$H_2BMe_2^-$	*1380*		
$H_2B^-(Me)CH_2P^+Ph_3$	*762*		
$H_2BEt_2^-$	*84*		
$H_2B^-(2\text{-}Bu)(CH_2P^+Ph_3)$	*762*		
$H_2B^-(2\text{-}Bu)(CH_2P^+Ph_3)$	*762*		
$H_2B^-(t\text{-}Bu)(CH_2P^+Ph_3)$	*762*		
$H_2B^-(Ph)(CH_2P^+Ph_3)$	*762*	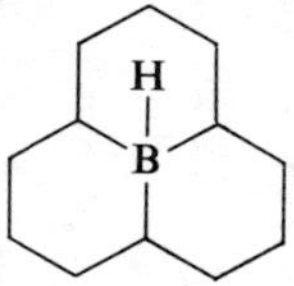	
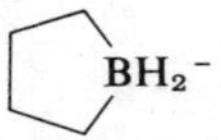 BH_2^- (cyclopentane ring)	*259*	H—B (tricyclic)	*202*
$HBMe_3^-$	*287*	BPh_4^-	*45, 550, 615, 645, 687, 862a, 868, 884, 886, 887, 912a, 986, 987, 1354, 1370, 1382, 1420, 1426, 1464, 1623, 1645b, 1709b, 1709c, 1766a, 1903, 1907, 1908, 1910, 1954b*
$HBEt_3^-$	*85, 130, 231, 231a, 287, 833*		
$HB(Et)_2(C{\equiv}CMe)^-$	*131*		
$HB(2\text{-}Bu)_3^-$	*230*		
$HB(n\text{-}Bu)(pinan\text{-}3\alpha\text{-}yl)_2^-$	*720*		
$HBPh_3^-$	*1876, 1903, 1908, 1913–1915*		
BMe_4^-	*730, 847, 1168a, 1771*	$Ph_3B(p\text{-}MeC_6H_4)^-$	*1596*

Compound	References
$Me_3B^-C{\equiv}N^+(t\text{-Bu})$	*382*
$B(CN)_4^-$	*1911*
Me_3BEt^-	*1646*
$Me_3BC{\equiv}CR^-$	*131*
$MeBPh_3^-$	*1597*
$NCBPh_3^-$	*729a, 747, 1069, 1871a, 1906, 1910*
$(RN^+C)B^-Ph_3$	*157, 801*
$NCB(o\text{-tolyl})_3^-$	*1906*
$NCB(m\text{-tolyl})_3^-$	*1906*
$NCB(p\text{-tolyl})_3^-$	*1906*
$NCB(p\text{-}R_2NC_6H_4)_3^-$	*1906*
BEt_4^-	*833, 1771*
Et_3BBu^-	*467*
$Et_3B(n\text{-}C_8H_{17})^-$	*833*
$Et_3B(C_{12}H_{25})^-$	*467*
$Et_3BC{\equiv}CR^-$	*131*
$EtB(\alpha\text{-naphthyl})_3^-$	*1597*
BVi_4^-	*1680, 1771*
$ViBPh_3^-$	*1680*
$B(C{\equiv}CH)_4^-$	*1954a*
$Pr_3BC{\equiv}CR^-$	*131*
$B(CH_2{=}CHCH_2)_4^-$	*1771*
$CH_2{=}CHCH_2BPh_3^-$	*1679*
BBu_4^-	*467, 730, 1552, 1714*
$(n\text{-Bu})_3BPh^-$	*865*
$(n\text{-Bu})B(CH_2Ph)_3^-$	*730*
$(n\text{-Bu})BPh_3^-$	*1913*
$B(C{\equiv}CPh)_4^-$	*999–1001, 1640*
$PhC_2BPh_3^-$	*1679, 1910*
$Ph_3CBPh_3^-$	*1909*
$Ph_3CCH_2CH{=}CHCH_2BPh_3^-$	*1777*
$Ph_3B(p\text{-}R_2NC_6H_4)^-$	*1906*
$Ph_3B^-C_6H_4N^+Me_3$	*1906*
$Ph_3B(\alpha\text{-}C_{10}H_7)^-$	*1596*
$Ph_2B(o\text{-tolyl})_2^-$	*1301, 1354*
$Ph_2B(o\text{-tolyl})(\alpha\text{-naphthyl})^-$	*1301*
$PhB(o\text{-tolyl})(p\text{-tolyl})(\alpha\text{-naphthyl})^-$	*1301*
$PhB(o\text{-tolyl})(p\text{-}ClC_6H_4)(\alpha\text{-naphthyl})^-$	*1301*
$PhB(o\text{-}PhC_6H_4)_3^-$	*1906*
$PhB(\alpha\text{-}C_{10}H_7)_3^-$	*1596*
$B(o\text{-tolyl})_4^-$	*1906*
$B(m\text{-tolyl})_4^-$	*1906*
$B(p\text{-tolyl})_4^-$	*1426, 1906*
$B(p\text{-}CF_3C_6H_4)_4^-$	*1831*
$B(p\text{-}R_2NC_6H_4)_4^-$	*1906*
$B(p\text{-}MeOC_6H_4)_4^-$	*1642*
$B(m\text{-}FC_6H_4)_4^-$	*1831*
$B(p\text{-}FC_6H_4)_4^-$	*1389*
$B(p\text{-}ClC_6H_4)_4^-$	*385*
$B(C_6F_5)_4^-$	*1164, 1167*
$B(C_4H_2O\text{-}R)_4^-$ (5-R-2-furyl)	*1422*
$B(C_4H_3S)_4^-$ (2-thienyl)	*1422, 1642*
$B(C_4H_3Se)_4^-$ (2-selenienyl)	*1421, 1422*
$Me_3BNH_2^-$	*826, 1714*

(continued)

TABLE 4–1 (*continued*)

Compound	Reference	Compound	Reference
$Me_3BNR_2^-$	*1714*	(2-pyridyl)$CH_2\bar{B}Ph(OR)_2$	*1303*
$(n\text{-}Bu)_3BNH_2^-$	*1714*	$PhB(aryl)Cl_2^-$	*133*
$Ph_3BNH_2^-$	*991*	$MeB(NR_2)Cl_2^-$	*1466*
Me_3BOH^-	*618, 679*	$MeB(OH)_3^-$	*679*
$Me_3BGeH_3^-$	*519a*	$(n\text{-}Bu)B(OH)_3^-$	*1310*
Ph_3BOH^-	*613, 991, 1040, 1910*	$PhB(OH)_3^-$	*1040, 1310*
$(p\text{-}Tolyl)_3BOH^-$	*1358*	$(o\text{-}Tolyl)B(OH)_3^-$	*1310*
Ph_3BOR^-	*988, 1908*	$(p\text{-}MeOC_6H_4)B(OH)_3^-$	*1310*
$Ph_2B(\alpha\text{-}C_{10}H_7)OR^-$	*1354*	$(p\text{-}BrC_6H_4)B(OH)_3^-$	*1310*
$PhB(\alpha\text{-}naphthyl)_2OR^-$	*1354*	$(1\text{-}Naphthyl)B(OH)_3^-$	*1310*
Ph_3BF^-	*612, 613*	$EtB(OR)_3^-$	*1955*
$Ph_3BSiPh_3^-$	*1675a*	(2-pyridyl)$CH_2B(OR)_3$	*1303*
$Ph_3BGeEt_3^-$	*343a*	(2-pyridyl)$B(OR)_3^-$	*1308, 1328*
Et_3BCl^-	*1841*	$CF_3BF_3^-$	*398–400*
$Ph_3BGePh_3^-$	*1675a*	$Ph_3P^+CH_2B^-F_3$	*1672*
$Et_2B(NR_2)_2^-$	*368*	$ViBF_3^-$	*1733*
$Ph_2B(NH_2)Cl^-$	*1293*	$EtBF_2Cl^-$	*168*
$Ph_2B(NR_2)Cl^-$	*1291*	$PhBCl_3^-$	*466, 1695*
$Me_2B(OH)_2^-$	*678, 679*	$PhBI_3^-$	*1699a*
$(n\text{-}Bu)_2B(OH)_2^-$	*1328*		
$Ph_2B(OH)_2^-$	*1310, 1343, 1352a, 1357*		
$(o\text{-}MeC_6H_4)_2B(OH)_2^-$	*1357*		
$(p\text{-}MeC_6H_4)_2B(OH)_2^-$	*1359*		
$(p\text{-}BrC_6H_4)_2B(OH)_2^-$	*1353*		
$(\alpha\text{-}C_{10}H_7)_2B(OH)_2^-$	*1353*		
$(n\text{-}Bu)_2B(OR)OH^-$	*1328*		
$(\alpha\text{-}naphthyl)_2B(OR)OH^-$	*1358*		
$(n\text{-}Pr)_2B(OR)_2^-$	*1661*		
$Ph_2B(OR)_2^-$	*64, 66, 1242, 1823*		

Hydrolysis of BR_4^- with 20% acetic acid results in rapid loss of 1 mole of alkane followed by slow hydrolysis of the resultant organoborane (*467*). Strong acid can decompose tetraphenylborate anion into boric acid and a mixture of organic fission products (*1583a*). In other instances, Ph_2BOH can be isolated (*1428a, 1428b*). The B—H bond is preferentially cleaved during the hydrolysis of triethylborohydride (*833*).

$$NaBEt_3H + H_2O \longrightarrow Et_3B + H_2 + NaOH \qquad (282)$$

The rate of $NCBH_3^-$ exchange with D_2O is about 15 times the hydrolysis rate, whereas exchange is barely perceptible for BH_4^- (*998*). The $ViBF_3^-$ anion exhibits greater hydrolytic stability than does $ViBF_2$ (*1733*). Ammonium ion can act as a proton source for C—B bond cleavage in tetraphenylborohydride (*466*).

$$RNH_3^+BPh_4^- \xrightarrow[-C_6H_6]{110\text{–}150^\circ C} RNH_2\cdot BPh_3 \xrightarrow[-C_6H_6]{200^\circ C} RNHBPh_2 \qquad (283)$$

The acid of tetraethynylborate, $HB(C{\equiv}CH)_4$, has been reported, but its structure is still in doubt (*1954a*).

Rearrangements and Organo Group Migration

Migration of an organic (or hydrogen) group from boron to an adjacent electroposition atom is observed in a number of instances (Section 3-2). A formal negative charge on boron, such as found in organoborohydrides, is obviously a factor that encourages the migrating group to part taking with it the bonded pair of electrons (*382, 802*).

$$3\ Ph_3\overset{+}{P}CH_2\overset{-}{B}R_3 \xrightarrow[(1972)]{\Delta} B(CH_2R)_3 + Ph_3P + 2\ Ph_3PBR_3 \qquad (284)$$

$$Me_3SiCH(\overset{+}{S}Me_2)\overset{-}{B}R_3 \xrightarrow[(1213)]{} Me_3SiCH(R)BR_2 + SMe_2 \qquad (285)$$

$$Na^+[R_3BC{\equiv}CR']^- + ClBR''_2 \xrightarrow[(131)]{-NaCl} \left[R_2\overset{-}{B}(R)\text{—}\overset{+}{C}{=}C(R')(BR''_2) \right] \longrightarrow (R_2B)(R)C{=}C(R')(BR''_2) \qquad (286)$$

Trialkylcyanoborates rearrange upon reaction with a protic acid (or other electrophile) to give, after a peroxide treatment, dialkyl ketones (*1558*).

$$R_3BCN^- \xrightarrow{H^+} R_2\overset{-}{B}(R)-C\equiv\overset{+}{N}H \longrightarrow R_2B-C(R)=\ddot{N}H$$

$$\longrightarrow [R_2\overset{-}{B}-C(R)=\overset{+}{N}H]_2 \text{ (cyclic dimer)} \longrightarrow (RB-CR_2-NH)_2 \text{ (cyclic)} \xrightarrow{(O)} R_2CO \qquad (287)$$

Hydride abstraction by a carbonium ion presumably accounts for the rearranged products in at least two instances, Eq. 288–289.

$$[(PhCH_2)_3B-n\text{-}Bu]^- \xrightarrow[(730)]{Et_3O^+BR_4^-} (PhCH_2)_2BCHPr(CH_2Ph) + (PhCH_2)_2BCHPh(Bu) \qquad (288)$$

$$n\text{-}Bu_4B^-Li^+ + ClCH_2Ph \xrightarrow[(859)]{} LiCl + Bu_3\overset{-}{B}-CHC_3H_7(H) \; {}^{\oplus}CH_2Ph$$

$$\longrightarrow Bu_2\overset{\ominus}{B}(Bu)-\overset{\oplus}{C}HC_3H_7 + PhCH_3 \longrightarrow Bu_2BC(H)(Bu)-C_3H_7 \qquad (289)$$

Sodium triethyl-1-propynylborate and acetyl chloride react to give both an alkynylketone and an oxaborole; the latter product is probably formed via a borenylide intermediate (**XLI**) (*127*).

$$Na^+Et_3\overset{-}{B}C\equiv CMe + RCOCl \longrightarrow RCOC\equiv CMe + \text{oxaborole (Me, R, Et, Et, B–Et, O)}$$

$$\xrightarrow{-NaCl} \left[Et-\overset{\ominus}{B}(Et)_2-\overset{\oplus}{C}=C(Me)COR\right] \text{ (XLI)} \longrightarrow \left[Et_2B-C(Et)=C(Me)-C(=O)R\right] \longrightarrow \text{oxaborole} \qquad (290)$$

(XLI)

In a related reaction, the cyclic compound (**XLII**) is formed from a cyanoborohydride (*158*), Eq. 291, and has also been proposed as an intermediate to

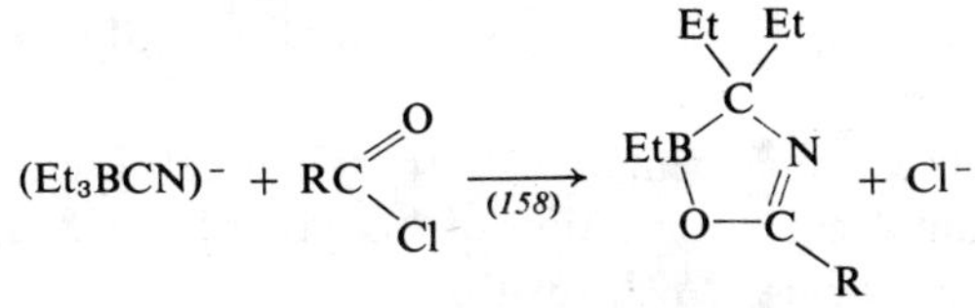

$$(Et_3BCN)^- + RC(=O)Cl \xrightarrow{(158)} \text{EtB}\langle\text{ring: } C(Et)_2\text{–N=C(R)–O}\rangle + Cl^- \quad (291)$$

(**XLII**)

account for the rearrangement products obtained from the acylation of R_3BCN^- with $(CF_3CO)_2O$ (*1559, 1560*).

Exchange Reactions

Ligand exchange between an organoborohydride and metal compounds can occur, although there are far fewer examples than with the transmetallation (Sections 3-1 and 3-2) and exchange (Section 3-1) reactions observed for trivalent organoboron compounds.

$$CrBPh_4 + HgCl_2 \xrightarrow[(912,\ 982)]{} PhHgCl + Ph_3B + [CrCl] \quad (292)$$

$$2\ NaBEt_4 + HgCl_2 \xrightarrow[(1956)]{} Et_2Hg + 2\ BEt_3 + 2\ NaCl \quad (293)$$

$$Me_3SnL_2BPh_4 \xrightarrow[(887)]{} Me_3SnPh + Ph_3B + 2\ L \quad (294)$$

$$(R_2B)_2CHCH_2CH_2Cl + NaBR_4 \xrightarrow[(130)]{\Delta} R_2B\text{–}\triangleleft + 2\ BR_3 + NaCl \quad (295)$$

$$2\ PhBI_2 + NaBPh_4 \xrightarrow[(1706)]{} 3\ Ph_2BI + NaI \quad (296)$$

$$RHgCl + NaBPh_4 \longrightarrow NaCl + RHgBPh_4 \xrightarrow[(876a)]{} RHgPh + Ph_3B \quad (297)$$

$$[H_2NBMe_3]^- \xrightarrow[(826)]{B_2H_6} [H_2NB_2H_6]^- \quad (298)$$

$$[Me_3BPMe_2BMe_3]^- \xrightarrow[(1770)]{B_2H_6} [H_3BPMe_2BH_3]^- \quad (299)$$

The last two reactions may, of course, be the result of boron–boron rather than ligand exchange. On standing, solutions of ethyltriphenylborate and methyltrinaphthylborate ions symmetrize to give tetraarylborates (*1597*). Tetraethyllead is a major product from a reaction of $NaBEt_4$ with lead metal and ethyl halides (*624*).

Oxidative and Photolytic Reactions

Anodic oxidation of *n*-butylboronate anion, $C_4H_9B(OH)_3^-$, produces boric acid, 1-butene, *cis*-2-butene, and *trans*-2-butene, with ratios of 3.3:1:1 for the last three compounds. It is suggested that a primary butyl radical R·

is formed which then loses a β-hydrogen, either directly, to give 1-butene,

$$RB(OH)_3^- \xrightarrow{-e^-} RB(OH)_3 \longrightarrow R\cdot + B(OH)_3 \qquad (300)$$

or after rearrangement to the secondary butyl radical, forming 2-butenes (*846*).

Photochemical rearrangement and electrooxidation of tetraarylborate salts lead to the intramolecular formation of biaryls (*657*, *658*, *1893*, *1894*) and 1-aryl-1,4-cyclohexadienes (*1893*, *1894*).

The quantitative oxidation of tetraphenylborate by hexachloroiridate(IV) to give diphenylborinic acid and biphenyl, Eq. 301, may proceed by a mechanism involving $BPh_4\cdot$ radicals, Eqs. 302a–c (*12*).

$$BPh_4^- + 2\,IrCl_6^{2-} + H_2O \longrightarrow Ph_2BOH + Ph_2 + 2\,IrCl_6^{3-} + H^+ \qquad (301)$$

$$BPh_4^- + IrCl_6^{2-} \longrightarrow BPh_4\cdot + IrCl_6^{3-} \qquad (302a)$$

$$BPh_4\cdot + IrCl_6^{2-} \xrightarrow{\text{fast}} BPh_4^+ + IrCl_6^{3-} \qquad (302b)$$

$$BPh_4^+ \xrightarrow{\text{fast}} [BPh_2^+] + Ph_2$$

$$\xrightarrow{H_2O} Ph_2BOH + H^+ \qquad (302c)$$

The electrochemical oxidation of BPh_4^-, which has also been found to yield Ph_2BOH and Ph_2 as products, is interpreted in terms of a primary two-electron oxidation to BPh_2^+ and Ph_2 (*657*), rather than a multistep process as above. The photochemical decomposition of BPh_4^-, which yields a mixture of phenylcyclohexadienes and Ph_2BOH in the absence of oxygen, as well as biphenyl in the presence of oxygen (*1892*, *1894*), can be interpreted in terms of a photochemical electron-detachment step (Eq. 303) followed by ground state reactions of the $BPh_4\cdot$ intermediate.

$$BPh_4^- \xrightarrow{h\nu} BPh_4\cdot + e^- \qquad (303)$$

In any case, it appears unlikely that an equilibrium amount of BPh_3 is responsible for the photolysis products, for the photolysis of this trivalent compound gives rise to phenol and phenylboric acid (*1895*).

Biphenyl can be generated by the ferric ion oxidation of tetraphenylborate

$$LiBPh_4 + 3\,FeCl_3 \longrightarrow$$

$$PhBCl_2 + 3\,FeCl_2 + LiCl + Ph\text{—}Ph + Ph\cdot$$

$$\xrightarrow{\text{acetone}} \text{benzene} \qquad (304)$$

(*1597*), and is also the product from the rearrangement of the same borate by the action of Rose Bengal-generated singlet oxygen (*562*). The primary step proposed for the latter reaction involves an electron transfer to the singlet oxygen molecule (*562*).

$$Ph_4B^- + {}^1O_2 \longrightarrow O_2^- + Ph_3\bar{B}\text{—}\overset{+}{C_6H_5}\cdot \qquad (305)$$

Examination of the photolysis products from potassium dimesityldiphenylborate in the presence of oxygen (*698*) indicates that a 1,3 shift of a diarylbora group has taken place (*702*). A possible mechanism involving bridged intermediates has been proposed to account for this rearrangement (*702*).

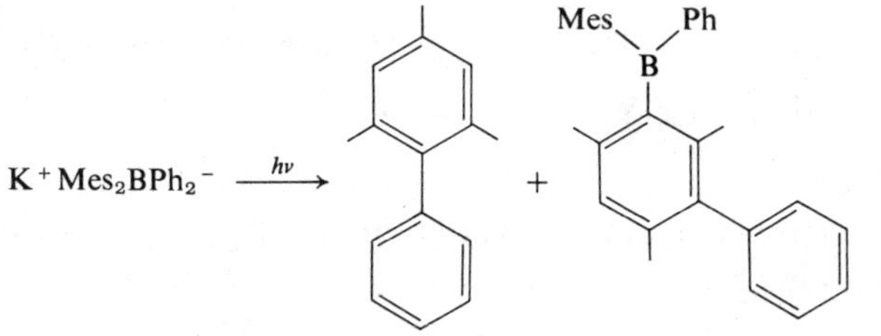

$$K^+Mes_2BPh_2^- \xrightarrow{h\nu} \quad + \quad \tag{306}$$

Mes Ph B

On exposure of chloroform solutions of tetraphenylborates to ultraviolet light it is presumed that a phenyl radical is formed and that this radical abstracts hydrogen from the solvent to form the observed benzene (*1597*).

Hydridic Reducing Agents

The cyanoborohydride ion offers a convenient reducing agent for alkyl halides and tosylates (*848a*, *850*), as well as imino esters, aldehydes, and ketones (*146*, *569*, *848a*), the latter two being reduced eventually to hydrocarbons with the same reagent (*849*). Triethylborohydride ion, Et_3BH^-, has been used to hydroborate olefins and reduce B—Cl bonds (*130*, *231a*, *833*).

$$NaBEt_3H + C_2H_4 \longrightarrow NaBEt_4 \tag{307}$$

Both lithium perhydro-9*b*-boraphenalylhydride (**XLIII**) (*202*) and $Li^+[2\text{-}Bu_3BH]^-$ (*230*) act as stereospecific reducing agents for the reduction of cyclic ketones.

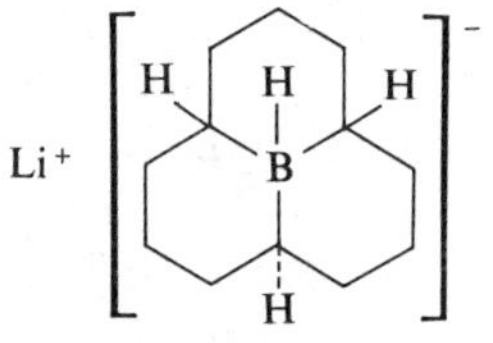

Li^+ H H H B H

(**XLIII**)

Asymmetric reductions of both imines (*36*) and ketones (*720*) have been accomplished with lithium alkyl(hydro)dipinan-3α-ylborates; and lithium triphenylborohydride has been employed to reduce acyl halides (*1913*).

$$LiBPh_3H + PhCOCl \xrightarrow{\text{pyridine}} PhCH_2OH + Ph_3B\cdot\text{pyridine} \tag{308}$$

Miscellaneous Reactions

Halogen cleavage of one tetraarylborohydride B—C bond produces a triarylboron and the aryl halide.

$$K^+[Ph_3Btolyl]^- + Br_2 \xrightarrow[(1597)]{} Ph_3B + tolylBr + KBr \tag{309}$$

$$R_2SBr^+Br^- + NaBPh_4 \xrightarrow[(143)]{} R_2SBPh_3 + PhBr + NaBr \tag{310}$$

$$[Ph_3BC{\equiv}CPh]^- + I_2 \xrightarrow[(1910)]{} Ph_3B + IC{\equiv}CPh + I^- \tag{311}$$

The reaction of lithium 1-alkynyltriorganoborates can be encouraged to take a different course (*1752a*) from that given in Eq. 311 whereby the net effect is the migration of an organo group from boron to the alkynyl group.

$$Li[R'_3BC{\equiv}CR] + I_2 \longrightarrow R'C{\equiv}CR + R'_2BI + LiI \tag{312}$$

Copper tetraphenylboron reacts with α,β-unsaturated carbonyl compounds to give what is presumed to be phenyl radical addition at the 4-position (*1641*).

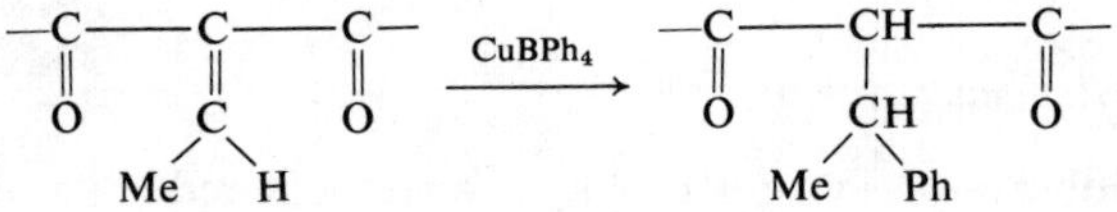

$$-\underset{\|}{\overset{}{C}}(=O)-C(=C(Me)H)-\underset{}{C}(=O)- \xrightarrow{CuBPh_4} -C(=O)-CH(CH(Me)Ph)-C(=O)- \tag{313}$$

Treatment of BAr_4^- with Hg and $HCCl_3$ gives B—C cleavage products that parallel those obtained from thermal decomposition of $NH_4^+BAr_4^-$ (see p. 141) (*1596*).

The cyclic ion (**XLIV**) undergoes Diels–Alder addition with a number of π-electron-poor dienophiles (*700*).

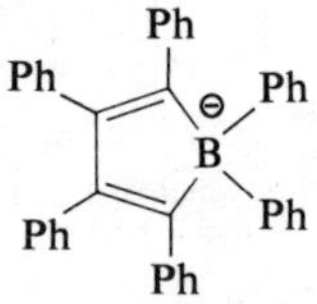

Ph, Ph, Ph, Ph, Ph, Ph; B⊖

(XLIV)

Both $NCBH_3^-$ and BPh_4^- have been incorporated into metal complexes, the former as a bridging ligand in copper complexes (*1120*), and the latter having one of the phenyl rings as a portion of a sandwich compound, $C_5H_5RuBPh_4$, in which one of the phenyl rings is π-bonded to the ruthenium atom (*731*). Similar π-complexes of tetraphenylborate ion with rhodium(I) (*1453*) and iridium are also known (*1662*).

Heating organoborohydrides can result simply in the reversal of the most common method of formation (Section 4-1, *Synthesis*).

$$NaBEt_3H \xrightarrow[(833)]{\Delta} NaH + Et_3B \tag{314}$$

$$HMgBR_3H \xrightarrow[(85)]{\Delta} MgH_2 + BR_3 \tag{315}$$

Properties

The greater stability of $CF_3BF_3^-$, as compared to CF_3BF_2, is attributed to filling of all the boron orbitals (*399*, *400*). Because BH_4^- contains four equivalent terminal hydrogen bonds about a tetrahedral boron, it is somewhat surprising to find the B—H infrared band for BEt_3H^- at 5.4 μm, a region generally attributed to bridge bonds (*833*). A bridge bond may also account for the observed ^{11}B nmr spectra (*1883*). An infrared spectrum of the monomethyl derivative of lithium borohydride also indicates a considerable change in structure relative to that of the parent compound (*1577*). It is now known that crystalline $LiBMe_4$ exists as a tetramer having bridging methyl groups, as in Me_6Al_2, with a short bridge C—B bond distance of 1.51 Å (*717*).

Infrared examination of $LiH \cdot 2BMe_3$ and $LiH \cdot 2BEt_3$ indicates the presence of single hydrogen bridges (*1169*), and this can be attributed to a $R_3BHBR_3^-$ structures for both compounds.

The high-field chemical shift of the *B*-methyl proton, τ10.12, of $Me_3N^+{\equiv}C{-}BMe_3^-$, suggests a strong C—B bond, and this suggestion is supported by the lack of chemical exchange with hexadecylamine (*382*). From a compliance matrix study a 26% s-character is estimated for the hybrid boron orbital bonding to CN in H_3BCN^- (*1580*). Tetraphenylborate ion is a powerful nmr shift reagent for pyridinium ions. The magnitude and direction of the observed shifts reflect the influence of aromatic ring currents in the anion (*1645a*).

4-2. Organoborane–Lewis Base Adducts, RBXYD (D = Donor Group)

Synthesis and Stability

The "vacant" *p*-orbital of trigonal a boron compound can react with an electron-pair donor to give a 1:1 adduct, Eq. 316, with the four groups

$$RBXY + D: \rightleftharpoons RXY\overset{\ominus}{B}:\overset{\oplus}{D} \qquad (316)$$

arranged about the boron in a nearly tetrahedral geometry. With a strong acceptor–donor combination such as $MeBF_2$ and Me_3N (*449*), the equilibrium is far to the right, whereas with a $MeB(OMe)_2$–Et_2O mixture, the equilibrium lies in the opposite direction. Table 4-2 lists a representative number of organoboron acceptor–donor compounds which are known to have a measurable degree of stability.

TABLE 4-2

Substituted Borane—Lewis Base Adducts ($BR_3\cdot LB$)

Compound	Reference	Compound	Reference
$H_2BMe\cdot NR_3$	*972, 1056, 1649*	$Me_3B\cdot NR_3$	*34, 59, 60, 94, 183, 189,*
$H_2BEt\cdot NR_3$	*1056*		*190, 205, 218, 286,*
$H_2BEt\cdot PR_3$	*641a, 872, 969, 1056,*		*300–303, 305, 444,*
	1062		*617, 618, 674, 679,*
$H_2B(n\text{-}Pr)\cdot NMe_3$	*759, 763*		*878, 1212, 1553, 1740,*
$H_2B(n\text{-}Pr)\cdot PR_3$	*641a*		*1878–1880*
$H_2B(i\text{-}Pr)\cdot NMe_3$	*759, 763*	$Me_3B\cdot N_2H_4$	*1553*
$H_2B(n\text{-}Bu)\cdot NMe_3$	*212, 759, 763*	$Me_3B\cdot HN{=}CPh_2$	*1554*
$H_2B(n\text{-}Bu)\cdot$pyridine	*756*	$Me_3B\cdot$pyridine	*188, 206, 207, 218, 1286*
$H_2B(s\text{-}Bu)\cdot NMe_3$	*212, 759*	$Me_3B\cdot OMe_3$	*688*
$H_2B(i\text{-}Bu)\cdot NMe_3$	*212, 763*	$Me_3B\cdot PR_3$	*359, 777, 878, 1837, 1838*
$H_2B(t\text{-}Bu)\cdot$pyridine	*756*	$Me_3B\cdot SMe_3$	*688*
$H_2B(t\text{-}Bu)\cdot NMe_3$	*763*	$Me_3B\cdot SeMe_2$	*688*
$H_2B(t\text{-}Bu)\cdot N_2H_4$	*1365*	$(CD_3)_3B\cdot NMe_3$	*1141*
$H_2BCH_2Ph\cdot$pyridine	*756*	$Me_2BCH_2CH_2SiMe_3\cdot NHMe_2$	*1213*
$H_2BCH_2Ph\cdot PPh_3$	*972*	$NCB(n\text{-}Bu)_2\cdot NR_3$	*1256*
$H_2BCH_2SiMe_3\cdot NMe_3$	*1213*	$NCB(i\text{-}Bu)_2\cdot NR_3$	*1256*
$H_2B(n\text{-pentyl})\cdot NMe_3$	*763*	$NCB(i\text{-}C_5H_{11})_2\cdot NR_3$	*1256*
$H_2B(\text{cyclopentyl})\cdot NR_3$	*212, 268*	$Et_3B\cdot NR_3$	*183, 616–618, 828, 952*
$H_2B(n\text{-hexyl})\cdot NMe_3$	*763*	$Et_3B\cdot P(NMe_2)Me_2$	*828*
$H_2B(3\text{-hexyl})\cdot NR_3$	*268*	$Et_3B\cdot P(NMe_2)_2Me$	*828*
$H_2BCMe_2CHMe_2\cdot NMe_3$	*226*	$Et_3B\cdot P(NMe_2)_3$	*828*
$H_2B(\text{cyclohexyl})\cdot NR_3$	*212, 268, 763*	$Et_3B\cdot PMe_3$	*828*
$H_2B(\text{exonorbornyl})\cdot NMe_3$	*212*	$Et_2BC{\equiv}CH\cdot NMe_3$	*962*
$H_2BCH_2Ph\cdot NMe_3$	*763*	$Et_2BC{\equiv}CMe\cdot NMe_3$	*962*
$H_2BPh\cdot NMe_3$	*589, 922, 1280–1282,*	$Et_2BC{\equiv}CMe\cdot$pyridine	*962*
	1876	$Et_2BC{\equiv}CEt\cdot NMe_3$	*962*
$H_2BPh\cdot$pyridine derivatives	*756, 757, 1385*	$Et_2BCp\cdot NMe_3$	*719*
$H_2BPh\cdot P(n\text{-}Bu)_3$	*1845*	$Et_2BCp\cdot$pyridine	*719, 719a*
$H_2B(o\text{-tolyl})\cdot NR_3$	*1845*	$Et_2BC{\equiv}CC_6H_{13}\cdot NMe_3$	*962*
$H_2B(o\text{-tolyl})\cdot$pyridine	*757*	$Et_2BC{\equiv}CEt\cdot NMe_3$	*962*

Compound	Reference
$H_2B(o\text{-tolyl})\cdot P(n\text{-Bu})_3$	*1845*
$H_2B(p\text{-tolyl})\cdot NR_3$	*1281, 1282*
$H_2B(p\text{-tolyl})\cdot$pyridine	*757*
$H_2B(p\text{-tolyl})\cdot P(n\text{-Bu})_3$	*1845*
$H_2B(p\text{-anisyl})\cdot NMe_3$	*1845*
$H_2B(p\text{-anisyl})\cdot$pyridine	*757*
$H_2B(p\text{-anisyl})\cdot P(n\text{-Bu})_3$	*1845*
$H_2B(p\text{-ClC}_6H_4)\cdot$pyridine	*757*
$H_2B(p\text{-BrC}_6H_4)\cdot NMe_3$	*1845*
$H_2B(p\text{-BrC}_6H_4)\cdot P(n\text{-Bu})_3$	*1845*
$H_2B(\text{mesityl})\cdot NMe_3$	*1845*
$H_2B(\text{mesityl})\cdot P(n\text{-Bu})_3$	*1845*
$H_2B(1\text{-naphthyl})\cdot NH_2R$	*1281, 1282*
$H_2B(1\text{-naphthyl})\cdot$pyridine	*757*
$HBMe_2\cdot NMe_3$	*1380, 1649*
$HBMe_2\cdot PHMe_2$	*359*
$HBMeCl\cdot NMe_3$	*1649*
$HB(CH_2SiMe_3)I\cdot NMe_3$	*1213*
$HBEt_2\cdot PHR_2$	*873*
$HB(n\text{-Bu})_2\cdot$pyridine	*210, 213*
$HB(i\text{-Bu})_2\cdot$pyridine	*210, 213*
$HB(s\text{-Bu})_2\cdot$pyridine	*210, 213*
$HB(CHMeCHMe_2)_2\cdot NMe_3$	*226*
$HB(\text{cyclopentyl})_2\cdot$pyridine	*213*
$HBPh_2\cdot NH_2Me$	*1027*
$HBPh_2\cdot$pyridine	*756, 758*
$HB(p\text{-tolyl})_2\cdot NH_2Me$	*1027*
$HB(p\text{-tolyl})_2\cdot$pyridine	*756, 758*
$HB(p\text{-CH}_3OC_6H_4)_2\cdot$pyridine	*756, 758*
$HB(p\text{-FC}_6H_4)_2\cdot NH_2Me$	*1027*
$HB(p\text{-ClC}_6H_4)_2\cdot NH_2Me$	*1027*
$HB(p\text{-ClC}_6H_4)_2\cdot$pyridine	*756, 758*
$HB(p\text{-BrC}_6H_4)_2\cdot NH_2Me$	*1027*
$HB(p\text{-BrC}_6H_4)_2\cdot$pyridine	*758*
$HB(CH_2CHMe_2)Cl\cdot THF$	*1541*
$EtB(C{\equiv}CMe)_2\cdot$pyridine	*962*
$EtB(C{\equiv}CC_6H_{13})_2\cdot NMe_3$	*962*
$EtBPh_2\cdot$pyridine	*1729, 1730*
$Vi_3B\cdot NR_3$	*670*
$(HC{\equiv}C)_3B\cdot NR_3$	*49*
$(HC{\equiv}C)_3B\cdot$pyridine	*49*
$(HC{\equiv}C)_2BPh\cdot$pyridine	*667*
$(HC{\equiv}C)BPr_2\cdot$pyridine	*962*
$HC{\equiv}CBPh_2\cdot$pyridine	*1729, 1730*
$HC{\equiv}CBPh(o\text{-tolyl})\cdot$pyridine	*667, 1729*
$HC{\equiv}CBPh(m\text{-tolyl})\cdot$pyridine	*667, 1729*
$HC{\equiv}CBPh(p\text{-tolyl})\cdot$pyridine	*667, 1729*
$HC{\equiv}CB(p\text{-tolyl})(p\text{-MeOC}_6H_4)$	*667, 1729*
Pyridine$\cdot Ph_2BC{\equiv}CBPh_2\cdot$pyridine	*1730*
$Me_3SiC{\equiv}CBPh_2\cdot$pyridine	*1730*
$EtSC{\equiv}CBPh_2\cdot$pyridine	*1730*
$(n\text{-Pr})_2B\cdot N_2H_4$	*1260*
$(i\text{-Pr})_3B\cdot NR_3$	*183*
$Pr_2BC{\equiv}CMe\cdot NMe_3$	*962*
$Pr_2BC{\equiv}CMe\cdot$pyridine	*962*
$PrB(C{\equiv}CMe)_2\cdot$pyridine	*962*
$(i\text{-Pr})B(C{\equiv}CMe)_2\cdot$pyridine	*962*
$(n\text{-Pr})BPh_2\cdot$pyridine	*1729*
$EtO(CH_2)_3BPh_2\cdot$pyridine	*1729*
$(\text{Cyclopropyl})_3B\cdot NMe_3$	*442*
$(\text{Cyclopropyl})_3B\cdot PMe_3$	*442*
$(\text{Allyl})_3B\cdot NR_3$	*330, 1341*
$(\text{Allyl})_3B\cdot$pyridine	*1783, 1785*
$MeCH{=}CHBPh_2\cdot$pyridine	*1729*
$MeC{\equiv}CBBu_2\cdot$pyridine	*962*
$(MeC{\equiv}C)_2B(i\text{-Bu})\cdot NMe_3$	*962*
$(MeC{\equiv}C)_2B(i\text{-Bu})\cdot$pyridine	*962*
$(MeC{\equiv}C)_2BPh\cdot$pyridine	*667, 1729*
$(MeC{\equiv}C)_2B(o\text{-tolyl})\cdot$pyridine	*1729*
$(MeC{\equiv}C)_2B(p\text{-tolyl})\cdot$pyridine	*1729*

(continued)

TABLE 4–2 (*continued*)

Compound	Reference	Compound	Reference
$MeC\equiv CBPh_2 \cdot$ pyridine	*1729*	$Ph_2B(\alpha$-naphthyl$) \cdot NR_3$	*1354*
$MeC\equiv CBPh(o$-tolyl$) \cdot$ pyridine	*667, 1729*	$Ph_2B(\alpha$-naphthyl$) \cdot$ pyridine	*1354, 1355*
$MeC\equiv CBPh(m$-tolyl$) \cdot$ pyridine	*667, 1729*	$PhB(p\text{-}ClC_6H_4)_2 \cdot NH_3$	*1355*
$MeC\equiv CBPh(p$-tolyl$) \cdot$ pyridine	*667, 1729*	$PhB(\alpha$-naphthyl$)_2 \cdot$ pyridine	*1354*
$ROCH_2C\equiv CBPh_2 \cdot$ pyridine	*1729*	$(o\text{-}MeC_6H_4)B(p\text{-}BrC_6H_4)_2 \cdot$ pyridine	*1355*
$Me_3SiCH_2C\equiv CBPh_2 \cdot$ pyridine	*1730*	$(o\text{-}MeC_6H_4)B(\alpha$-naphthyl$)_2 \cdot$ pyridine	*1354, 1355*
$(n\text{-}Bu)_3B \cdot N_2H_4$	*1260*	$(p$-Tolyl$)_3B \cdot NR_3$	*990, 996*
$(n\text{-}Bu)_3B \cdot$ pyridine	*493*	$(p$-Tolyl$)_3B \cdot$ pyridine	*990, 996*
$(n\text{-}Bu)_2B(s\text{-}Bu) \cdot$ pyridine	*493*	$(p$-Xylyl$)_3B \cdot NR_3$	*997*
$(t\text{-}Bu)_3B \cdot NR_3$	*183*	$(p$-Xylyl$)_3B \cdot$ pyridine	*997*
$(i\text{-}Bu)B(C\equiv CEt)_2 \cdot$ pyridine	*962*	$(o\text{-}ROC_6H_4)_3B \cdot NR_3$	*1038a*
$CH_2{=}CHC\equiv CBPh_2 \cdot$ pyridine	*667, 1729*	$(C_6F_5)_3B \cdot NR_3$	*1164, 1165, 1167*
$Cp_3B \cdot$ pyridine	*1579*	$(C_6F_5)_3B \cdot PPh_3$	*1167*
$(CH_2{=}CMeC\equiv C)_2B(o$-tolyl$) \cdot$ pyridine	*1729*	(1-Naphthyl)$_3B \cdot NR_3$	*299, 997*
$CH_2{=}CMeC\equiv CBPh_2 \cdot$ pyridine	*667, 1729*	(1-Naphthyl)$_3B \cdot$ pyridine	*997*
$MeC\equiv CC\equiv CBPh_2 \cdot$ pyridine	*1729*	(5-R-2-furyl)$_3$ B · pyridine [structure: R, O]	*1422*
(Cyclohexyl)$_3B \cdot NR_3$	*996*	(2-thienyl)$_3$ B · pyridine [structure: S]	*1422*
(Cyclohexyl)$_3B \cdot$ pyridine	*996*	(2-selenienyl)$_3$ B · pyridine [structure: Se]	*1422*
$(n\text{-}BuC\equiv C)_3B \cdot$ pyridine	*49*	$(n\text{-}Pr)_2BN_3 \cdot$ pyridine	*1511*
$EtC\equiv CC\equiv CBPh_2 \cdot$ pyridine	*1729*	$Ph_2BN_3 \cdot$ pyridine	*1509, 1511*
Cyclopenten-1-yl$C\equiv CBPh_2 \cdot$ pyridine	*667, 1729*	$PhB(o$-tolyl$)N_3 \cdot$ pyridine	*1514*
$(PhCH_2)_3B \cdot NR_3$	*997*	$PhB(p$-tolyl$)N_3 \cdot$ pyridine	*1514*
$(PhCH_2)_3B \cdot$ pyridine	*951, 997, 1891*		
(Cyclohexene-1-yl$C\equiv C)_2BPh \cdot$ pyridine	*1729*		
(Cyclohexene-1-yl$C\equiv C)_2B(p$-tolyl$) \cdot$ pyridine	*1729*		
Cyclohexyl$CH_2CH_2BPh_2 \cdot$ pyridine	*1729*		
Cyclohexene-1-yl$C\equiv CBPh_2 \cdot$ pyridine	*667, 1729*		
$PhCH_2CH_2BPh_2 \cdot$ pyridine	*1729*		
$PhCH_2CH_2BPh(o$-tolyl$) \cdot$ pyridine	*1729*		
$PhCH_2CH_2BPh(m$-tolyl$) \cdot$ pyridine	*1729*		

Compound	References
$PhCH_2CH_2BPh$(*p*-tolyl)·pyridine	*1729*
$PhCH_2CH_2BPh$(*p*-$MeOC_6H_4$)·pyridine	*1729*
PhCH=$CHBPh_2$·pyridine	*1729*
$(PhCH{=}CH)_3B$·pyridine	*1784*
$(PhC{\equiv}C)_3B\cdot NR_3$	*1000*
$(PhC{\equiv}C)_3B$·pyridine	*49, 1000*
$(PhC{\equiv}C)_3B$·tetrahydrofuran	*1000*
$(PhC{\equiv}C)_2BPh$·pyridine	*667, 1729*
$(PhC{\equiv}C)_2B$(*o*-tolyl)·pyridine	*1729*
$(PhC{\equiv}C)_2B$(*p*-tolyl)·pyridine	*1729*
$(PhC{\equiv}C)_2B$(*p*-$MeOC_6H_4$)·pyridine	*1729*
$PhC{\equiv}CBPh_2$·pyridine	*1729, 1730*
$PhC{\equiv}CBPh$(*o*-tolyl)·pyridine	*667, 1729*
$PhC{\equiv}CBPh$(*m*-tolyl)·pyridine	*667, 1729*
$PhC{\equiv}CBPh$(*p*-MeC_6H_4)·pyridine	*667, 1729*
$PhC{\equiv}CB$(*p*-tolyl)(*p*-$MeOC_6H_4$)·pyridine	*667, 1729*
Anthranyl-*9*-$C{\equiv}CBPh_2$·pyridine	*1729*
[boron-centred tricyclic structure: B←LB] LB = NH_3, pyridine	*692* *929*
$Ph_3B\cdot NR_3$	*466, 988, 1354, 1448–1450, 1596, 1597*
Ph_3B·pyridine	*988, 990*
$Ph_3B\cdot PPh_3$	*972, 1905*
$Ph_3B\cdot Ph_3PNR$	*35, 1957*
$Ph_3B\cdot SR_3$	*143, 1471*
Ph_2B(*o*-$CH_3C_6H_4$)·pyridine	*1355*
Ph_2B(*p*-MeC_6H_4)·NH_3	*1355*
Ph_2B(*p*-MeC_6H_4)·pyridine	*1355*
Ph_2B(*p*-BrC_6H_4)·pyridine	*1729, 1730*
PhB(*p*-ClC_6H_4)N_3·pyridine	*1514*
PhB(α-naphthyl)$_2$·NH_3	*1596*
(*p*-Tolyl)$_2BN_3$·pyridine	*1514*
(*p*-$MeOC_6H_4$)$_2BN_3$·pyridine	*1514*
(*p*-ClC_6H_4)$_2BN_3$·pyridine	*1514*
$Me_2BOH\cdot NH_3$	*678*
$Me_2BOMe\cdot NHMe_2$	*421*
(*n*-Bu)$_2$BOR·pyridine	*8*
$Ph_2BOR\cdot NR_3$	*1107, 1449, 1450*
Ph_2BOR·pyridine	*8*
PhB(*p*-tolyl)OR·NR_3	*1292*
PhB(α-naphthyl)OR·NR_3	*1292, 1450*
$Me_2BF\cdot NR_3$	*1882*
$Et_2BF\cdot NMe_3$	*79, 354, 962, 1812, 1813*
Et_2BF·pyridine	*1812*
(*n*-Pr)$_2$BF·NMe_3	*962, 1812, 1813*
(*n*-Pr)$_2$BF·pyridine	*1812*
(*n*-Bu)$_2$BF·NMe_3	*1812, 1813*
(*n*-Bu)$_2$BF·pyridine	*1812*
$Me_2BCl\cdot NMe_3$	*96, 1649*
Et_2BCl·pyridine	*962*
Pr_2BCl·pyridine	*962*
(*n*-Bu)$_2$BCl·$N{\equiv}CBr$	*1224*
(*n*-Bu)$_2$BCl·(2)pyridine	*650*
$Ph_2BCl\cdot NR_3$	*1289*
Ph_2BCl·pyridine	*1289, 1295*
$Ph_2BCl\cdot OR_2$	*1292*
PhB(*p*-tolyl)Cl·OR_2	*1292*
PhB(*p*-ClC_6H_4)Cl·pyridine	*1289*
PhB(α-naphthyl)Cl·OR_2	*1292*
(α-Naphthyl)$_2$BCl·OR_2	*1292*
(α-Naphthyl)$_2$BCl·pyridine	*1289*
$Me_2BBr\cdot HN{=}CR_2$	*1226*
EtB(*s*-Bu)Br·pyridine	*309*

(continued)

TABLE 4-2 (*continued*)

Compound	Reference	Compound	Reference
$NCB(NR_2)_2 \cdot NMe_3$	*115*	$EtBCl_2 \cdot NR_3$	*179*
$PhB(N_3)_2 \cdot$ pyridine	*1511*	$ClCH_2CH_2BCl_2 \cdot OMe_2$	*771*
$(n\text{-}Bu)B(OPh)_2 \cdot$ pyridine	*8*	$(n\text{-}Bu)BCl_2 \cdot$ pyridine derivatives	*173, 1387*
$PhB(OEt)_2 \cdot NEt_3$	*589*	$PhBCl_2 \cdot NR_3$	*133, 179, 466, 1128, 1131*
$PhB(OR)_2 \cdot$ pyridine derivatives	*8, 1386*	$PhBCl_2 \cdot NCCH_3$	*1131*
$ArB(OR)_2 \cdot NR_3$	*527*	$PhBCl_2 \cdot OEt_2$	*1131*
$MeBF_2 \cdot NMe_3$	*168, 354, 449*	$PhBCl_2 \cdot$ pyridine derivatives	*173, 1387*
$EtBF_2 \cdot NMe_3$	*79, 449, 962, 1742*	$PhBCl_2 \cdot PhPH_2$	*419*
$EtBF_2 \cdot$ pyridine	*962*	$PhBCl_2 \cdot SMe_2$	*1657a*
$ViBF_2 \cdot NMe_3$	*449*	$C_6F_5BCl_2 \cdot NCX$	*1224*
$(n\text{-}Pr)BF_2 \cdot NMe_3$	*449, 1568*	$C_6F_5BCl_2 \cdot$ pyridine	*397*
$(n\text{-}Pr)BF_2 \cdot$ pyridine	*962*	Ferrocenyl$BCl_2 \cdot OR_2$	*984*
$(i\text{-}Pr)BF_2 \cdot$ pyridine	*962*	Ferrocenyl$BCl_2 \cdot NR_3$	*984*
$(n\text{-}Bu)BF_2 \cdot NR_3$	*1329, 1812, 1813*	Ferrocenyl$BCl_2 \cdot$ pyridine	*984*
$(n\text{-}Bu)BF_2 \cdot$ pyridine	*173, 1812*	$MeB(SR)Br \cdot HN{=}CR_2$	*1226*
$(i\text{-}Bu)BF_2 \cdot NMe_3$	*962*	$PhB(SR)Br \cdot HN{=}CR_2$	*1226*
$(i\text{-}Bu)BF_2 \cdot$ pyridine	*962*	$MeBBr_2 \cdot HN{=}CR_2$	*1226*
$(n\text{-}C_5H_{11})BF_2 \cdot NR_3$	*1812, 1813*	$PhBBr_2 \cdot NCCH_3$	*1131*
$(n\text{-}C_5H_{11})BF_2 \cdot$ pyridine	*1812*	$PhBBr_2 \cdot HN{=}CR_2$	*1226*
$(n\text{-}C_6H_{13})BF_2 \cdot NR_3$	*1329, 1812, 1813*	$PhBBr_2 \cdot SMe_2$	*1657a*
$(n\text{-}C_6H_{13})BF_2 \cdot$ pyridine	*1812*	$n\text{-}BuBI_2 \cdot SMe_2$	*1657a*
$MeBCl_2 \cdot NMe_3$	*168, 1879*	$PhBI_2 \cdot SR_2$	*1657a*

Because the adduct stabilities are dependent upon the nature of both the organoborane acceptor and the donor compound, it is hazardous to give relative acceptor strengths of the borane without making reference to the specific donor molecule. A general trend can be noted, however, that holds for adducts with simple bases: acceptor strength, Z = I > Br > Cl > H > CF_3 > F > aryl > alkyl > OR for BZ_3 compounds (*8, 184, 190, 302, 354, 449, 1039, 1131, 1165, 1526, 1533, 1649*). The pentafluorophenyl group lies somewhere between Cl and F (*1165*); and $B(CD_3)_3$ > $B(CH_3)_3$ (*1141*). Tricyclopropylborane is a stronger Lewis acid toward Me_3N and Me_3P than is triisopropylboron, in that the former forms complexes whereas the latter does not (*442*). Factors that influence the above orders includes: (a) electron withdrawal properties of the boron attached group; atoms having greater electron withdrawing strength should improve the boron-donor bond strength; (b) π-orbital interaction between the "vacant" *p*-orbital of boron with a neighboring *p*-orbital of the attached group; the greater the interaction the less likely the boron shall form an addition compound; (c) rehybridization energy ($sp^2 \rightarrow sp^3$); and (d) steric effects.

Lewis acidity toward amines decreases in the sequence Me_3Al > Me_3Ga > Me_3In > Me_3B > Me_3Tl (*423, 519, 1041*), and although steric effects between donor and acceptor may play some part in the position in which the Me_3B is found, the weakness of Me_3B as an acid must also be due to the relatively large reorganizational energy in forming an adduct.

Properties of the Lewis base which influence the stability of an adduct are reviewed elsewhere (*184b*), but it can be safely generalized that they are greatly dependent on steric (*184, 188–190, 206, 207, 218, 286, 299–301, 303, 828, 1038*), as well as electronic factors. Addition compounds of tri-α-naphthylboron and amines exist in two polymorphic modifications. This is probably a result of restricted rotation of the α-naphthyl group (*229*). The reasonably stable complexes formed from dialkylboranes and aluminum methoxide, $(MeO)_3Al\cdot 3R_2BH$, permit the synthesis of pyridine–dialkylboranes by a simple displacement (*210*).

$$\begin{array}{ccc} (Me_3O)_3Al\cdot 3R_2BH + 3\ py & \longrightarrow & (Me_3O)_3Al + 3\ R_2BH\cdot py \\ \uparrow & & \\ 3\ R_2BOMe + AlH_3 & & \end{array} \qquad (317)$$

Ammonium ions can act as a proton source for the cleavage of organoborohydride ions providing another route to amine–organoborane adducts (Section 3-2).

$$RNH_3{}^+BPh_4{}^- \xrightarrow[(466)]{110\text{–}150^\circ C} RNH_2\cdot BPh_3 + C_6H_6 \qquad (318)$$

Reactions

Displacement of trimethylamine from trimethylamine–alkylboranes and trimethylamine–phenylborane is observed with methylenetriphenylphosphorane (**XLV**). The qualitative order of reactivity, R = *t*-butyl > 2-butyl > methyl, suggests that these reactions occur by the combination of the free borane and substrate (*762*).

$$(C_6H_5)_3\overset{+}{P}\overset{-}{C}H_2 + RBH_2 \cdot N(CH_3)_2 \xrightarrow[80°C]{C_6H_6} (C_6H_5)_3\overset{+}{P}CH_2\overset{-}{B}H_2R + N(CH_3)_3 \qquad (319)$$

(XLV)

Displacement at boron of trimethylamine–butylboranes, and also of various para- and ortho-substituted trimethylamine–monoarylboranes, with tri-*n*-butylphosphine as the nucleophile occurs largely by an S_N2 mechanism, whereas a moderately bulky group on the boron (e.g., *t*-butyl) leads to a predominantly first-order S_N1 displacement reaction (*769*, *770*, *1845*). Not unexpectedly, S_N1 is favored over the alternant S_N2 mechanism in those amine–arylborane [e.g., di(para-substituted phenyl)boranes] nucleophilic displacement reactions in which the reactant molecules are not only sterically constrained but are also capable of yielding borane intermediates stabilized by the additional electron delocalization of two phenyl groups (*1027*).

Concentration dependence studies together with energies and entropies of activation indicate that the exchange reaction between Me_3NBMe_3 and BMe_3 and also the exchange between Me_3NBMe_3 and Me_3N proceed by dissociative mechanisms (*444*). The gas-phase dissociation energy for Me_3NBMe_3 is 17.6 kcal mole^{-1}, and at -24°C, the entropy of activation is $+15.0$ eu (*444*). The apparent displacement reaction between $Me_3P \cdot BMe_3$ and Me_3P in toluene proceeds by a dissociative mechanism with activation parameters: $\Delta H^{\ddagger}$ 63 kJ mole^{-1} and $\Delta S^{\ddagger}$ 36 JK^{-1} mole^{-1} (*25*). In an apparent displacement–ligand exchange sequence, diborane reacts with $Me_3B \cdot N_2H_4$ to give 1,1-dimethyldiborane as one of the major products (*1553*). In an ammonium chloride-catalyzed reaction between trimethylamine–alkylborane and ammonia at 100–150°C, formation of the observed product, *B,B,B*-trialkylborazine (**XLVI**) (*759*, *764*), is probably preceded by a displacement reaction (Eq. 320a).

$$R\overset{-}{B}H_2\overset{+}{N}(CH_3)_3 + NH_3 \longrightarrow R\overset{-}{B}H_2\overset{+}{N}H_3 + (CH_3)_3N \qquad (320a)$$

$$3\,R\overset{-}{B}H_2\overset{+}{N}H_3 \longrightarrow 6\,H_2 + \text{(XLVI)} \qquad (320b)$$

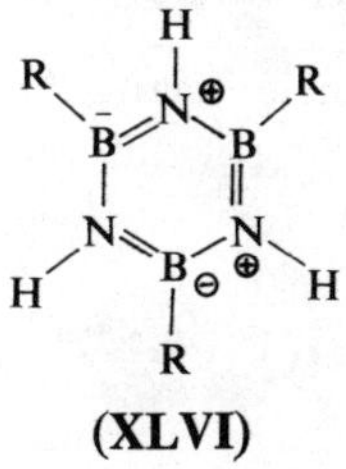

(XLVI)

In a similar fashion, *o*-phenylenediamine reacts with trimethylamine–alkylboranes with eventual loss of trimethylamine and hydrogen and the formation of 2-alkylborabenzimidazolines (*759*, *763*).

$$C_6H_4(NH_2)_2 + RBH_2N(CH_3)_3 \xrightarrow[80°C]{C_6H_6} C_6H_4(NH)_2BR + (CH_3)_3N + 2\,H_2 \qquad (321)$$

(XLVII)

The action of tetraalkyldiboranes on hydrazine at 100–150°C affords *B*-substituted hydrazinoboranes in good yield (*1460*).

$$(R_2BH)_2 + N_2H_4 \longrightarrow R_2BNHNHBR_2 + 2\,H_2 \qquad (322)$$

Trimethylamine–alkylboranes react with two equivalents of alkanethiols at 60–100°C, affording the corresponding dialkyl(alkylthio)boronate esters (*760*, *765*).

$$RBH_2 \cdot N(CH_3)_3 + 2\,R'SH \longrightarrow RB(SR')_2 + 2\,H_2 + (CH_3)_3N \qquad (323)$$

Dimethylphosphine–dimethylborane prepared from dimethylphosphine and tetramethyldiborane, undergoes disproportionation above 40°C, and upon heating to 165°C, there is some evidence that the main reaction is (*359*)

$$2\,(CH_3)_2PH \cdot HB(CH_3)_2 \longrightarrow H_2 + (CH_3)_2PH \cdot B(CH_3)_3 + \underset{\text{(polymer)}}{(CH_3)_2PBHCH_3} \qquad (324)$$

As the above examples indicate, Lewis bases, which also serve as proton sources, may be used to cleave B—H bonds (Eqs. 320–324) or as shown in Eq. 325, and mentioned earlier in Section 3-2, can sever B—C linkages.

$$RNH_2 \cdot BPh_3 \xrightarrow[(466)]{200°C} RNHBPh_2 + C_6H_6 \qquad (325)$$

Amine complexes of the triarylboranes can be converted into the triarylboranes themselves by the action of mineral acids and by pyrolysis (*1355*).

$$Ar_3B \cdot NH_3 + HCl \longrightarrow Ar_3B + NH_4Cl \qquad (326)$$

As with amine–boranes the B—H bonds in the aminoorganosubstituted boranes exhibit a characteristic infrared absorption at 4.35 μm, and reduce silver ion, iodine, and hydrogen ion to free silver, iodide ion, and molecular hydrogen, respectively (*756–758*, *763*). From earlier observations, methylation of trimethylamine–borane was found to increase the rate at which hydrogen chloride attacks the remaining B—H links to produce hydrogen (*1649*). A corresponding increase of B—H activity is also observed in the hydrolysis of the phenyl-substituted compounds, e.g., pyridine–diphenylborane > pyridine–phenylborane (*772*). In a reaction of iodine with pyridine–phenylborane in the presence of excess pyridine, the iodide of the bispyridine complex of the phenylboronium ion is produced (*565*).

The hydrolysis of pyridine–diphenylborane follows second-order kinetics, first-order both in water and amine–borane (*772*). An observed kinetic isotope effect is small and temperature-independent for substitution of deuterium for protium on boron, whereas, the effect is large and temperature-dependent for a similar isotopic substitution in water (*772, 1115*). A mechanism consistent with the isotope effect involves an initial attack of water on the boron–hydrogen bond in a nonlinear transition state (*772*).

$$\underset{-\ +}{(Ph)_2\overset{H}{\overset{|}{B}}Py} + H_2O \longrightarrow \left[\begin{array}{c} H \\ O{-}H \cdots B(Ph)_2Py \\ \cdots H \end{array} \right] \longrightarrow (Ph)_2B\overset{+}{P}y + H_2 + OH^- \text{(slow)} \tag{327a}$$

$$(Ph)_2B\overset{+}{P}y + OH^- \longrightarrow (Ph)_2BOH + Py \text{ (fast)} \tag{327b}$$

Increasing the electron density on boron by suitable substitution on the benzene rings promotes the rate of hydrolysis, indicating a slightly electron-attracting boron atom in the transition state (*772*). The above mechanism is to be contrasted with the observation that hydrolysis of amine–boranes probably proceeds by an initial displacement of BH_3 from the amine nitrogen with protons.

Both normal and secondary kinetic isotope effects have been observed in the hydrolysis of arylborane–amine adducts with k_H/k_D for R = Me > H > Cl (*513, 772*).

$$p\text{-}RC_6H_4BH_2 \cdot C_5H_5N + 2\,H_2O \longrightarrow p\text{-}RC_6H_4B(OH)_2 + 2\,H_2 + C_5H_5N \tag{328}$$

Cleavage of the B—C bond of complex (**XLVIIa**) with bromine gives the ω-bromide (**XLVIIb**) (*1836a, 1836b*).

$$\underset{\textbf{(XLVIIa)}}{\text{cyclo-}(CH_2)_4B(n\text{-Bu}) \leftarrow Py} \xrightarrow{Br_2} \underset{\textbf{(XLVIIb)}}{Br{-}(CH_2)_4\overset{n\text{-Bu}}{\overset{|}{B}}{-}Br \leftarrow Py} \tag{329}$$

The relative ease with which trialkylamine–trialkylboron compounds undergo hydrogenolysis, as compared to other organometallic systems, is partially attributed to relief of B-strain in the transition state (*1573*).

ω-Dialkylaminoalkylboron compounds of the type $R_2N(CH_2)_nBR'R''$ (n = 3, 4) (R′ = n-Bu, R″ = MeO, EtS) are equilibrium mixtures of coordinated cyclic and noncoordinated open forms that interconvert at a rate exceeding 10^3 sec^{-1} (*141c, 1836b*).

Donor compounds tend to be more stable toward disproportionation than do the organoboranes themselves (*268*) (cf. Section 3-1), a fact which can be attributed to the lack of the empty boron orbital thought necessary to participate in a bridge intermediate during organo group transfer.

A ring compound with a repeating BCN sequence (**XLVIII**) is obtained in low yield from sodium hydride and $H_2B[N(CH_3)_3]_2{}^+Cl^-$ (*1367*).

$$\begin{array}{ccccc} & & H_2 & & \\ & & B & & \\ (CH_3)_2N & \nearrow & & \searrow & CH_2 \\ | & & & & | \\ H_2C & \searrow & & \swarrow & N(CH_3)_2 \\ & & B & & \\ & & H_2 & & \end{array}$$

(XLVIII)

The use of Lewis base adducts of organoboranes as hydroborating agents is reviewed in Section 3-1.

4-3. Organoboronium Ions

From the large relative abundance of Me_2B^+ and Et_2B^+ ions in the mass spectra of Me_3B and Et_3B (*1065*) (Section 2-6), perhaps it is not surprising that solvated species of the ions are reasonably stable under normal bench-top laboratory conditions. A number of solvated organoboronium ions of the two types $RR'B(LB)_2{}^+$ (Table 4-3) and $R_2B(LB)^+$ (*475–477*, *694*) have been characterized or implied by conductance measurements of solutions of trigonal boron compounds containing good leaving groups and appropriate donor groups. The reaction of iodine with either pyridine(1,1,3-trimethylpropyl)borane or pyridinedicyclohexylborane in the presence of excess pyridine produces the corresponding bispyridineboron cation, Eq. 330 (*564*).

$$2\ RR'BH \cdot NC_5H_5 + I_2 + 2\ C_5H_5N \longrightarrow 2\ RR'B(NC_5H_5)_2{}^+I^- + H_2 \qquad (330)$$

The course of the reaction is slightly different when reactants with less bulky substituents are used, Eq. 331 (*565*, *566*), but still producing an organoboronium ion.

$$RR'BH \cdot \text{amine} + I_2 + 2\ \text{amine} \longrightarrow RR'B(\text{amine})_2{}^+I^- + \text{amine} \cdot H^+I^- \qquad (331)$$

The solvated boron cation, Ph_2B^+ may be produced several ways, two of which involve the addition of a Lewis acid to help pull off the boron-attached halogen. A comparison of calculated (Pariser–Parr–Pople technique) with observed energies indicates that, in solution, the Ph_2B^+ ion is coordinated, whereas the 9-borafluorene cation is not (*40*).

$$2\ Ph_2BCl + SnCl_4 \xrightarrow[(1294)]{EtNH_2} [Ph_2B(EtNH_2)_2{}^+]_2SnCl_6{}^{2-} \qquad (332)$$

$$Ph_2BCl + AlCl_3 \xrightarrow[(475)]{MeCOEt} [Ph_2B(solv)_{2\ or\ 1}]^+ + AlCl_4{}^- \qquad (333)$$

TABLE 4-3

Organoboron Cationic Species

Compound	Reference
$[Me_3SiCH_2B(H)(\leftarrow PMe_3)(\leftarrow NMe_3)]^+$	*1213*
$[Me_3SiCH_2B(H)(\leftarrow Py)(\leftarrow NMe_3)]^+$	*1213*
$CH_3CH(CH_3)C(CH_3)_2BH(C_5H_5N)_2^+$	*564*
$(Cyclohexyl)BH(C_5H_5N)_2^+$	*564, 566*
$PhBH(Py)_2^+$	*565, 566*
$Me_2B(NH_3)_2^+$	*1380*
$Me_2B(ethylenediamine)^+$	*1380*
$Pr_2B(N_2H_4)_2^+$	*1475*
n-$Bu_2B(NH_2R)_2^+$	*1690*
$Bu_2B(N_2H_4)_2^+$	*1460*
$(n$-$Bu_2BPy_2)^+$	*650*
$(Cyclohexyl)_2B(C_5H_5N)_2^+$	*564*
$(CH_2)_6B(NR_2)_2^+$ (ring)	*1307*
$Ph_2B(N_2H_4)_2^+$	*1475*
$Ph_2B(bipyridyl)^+$	*69, 476, 477*
$[O(C_6H_4)_2B \leftarrow (N,N')bipyridyl]^+$	*476*
$Me(Cl)B(NR_3)_2^+$	*1466*
$Me(Br)B(NR_3)_2^+$	*1466*
$Me(Cl)B(ethylenediamine)^+$	*1466*
$Ph(Cl)B(NR_3)_2^+$	*1470*
$Ph_2B(NR_3)_2^+$	*1294, 1296*
$[Me_3NBH_2CH_2NR_3]^+$	*1366, 1369*
$Me_3NBH_2CNMe^+$	*1601*
$Me_3NBH_2CNBH_2NMe_3^+$	*1601*

A reasonably stable bisborane cation, bridged by a CN group, is prepared by the following scheme (*1601*):

$$B_2H_6 \xrightarrow{CN^-} H_3BCNBH_3^- \xrightarrow{I_2} IH_2BCNBH_2I^- \xrightarrow[(-2\,I^-)]{Me_3N} Me_3NBH_2CNBH_2NMe_3^+ \quad (334)$$

For additional routes to organoboronium ions see Section 5-2, *Coordination Compounds.*

The dicyclohexylbispyridineboron cation undergoes dehydroboration under relatively mild conditions to give the cyclohexylhydridobispyridineboron cation (*564*).

4-4. Carbon Monoxide–Borane; H_3BCO

Synthesis

A borane carbon monoxide adduct containing a B—C bond is obtained from diborane and carbon monoxide at elevated pressures (*356*) in a reaction that is catalyzed by ethers (*1199*) or from a facile reaction between boroxine and carbon monoxide (*1853*).

$$B_2H_6 + 2\,CO \longrightarrow 2\,H_3BCO \quad (335)$$

$$H_3B_3O_3 + CO \longrightarrow H_3BCO + B_2O_3 \quad (336)$$

The products isolated from the reaction of carbon monoxide with alkylated diboranes do not contain the B—H bonds expected of a partially alkylated BH_3CO. It is suspected (*1731*) that these compounds contain $(BCO)_2$ hexatomic rings similar to the rearranged products obtained from trialkylborons and carbon monoxide (*807*) (Section 3-2).

When CO is passed over liquid B_8F_{12} at $-70°C$, the white solid $(BF_2)_3BCO$, a structural analog of H_3BCO, is prepared in 90% yield (*885, 1792*).

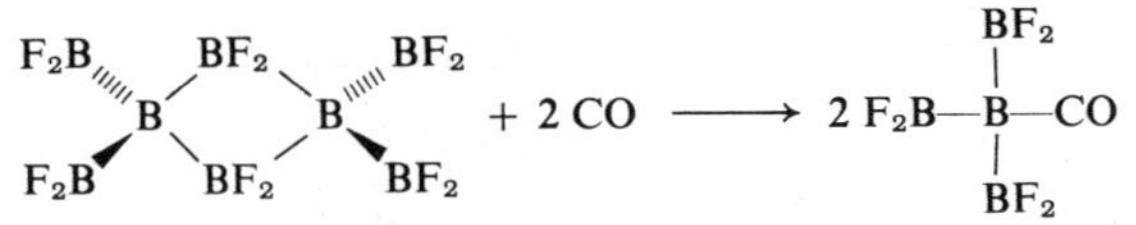

$$(F_2B)_2B(\mu\text{-}BF_2)_2B(BF_2)_2 + 2\,CO \longrightarrow 2\,F_2B\text{—}B(BF_2)_2\text{—}CO \quad (337)$$

TABLE 4-4

Borane–Carbonyl Compounds

Compound	Reference
BH_3CO	*356, 629, 638, 1027a, 1152, 1199, 1731, 1750, 1853*
$(BF_2)_3BCO$	*885, 1772*
$(SiCl_3)_2(BCl_2)BCO$	*885*
B_3H_7CO	*1922*

A related complex, $(SiCl_3)_2(BCl_2)BCO$ is formed when the condensate from a silicon dichloride–B_2Cl_4 mixture is allowed to warm from −196°C in the presence of CO (*885*) (Table 4-4).

Reactions

Decomposition of H_3BCO occurs readily at room temperature, although equilibrium is not easily reached unless the mixture is heated to 100°C (*356, 1154*).

$$H_3BCO \rightleftharpoons BH_3 + CO \text{ (rapid equilibrium)} \quad (338a)$$

$$BH_3 + H_3BCO \rightleftharpoons B_2H_6 + CO \quad \text{(slow)} \quad (338b)$$

The rapid decrease in rate at the initial stage of the reaction is attributed to an inhibitory effect of carbon monoxide. The other product, diborane, has no appreciable effect upon the rate in the early stages of decomposition. From a kinetic study a mechanism has been proposed, Eqs. 338a–b, in which the rate-limiting step is the displacement of carbon monoxide by a borane fragment (Eq. 338b) (*349*). This mechanism was challenged (*83, 85a, 629*) but has been vindicated (*644*).

Advantage of the BH_3CO decomposition mechanism has been taken in the isolation of the elusive monomeric borane species, BH_3, achieved by trapping the products from the initial step, Eq. 338a. This is accomplished by low-temperature quenching (*642, 879*) or by dilution with an inert carrier gas (*1153*). It has been verified that the first step, Eq. 338a, is first order with respect to BH_3CO, with an estimated bond dissociation energy of 23.1 ± 2 kcal $mole^{-1}$ (*595*).

In a reaction reminiscent of the ether-catalyzed hydroboration reaction (Sections 3-1, 5-1, and 5-2), the B—H bonds of BH_3CO add to ethylene at ambient temperatures to form triethylboron with the release of carbon monoxide (*1731*).

$$H_3BCO + 3\,H_2C{=}CH_2 \longrightarrow BEt_3 + CO \quad (339)$$

In a reaction with trimethylamine, BH_3CO merely loses carbon monoxide with the simultaneous formation of trimethylamine–borane (*356*).

$$BH_3CO + Me_3N \longrightarrow Me_3NBH_3 + CO \quad (340)$$

The displacement of CO from BH_3CO with Me_3N is second order at 207°K, but at 273°K, the unimolecular rupture of the B—C bond becomes important as a rate-controlling step (*715*). With the bulkier Et_3N, the reaction is almost exclusively unimolecular, with E_a = ca. 21.4 ± kcal $mole^{-1}$ for the dissociation of BH_3CO and D(B—C) is assumed to be about the same value (*716*). This agrees well with a previous value of 23.1 ± 2 kcal $mole^{-1}$ reported earlier (*595*).

With ammonia, as well as with primary and secondary amines, BH_3CO forms an ionic solid in a reaction that is analogous to that between carbon dioxide

$$H_3BCO + 2\ HNR_2 \longrightarrow H_2NR_2^+\left[H_3BC\begin{matrix}{}^{\diagup\!\!\diagup O}\\{}_{\diagdown NR_2}\end{matrix}\right]^- \qquad (341)$$

and bases (*378*, *811*, *1525*). In a similar fashion, *O*-methylboranocarbonate ion and potassium boranocarbonate are formed from the treatment of BH_3CO with methoxide ion and potassium hydroxide, respectively (*1152*).

$$H_3BCO + NaOMe \longrightarrow Na^+H_3BCO_2Me^- \qquad (342)$$

$$H_3BCO + 2\ KOH \longrightarrow K_2H_3BCO_2 + H_2O \qquad (343)$$

High-purity H_3BCO can be regenerated from the $K_2H_3BCO_2$ salt by the action of 85% H_3PO_4 (*1152*). The free acid-ion $H_3BCO_2H^-$ can be obtained from the hydrolysis of *N*-methylboranocarbamate (*1152*).

$$Na^+H_3BC(O)NHCH_3^- + 2\ H_2O \longrightarrow Na^+H_3BCO_2H^- + MeNH_3^+OH^- \qquad (344)$$

Attempts to prepare $H_3BCO_2H^-$ directly from H_3BCO and water have resulted, instead, in the production of $B(OH)_3$, CO, and H_2 (*1151*). Analogous reactions occur between H_3BCO and alcohols (*1150*). Reaction 345 is favored at 0°C, whereas reaction 346 occurs exclusively at -78°C. This is

$$H_3BCO + 3\ ROH \longrightarrow B(OR)_3 + CO + 3\ H_2 \qquad (345)$$

$$H_3BCO + 2\ ROH \longrightarrow H_2 + (RO)_2BCH_2OH \longrightarrow ROB\begin{matrix}{}^{\diagup CH_2O \diagdown}\\{}_{\diagdown OCH_2 \diagup}\end{matrix}BOR \qquad (346)$$

(XLIX)

consistent with the suggestion that reaction 345 occurs as a result of the initial dissociation of H_3BCO and subsequent solvolysis of the produced borane (or diborane). At -78°C, dissociation of H_3BCO is assumed to be negligible and accounts for a product in which the B—C bond is retained. The ester (**XLIX**) is also prepared, in high yield, by the action of HCl with ethanolic solutions of $K^+H_3BCO_2Et^-$ (*1150*).

With glycinate ion, the product from a reaction with H_3BCO stops with adduct formation, Eq. 347; the boranocarbamate can be formed from the boranocarbonate, Eq. 348 (*1952*).

$$K^+(H_2NCH_2CO_2)^- + H_3BCO \longrightarrow K^+\left[O_2CCH_2NH_2C\begin{matrix}{}^{\diagup\!\!\diagup O}\\{}_{\diagdown BH_3}\end{matrix}\right]^- \qquad (347)$$

$$K^+(H_2NCH_2CO_2)^- + K^+\left[EtOC\begin{matrix}{}^{\diagup\!\!\diagup O}\\{}_{\diagdown BH_3}\end{matrix}\right]^- \longrightarrow K_2^{2+}\left[O_2CCH_2NHC\begin{matrix}{}^{\diagup\!\!\diagup O}\\{}_{\diagdown BH_3}\end{matrix}\right]^{2-} + EtOH \qquad (348)$$

Mass spectroscopic monitoring of the oxidation of BH_3CO with O_2 gives evidence of the formation of $H_2B_2O_3$ (*81*). Nitric oxide reacts with borane carbonyl below $-130°C$ in a redox reaction yielding nitrous oxide, nitrogen, carbon monoxide, and boric acid (*811a*).

Physical Properties

Carbon monoxide-(C—B)borane has been the subject of intensive spectroscopic investigations. Electron diffraction measurements (*82*) indicate that the molecule has its B—C—O atoms linearly arranged with the three hydrogens attached to the boron, giving overall C_{3v} symmetry. Microwave studies (*671–673*) have confirmed this structure and, in addition, provided more precise values for the molecular dimensions ($d_{BH} = 1.194$ Å, $d_{BC} = 1.540$ Å, $d_{CO} = 1.131$ Å, angle HBH $= 113°52'$). To explain the nature of the bonding in BH_3CO the following resonance contributions have been suggested (*82*, *673*) and weighted (*673*) on the basis of the microwave data and dipole moment measurements (*1744*):

H_2H—$\overset{-}{B}$—C≡$\overset{+}{O}$	H_2H—$\overset{-}{B}$—$\overset{+}{C}$=O	$\overset{+}{H}$ $H_2\overset{-}{B}$=C=O	$\overset{+}{H}$ $H_2\overset{-}{B}$=$\overset{+}{C}$—$\overset{-}{O}$	H_2H—B CO
(La)	**(Lb)**	**(Lc)**	**(Ld)**	**(Le)**
ca. 30% (La + Lb)		ca. 20% (Lc + Ld)		ca. 50%

The predominance of the contribution that lacks a formal B—C bond (**Le**) is consistent both with the low chemical stability of the molecule and the large H—B—H angle which is nearly midway between the sp^3 (109°+) and sp^2 (120°) hybrid angles. From a compliance matrix study an estimated 16% s-character is estimated for the boron hybrid bond to carbon (*1580*). Additionally, the ionic character of the B—C bond has been estimated to be about 20–30% (*472*) from the observed ^{11}B quadrupole coupling constant (*671*, *672*, *1867*). Both photoelectron studies (*1124*, *1126*) and compliance matrix studies (*1580*) support some hyperconjugative back donation of BH_3 to CO (**Lc**, **Ld**). Calculated bond orders from Gaussian-type atomic orbitals for BH_3CO molecule are: B—H 0.516, C—O (σ) 0.771, C—O (π) 0.376, B—C (σ) 0.425, B—C (π) 0.049 (*42*); 0.236 electrons are donated from CO to the BH_3 group. Electron donation stems mainly from the carbon $2s$ orbital and is transferred to the empty $2p_z$ orbital of the boron atom (*42*, *1023a*); the boron assumes a negative charge ($0.281\ e^-$) and the carbon is strongly positive (0.423 units) (*42*). The calculated dipole moment 1.541D (*42*) is also in reasonable agreement with the experimental value of 1.795 D (*1744*).

Infrared (*116*, *117*, *439*, *440*, *1155*) and Raman (*116*, *1764*, *1765*) spectroscopic results for BH_3CO are in essential agreement with the assigned structure. The BC force constant, 2.97×10^5 dynes cm^{-1}, is reasonably consistent with the contribution of the double bond resonance structures (**Lc, Ld**). A reasonably good correlation between bond length versus stretching frequency of the carbonyl group is obtained, with BH_3CO included among several compounds studied (*1155*); CO 1.31 $\pm$ 0.01 Å (*673*), 2164 cm (*440*). All of the molecular vibrational constants for $^{11}BH_3CO$, $^{10}BH_3CO$, $^{11}BD_3CO$, and $^{10}BD_3CO$ have been calculated using a standardized set of symmetry coordinates (*520*).

From structural and spectroscopic properties a computed $\Delta S^\circ = 32.1$ eu (*1694*) for the reverse of Eq. 335 compares favorably with the experimental entropy of 32.5 eu (*349*). The calculated heat of formation, 22.7 (0°K) kcal $mole^{-1}$, for the reverse of Eq. 338a using basis sets of Gaussian-type atomic orbitals (*42*) is in reasonable agreement with an adjusted experimental value of ca. 16.6 kcal $mole^{-1}$ (*42, 1200*). A study of the potential surface for the reverse of Eq. 338a shows that the reaction does not need any activation energy (*1023a*). From mass spectroscopic studies a B—C bond dissociation energy ranging from 23 to 34 kcal $mole^{-1}$ has been reported for Eq. 338a (*595, 596, 643*).

The molecular beam mass spectrum of BH_3CO is shown to have fragments of the type BH_xC^+, BH_yCO^+, and BH_x^+ ($y = 0, 1, 2, 3$; $x = 0, 1, 2$) (*788*).

Chapter 5

Organodiboranes

Historically, the discovery that not more than four of the six hydrogen atoms are replaced by alkyl groups provided chemical evidence in favor of the bridge model of diborane (*1651, 1654*). Thus far, no compound has been described in the literature in which an alkyl group bridges two boron atoms, as is observed in some aluminum compounds, e.g., $Al_2(CH_3)_6$. However, a related bridging species may well be present as a high-energy intermediate during alkyl–boron exchange reactions (*1780*) (Section 3-1).

An S.C.M.O. method has been employed to study the valence electronic structures of hypothetical methyl bridging isomers of methylated diboranes. Hydrogen is found to be energetically favored over methyl as a bridging group; and this stems mainly from the different nuclear repulsion energies of the isomers (*1112*).

5-1. Synthesis

Exchange Reactions

Exchange between Diborane and Trialkyl and Triarylboranes

Four of the five possible methyl- and ethyldiboranes can be prepared by the room temperature equilibration of diborane with the corresponding trialkylborane (*1651, 1654*).

$$B_2H_6 + BR_3 \longrightarrow RB_2H_5,\ 1,1\text{-}R_2B_2H_4,\ R_3B_2H_3,\ \text{and}\ R_4B_2H_2 \qquad (349)$$

Approximate equilibrium constants have been determined for the trimethylborane exchange from equilibration experiments (*1654*) and from thermodynamic data (*1200*).

$$2\ CH_3B_2H_5 = 1,1\text{-}(CH_3)_2B_2H_4 + B_2H_6 \qquad (350)$$

$$3\ [1,1\text{-}(CH_3)_2B_2H_4] = 2\ (CH_3)_3B_2H_3 + B_2H_6 \qquad (351)$$

$$4\ (CH_3)_3B_2H_3 = 3\ (CH_3)_4B_2H_2 + B_2H_6 \qquad (352)$$

$$6\ CH_3B_2H_5 = 2\ (CH_3)_3B + 5\ B_2H_6 \qquad (353)$$

Equation	K_{expt}	K_{calc}
350	2.8	3.1
351	2.7×10^{-4}	5.2×10^{-4}
352	6.7×10^{-3}	3.5×10^{-3}
353	—	$\sim 10^{-12}$

The alkylation follows pseudo-second-order kinetics ($k_f = 0.40 \pm 0.02$ liter minute^{-1} mole^{-1}) after an induction period that is suppressed with an excess of diborane (*1665*, *1830*). From exchange studies carried out in ether solution (*304*, *305*, *1240*), alkyl groups with large steric requirements were found to reduce the proportion of tetraalkyldiborane relative to the less alkylated products (*305*). 1,2-Dialkyldiborane, not detectably present in the above exchange, is obtained by the reversible disproportionation of the monoalkylated diborane (*1649*, *1723*), Eq. 354.

$$2\,RB_2H_5 \rightleftharpoons B_2H_6 + (RBH_2)_2 \qquad (354)$$

R	$K \times 10^2$	Reference
CH_3	7.01	*1830*
C_2H_5	3.9 (g)	*1723*
	16.9 (liq)	

A related exchange involving a cyclic tetrasubstituted diborane (**LI**) has been used to synthesize the corresponding asymmetrical derivative (**LII**) (*963*); however, the infrared spectrum given for this compound is suspiciously like that of 1,2-tetramethylenediborane.

$$(C_4H_8)B(\mu\text{-}H)_2B(C_4H_8) + B_2H_6 \xrightleftharpoons{120°C} H_2B(\mu\text{-}H)_2B(C_4H_8) \qquad (355)$$

(**LI**) (**LII**)

When a mixture of triphenylborane and diborane is heated to about 80°C in both the presence (*1279*) and absence (*1876*) of ether, 1,2-diphenyldiborane is the only phenyl-substituted diborane obtained. This is in striking contrast to the above alkyl exchange in which all diborane derivatives except the 1,2-dialkyldiborane are found.

When alkyl exchange is carried out in tetrahydrofuran, the n-$Pr_3B/THF \cdot BH_3$ reaches equilibrium significantly faster than the i-$Pr_3B/THF \cdot BH_3$ system, reflecting an expected increased steric demand in the transition state for the latter (*1534*).

$$R_3B + BH_3 \xrightleftharpoons{THF} \underset{(dimer)}{R_2BH} + \underset{(dimer)}{RBH_2} \qquad (356)$$

$$2\,R_2BH\,(\text{dimer}) \underset{}{\overset{THF}{\rightleftharpoons}} R_3B + RBH_2\,(\text{dimer}) \tag{357}$$

$$2\,RBH_2\,(\text{dimer}) \underset{}{\overset{THF}{\rightleftharpoons}} R_2BH\,(\text{dimer}) + BH_3 \tag{358}$$

Equilibrium constants (Table 5-1) were calculated both on the basis of (a) total monomer plus dimer (Section 5-2) borane concentrations and (b) free monomer concentration (*1534, 1544*).

The dimethylamino bridged diborane (**LIII**) exchanges with either trimethylboron or methylmagnesium iodide to give a *B*-monomethyl derivative (**LIV**) (*551*). The latter is cyclic below $-30°C$ but opens to $H_3B—NMe_2—BHMe$ above this temperature (*551*).

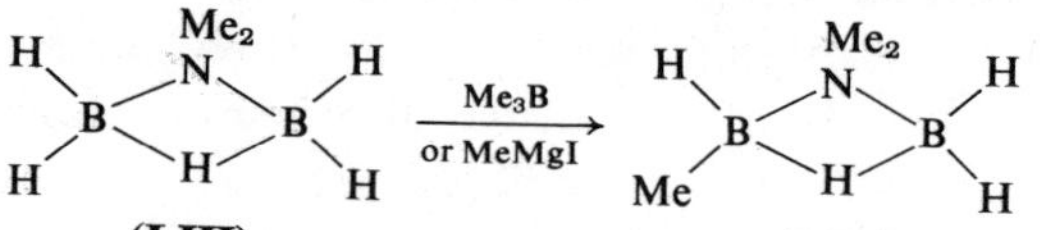

$$(\textbf{LIII}) \xrightarrow[\text{or MeMgI}]{Me_3B} (\textbf{LIV}) \tag{359}$$

The cyclic borane (**LV**) undergoes alkyl exchange with tetraalkyldiboranes to give the unsymmetrical (**LVI**) (*937*).

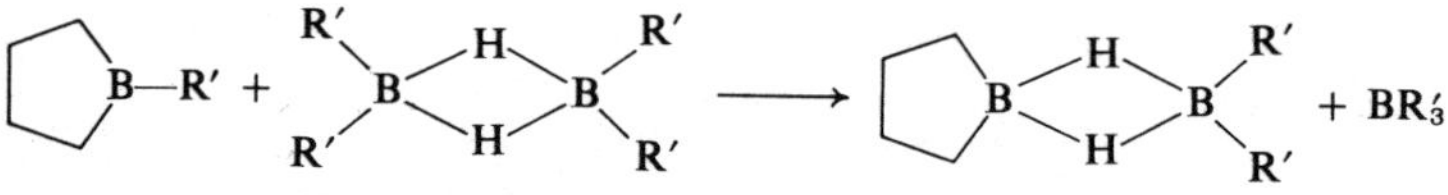

$$\text{(cyclo)B—R'} + (\textbf{LV}) \longrightarrow (\textbf{LVI}) + BR'_3 \tag{360}$$

In contrast, the following order of reactivity of various types of B—C bonds toward exchange with borane ($THF \cdot BH_3$) has been established (*263, 265*)

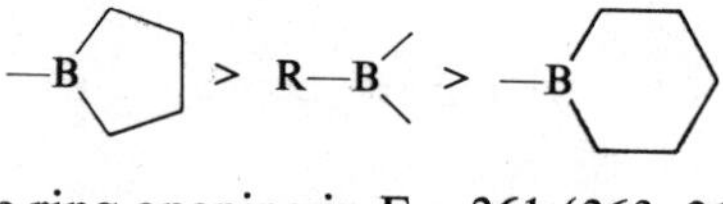

—B(cyclopentyl ring) > R—B< > —B(cyclohexyl ring)

which accounts for the ring opening in Eq. 361 (*263, 264*) and exchange with bonds exo to the ring in Eq. 362 (*1418*).

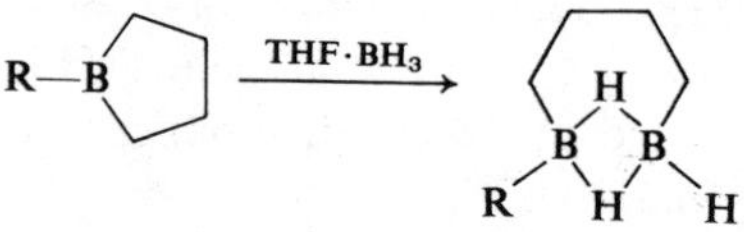

$$\text{R—B(ring)} \xrightarrow{THF \cdot BH_3} \text{(cyclic H-bridged diborane)} \tag{361}$$

$$2\,\text{(ring)B—}(CH_2)_5\text{—B(ring)} + 2\,THF \cdot BH_3 \longrightarrow 3\,\text{(ring)B(H)}_2\text{B(ring)} \tag{362}$$

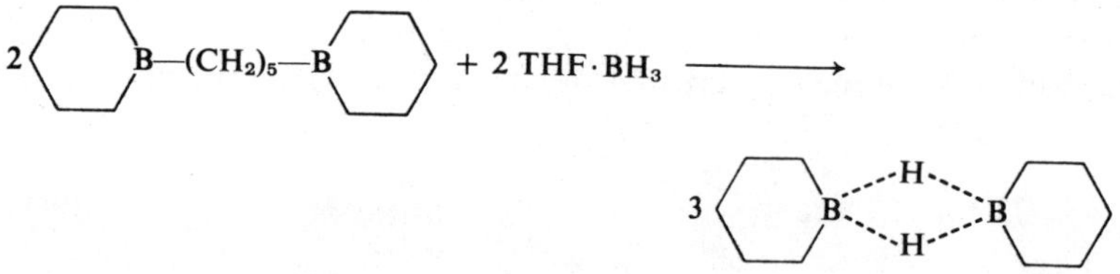

TABLE 5-1

Equilibrium Constants for Eqs. 356–358

	R = *n*-propyl		R = isopropyl	
Equation	(a)	(b)	(a)	(b)
356	82	31	189	165
357	0.098	0.33	0.87	1.65
358	0.13	0.10	0.60	0.04

Reductions with Metal Hydrides

A mixture of trialkylborane and hydrogen halide reacts at elevated temperatures under pressure with an inorganic complex hydride such as $NaBH_4$, $LiBH_4$, or $LiAlH_4$ to give a mixture of alkylated diboranes (*1137*, *1139*).

$$(6-n)NaBH_4 + (6-n)HCl + (n+2)BR_3 \longrightarrow 4\,B_2H_{6-n}R_n + (6-n)NaCl + (6-n)RH \quad (363)$$

$$(6-n)LiBH_4 + (6-n)HCl + nBR_3 \longrightarrow 3\,B_2H_{6-n}R_n + (6-n)LiCl + (6-n)H_2 \quad (364)$$

Alternatively, boron trichloride may be used in place of the hydrogen halide, or the hydride may be treated with an alkylboron halide, R_2BX or RBX_2 (*1137–1139*), or corresponding aluminum derivatives, AlR_3, R_2AlX, $RAlX_2$ (*1136*). As implied above (Eqs. 363 and 364) the extent of diborane alkylation is controlled by the ratio of reactants. It is interesting to note that not only does the reaction proceed more readily with $LiBH_4$ than with $NaBH_4$, but it also follows a slightly different course. The former hydride releases no hydrocarbon, hydrogen being the only gas formed, whereas the reverse is observed with $NaBH_4$ (*1137*).

Other reported reductions (Eqs. 365–371) of alkyl- or arylboron halides as well as boron–oxygen bonds with ether solutions of complex metal hydrides imply convenient general routes to the 1,2-di- or tetrasubstituted diboranes.

$$2\,C_6H_5BCl_2 + 4\,LiBH_4 \xrightarrow[(1877)]{} (C_6H_5BH_2)_2 + 2\,B_2H_6 + 4\,LiCl \quad (365)$$

$$2\,C_6H_5B(OR)_2 + LiAlH_4 \xrightarrow[(1877)]{} (C_6H_5BH_2)_2 + LiAl(OR)_4 \quad (366)$$

$$2\,R_2BOAr \xrightarrow[(213)]{LiAlH_4} R_4B_2H_2 \quad (367)$$

$$2\,(C_4H_8)B{-}Cl \xrightarrow[(954)]{LiAlH_4} (C_4H_8)B(\mu\text{-}H)_2B(C_4H_8) \quad (368)$$

$$\text{9-chloro-9-borafluorene} + NaB(C_2H_5)_3H \xrightarrow{(947)} \tfrac{1}{2}(\text{9-borafluorene-H})_2 + NaCl + B(C_2H_5)_3 \quad (369)$$

$$2\ \text{(3-methylborolane)B—OBu} \xrightarrow[(1932)]{LiAlH_4} (\text{(3-methylborolane)BH})_2 \quad (370)$$

$$\text{1,2,3,4-tetrakis}(BCl_2)\text{tetrahydronaphthalene} \xrightarrow[(1950)]{LiAlH_4} \text{1,2,3,4-tetrakis}(BH_2)\text{tetrahydronaphthalene} \quad (371)$$

When the reduction of phenylboron dichloride is carried out with lithium aluminum hydride at the reflux temperature of dioxane (100°C), only the disproportionated products, triphenylborane and diborane, are found (*1448*). In contrast, a similar reduction of diethylphenylboronate in diethyl ether at room temperature gives the phenylborane (dimer) which is isolated as the pyridine adduct (*757*). Also, 1-butoxy-3-methylboracyclopentane (**LVII**) is reduced without difficulty by lithium aluminum hydride in ether solution to the corresponding tetrasubstituted diborane (**LVIII**) (*1932*).

$$\underset{\textbf{(LVII)}}{CH_3\text{-}C_4H_7BOC_4H_9} \xrightarrow{LiAlH_4} \underset{\textbf{(LVIII)}}{(CH_3\text{-}C_4H_7BH)_2} \quad (372)$$

A few reductions involving metals have led to useful routes for preparing alkylated diboranes, Eqs. 373–375.

$$C_4H_8BCl \xrightarrow{K} \tfrac{1}{2}(C_4H_8B\text{—}C_4H_7BH)_2 \xrightarrow[(945)]{heat} \tfrac{1}{2}(C_4H_8BH)_2 + \tfrac{1}{3}C_{12}H_{21}B_3 \quad (373)$$

$$6\,BX_3 + 6\,H_2 + 6\,RX + 8\,Al \xrightarrow[(1394)]{\text{heat, pressure}} 3\ 1,1\text{-}R_2B_2H_4 + 8\,AlX_3 \quad (374)$$

$$6\ BX_3 + 3\ H_2 + 12\ RX + 10\ Al \xrightarrow[(1394)]{\text{heat, pressure}} 3\ R_4B_2H_2 + 10\ AlX_3 \tag{375}$$

The dialkylated diborane product from the room temperature reduction of olefinic halides with $NaBH_4$ in diglyme (*1848*) has been identified as the 1,1-isomer (*1127*).

$$CH_2{=}CHBr + NaBH_4 \longrightarrow \tfrac{1}{2}(C_2H_5)_2B_2H_4 + NaBr \tag{376}$$

$$CH_2{=}CH{-}CH_2Br + NaBH_4 \longrightarrow \tfrac{1}{2}(CH_3CH_2CH_2)_2B_2H_4 + NaBr \tag{377}$$

This suggests rapid formation of trialkylboranes followed by an alkyl exchange as discussed in the previous section.

Reduction of the B_2Cl_4–cyclohexene addition product with $LiBH_4$ gives an internally bridged (**LIX**) (*122*).

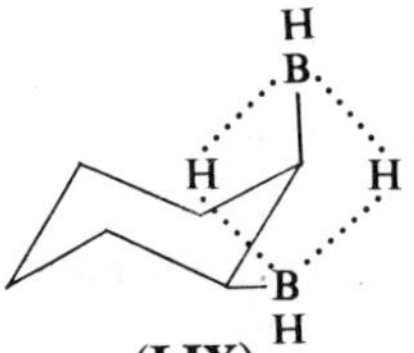

(LIX)

Hydroboration

Hydroboration of the great majority of olefins with diborane at elevated temperatures (*848*) or at ambient temperatures in the presence of ethers (*184a*, *184b*, *294*, *297*) is usually allowed to proceed to the corresponding trialkylborane. However, it is recognized that alkyldiboranes are obvious intermediates in this reaction (*1873*). In fact, equilibrium concentrations of the various alkylated diboranes are found when low ratios of olefin–diborane are used (*304*, *305*, *1118*). Further, when sterically hindered olefins are used, the reaction is conveniently stopped at the di- or tetraalkyldiborane stage. As a general rule, trisubstituted olefins, such as 2-methyl-2-butene and 2,4,4-trimethyl-2-pentene, usually proceed to the tetraalkyldiborane, whereas tetrasubstituted olefins, such as 2,3-dimethyl-2-butene, proceed only to the

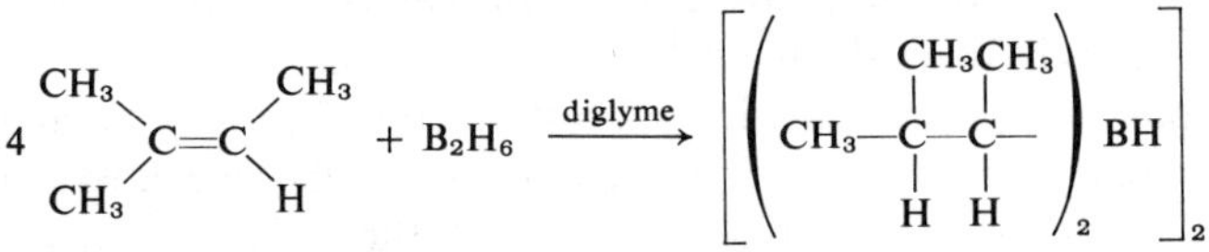

$$4\ (CH_3)_2C{=}C(CH_3)H + B_2H_6 \xrightarrow{\text{diglyme}} \left[\left(CH_3{-}\underset{H}{\overset{CH_3}{C}}{-}\underset{H}{\overset{CH_3}{C}}{-}\right)_2 BH\right]_2 \tag{378}$$

$$2\ (CH_3)_2C{=}C(CH_3)_2 + B_2H_6 \xrightarrow{\text{diglyme}} \left[\left(CH_3{-}\underset{H}{\overset{CH_3}{C}}{-}\underset{CH_3}{\overset{CH_3}{C}}{-}\right) BH_2\right]_2 \tag{379}$$

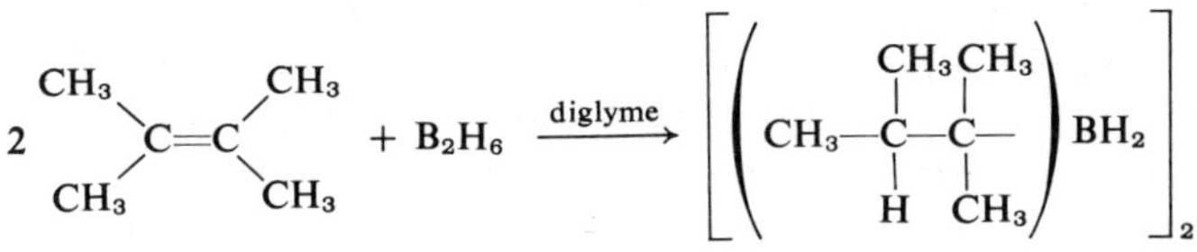

1,2-dialkyldiborane stage (*186, 226, 243, 296, 297, 319, 322, 1966, 1968, 1969*). This rule is, of course, subject to modification when extraordinarily large groups are attached to the double bond. For instance, *trans*-di-*t*-butylethylene is reported to stop at the dialkyldiborane stage (*1133*).

A vapor-phase reaction between butadiene and diborane at 100°C with hydrogen as a diluent produces, in addition to polymeric materials, an appreciable yield of the cyclic 1,2-tetramethylenediborane (**LX**) and a smaller amount of 1,2-(1′-methyltrimethylene)diborane (**LXI**) (*1866*). A similar cyclic

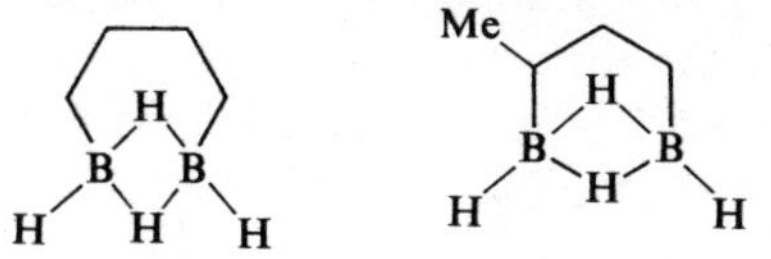

H H H B B H Me H B B H H H

(LX) **(LXI)**

structure (**LXII**) has been strongly suggested (*162, 259, 265, 1968, 1973*) for the 2:1 butadiene–diborane product that was originally proposed to be bis(bora-cyclopentane) (**LXIII**) (*937*). Both the formation of (**LX**) and (**LXII**) are catalyzed by ethers (*1929*).

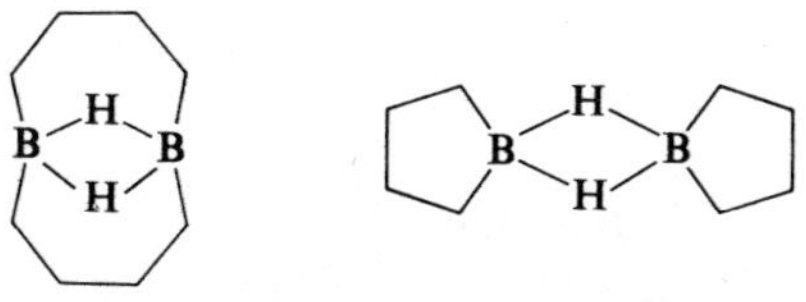

B H H B B H H B

(LXII) **(LXIII)**

Terminal addition of diborane to allene in the gas phase results in the formation of the cyclic 1,2-trimethylenediborane (**LXIV**). Polymerization of this compound occurs rapidly in liquid phase and is attributed in part to release of ring strain. Poly(μ-trimethylene)diborane depolymerizes quantitatively at 60°C *in vacuo*, thus establishing the reversibility of the reaction $nB_2C_3H_{10} \rightleftharpoons [(CH_2)_3—B_2H_4]_n$, which probably proceeds via mobile bridge hydrogens (*1118*).

Only a minimum of mechanistic attention has been paid to the catalytic effect that ethers have on the hydroboration reaction. After determining that the hydroboration of olefins with 1,2-bis(3-methyl-2-butyl)diborane is first order in olefin and also first order in the substituted diborane, it was suggested that the role of the solvent is to coordinate with the dialkylborane monomer, the "leaving group" in this particular reaction (*242*). However, the room temperature hydroboration of ethylene with borine carbonyl (*1731*) intimates, by analogy, that a rather loosely held adduct may account for acceleration of B—H addition to unsaturated systems in ether solvents.

There is indirect evidence that $H_2BCH_2CH_2BH_2$—dimer or polymer(?)—is

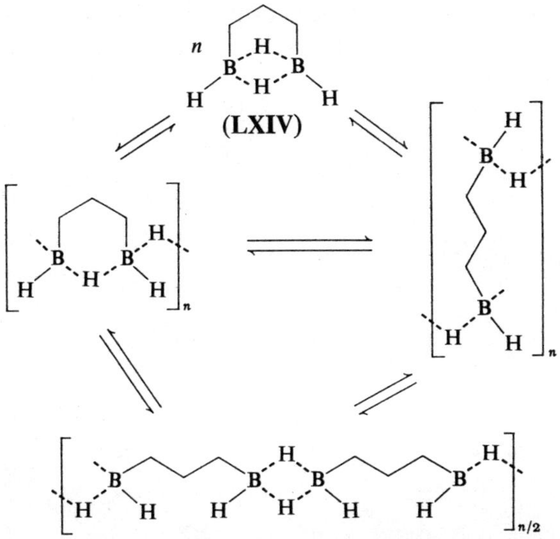

(LXIV)

formed from acetylene and excess diborane in the presence of 1,2-dimethoxyethane (*414*). Hydroboration of 1,4-pentadiene produces a significant amount of the biscyclic diborane (**LXV**) (*260, 261*) which can be isomerized to the

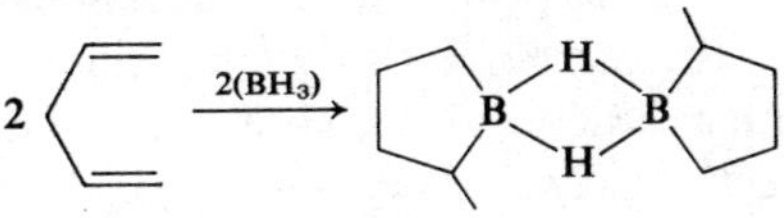

$$2 \text{ (1,4-pentadiene)} \xrightarrow{2(BH_3)} \text{(LXV)} \quad (380)$$

(LXV)

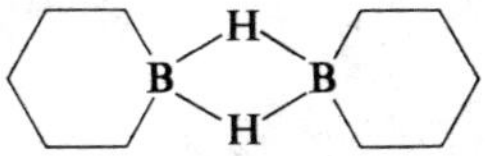

(LXVI)

bisborinane (**LXVI**) (*261*). Apparently, ring closure, Eq. 380, proceeds to a considerable extent with boron attack at a secondary carbon. This can be circumvented by utilizing the greater terminal directive effect of 2,4-dimethyl-1,4-pentadiene during hydroboration (*260*). Bisborepane can be prepared in

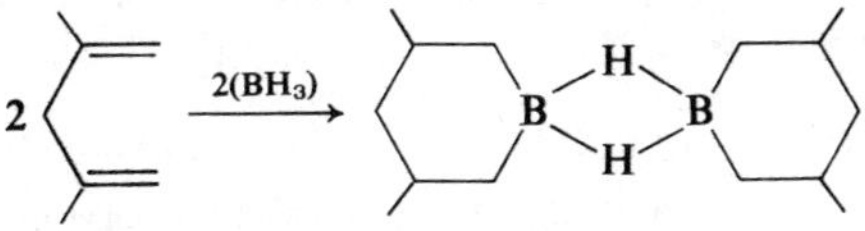

$$2 \text{ (2,4-dimethyl-1,4-pentadiene)} \xrightarrow{2(BH_3)} \quad (381)$$

$$2 \xrightarrow{2(BH_3)} \quad (382)$$

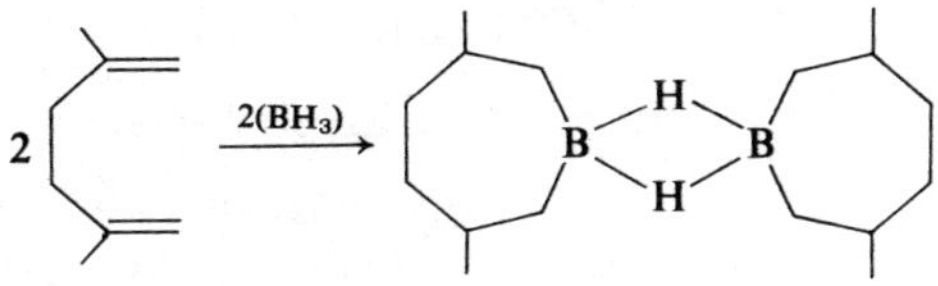

an analogous fashion (*260*); and the unusually stable bis-9-borabicyclo[3.3.1]-nonane (**LXVII**) is obtained from the hydroboration–isomerization of 1,5-cyclooctadiene (*892, 937*).

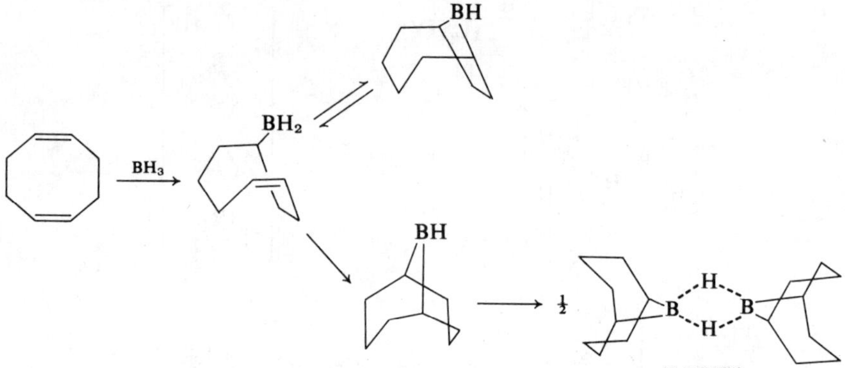

BH_3 BH_2 BH BH ½ H B B H

(**LXVII**)

(383)

The addition of cyclohexene or *d*-α-pinene to a tetrahydrofuran·BH_3 (in a 1:1 ratio) solution yields the unsymmetrical 1,1-$R_2B_2H_4$ in equilibrium with relatively small amounts of the symmetrical $R_2BH_2BR_2$ (*226*). Increasing the olefin/THF·BH_3 ratio to 2 gives primarily tetraalkyldiboranes in equilibrium with small quantities of the unsymmetrical dialkyldiboranes and THF·BH_3 (*226*) Eq. 384. For other kinetic and equilibrium studies on the hydroboration reaction see Section 3-1.

$$2\ R_2B_2H_4 \rightleftharpoons R_4B_2H_2 + 2\ THF \cdot BH_3 \tag{384}$$

Reduction of Organoboron Esters and Halides with Diborane

The interaction of diborane and esters of dialkylborinic acids at room temperature yields predominantly tetraalkyldiboranes, 1,1-dialkyldiboranes, and esters of alkylboronic acids (*1344, 1345*). If the reaction mixture is allowed to stand for 1 week, the unsymmetrical dialkyldiborane isomerizes into the

$$4\ R_2BOR' + 2B_2H_6 \rightleftharpoons 3\ R_2B_2H_4 + 2\ RB(OR')_2 \tag{385}$$

$$4\ R_2BOR' + R_2B_2H_4 \rightleftharpoons 2\ RB(OR')_2 + 2\ R_4B_2H_2 \tag{386}$$

symmetrical 1,2 derivative (*1345*). Similar attempts to reduce diphenylboron chloride (*1279*) and esters of diarylborinic or arylboronic acid (*1277, 1279, 1281*) with diborane all result in formation of the 1,2-diaryldiborane.

$$3\ (C_6H_5)_2BOR + 2\ B_2H_6 \longrightarrow 3\ (C_6H_5BH_2)_2 + B(OR)_3 \tag{387}$$

The absence of diphenylborane (as either dimer or 1,1-diphenyldiborane) in the product mixture is attributed to a disproportionation, which must be more rapid than with the corresponding alkyl compounds, to give the 1,2-diaryl-diborane and triphenylborane. Undoubtedly, the latter compound reacts further with diborane under these conditions to form more of the observed product.

Dehydroboration

The reversibility of the hydroboration reaction was first realized when dibutyldiborane and olefin were obtained from tri-*n*-butylborane at elevated temperatures (*1613*).

$$\underset{\text{H}\quad\text{B}-}{-\overset{|}{\underset{|}{\text{C}}}-\overset{|}{\underset{|}{\text{C}}}-} \rightleftharpoons \text{>C=C<} + \text{H}-\underset{|}{\text{B}}- \qquad (388)$$

Other dehydroboration reactions have since been observed (*48*, *932*, *1901*) (Section 3-2), and have occasionally been used to synthesize alkylated diboranes (*187*, *978*). As with hydroboration, dehydroboration is catalyzed by ethers (*226*, *298*). Thus, in diglyme solution, a marked enhancement of organoborane isomerization is observed (*298*, *781*); and tetraisopinocampheyldiborane appears to be in equilibrium with substantial amounts of triisopinocampheyldiborane and α-pinene (*187*, *226*, *243*).

A dehydroboration–hydroboration sequence is probably operating when thermal action converts seven- or eight-membered rings of bis(boracycloalkanes) into the more stable six-membered ring (*937*).

Insertion

Dimethylboryl chloride reacts with an equimolar quantity of sodium borohydride to produce 1,1-dimethyldiborane (*636*).

$$Me_2BCl + NaBH_4 \longrightarrow NaCl + 1,1\text{-}Me_2B_2H_4 \qquad (389)$$

It is proposed (*636*) that the high-temperature reaction of Me_3B and $NaBH_4$ in the presence of HCl to produce mixed methylated diboranes (*1139*) proceeds through the intermediate formation of Me_2BCl and subsequent reaction of this with borohydride ion.

Miscellaneous Preparative Reactions

Alkylated diboranes are produced by exchange of diborane with a few metal alkyls, e.g., ethyllithium (*1646*) and trimethylgallium (*1648*). Similarly,

$$GaMe_3 + 3\,B_2H_6 \longrightarrow Ga + 3\,MeB_2H_5 + \tfrac{3}{2}\,H_2 \qquad (390)$$

alkyl exchange has been observed between aluminum borohydride and both tin and lead tetramethyl, with formation of, among other products, methylated diboranes (*816*).

A reported preparation of phenylborane (dimer) from a reduction of $C_6H_5BI \cdot 2HI$ with hydrogen in ethanol (*1507a*) is undoubtedly in error (*1448*, *1876*). Perhaps the first unquestionable evidence for the existence of this compound comes from the reaction of diborane with benzene at elevated temperatures (*848*).

Hydrogenation of a trialkylboron, used occasionally to prepare tetra-alkyldiboranes (*928*, *942*, *970*), probably occurs through direct addition to the carbon–boron bond (*890*, *1584*) rather than a dehydroboration followed by hydrogenation of the resulting olefin.

$$\text{(1-ethyl-3-methyl-benzoborole)} \xrightarrow{\text{heat, } H_2} \tfrac{1}{2}\left[\text{(3-methyl-benzoborole, B–H)}\right]_2 + C_2H_6 \qquad (391)$$

A mixture of *n*-propyldiboranes is obtained along with tri-*n*-propyl borane from the reaction of diborane with cyclopropane at about 95°C. Apparently the "banana" bonds of cyclopropane can provide sufficient electron density to form the expected π-complex intermediate (*689*).

Several organoborohydrides provide sources of organodiboranes, Eq. 392,

$$Li(PhBH_3) + \tfrac{1}{2} B_2H_6 \xrightarrow[(1879)]{\text{ether}} LiBH_4 + \tfrac{1}{2} (PhBH_2)_2 \qquad (392)$$

$$C_5H_9BH_2Li^+ \xrightarrow[(259)]{MeSO_3H} \left[C_5H_9B(\mu\text{-}H)_2BC_5H_9\right] \longrightarrow \text{(bicyclic diborane)} + \left[\text{B(}\mu\text{-H)}_2\text{B}-C_4H_8\right]_n \qquad (393)$$

393. At elevated temperatures (>80°), magnesium alkylborohydrides are sources of alkyldiboranes (*84*).

1,1-Dialkyldiboranes can be generated from alkylborane adducts upon treatment with boron trifluoride etherate (*212*).

$$RBH_2 \cdot py + BF_3 \cdot Et_2O \longrightarrow \underset{\text{dimer}}{RBH_2} + BF_3 \cdot py + Et_2O \qquad (394)$$

Diborane reacts with the silyl-substituted sulfonium ylide (**LXVIII**) to form the substituted diborane (**LXIX**). The formation of (**LXIX**) can be explained in terms of initial adduct formation followed by a hydride shift with concomitant liberation of methyl sulfide (*1213*) (Section 3-2).

$$\underset{\text{(LXVIII)}}{Me_3SiCHSMe_2} \xrightarrow{B_2H_6} Me_3Si\underset{\underset{+}{SMe_2}}{C}H(BH_3^-) \longrightarrow \underset{\text{(LXIX)}}{\underset{\text{(dimer)}}{Me_3SiCH_2BH_2}} + SMe_2 \quad (395)$$

Known organodiboranes are listed in Table 5-2.

5-2. Reactions

Hydrolysis, Alcoholysis, and Related Reactions

As with diborane, the bridge and terminal boron-bonded hydrogens of alkylated diboranes are normally quite labile at room temperature toward hydrolysis and alcoholysis (*304, 305, 1240, 1650, 1654, 1723, 1848*). Since the alkyl–boron bonds are not as easily cleaved, the alkylboronic or alkylborinic acids (or esters in the case of alcoholysis) produced are often quite useful in determining the structure of the original alkylated diborane. Similarly, dialkylthioborinic esters are obtained from the interaction of

$$(CH_3)_2B(\mu\text{-}H)_2BH_2 + 4\,H_2O \longrightarrow (CH_3)_2BOH + B(OH)_3 + 4\,H_2 \quad (396)$$

$$(CH_3)HB(\mu\text{-}H)_2BH(CH_3) + 4\,H_2O \longrightarrow 2\,CH_3B(OH)_2 + 4\,H_2 \quad (397)$$

tetraalkyldiboranes with thiols (*360, 1240*). 1,2-Diaryldiboranes react with hydroxy compounds in a manner analogous to the 1,2-dialkyldiboranes (*848, 1277, 1278, 1876*).

Notable exceptions to the rather labile character of B—H bonds of alkyldiboranes are bis(boracyclopentane) (*937, 963*) and bis(1,2-tetramethylene)-diborane, which are both unusually stable to both hydrolysis and alcoholysis (*162, 937, 963*).

Oxidation and Oxidative Hydrolysis

As with diborane and trialkylborons, alkylderivatives of diborane react with air (*1240*), with the formation of B—O and R—O bonds. In a study on the nonexplosive partial oxidation of 1,2-diethyldiborane the rate dependence

TABLE 5-2

Organodiboranes

Compound	Reference
MeB_2H_5	*1083, 1139, 1648, 1654*
1,2-$Me_2B_2H_4$	*1086, 1139, 1649, 1901a*
1,1-$Me_2B_2H_4$	*636, 1087, 1139, 1654, 1683, 1901a* 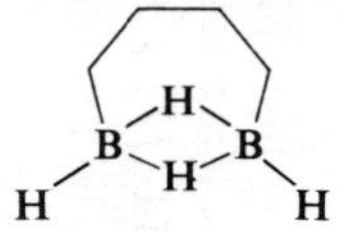
$Me_3B_2H_3$	*952, 1088, 1139, 1654*
$Me_4B_2H_2$	*952, 1088, 1137, 1139, 1602, 1603, 1654*
EtB_2H_5	*1085, 1118, 1650* 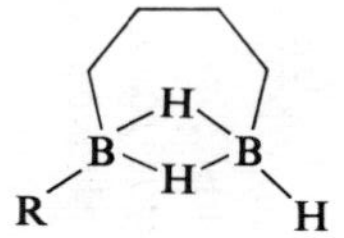
1,2-$Et_2B_2H_4$	*1086, 1723, 1848*
1,1-EtB_2H_4	*1087, 1118, 1650, 1683, 1723*
$Et_3B_2H_3$	*952, 1088, 1650, 1683*
$Et_4B_2H_2$	*952, 954, 957, 961, 973, 1088, 1650, 1683* 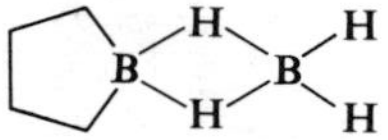
n-PrB_2H_5	*689, 1118, 1650*
1,2-*n*-$Pr_2B_2H_4$	*1345, 1346, 1848*
1,1-*n*-$Pr_2B_2H_4$	*1118, 1240, 1345, 1650*
$Pr_3B_2H_3$	*952* 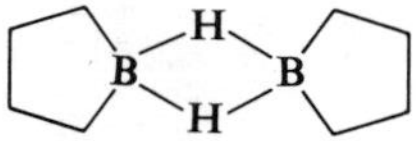
$Pr_4B_2H_2$	*952, 1137, 1138, 1240, 1345, 1346*
n-BuB_2H_5	*1118*
n-$Bu_3B_2H_3$	*952* 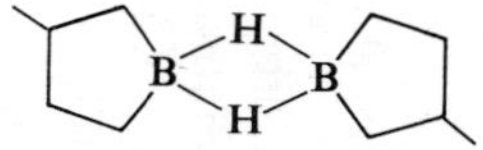
i-$Bu_3B_2H_3$	*952*
n-$Bu_4B_2H_2$	*213, 952, 1240, 1312*
s-$Bu_4B_2H_2$	*213*
i-$Bu_4B_2H_2$	*213, 1312*
1,2-$(Me_2CHCHCl)_2B_2H_4$	*1541*
1,2-$(Me_2CHCHMe)_2B_2H_4$	*1391* 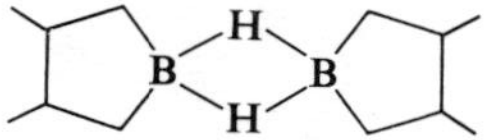

Compound	Reference
	1118, 1866, 1929
R = *n*-Bu, *n*-Oct	*263, 264*
	937, 963
	259, 937, 945, 963
(mixture of isomers)	*605, 937, 945, 949, 963, 1932*
	963

Compound	References
1,1-$(Me_2CHCHMe)_2B_2H_4$	*226, 321*
$(Me_2CHCHMe)_4B_2H_2$	*226*
$Me_2CHCMe_2B_2H_5$	*226*
1,2-$(Me_2CHCMe_2)_2B_2H_4$	*226, 243, 1419, 1968*
1,2,2-$(Me_2CHCMe_2)_3B_2H_3$	*1419*
$(Cyclopentyl)_4B_2H_2$	*213*
1,1-$(Cyclohexyl)_2B_2H_4$	*226*
$(Cyclohexyl)_3B_2H_3$	*952*
$(Cyclohexyl)_4B_2H_2$	*226, 243, 952*
$(Methylcyclohexyl)_4B_2H_2$	*226, 243*
1,2-(2,2,4-Me_3-3-pentyl)$_2B_2H_4$	*226, 243*
$(n\text{-}C_{10}H_{21})_4B_2H_2$	*932*
1,1-$(Isopinocampheyl)_2B_2H_4$	*226*
$(Isopinocampheyl)_3B_2H_3$	*186, 226*
$(Isopinocampheyl)_4B_2H_2$	*186, 187, 226*
1,2-$(n\text{-}C_{12}H_{25})_2B_2H_4$	*932*
1,2-$(Me_3SiCH_2)_2B_2H_4$	*1213*
1,2-$Ph_2B_2H_4$	*1277, 1279, 1876*
$Ph_3B_2H_3$	*952*
$Ph_4B_2H_2$	*952*
1,2-$(o\text{-}MeC_6H_4)_2B_2H_4$	*1278*
1,2-$(p\text{-}MeC_6H_4)_2B_2H_4$	*1281*
1,2-$(p\text{-}ClC_6H_4)_2B_2H_4$	*1277*
1,2-$(\alpha\text{-}Naphthyl)_2B_2H_4$	*1277*
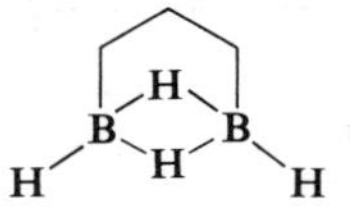	*1118*
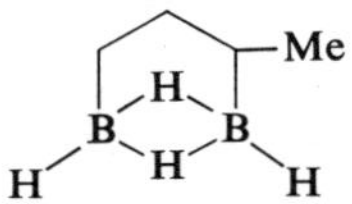	*1118, 1866, 1929*
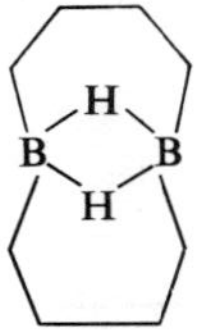	*162, 259, 963, 1929*
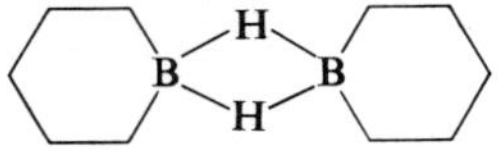	*261, 263, 937, 1418*
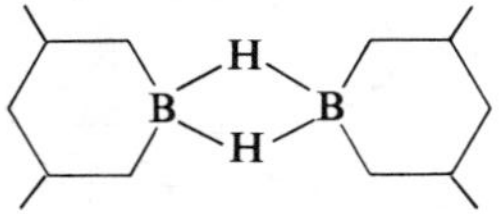	*260*
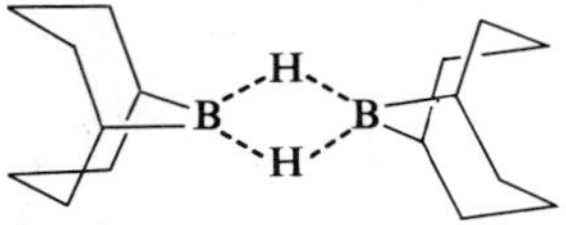	*892* *937*
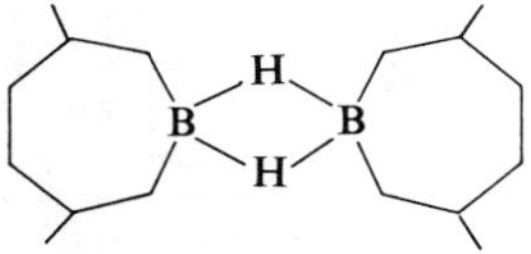	*260*
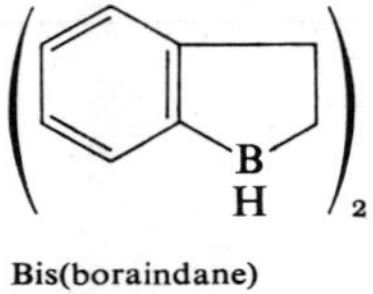 Bis(boraindane)	*952, 970*

(*continued*)

TABLE 5-2 (*continued*)

Compound	Reference	Compound	Reference
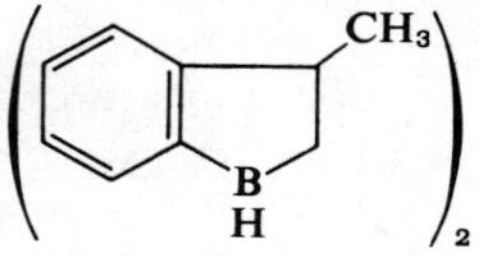 Bis(3-methylboraindane)	*970*	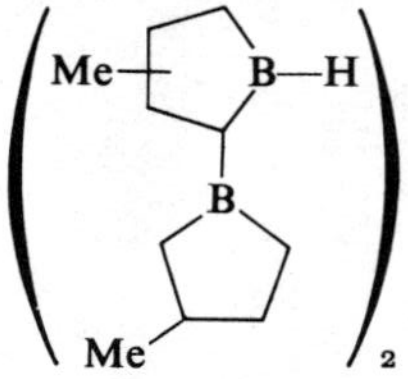	*949*
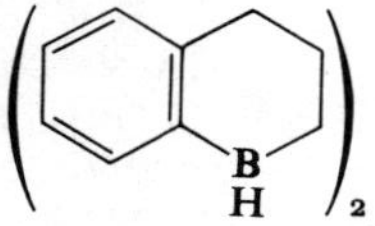 Bis(boratetralin)	*952, 970*	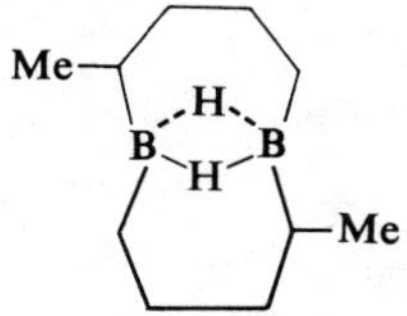	*1418*
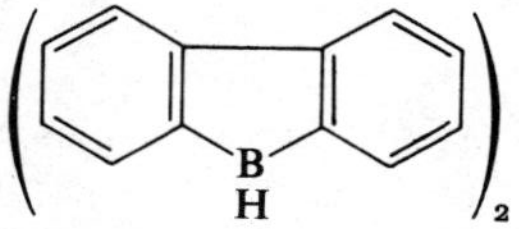 Bis(borafluorene)	*947*		

was found to be zero order in oxygen and first to three-halves order in 1,2-diethyldiborane, with an activation energy of 32.5 ± 1.5 kcal mole^{-1} (*1634a*).

$$(C_2H_5BH_2)_2 + 2\ O_2 \xrightarrow{40-80°C} 2\ C_2H_5OBO + 2\ H_2 \tag{398}$$

Tetramethyldiborane, 1,1-dimethyldiborane, and diborane are all much more stable than trimethylborane toward oxidation with molecular oxygen in the low temperature region of 77–170°K. This work tends to support a primary oxidation step involving formation of a bond between oxygen and the empty orbital of the boron atom of a monomeric trivalent boron species (Section 3-2). Diborane and its methyl derivatives do not possess energetically accessible vacant orbitals; and further, the heat of dissociation of the BH_2B bridging unit (see section on diborane–borane equilibrium below, p. 186) is sufficiently high to prevent dissociation to a monomeric borane species at the low temperatures studied (*1530*). Both the bulky bis-9-borabicyclo[3.3.1]-nonane,9-BBN (**LXVII**), (*892*), and bis(1,2-tetramethylene)diborane (**LXII**), exhibit a high degree of stability toward oxygen (*937*, *963*, *1621*). In both of these compounds, it is probable that the difficulty with which the bridge linkage is broken is responsible for their low reactivity. Thus, the resistance to widening of the C—B—C angle to 120° prevents 9-BBN from dissociation into the potentially reactive monomer; and the rather rigid bicyclic framework of the bis(1,2-tetramethylene)diborane does not allow the bridge to break without considerable expenditure of energy.

Oxidative hydrolysis with hydrogen peroxide has been used extensively in sequence with the hydroboration reaction in the synthesis of alcohols from olefins via alkylated diboranes (*184a*) (Sections 3-1 and 3-2).

$$R_4B_2H_2 \xrightarrow[NaOH]{H_2O_2} 4\ ROH + B(OH)_3 + H_2 \tag{399}$$

Disproportionation and Exchange Reactions

The reversible nature of the alkyl exchange between trialkylboranes and diborane (Section 5-1) is implied in several studies (*1654*, *1830*, *1883*). This disproportionation behavior may partially account for the production of trialkylboranes by the thermal interaction of diborane with olefins (*848*) after the initial formation of an alkyldiborane by an addition step (*1873*). Alternatively, of course, the trialkylborane may arise solely through a series of B—H additions (1873).

1,2-Diphenyldiborane appears to be moderately stable toward disproportionation at room temperature (*757*, *1277*, *1876*), although at 100°C in dioxane, a reaction that should have yielded this phenyl derivative resulted instead in the formation of triphenylborane and diborane (*1448*). Attempts to obtain pure diphenylborane (dimer) have met with difficulties (*952*, *1277*, *1279*, *1876*), leading rather to 1,2-diphenyldiborane and triphenylborane. However, diarylboranes have been isolated as relatively stable pyridine adducts (*758*). Köster and Benedikt (*947*) have prepared bis(9-borafluorene), which represents the first example of a stable tetraaryldiborane.

The dimer (**LXX**) slowly disproportionates at room temperature to the triorganoborane (**LXXI**) (*1213*).

$$3\,(Me_3SiCH_2BH_2)_2 \longrightarrow 2\,(Me_3SiCH_2)_3B + 2\,B_2H_6 \quad (400)$$

(**LXX**) (**LXXI**)

Bisborolane (**LXXII**) will rearrange to (**LXXIII**) and a polymer (**LXXIV**) (*259*, *1418*), whereas bisboracyclane (**LXXV**) does not interconvert with (**LXXVI**) and (**LXXVII**) (*261*, *1418*). This points to the greater stability of the six-membered (**LXXV**) over the five-membered (**LXXII**) ring system.

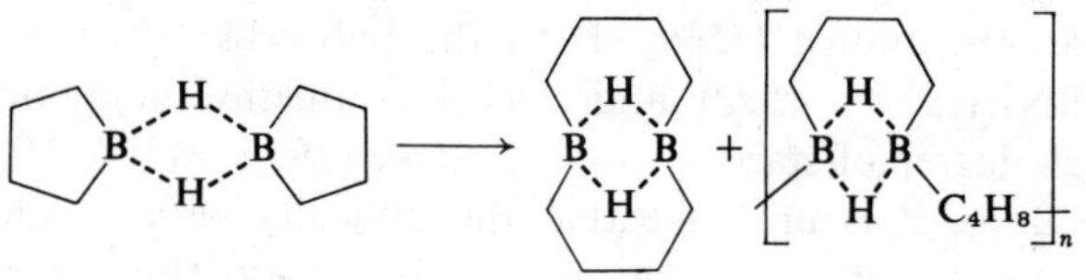

(401)

(**LXXII**) (**LXXIII**) (**LXXIV**)

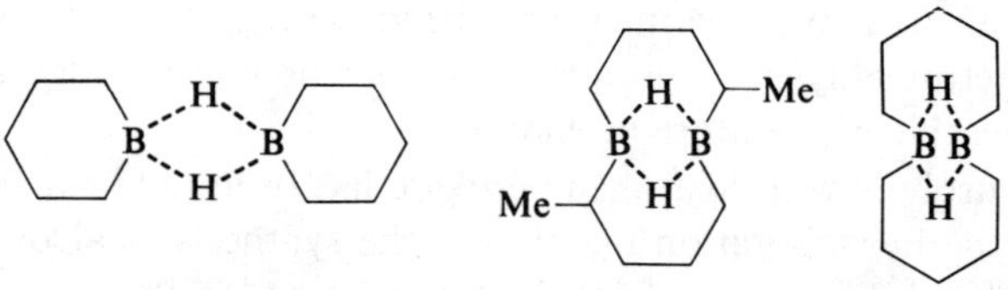

(**LXXV**) (**LXXVI**) (**LXXVII**)

Dialkyl and tetraalkyldiboranes have been shown to equilibrate with boric, alkylboronic, and dialkylborinic esters as illustrated in Eqs. 402 404 (*1344–1346*).

$$3\,R_2B_2H_4 + 4\,(R'O)_3B \rightleftharpoons 6\,RB(OR')_2 + 2\,B_2H_6 \quad (402)$$

$$2\,R_2B_2H_4 + 4\,(R'O)_3B \rightleftharpoons 6\,RB(OR')_2 + R_2B_2H_4 \quad (403)$$

$$R_2B_2H_4 + 4\,R_2BOR' \rightleftharpoons 2\,RB(OR')_2 + 2\,R_4B_2H_2 \quad (404)$$

A related equilibrium probably operates in the reaction between tetraalkyldiboranes and sodium methoxide (*929*).

$$2\,R_4B_2H_2 + NaOCH_3 \longrightarrow NaBH_4 + 2\,BR_3 + R_2B(OCH_3) \quad (405)$$

The ready ability of alkylated diboranes to enter into a number of interchange reactions has led to their use as catalysts in other transboronation transformations (*938*, *958*, *985*, *1343*, *1344*, *1347*, *1780*).

$$BR_3 + 2\,B(OR')_3 \xrightarrow[\text{(catalytic amounts)}]{R_4B_2H_2} 3\,RB(OR')_2 \quad (406)$$

Exchange reactions have been observed between alkyldiboranes and aminoboranes or compounds with B—N—B linkages (*963*).

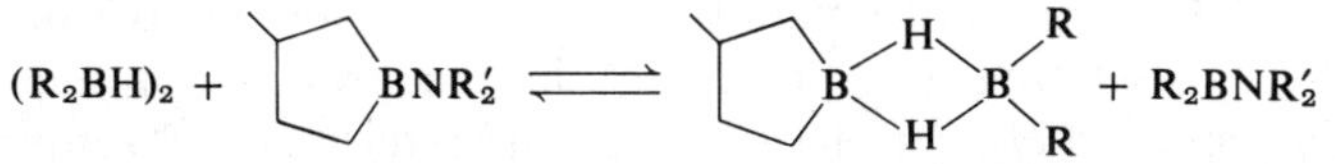

$$(R_2BH)_2 + \text{[boracyclopentane]}BNR'_2 \rightleftharpoons \text{[boracyclopentane]}B(\mu\text{-}H)_2BR_2 + R_2BNR'_2 \qquad (407)$$

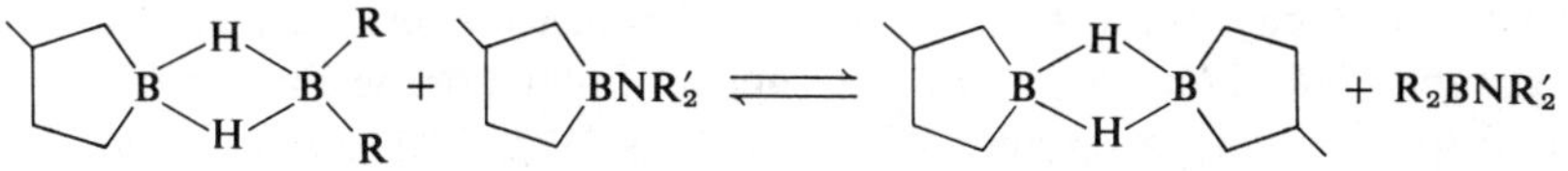

$$\text{[boracyclopentane]}B(\mu\text{-}H)_2BR_2 + \text{[boracyclopentane]}BNR'_2 \rightleftharpoons \text{[boracyclopentane]}B(\mu\text{-}H)_2B\text{[boracyclopentane]} + R_2BNR'_2 \qquad (408)$$

$$(R_2BH)_2 + \text{[boracyclopentane]}B\text{—}N(R')\text{—}B\text{[boracyclopentane]} \rightleftharpoons \text{[boracyclopentane]}B\text{—}N(R')\text{—}BR_2 + \text{[boracyclopentane]}B(\mu\text{-}H)_2BR_2$$

$$\rightleftharpoons \text{[boracyclopentane]}B(\mu\text{-}H)_2B\text{[boracyclopentane]} + R_2B\text{—}N(R')\text{—}BR_2 \qquad (409)$$

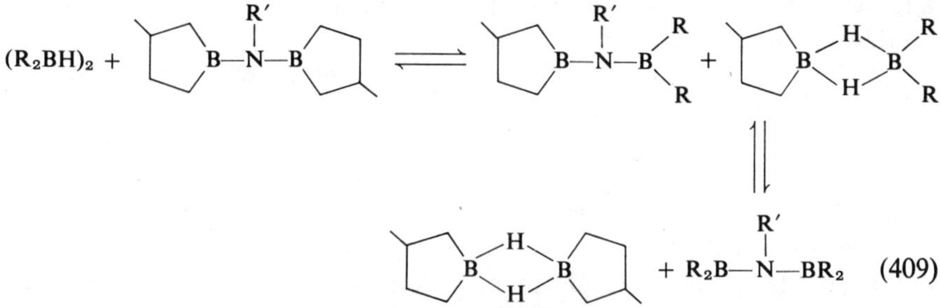

Studies on the disproportionation of methyl(dimethylamino)borane (*351*) tend to support an exchange mechanism that does not involve breaking of the B—N bond in the above bis(boracyclopentane) transformations. For the same reason, B—N cleavage probably does not occur in a related exchange between diborane and hydrazino-1,2-bis(diethylborane) (*1475*).

$$Et_2B\text{—}NHNH\text{—}BEt_2 + B_2H_6 \longrightarrow H_2B\text{—}NHNH\text{—}BH_2 + (Et_2BH)_2 \qquad (410)$$

Exchange reactions of alkylated diboranes with higher boron hydrides are discussed in Sections 6-2 and 6-4.

Hydroboration

For an extensive discussion of the role of alkylated diboranes in the hydroboration reaction, one is advised to consult the authoritative books by H. C. Brown (*184a*, *184b*); only a brief summary of pertinent chemistry is included here (see also Section 3-1).

Alkylated diboranes with large steric requirements [e.g., 1,2-bis(2,3-dimethyl-2-butyldiborane, tetra(3-methyl-2-butyl)diborane, and tetracyclohexyldiborane] are used extensively as selective reducing agents. Applications have included: (a) the competitive hydroboration of olefins and dienes (*185*, *186*, *242*, *244*, *290*, *318*, *321*, *1966*, *1968*, *1974*); (b) steric control over the mode of addition to olefins and alkynes (*186*, *224*, *313*, *317*, *321*, *592*, *893*, *1966*, *1968*); (c) use us a monohydroborating agent of alkynes and dienes (*313*, *323*, *1391*, *1968*, *1974*, *1975*); (d) the selective reduction of various allyl derivatives (*200*, *224*, *225*, *227*); (e) the asymmetric synthesis of alcohols from the hydroboration of olefins with optically active alkylated diboranes

(*185–187*, *320*, *794*, *1965*, *1967*, *1969*). In addition, alkylated diboranes have found use as a reagent for the sterically controlled reductions of cyclanones to alcohols (*195*), for the formation of cyclic organoboranes from dienes (*262*, *269*), for selective hydroborations in the presence of unprotected carboxylic groups (*194*), for the stereospecific reductions of ketones (*307*) and of lactones (*668*), and for the reduction of numerous unsaturated functional groups (*196*, *216*, *593*). Optically active dialkylborane dimers have been used for a variety of asymmetric reductions (*384*, *1389b*, *1746*, *1836*), and diisopinocampheylborane (dimer) has been effective for the asymmetric synthesis of α-amino acids from nitriles (*547a*).

A comparison of relative hydroboration rates of olefins with tetra(3-methyl-2-butyl)diborane indicates the following general order: alkynes > terminal olefins > cis internal olefins > trans internal olefins (*244*). Kinetic data establish the hydroboration of olefins with this tetraalkyldiborane to be second order, first order in olefin and first order in the substituted diborane (*242*). Presumably, the rate-determining step gives, in addition to the trialkylborane product, one molecule of a dialkylborane monomer, which may dimerize or react with a second molecule of olefin in a rapid second step. Since the dialkylborane dimer rather than monomer participates in the

$$R_2B(\mu\text{-}H)_2BR_2 + \gt C{=}C\lt \longrightarrow \left[\begin{array}{c} -\overset{|}{C}\text{---}\overset{|}{C}- \\ \vdots \quad\ \vdots \\ R_2B\text{---}H \\ \vdots \quad\ \vdots \\ H\text{---}BR_2 \end{array}\right] \longrightarrow R_2B\overset{|}{\underset{|}{C}}-\overset{|}{\underset{|}{C}}H + R_2BH \qquad (411)$$

activated complex, it is not surprising that fairly selective hydroborations are observed with tetraethyldiborane (*958*).

A stepwise reduction of the triple bond of propargyl chloride with tetraethyldiborane is observed with the formation of 1,1-bis(diethylboryl)-3-chloropropane (*130*).

$$[(C_2H_5)_2BH]_2 + 2\,HC{\equiv}CCH_2Cl \longrightarrow 2\,(C_2H_5)_2BCH{=}CHCH_2Cl \qquad (412)$$

$$[(C_2H_5)_2BH]_2 + 2\,(C_2H_5)_2BCH{=}CHCH_2Cl \longrightarrow 2\,[(C_2H_5)_2B]_2CHCH_2CH_2Cl \qquad (413)$$

Unsymmetrical trialkylboranes are isolated by hydroboration of alkenes with tetraalkyldiboranes (*132*, *959*, *1240*). In contrast, phenyldialkylboranes, prepared from 1,2-diphenyldiborane, undergo a rapid redistribution at room temperature into triphenylboron and a trialkylboron (*1278*). 1,2-Diaryldiboranes add to 1,3-dienes in toluene with the formation of 1-arylboracyclopentanes which appear to be more stable toward disproportionation than the open chain analogs (*1278*).

$$(ArBH_2)_2 + 2\,CH_2{=}CH{-}CH{=}CH_2 \longrightarrow 2\,Ar{-}B\langle\begin{matrix} CH_2{-}CH_2 \\ | \\ CH_2{-}CH_2 \end{matrix} \qquad (414)$$

Coordination Compounds with Lewis Bases

The behavior of methylated diboranes toward ammonia at low temperatures is similar to diborane in that white saltlike "diammoniates" (Table 5-3) are formed from an unsymmetrical cleavage of the bridge hydrogens (*352, 1380, 1649*).

$$Me_2BH_2BMe_2 + 2\,NH_3 \longrightarrow [Me_2B(NH_3)_2]^+[H_2BMe_2]^- \qquad (415)$$

Although the stability of the adducts decreases with increasing number of methyl groups, there appears to be no disproportionation of one into the

TABLE 5-3

Organodiborane—Lewis Base Compounds

Compound	Reference
$MeB_2H_5 \cdot 2NH_3$	*1651*
$Me_2B_2H_4 \cdot 2NH_3$	*1649, 1651*
$Me_3B_2H_3 \cdot 2NH_3$	*1651*
$Me_4B_2H_2 \cdot 2NH_3$	*1651*
Me(H)B(μ-NMe_2)(μ-H)BH_2 (< −30°)	*551*
$MeHBN(Me)_2 \cdot BH_3$ (> −30°)	*551*
$Me_2NHBH_2(CH_2)_4BH_2NHMe_2$	*1929*
$Me_3NBH_2(CH_2)_4BH_2NMe_3$	*1929*
$^+L_2HB(CH_2)_4BH_3^-$ L = NH_3, $MeNH_2$	*1929*
$(H_3N)_2B^+\ {}^-BH_2$ (bicyclic)	*1929*
(L)HB BH(L) (bicyclic) L = Me_3N, Me_2NH, $MeNH_2$	*1929*

others. Aminodimethylborane is produced by the mild decomposition of the diammoniates of 1,1-dimethyl-, trimethyl-, and tetramethyldiboranes, whereas heating the diammoniates to 200°C in a closed tube affords *B*-methyl derivatives of borazine (*1649*, *1651*).

A simple $Me_2BH \cdot NH_3$ adduct can be obtained from a reaction of ammonia with tetramethyldiborane carried out in ether solution (*1380*); and the direct interaction of Me_2PH with $Me_4B_2H_2$ gives the $Me_2BH \cdot HPMe_2$ adduct (*359*). Both symmetrical and unsymmetrical bridge cleavage with amines occurs with the cyclic organodiboranes (**LXXVIII**) and (**LXXIX**), the latter process favored by the less substituted amines (*1929*).

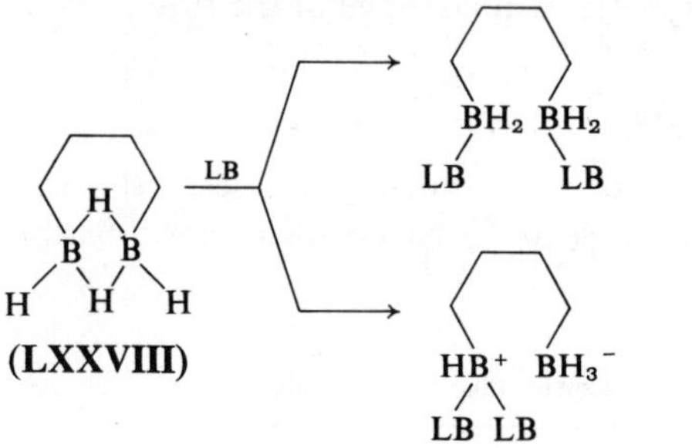

(**LXXVIII**) $\xrightarrow{LB}$ ring-$BH_2(LB)$ $BH_2(LB)$ (416a)

(**LXXVIII**) $\xrightarrow{LB}$ ring-$HB^+(LB)_2$ BH_3^- (416b)

(**LXXIX**) $\xrightarrow{LB}$ $LB \rightarrow BH$ $BH \leftarrow LB$ (417a)

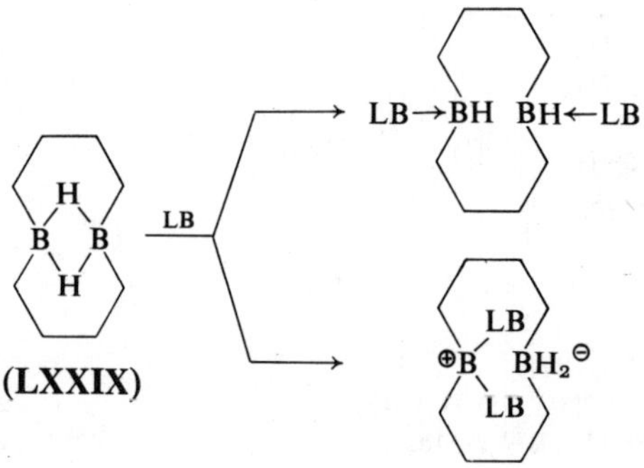

(**LXXIX**) $\xrightarrow{LB}$ $^{\oplus}B(LB)_2$ $BH_2^{\ominus}$ (417b)

Primary and secondary amines react with 1,2-alkyl or aryldiboranes to give amine–borane adducts, which, with appropriate thermal encouragement,

$$\tfrac{1}{2}(CH_3BH_2)_2 + (CH_3)_2NH \xrightarrow{-78^\circ C} CH_3BH_2 \cdot NH(CH_3)_2$$

$$\downarrow \Delta, -H_2$$

$$\underset{\text{gas}}{CH_3BHN(CH_3)_2} \rightleftharpoons \tfrac{1}{2}\underset{\text{liquid}}{[CH_3BHN(CH_3)_2]_2}$$

$$\updownarrow$$

$$\tfrac{1}{2}(CH_3)_2BN(CH_3)_2 + \tfrac{1}{4}[H_2BN(CH_3)_2]_2 \qquad (418)$$

evolve hydrogen and yield the corresponding aminoboranes. Aminoboranes generally exhibit a tendency to disproportionate by reversibly exchanging R (or Ar) for H on boron.

$$\tfrac{1}{2}(C_6H_5BH_2)_2 + (C_2H_5)_2NH \xrightarrow[(1280,\ 1282)]{-30°C} C_6H_5BH_2HN(C_2H_5)_2$$

$$C_6H_5BH_2HN(C_2H_5)_2 \xrightarrow{\text{ca. } 110°C,\ -H_2} C_6H_5BHN(C_2H_5)_2$$

$$C_6H_5BHN(C_2H_5)_2 \longrightarrow \tfrac{1}{2}(C_6H_5)_2BN(C_2H_5)_2 + \tfrac{1}{4}[H_2BN(C_2H_5)_2]_2 \quad (419)$$

The complexes of 1,2-diaryldiboranes with primary amines behave analogously when heated, the only difference being that the alkylaminoborines formed from the symmetrization step are further converted to *N*-trialkylborazines (*1281*).

$$ArBH_2\cdot NH_2R \xrightarrow[-H_2]{\text{ca. } 100°C} [ArBHNHR] \longrightarrow \tfrac{1}{2}Ar_2BNHR + \tfrac{1}{2}H_2BNHR$$

$$\tfrac{1}{2}H_2BNHR \xrightarrow{-\frac{1}{2}H_2} \tfrac{1}{6}(HBNR)_3 \quad (420)$$

On heating the complexes in the presence of excess amine, or with thiols, high yields of bis(alkylamino)arylboranes (*1281*) or (alkylthio)(dialkylamino)-arylboranes (*1282*), respectively, are obtained.

$$ArBH_2\cdot NH_2R + RNH_2 \xrightarrow{50\text{–}150°C} ArB(NHR)_2 + 2\ H_2 \quad (421)$$

$$ArBH_2\cdot NHR_2 + R'SH \xrightarrow{50\text{–}150°C} ArB(SR')(NR_2) + 2\ H_2 \quad (422)$$

Similar chemistry is observed with the primary amine complexes of dialkylboranes (*922, 923, 942, 963, 1240*), for example, Eq. 423.

$$\tfrac{1}{2}\,[\text{(3-methylcyclopentyl)B}(\mu\text{-H})_2\text{B(3-methylcyclopentyl)}] \xrightarrow[\text{ca. } 30°C]{H_2NR} \text{(3-methylcyclopentyl)}BH\cdot NH_2R$$

$$\text{(3-methylcyclopentyl)}BH\cdot NH_2R \xrightarrow{80°C,\ -H_2} \text{(3-methylcyclopentyl)}BNHR$$

$$\text{(3-methylcyclopentyl)}BNHR \xrightarrow[150°]{(\text{cyclopentyl-}BH)_2} \text{(3-methylcyclopentyl)}B\text{—}N(R)\text{—}B\text{(3-methylcyclopentyl)} + H_2 \quad (423)$$

The reactions of 1,2-diphenyldiborane with alkyl (and aryl) diaminoboranes (*922*) and with diamines (*923*) have been investigated.

A general method of obtaining tertiary amine or pyridine complexes of organo-substituted boranes has been developed (*756*), which avoids the necessity of preparing intermediate alkylated diboranes in a separate step (*1876*). This route involves reduction of an appropriately substituted boronate or borinate (or boroxine) with lithium aluminum hydride in the presence of the desired amine (*565, 589, 757, 758, 763*).

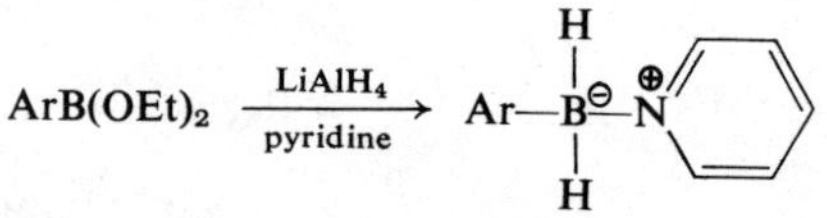

$$ArB(OEt)_2 \xrightarrow[\text{pyridine}]{LiAlH_4} Ar{-}\overset{\ominus}{B}H_2{-}\overset{\oplus}{N}C_5H_5 \tag{424}$$

$$Ar_2B(OEt) \xrightarrow[\text{pyridine}]{LiAlH_4} Ar_2\overset{\ominus}{B}H{-}\overset{\oplus}{N}C_5H_5 \tag{425}$$

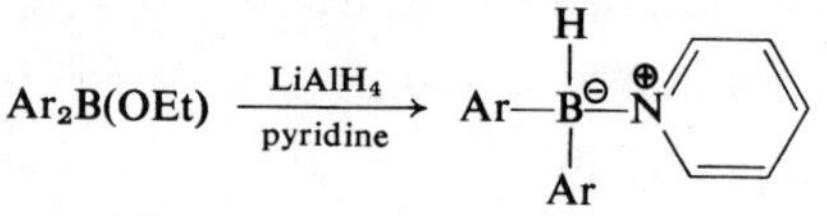

For additional related chemistry see Section 4-2.

Diborane–Borane Equilibrium

Nearly all mono- and dialkylboranes are known as dimeric species in which each boron is tetracoordinated and bonded to the adjacent boron through two bridge hydrogens (Table 5-2). Apparently, the presence of large bulky alkyl groups such as 3-methyl-2-butyl does not lead to any observable dissociation even in the tetraalkyldiboranes (*226*). However, the recently proposed bis(2,3-dimethyl-2-butyl)borane (*1419*), and also the tetraboryl compound (**LXXX**) (*1950*), have been reported to exist in monomeric form.

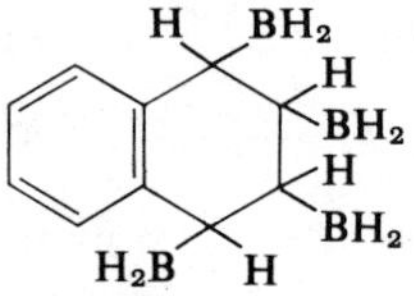

H BH2 H BH2 H BH2 H2B H

(LXXX)

There is no *direct* indication that the dimeric alkylated diboranes are dissociated in tetrahydrofuran (*226*), a solvent that readily dissociates diborane itself. Steric influences may be responsible for the instability of the alkylborane addition compounds relative to those formed by borane. Moreover, the bridge structures in the alkyl-substituted diboranes might be expected, on the basis of compensating polar effects (*226*), to possess stabilities comparable to that of diborane.

The trends in the calculated monomer–dimer equilibria (Eqs. 426–430) in tetrahydrofuran can be rationalized both in terms of a steric effect imposed by the alkyl groups on the extent of complexing with solvent molecules and by a simple electrostatic interaction between $B^{\delta+}$—$H^{\delta-}$ bonds (*1534, 1544*). Electron densities calculated by an extended Hückel method indicate greater positive charge on the boron and greater negative charge on the boron-attached hydrogen of boranes with increasing alkyl substitution (*1534*).

		Equilibrium Constants at 25°C		
		R = *n*-Pr	R = *i*-Pr	
$BH_3 + RBH_2 \rightleftharpoons$	RB_2H_5	0.6	0.6	(426)
$BH_3 + R_2BH \rightleftharpoons$	1,1-$R_2B_2H_4$	0.8	3.2	(427)
$2\ RBH_2 \rightleftharpoons$	1,2-$R_2B_2H_4$	1.0	1.8	(428)
$RBH_2 + R_2BH \rightleftharpoons$	1,1,2-$R_3B_2H_3$	1.4	3.5 (1.8)	(429)
$2\ R_2BH \rightleftharpoons$	$R_4B_2H_2$	2.3	20	(430)

An interesting observation was made some years ago to the effect that bridge bonding occurs only when the hydrogen atoms have a partial negative charge (*1634*). On the basis of the performed calculations a number of known compounds, including alkylated diboranes, fitted well within this scheme. However, the hydrogen atoms of both phenylborane and diphenylborane were given positive charges, implying predominantly monomeric character for these compounds. Since then, phenylborane has been synthesized and found to occur as a dimer (*1277*, *1876*), and a diarylborane, 9-borafluorene (**LXXXI**), has been reported to exist as a colorless dimer, which, however,

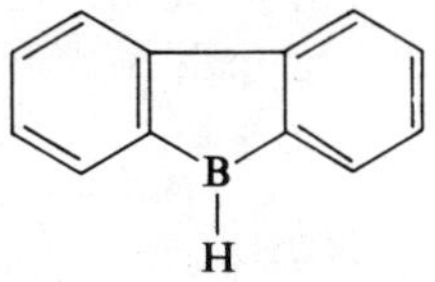

(**LXXXI**)

dissociates into the yellow monomer when a benzene solution is heated to 80°C (*947*).

The bridge-breaking energies for the various methyldiboranes have been estimated to fall in the 25–28.5 kcal range (*1200*); however, these figures may have to be revised upward by about 4–10 kcal (*644*). The rate constant for the combination of two ethylborane molecules is one order of magnitude less than estimated for the combination of two borane molecules (*594*), suggesting that the ethyl substituent either sterically hinders the reaction or lowers the reactivity by inductive or hyperconjugative effects on the vacant *p* orbital of the boron.

The only alkyldiborane capable of geometrical isomerism is the 1,2-dialkyldiborane. Of the two possible forms, the trans configuration should predominate under equilibrium conditions on steric grounds. This has not received experimental verification, and in fact there is some indication that

cis 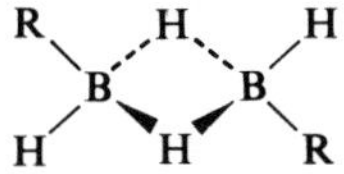　　　　trans

the 1,2-dimethyldiborane exists primarily in the cis configuration (Section 5-3, *Vibrational Spectroscopy*).

Miscellaneous Reactions

Triphenylphosphine oxide can be reduced to triphenylphosphine with either alkyldiboranes or by trialkylboranes. The triorganoboranes requires

$$Ph_3PO + (R_2BH)_2 \xrightarrow{100°C} Ph_3P + R_2BOBR_2 + H_2 \rightarrow R_3B + (RBO)_3 \quad (431)$$

higher temperatures (220°C) than do the organodiboranes, Eq. 431, and therefore probably undergo a dehydroboration (Section 3-2) to R_2BH and olefin before reacting (*969*).

Tetraethyldiborane reacts with benzene at about 200°C to produce triphenylborane in 25% yield (*971*). Similar reactions have been effected intramolecularly in the syntheses of boron heterocycles (*939*, *965*, *970*, *974*, *977*, *978*, *1123*) and probably occur through a concerted four-center transition state mechanism (Section 3-1).

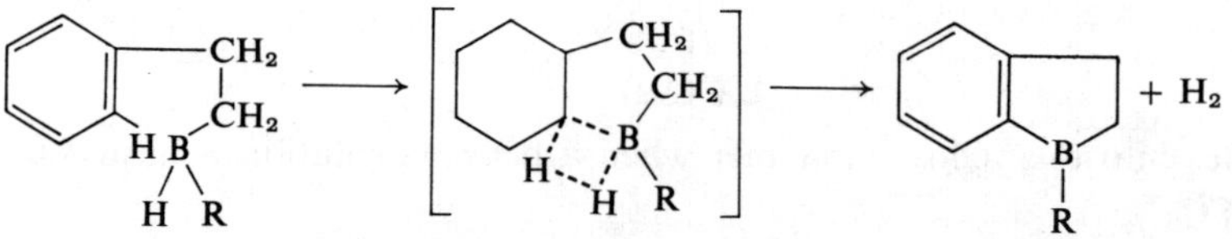

$$C_6H_5CH_2CH_2BH(H)R \longrightarrow [\text{transition state}] \longrightarrow C_6H_4(CH_2CH_2)BR + H_2 \quad (432a)$$

The rearrangement of the α-chloro-substituted alkyldiborane (**LXXXII**) to the *B*-chloro compound (**LXXXIII**) is facilitated by the presence of Lewis

$$(Me_2CHCH(Cl)BH_2)_2 \xrightarrow{THF} 2\ Me_2CHCH_2BH(Cl)\cdot THF \quad (432b)$$

(**LXXXII**) (**LXXXIII**)

acids; and the first-order kinetics with respect to (**LXXXII**) is interpreted in terms of an intramolecular hydride transfer (*1539–1541*).

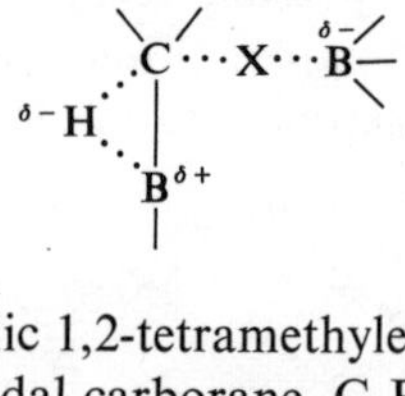

$$\delta^{-}H \cdots C \cdots X \cdots B^{\delta-}, \quad H \cdots B^{\delta+}$$

Flash thermolysis of the cyclic 1,2-tetramethylenediborane produces a low yield of the pentagonal pyramidal carborane, $C_4B_2H_6$ (*706*, *1506*).

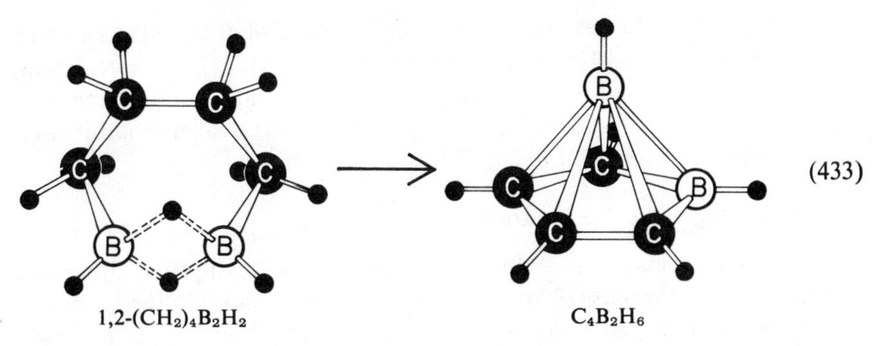

(433)

A reaction of tetramethyldiborane with sodium in ammonia at low temperatures produces the rather interesting salt, $Na_2HB(CH_3)_2$ (*352*, *370*).

$$(CH_3)_4B_2H_2 + 2\,Na + NH_3 \xrightarrow[-75^\circ C]{NH_3} Na_2HB(CH_3)_2 + (CH_3)_2BHNH_3 \quad (434)$$

It is stable as a white solid *in vacuo* up to 90°C; however, it hydrolyzes rapidly and quantitatively to $(CH_3)_2BOH$, $2\,H_2$, and $2\,NaOH$. Dissolved in liquid ammonia, it is yellow, diamagnetic, well ionized, and unstable relative to disproportionation, a reaction that is promoted by tetramethyldiborane. In dimethyl ether, in which it is highly aggregated, Na_2HBMe_2 is a ready source of hydride ion, easily converting chlorosilane to silane. In liquid ammonia the anion of the salt acts as a Lewis base, bonding with trimethylborane to form the compound $Na_2HB(CH_3)_2 \cdot B(CH_3)_3$ which is stable up to 100°C *in vacuo* (*352*). The calcium salt, $CaHB(CH_3)_2NH_3$, prepared in a manner analogous to the sodium salt, curiously does not form an adduct with trimethylborane (*369*, *370*).

5-3. Physical Properties

Structure

An electron diffraction study of tetramethyldiborane indicates that the methyl groups are staggered with respect to the bonds radiating from the boron atoms with a barrier to rotation of ca. 1 kcal mole^{-1}. The B—C bond is 0.012 ± 0.003 Å longer than that found in Me_3B, and the B—B distance

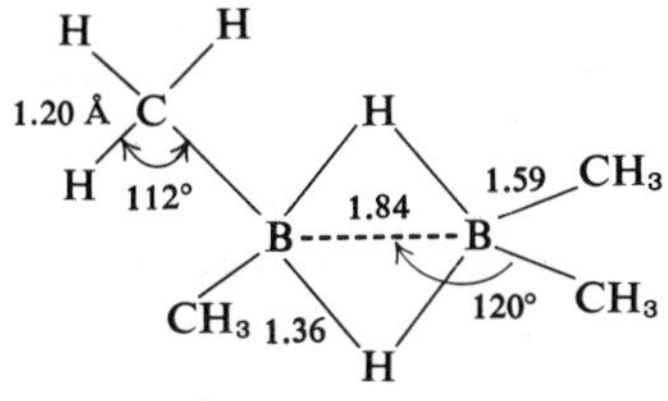

H H
1.20 Å C H
H 112° 1.84 1.59 CH_3
B - - - - - - - - B
CH_3 1.36 120° CH_3
H

is 0.065 ± 0.01 Å longer than that in B_2H_6. Both of these effects can be rationalized in terms of hyperconjugative (Chapter 2) CH_3—B interaction, which should be greater in Me_3B, or in terms of a steric effect in which non-bonded repulsions between ligands decreases with smaller coordination number about boron (*377*).

Nuclear Magnetic Resonance

The ^{11}B nmr chemical shifts (Table 5-4) of the methylated diboranes exhibit two major trends (*1898*). First, alkyl groups tend to shift the resonance absorption of attached boron nuclei to lower field (about 4 ppm per alkyl group). Second, unsymmetrical substitution promotes a more pronounced shift to lower field for the more substituted boron, as well as a shift to higher field for the less substituted boron. Quantitatively, about 6 ppm is added or subtracted when the two parts of the molecule differ by one alkyl group; this value is doubled to about 12 ppm for a difference of two alkyl groups.

Proton nmr analyses of the diboranes B_2H_6, $CH_3B_2H_5$, 1,1-$(CH_3)_2B_2H_4$, 1,2-$(CH_3)_2B_2H_4$, 1,1,2-$(CH_3)_3B_2H_3$, and 1,1,2,2-$(CH_3)_4B_2H_2$ reveal that the bridge hydrogen shifts −0.40 ppm per methyl group on the diborane molecule. Both the terminal proton and the methyl proton chemical shifts are moved significantly upfield with substitution on the far boron atom (*1068*, *1119*). From ^{11}B-decoupled proton spectra, long-range homonuclear proton coupling can be assessed: $J_{H_T—H\mu} = 7.5–8.7$ Hz, $J_{CH_3—H\mu} = 2.4–3.5$ Hz, $J_{CH_3—H_T} = 5.0–5.3$ Hz (*1068*). The partially collapsed H_T—B quartet and the narrowing of H_μ for (**LXXXIV**) is attributed to the boron-11 quadrupole relaxation associated with a polymeric structure (*1119*).

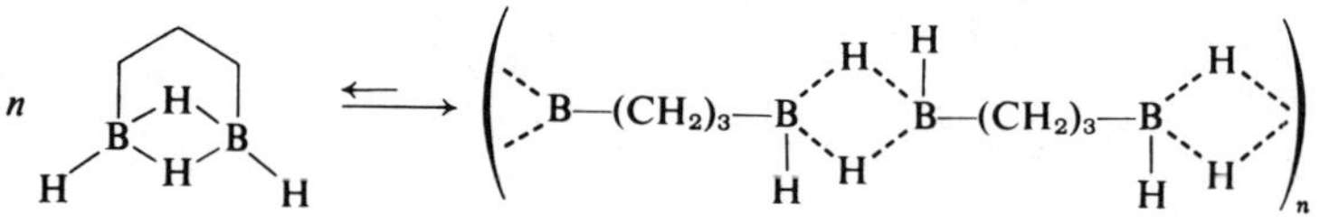

$$n\ \text{(LXXXIV)} \rightleftarrows \left(\text{B—(CH}_2)_3\text{—B(H)(H}_\mu)_2\text{B—(CH}_2)_3\text{—B(H)(H}_\mu)_2 \right)_n \qquad (435)$$

(LXXXIV)

Vibrational Spectroscopy

From the many infrared and Raman studies on alkylated diboranes (*371–373*, *439*, *879a–879c*, *1076–1079*, *1083*, *1085–1088*, *1685*), it is found that BH_2 stretching frequencies are lowered by about 10–15 cm^{-1} for each alkyl group substituted on the other side of the bridge and lowered by about 25 cm^{-1} by an alkyl group attached to the same side. An intense band assigned to the asymmetric in-phase stretching of the BH_2B bridge is located in the 1580–1610 cm^{-1} region of all alkyldiboranes except 1,1-dialkyldiboranes, where it is characteristically lowered by about 50 cm^{-1} (Table 5-5). This general rule is not applicable to strained cyclic systems (*1118*, *1119*, *1866*).

TABLE 5-4

Chemical Shifts and Coupling Constants for Diborane and Methyl Derivatives

	B_2H_6[a]	$CH_3B_2H_5$	1,1-$(CH_3)_2B_2H_4$	1,2-$(CH_3)_2B_2H_4$	1,1,2-$(CH_3)_3B_2H_3$	1,1,2,2-$(CH_3)_4B_2H_2$
$\delta_{^{11}B}$ (ppm)[b]	−16.6	−26.7(1) −8.8(2)	−36.4(1) −3.6(2)	−20.5	−29.2(1) −13.6(2)	−24.8
$J_{^{11}B-H_T}$ (Hz)	128 ± 4	127 ± 5 (all)	125.5	131.2	133.7	N/A
$J_{^{11}B-H_\mu}$ (Hz)	48	41 ± 6(1) 48 ± 5(2)	36.4 49.7	47.5	38.9(1) 45.2(2)	39.7
τ_{H_μ} (ppm)[c]	10.49	10.09	9.55	9.73	9.27	8.90
τ_{H_T} (ppm)[c]	5.99	5.60(1) 6.40(2)	6.74	6.02	6.24	N/A
τ_{CH_3} (ppm)[c]	N/A	9.47	9.48	9.64	9.59(1) or (1′) 9.60(1′) or (1) 9.70(2)	9.66
$J_{H_\mu-^{11}B}$ (Hz)	46.2	~45	?	44.0	40–45 (?)	?
$J_{H_T-^{11}B}$ (Hz)	133.0	129(1) 129(2)	128	129.0	126.0	N/A
$J_{H_T-H_\mu}$ (Hz)	7.5 ± 0.1	?	8.7 ± 0.1	?	8.0 ± 0.1	N/A
$J_{CH_3-H_\mu}$ (Hz)	N/A	3.2	2.4 ± 0.4	?	2.65(1 or 1′) 2.95(1′ or 1) 3.45(2)	3.0
$J_{CH_3-H_T(gem)}$ (Hz)	N/A	5.0	N/A	?	5.3 ± 0.1	N/A

[a] The ^{11}B data from Ref. *1503*.
[b] Relative to $BF_3 \cdot Et_2O$, $\delta = 0$; the first three rows of data are from ^{11}B nmr spectra, the remainder are obtained from proton spectra.
[c] Relative to TMS, $\tau = 10.00$.

TABLE 5-5

Infrared Assignments for Methylated Diboranes

Compound	Selected infrared frequencies	
	B—H$_T$ stretch (cm^{-1})	Asym in-phase B—H$_B$ stretch (cm^{-1})
$(CH_3)(H)\overset{1}{B}(\mu\text{-}H)_2\overset{2}{B}H_2$	Sym 2513 Asym 2571	1592
$(CH_3)_2B(\mu\text{-}H)_2BH_2$	Sym 2494 Asym 2571	1546
$(CH_3)(H)B(\mu\text{-}H)_2B(H)(CH_3)$ (cis and/or trans)	2519	1610
$(CH_3)_2B(\mu\text{-}H)_2B(CH_3)(H)$	2506	1605
$(CH_3)_2B(\mu\text{-}H)_2B(CH_3)_2$	—	1605

In addition to the above trends, other characteristic absorptions found for the series $Me_nB_2H_{6-n}$ (n = 1–4) include peaks at around 2950 cm^{-1} asymmetric methyl stretching vibration), 2900 (symmetric methyl stretch), 1970–2150 (symmetric bridge hydrogen–boron stretch), 1420–1460 (asymmetric methyl deformation), 1320 (symmetric methyl deformation), and 500–700 cm^{-1} (B—B stretch).

The sharpness of the Raman lines in the methyldiboranes as compared with those in trimethylborane does support the contention that there is less internal rotation of the methyl group in the former than in the latter (*373*). A vibrational analysis of 1,2-$Me_2B_2H_4$ leads to the conclusion that the observed spectrum can be attributed to the cis isomer (*879b*, *1086*).

Mass Spectrometry

The mass spectra of isotopically labeled methyl- and ethyldiboranes have been studied and, as found for trialkylboranes (Section 2-6), prominent peaks are exhibited at masses corresponding to R_2B^+ ions (*1683*, *1901a*).

Chapter 6

Other Organopolyboranes

6-1. Carbon Monoxide–Triborane

The displacement of dimethyl ether from ether–triborane(7) with carbon monoxide is facilitated with a Lewis acid such as boron trifluoride.

$$Me_2OB_3H_7 + CO + BF_3 \longrightarrow B_3H_7CO + Me_2OBF_3 \qquad (436)$$

The structure of the product, carbon monoxide–triborane, is believed to be that depicted by (**LXXXV**) from ^{11}B nmr data (*1522*), with a three-center, two-electron bond joining the three-boron atoms.

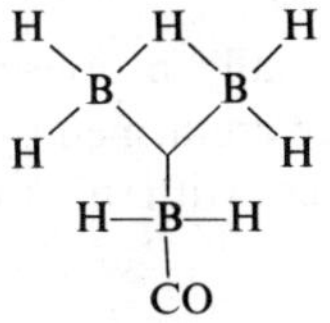

```
H     H     H
 \   / \   /
  B       B
 /   \ /   \
H     |     H
   H—B—H
      |
     CO
```

(**LXXXV**)

6-2. Organotetraboranes

Alkyltetraborane(10) Compounds

Methyl-substituted tetraborane(10) derivatives (Table 6-1) have been obtained, variously, by an exchange reaction between tetraborane and monomethyldiborane (or dimethyldiborane) (Eq. 437) (*1142*), by alkyl–hydrogen exchange between dimethylmercury and B_4H_{10} (*1363*), by boron insertion into a triboro compound (Eqs. 438–439) (*515*, *636*), and by insertion of

TABLE 6-1

Tetraborane Derivatives

Compound	Reference	Compound	Reference
2-MeB_4H_9	*513a, 514, 515, 1142, 1363*	B_4H_8CO	*155, 357, 1731*
		$B_4H_8CONH_2^-$	*1525*
1,2-$Me_2B_4H_8$	*514, 515*	$Me_3NB_4H_8CO$	*1731*
2,2-$Me_2B_4H_8$	*514, 515, 636*	$Me_2OB_4H_8CO$	*1731*
2,4-$Me_2B_4H_8$	*514, 515*	$(C_2H_4)_4B_4H_8CO$	*1731*
$(CH_2)_2B_4H_8$	*714, 1899*		

1,2-dimethyldiborane into bis(trifluorophosphine)diborane(4) (*514*). The last two mentioned procedures lead to isomeric 1,2-, 2,2-, and 2,4-dimethyl derivatives.

$$B_4H_{10} + MeB_2H_5 \longrightarrow 2\text{-}MeB_4H_9 + B_2H_6 \quad (437)$$

$$Me_2BCl + NaB_3H_8 \longrightarrow 2,2\text{-}Me_2B_4H_8 + NaCl \quad (438)$$

$$\tfrac{1}{2}(MeBH_2)_2 + Me_2OB_3H_7 \longrightarrow 2\text{-}MeB_4H_9 \quad (439)$$

Attempts to prepare ethyltetraborane through an exchange between ethyldiboranes and tetraborane (*1724*) or from ethylene and tetraborane have been unsuccessful. However, a reaction between the latter two compounds in a hot–cold reactor has been reported to give 2,4-dimethylenetetraborane (**LXXXVI**) in good yield (*741*).

$$C_2H_4 + B_4H_{10} \xrightarrow{100°C} (CH_2)_2B_4H_8 + H_2 \quad (440)$$

The cyclic bridge structure originally proposed for this compound on the basis of infrared evidence has been confirmed by ^{11}B and ^{1}H nuclear magnetic resonance studies (*1684*). Consistent with this structural assignment, oxidative

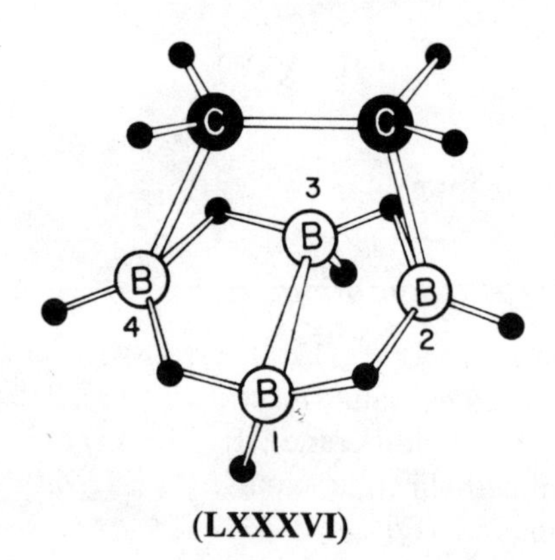

(LXXXVI)

hydrolysis and methanolysis of 2,4-dimethylenetetraborane produces ethylene glycol and 1,2-bis(dimethoxybora)ethane, respectively (*741*).

$$(CH_2)_2B_4H_8 + 12\ H_2O \longrightarrow HOCH_2CH_2OH + 4\ B(OH)_3 + 9\ H_2 \qquad (441)$$

$$(CH_2)_2B_4H_8 + 10\ CH_3OH \longrightarrow (CH_3O)_2BCH_2CH_2B(OCH_3)_2 + 2\ B(OCH_3)_3 + 9\ H_2 \qquad (442)$$

Support for the proposed mechanism of 2,4-dimethylenetetraborane formation, Eq. 443, is obtained from deuterium isotope studies (*1899*):

$$B_4H_{10} \xrightarrow{-H_2} [B_4H_8] \xrightarrow{C_2D_4} (CD_2)_2B_4H_8 \qquad (443)$$

Symmetrical bridge hydrogen cleavage of $(CH_2)_2B_4H_8$ with 2 moles of trimethylphosphine occurs with the formation of $Me_3P \rightarrow BH_2—CH_2CH_2—B_3H_6 \leftarrow PMe_3$ (*149*).

Tetraborane(8)carbonyl

The compound tetraborane(8)carbonyl, B_4H_8CO, is prepared by the action of carbon monoxide on either B_5H_{11} (*357*, *1731*) or B_4H_{10} (*155*). The latter reaction is first order in tetraborane(10) and zero order with respect to carbon monoxide, suggesting a slow step, Eq. 444, followed by a fast combination of the resulting B_4H_8 with carbon monoxide, Eq. 445 (*155*, *1731*). Similarly,

$$B_4H_{10} \longrightarrow B_4H_8 + H_2 \quad \text{(slow)} \qquad (444)$$

$$B_4H_8 + CO \longrightarrow B_4H_8CO \quad \text{(fast)} \qquad (445)$$

the reaction of carbon monoxide with pentaborane(11) probably involves a slow cleavage Eq. 446, with subsequent formation of the appropriate carbonyl adducts of the intermediate boron hydride species, Eqs. 447–448.

$$B_5H_{11} \longrightarrow B_4H_8 + BH_3 \quad \text{(slow)} \qquad (446)$$

$$B_4H_8 + CO \longrightarrow B_4H_8CO \quad \text{(fast)} \qquad (447)$$

$$BH_3 + CO \longrightarrow BH_3CO \quad \text{(fast)} \qquad (448)$$

The reversibility of these reactions is evidenced by the high yields of B_4H_{10} and B_5H_{11} when H_2 or B_2H_6 reacts with B_4H_8CO (*1731*).

The infrared spectrum of B_4H_8CO includes a carbon–oxygen stretching band very similar to that of BH_3CO, suggesting a parallel kind of B—C—O bonding. The structure of B_4H_8CO is probably related to $B_4H_8PF_2NMe_2$ (*1050*) to which it can be converted by CO displacement with Me_2NPF_2 (*1766*) and also to $B_4H_8PF_3$, which is obtained by a reversible displacement of B_4H_8CO with PF_3 (*1643a*, *1731*).

```
            H
            |
      H   H-3B-H   H
       \  |  |  |  /
          B  |  B
      H  2\  | /4  H
            B
          /  1 \
         H      CO
```

The carbonyl group most likely lies in the perpendicular mirror plane through B(1) and B(3) and bisecting the line B(2) and B(4). Support for this structure is obtained from boron-11 nmr studies (*1454*, *1731*). A molecular beam mass spectrometric study (827) on this compound indicates that the fragmentation pattern parallels the known chemistry (*1731*) in that the B—C bond is rather easily broken.

A number of basic reagents attack B_4H_8CO to form nonvolatile materials without liberation of carbon monoxide (*1525*, *1731*).

$$B_4H_8CO + (CH_3)_3N \longrightarrow (CH_3)_3NB_4H_8CO \tag{449}$$

$$B_4H_8CO + (CH_3)_2O \longrightarrow (CH_3)_2OB_4H_8CO \tag{450}$$

$$B_4H_8CO + x\,H_2O \longrightarrow 8.8\,H_2 + y\,B(OH)_3 + \text{residue} \tag{451}$$

$$B_4H_8CO + 2\,NH_3 \longrightarrow B_4H_8CO\cdot 2\,NH_3 \tag{452}$$

On the basis of largely chemical evidence, the structure of $B_4H_8CO\cdot 2NH_3$ appears to be analogous to that of $BH_3CO\cdot 2NH_3$ (see Section 4-4) (*152*).

$$[NH_4]^+ \left[\begin{array}{c} H_2N \diagdown \;\; \diagup\!\!\!\diagup O \\ C \\ | \\ B_4H_8 \end{array} \right]^-$$

It is interesting that ethylene does not displace carbon monoxide from tetraborane(8)carbonyl with formation of $B_4H_8C_2H_4$. Instead, the two reagents combined in a ratio of 4:1 to form $(C_2H_4)_4B_4H_8CO$ of undetermined structure (*1731*).

6-3. Organopentaborane(9) Compounds

Synthesis

Friedel–Crafts-Type Alkylation

Nearly all substitutions at the apex of pentaborane have been effected using conditions generally reminiscent of aromatic electrophilic substitution reactions. Alkylation of the apical boron atom in pentaborane(9) occurs with olefins or alkyl halides in the presence of a Lewis acid catalyst such as aluminum chloride (*30*, *135*, *599*, *1500*, *1628*, *1815*), Eq. 453. Alternatively, trialkylborates (*1896*), alkyl ethers (*1029*, *1030*), and alkyl silicates (*1031*) have been utilized as alkylating agents. Multiple alkylation of pentaborane does not occur under the conditions of this modified Friedel–Crafts reaction (*1628*) even when drastic conditions are imposed (*1502*). This can be considered as supporting evidence that the acid-catalyzed alkylation reaction results in substitution at the chemically unique apical boron atom. However,

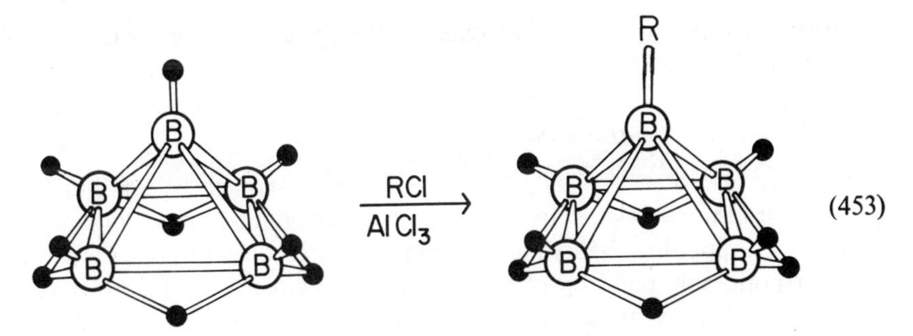

(453)

the site of substitution is unambiguously established by ^{11}B and ^{1}H nuclear resonance (*599*).

Other known acid-catalyzed pentaborane substitution reactions also result in apically substituted pentaboranes (*31*, *599*, *1505*). This is in agreement with the predicted site of electrophilic substitution based on charge distributions as determined from valence bond and molecular orbital treatments (*1759*). Furthermore, the position of observed attachment of boron on the carbon skeleton (i.e., the most substituted carbon when unsymmetrical alkenes are used) is consistent with this predicted mechanism (*1628*). The position of boron attachment is kinetically rather than thermodynamically controlled, for basal substituted alkylpentaboranes are found to be about 3 kcal mole^{-1} more stable than the corresponding apically substituted pentaboranes (*744*).

Synthesis from Unsaturated Compounds

The reaction of pentaborane at elevated temperatures with olefins in the

$$B_5H_9 + CH_2{=}CH{-}CH_2CH_3 \xrightarrow{150°C} 2\text{-}(CH_3CH_2CH_2CH_2)B_5H_8 \qquad (454)$$

$$1\text{-}RB_5H_8 \xrightarrow[200°C]{\text{Lewis base, room temp.}} 2\text{-}RB_5H_8 \qquad (455)$$

$$\begin{matrix} CH_2{=}CH_2 + B_2H_6 \\ \text{or} \\ R_3B + B_2H_6 \end{matrix} \xrightarrow{\text{heat}} \begin{matrix} \text{Complex mixtures containing} \\ \text{monoalkylated and} \\ \text{polyalkylated pentaboranes} \end{matrix} \qquad (456)$$

absence of a catalyst (*1630*) gives good conversions but poor yields of the appropriate 2-alkylpentaboranes, Eq. 454. The position of attachment of the boron atom on the carbon skeleton (least substituted when unsymmetrical olefins are used), as well as the basal substitution on the pentaborane pyramid, are in accord with a proposed mechanism of nucleophilic attack by the olefin (*1630*). The small positive charge assigned to each basal boron atom in pentaborane from valence bond and molecular orbital treatments (*1759*), in contrast to the negative charge assigned to the apical boron, gives theoretical support to this mechanism.

Base-Catalyzed Rearrangement

With the use of a sterically hindered Lewis base such as 2,6-dimethylpyridine or hexamethylenetetramine, a rearrangement of 1-alkylpentaboranes to the corresponding 2-alkylpentaboranes is effected at room temperature without noticeable build-up of an intermediate adduct, Eq. 455 (*355*, *1496*, *1815*). On the other hand, use of a relatively strong base such as trimethylamine reportedly converts the 1-alkylpentaborane to an intermediate salt,

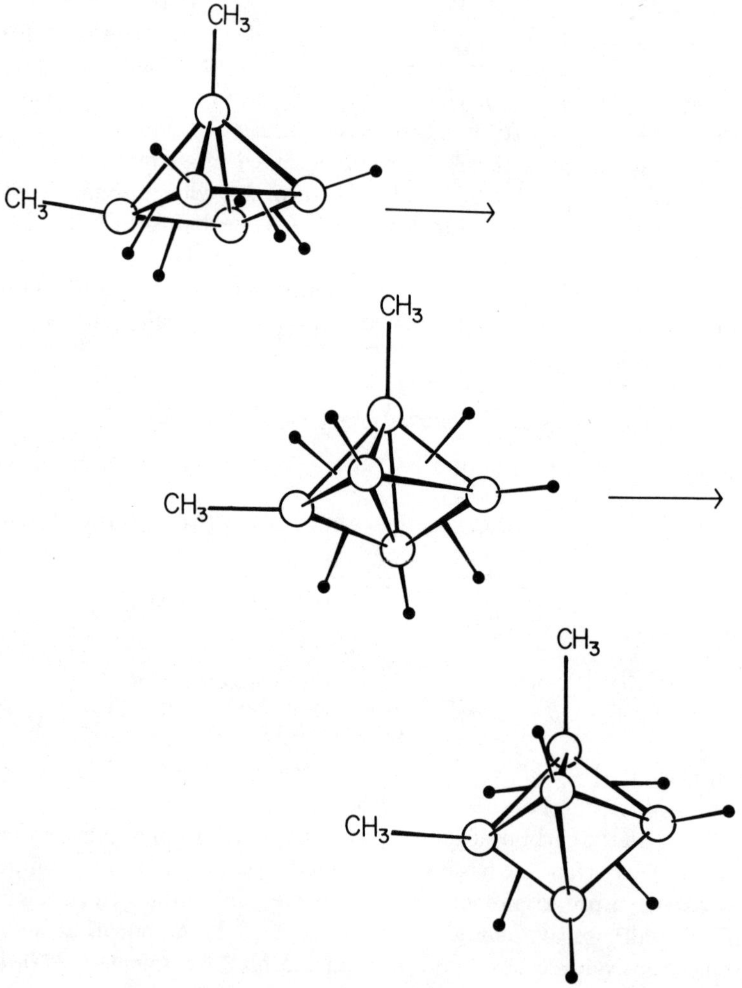

CH_3 CH_3 ⟶ CH_3 CH_3 ⟶ CH_3 CH_3 (457)

$[Me_3NH]^+[RB_5H_7]^-$, which then yields the more stable 2-alkylpentaborane isomer upon treatment with an appropriate acid (*841*). But on the basis of more recent investigations it now appears doubtful that this intermediate $RB_5H_7^-$ anionic species is solely responsible for the apparent base-catalyzed rearrangement of alkylpentaboranes (*167*, *905*, *1500*) and that adduct formation is a reasonable competing mechanism. Product distribution analyses of the base-catalyzed rearrangements of polysubstituted pentaboranes (*1815*) lends experimental support for a minimum atomic motion mechanism (*625*, *1121*, *1501*), which involves only minor boron skeletal movement and the migration of adjacent bridge hydrogens at a particular basal boron atom (incipient apex) to positions about the apical (incipient basal) boron.

A schematic diagram that summarizes the preparation of various methyl derivatives of pentaborane is shown in Fig. 6-1 (*1815*). Of particular interest is the base-catalyzed rearrangement of 1,2-Me_2-3-ClB_5H_6, which gives a mixture of dimethylchloro products. It is not surprising to find (**LXXXVII**) and (**LXXXVIII**) produced in greater yield than (**LXXXIX**), for a basal preference of substituents has already been observed in the monosubstituted series. The isomer (**XC**) is produced in very low yield, and 2-Cl-1,4-$Me_2B_5H_6$ is absent altogether. Neither of these two isomers are allowed from a single rearrangement step proceeding through the minimum atomic motion mechanism. If this mechanistic scheme is hypothetically extended to include further rearrangement of (**LXXXVII**), (**LXXXVIII**), and (**LXXXIX**), compound (**LXXXVII**) is a potential precursor to the low yield isomer (**XC**), and (**LXXXVIII**) would be a precursor to the absent 2-Cl-1,4-$Me_2B_5H_6$. The latter two rearrangements involve a net uphill energy change based on the substituent positional preference for the base of the pyramid and would account for the observed results.

Miscellaneous Preparative Reactions

An intermolecular hydrogen exchange presumably accompanies, or perhaps is responsible for, the thermal isomerization of 1-alkylpentaboranes to the corresponding 2-isomer, Eq. 455 (*1502*, *1502a*). Pyrolysis of ethylene–diborane and trimethylborane–diborane mixtures (*137*) yields 2-alkylpentaboranes, as well as mixtures of polyalkylated pentaboranes and alkylated decaboranes. The apically substituted monomethyl derivative of pentaborane(9) is a product of ether cleavage of 1-bromopentaborane (*355*). Apparently, dimethyl

$$\text{1-BrB}_5\text{H}_8 \xrightarrow{\text{Me}_2\text{O}} \text{1-MeB}_5\text{H}_8 \qquad (458)$$

ether is too weak to act as a rearrangement catalyst, for no 2-methyl isomer is detected. Pentaborane(9) reacts readily with NaCN to form a substituted pentaborate salt, $B_5H_9CN^-$ (*23*).

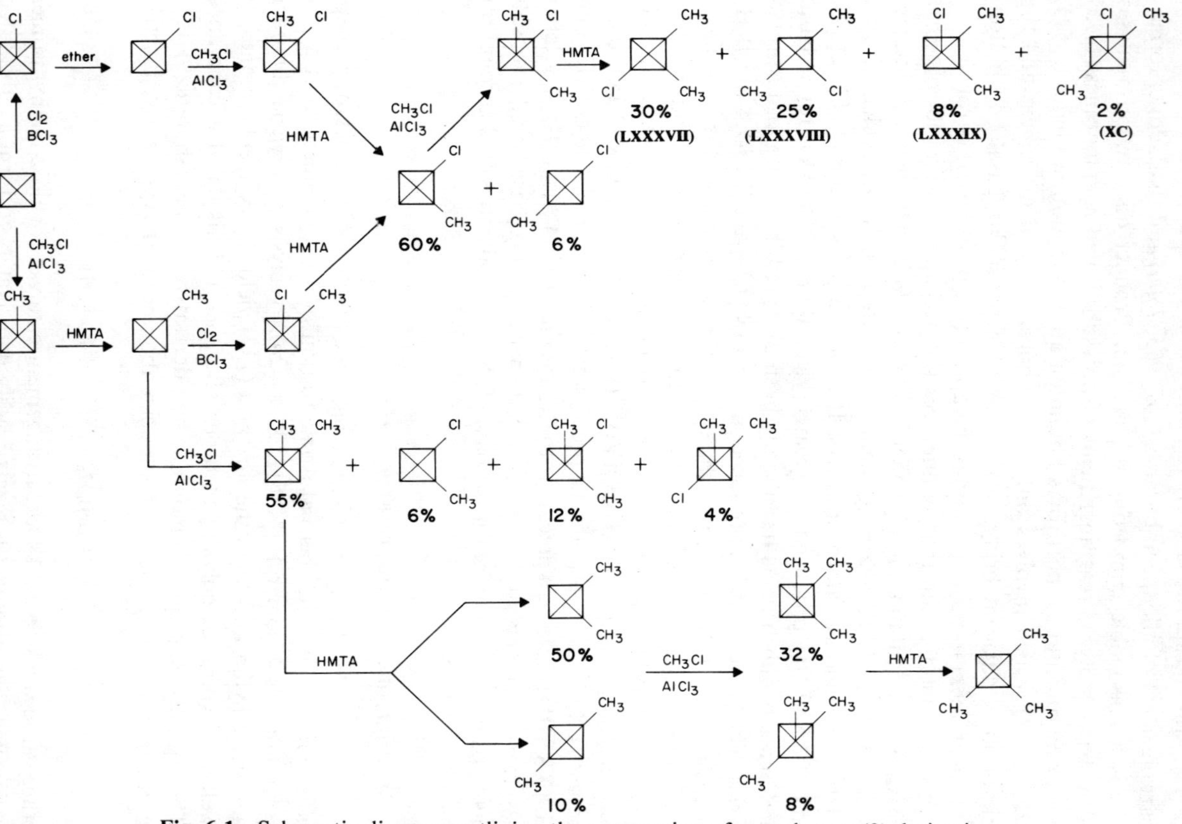

Fig. 6-1. Schematic diagram outlining the preparation of pentaborane(9) derivatives.

TABLE 6-2

Pentaborane(9) Derivatives

Compound	Reference	Compound	Reference
1-MeB_5H_8	*355, 599, 905, 1028, 1141a, 1496, 1500, 1628, 1815*	1-Me-2-ClB_5H_7	*1815*
2-MeB_5H_8	*355, 905, 1028, 1496, 1500, 1502, 1815*	2-Me-1-ClB_5H_7	*1815*
1-EtB_5H_8	*135, 599, 1141a, 1496, 1500, 1628*	2-Me-3-ClB_5H_7	*1815*
2-EtB_5H_8	*549, 599, 1496, 1500, 1502, 1628, 1630*	2-Me-4-ClB_5H_7	*1815*
1-*i*-PrB_5H_8	*1141a, 1628*	2-Me-1-BrB_5H_7	*1500, 1815*
1-*s*-BuB_5H_8	*1628*	2-Me-3-BrB_5H_7	*1815*
2-*n*-BuB_5H_8	*1630*	2-Me-4-BrB_5H_7	*1815*
2-*s*-BuB_5H_8	*1630*	1,2,3-$Me_3B_5H_6$	*1500, 1815*
2-*i*-BuB_5H_8	*1630*	1,2,4-$Me_3B_5H_6$	*1815*
1-$Cl_2BCH_2B_5H_8$	*31*	2,3,4-$Me_3B_5H_6$	*1815*
1-$B_5H_8CH_2B_5H_8$	*31*	1,2-Me_2-3-ClB_5H_6	*1815*
1,2-$Me_2B_5H_7$	*1500, 1502, 1814, 1815*	1,2-Me_2-4-ClB_5H_6	*1815*
2,3-$Me_2B_5H_7$	*1500, 1501, 1815*	2,3-Me_2-1-ClB_5H_6	*1815*
2,4-$Me_2B_5H_7$	*1815*	2,4-Me_2-1-ClB_5H_6	*1815*
1-Me-2-*s*-BuB_5H_7	*1628*	2,3-Me_2-4-ClB_5H_6	*1815*
		1,2,3,4-$Me_4B_5H_5$	*1067*
		μ-(Me_2B)-B_5H_8	*637*

The formation of carboranes from the reaction of pentaborane with acetylenes is reviewed in detail elsewhere (*696a*, *1497*). Known alkylpentaboranes are listed in Table 6-2.

Reactions

Thermal and base-catalyzed rearrangements, as well as Lewis-base adduct chemistry, of alkylpentaboranes are discussed in the above section.

The alkylpentaboranes are sensitive to air oxidation and to hydrolysis, the derivatives with smaller alkyl groups resembling pentaborane(9) more closely in this respect (*1628*, *1830*).

Bridge deprotonation of 1- and 2-$CH_3B_5H_8$ by potassium hydride gives the mononegative ions 1-$CH_3B_5H_7^-$ and 2-$CH_3B_5H_7^-$, respectively (*167*). Low-temperature pmr studies indicate a partial tautomeric quenching of the bridging protons in 2-$CH_3B_5H_7^-$, but not in the case of the apically substituted anion (*167*).

Flash thermolysis or electric discharge of 1-alkylpentaboranes can give monocarbon carboranes in low yield (*706*, *1498*, *1504*).

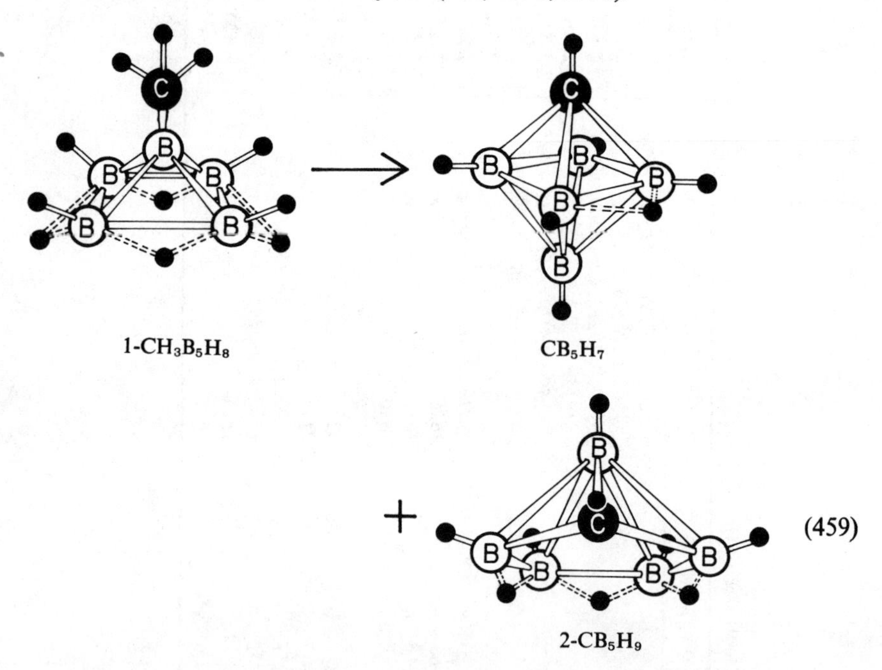

The modestly high-yield conversion of 1,2-dimethylpentaborane(9) to 1,5-dicarbaclosopentaborane(5) is accompanied by the production of many other carboranes. Together, the combined yield of carborane products

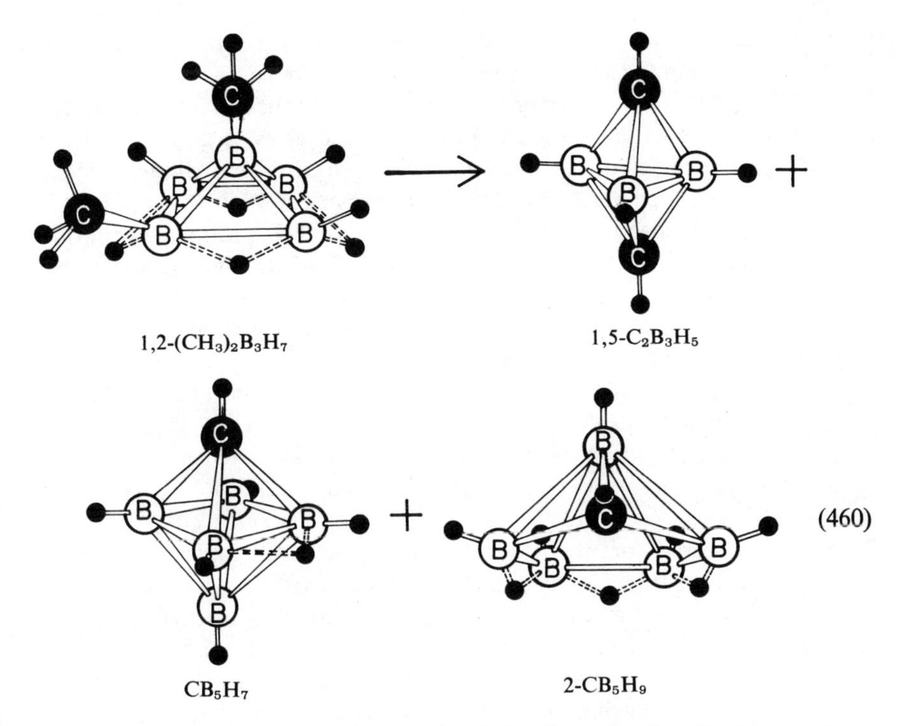

approaches 40%, which is strikingly high when the drastic temperatures needed for the conversion are considered. The formation of a two-boron-containing product, diborane, as the second most important product to the three-boron carborane does tend to suggest that a significant fraction of the pyrolysis can be summarized by Eq. 461.

$$1,2\text{-}(CH_3)_2B_5H_7 \longrightarrow 1,5\text{-}C_2B_3H_5 + B_2H_6 + H_2 \qquad (461)$$

A boron–boron exchange has been observed on treating the 1:2 adduct of 1-methylpentaborane with trimethylamine with ^{10}B-enriched diborane. The more stable 2-methylpentaborane is liberated and exhibits ^{10}B enrichment in all positions except B-2 (*1499*).

Physical Properties

An X-ray diffraction study of $2,3\text{-}Me_2B_5H_7$ gives a B—C bond distance of 1.55 Å (*625*, *1501*) whereas a longer such bond, 1.62 Å, is suspected for the monosubstituted $1\text{-}MeB_5H_8$ (*426*). In the latter study, a dipole moment of 1.93 ± 0.05 D is found for this methyl derivative; this may be compared with 3.3 D in the parent pentaborane. The microwave data for $1\text{-}MeB_5H_8$ are consistent with a model having free internal rotation and vibration–

internal–rotation interactions. Lack of barrier perturbations leads to an upper limit of 1 cm^{-1} for the twelvefold barrier (*426*).

In both ^{11}B and proton magnetic resonance chemical shift studies, the effect of alkyl substitution on the pentaborane cage is largely felt by the contiguously attached boron atoms and neighboring bridge hydrogens. Thus, an apically substituted boron is shifted ca. 8–9 ppm downfield, basal substitution shifts the attached boron ca. 14–15 ppm downfield, and the neighboring bridge hydrogens ca. 0.6 ppm downfield (*1496*, *1814*, *1815*). The effects are additive for polysubstituted derivatives. In this regard, the ^{11}B nmr data for the alkylated pentaboranes obtained from the ethylene–diborane reaction (*137*) should be reinterpreted in favor of a downfield chemical shift effect of the alkyl group. The methylene protons of 1-ethylpentaborane are located 0.23 ppm to higher field than the methyl protons of the ethyl group (*1067*), which may be a reflection of the negative charge associated with the apex boron of the pentaborane pyramid. In this same compound, $J_{B-H_{Me}} = 5.6$ Hz is slightly smaller than $J_{B-H_{CH_2}} = 6.8$ Hz. About the same coupling, 5.4–7.5 Hz, is observed for methyl proton–apical boron(11) spin interactions in 1-methylpentaborane and several derivatives (*1067*). Also, long-range coupling of ca. 0.8 Hz is found between apically attached methyl hydrogens and basal borons. Coupling between a basal attached methyl hydrogen and neighboring bridge hydrogens falls between 2.9 and 3.3 Hz (*1067*).

Although charge distribution and some other effects may account for the chemical shift difference, $\Delta\delta_{apex-base} = 0.35$ ppm, between apically situated methyl hydrogens and their basal counterparts (*1500*, *1814*), it is also possible to account for this observation by adopting a classic free-electron model with current loops parallel to the base plane of the pyramidal boron framework (*1161*). Boron(11)–carbon(13) coupling of 73 Hz has been observed for 1-MeB_5H_8 (*580*) and corresponds to sp^3-hybridized carbon bonded to approximately sp^2-hybridized boron.

Mass spectra of several alkyl derivatives of pentaborane(9) have been reported (*548*, *1397*); and appearance potentials of these compounds indicate lower ionization energies for the apical substituted alkylpentaboranes than for the basal isomers (*1397*). From electron impact methods and extended Hückel calculations, the 2-MeB_5H_8 isomer is the more thermodynamically stable species (*1398*).

6-4. Organopentaborane(11) Compounds

Reversible exchange at ambient temperatures between pentaborane(11) and monoalkyldiboranes, or 1,2-dialkyldiboranes, provides a convenient

TABLE 6-3

Pentaborane(11) Derivatives

Compound	Reference
2-MeB_5H_{10}	*1142*
2,5-$Me_2B_5H_9$	*1142*
2-EtB_5H_{10}	*1146, 1724*
2,5-$Et_2B_5H_9$	*1146, 1724*

method for obtaining monoalkyl and dialkylpentaborane(11) derivatives (Table 6-3) (*1142*, *1724*).

$$B_5H_{11} + 1{,}2\text{-}R_2B_2H_4 \rightleftharpoons RB_5H_{10} + RB_2H_5 \quad (462)$$

$$RB_5H_{10} + 1{,}2\text{-}R_2B_2H_4 \rightleftharpoons R_2B_5H_9 + RB_2H_5 \quad (463)$$

Other methods for alkylpentaborane(11) preparation include the elevated temperature (100°C) reaction between tetraborane and a mixture of methylated diboranes (*1142*) and a direct interaction of ethylene with pentaborane(11) (*1146*). The latter reaction gives dimethylenetetraborane [Section 6-2, *Alkyltetraborane(10) Compounds*] as well as monoethyldiborane and ethylpentaborane(11). A plausible mechanism for the formation of all products involves the initial cleavage of pentaborane(11):

$$B_5H_{11} \rightleftharpoons [BH_3] + [B_4H_8] \quad (463a)$$

$$C_2H_4 + [BH_3] \rightleftharpoons [C_2H_5BH_2] \quad (463b)$$

$$[C_2H_5BH_2] + [B_4H_8] \rightleftharpoons C_2H_5B_5H_{10} \quad (463c)$$

$$C_2H_4 + [B_4H_8] \rightleftharpoons (CH_2)_2B_4H_8 \quad (463d)$$

$$[C_2H_5BH_2] + [BH_3] \rightleftharpoons C_2H_5B_2H_5 \quad (463e)$$

From ^1H nmr and infrared evidence (*1142*, *1146*), the monoalkylated product is most probably 2-alkylpentaborane(11). The presence of two methyl resonances in the ^1H nmr of 2-methylpentaborane(11) is attributed to a mixture of the possible exo and endo forms, (**XCI**) and (**XCII**) (*1142*, *1161*).

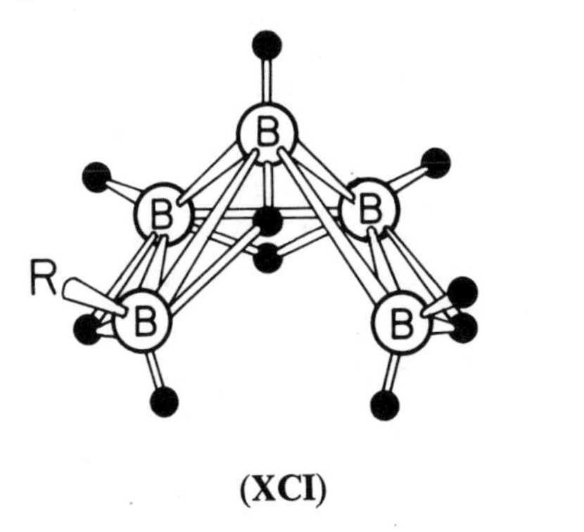

(XCI)

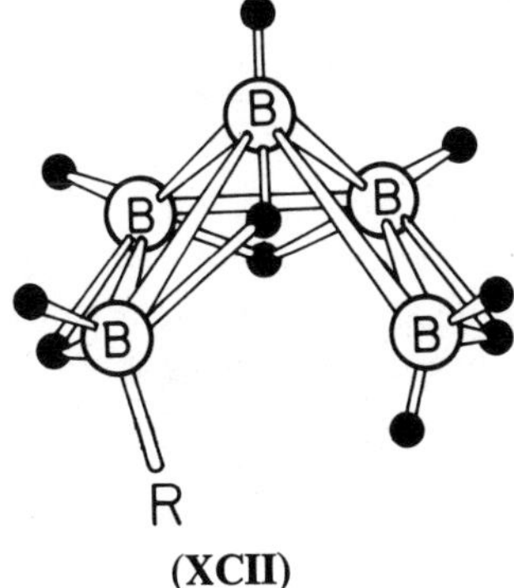

(XCII)

Although the structure of the dialkyl derivative of pentaborane(11) has not been unambiguously determined, it is believed that the two alkyl groups are attached to different boron atoms. This assignment is partially supported by the inability of B_5H_{11} to exchange with 1,1-diethyldiborane. However, this unobserved exchange may involve an unfavorable rate rather than an unfavorable equilibrium (*1724*).

Products that can be predicted on the basis of the above equilibria Eqs. 462–463 and from other known pentaborane(11) chemistry have been isolated from decomposition studies on the alkylated pentaboranes(11). One such decomposition gave evidence for formation of a trialkylated pentaborane(11) (*1142*).

6-5. Organohexaborane(10) Compounds

Alkyl derivatives of hexaborane(10) (Table 6-4) have been successfully prepared by borane insertion into the pentaborane(9) cage. Lithium octahydropentaborate(1−) reacts with dimethyl boron chloride at low temperatures to produce μ-dimethylborylpentaborane(9), in which the dimethylboryl moiety occupies a bridging position between two boron atoms in the base of the B_5H_9 tetragonal pyramidal framework. In the presence of dimethyl

TABLE 6-4

Hexaborane(10) Derivatives

Compound	Reference
$3\text{-}CH_3B_6H_9$	*166, 863, 863a*
$4,5\text{-}(CH_3)_2B_6H_8$	*637*

$$Me_2BCl + LiB_5H_8 \longrightarrow \mu\text{-}(Me_2B)B_5H_8 + LiCl \xrightarrow{\text{ether}} 4,5\text{-}Me_2B_6H_8 \quad (464)$$

ether, this compound isomerizes to 4,5-dimethylhexaborane(10) (*637*). In a presumably analogous reaction, 2-methylhexaborane(10) has been reportedly prepared (*166*). Protonation of the basal 4,5-boron–boron bond of $2\text{-}MeB_6H_9$ (*863a*) has been established by boron-11 and proton nmr (*863*).

$$2\text{-}MeB_6H_9 + HCl + BCl_3 \longrightarrow [2\text{-}MeB_6H_{10}^+][BCl_4^-] \quad (465)$$

6-6. Organodecaborane(14) Compounds

Synthesis

In contrast to the single apical alkylation of pentaborane(9), Friedel–Crafts alkylation of decaborane(14) with alkyl halides or olefins yields a mixture of derivatives (*135*, *462*, *1900*).

$$B_{10}H_{14} \xrightarrow[AlCl_3]{CH_2Br} \underset{5\%}{1\text{-}MeB_{10}H_{13}} + \underset{15\%}{2\text{-}MeB_{10}H_{13}} + \underset{19\%}{1,2\text{-}Me_2B_{10}H_{12}} + \underset{13\%}{2,4\text{-}Me_2B_{10}H_{12}}$$
$$+ \underset{5\%}{1,2,3\text{-}Me_3B_{10}H_{11}} + \underset{17\%}{1,2,4\text{-}Me_3B_{10}H_{11}} + \underset{22\%}{1,2,3,4\text{-}Me_4B_{10}H_{10}}$$
$$+ \underset{3\%}{1,2,3,5(\text{or } 8)\text{-}Me_4B_{10}H_{10}} \qquad (466)$$

The preferred boron sites of this presumed electrophilic substitution are, however, in accordance with the regions of high charge density, as computed by molecular orbital calculations (*1065d*).

Position	2,4	1,3	5,7,8,10	6,9
Charge	−0.0101	+0.0294	+0.0439	+0.0920

Conversely, the 6,9 and 5,7,8,10 positions are the preferred sites of nucleophilic substitution (*575*).

$$B_{10}H_{14} \xrightarrow{CH_3Li} \underset{45\%}{6\text{-}MeB_{10}H_{13}} + \underset{17\%}{6,5(\text{or } 8)\text{-}Me_2B_{10}H_{12}} + \underset{8\%}{6,9\text{-}Me_2B_{10}H_{12}} + \underset{1.5\%}{5\text{-}MeB_{10}H_{13}} \qquad (467)$$

Grignard reagents also give the expected 6-alkylated decaboranes (Fig. 6-2), but in low yield (*573*). The major product, instead, is a rather novel Grignard

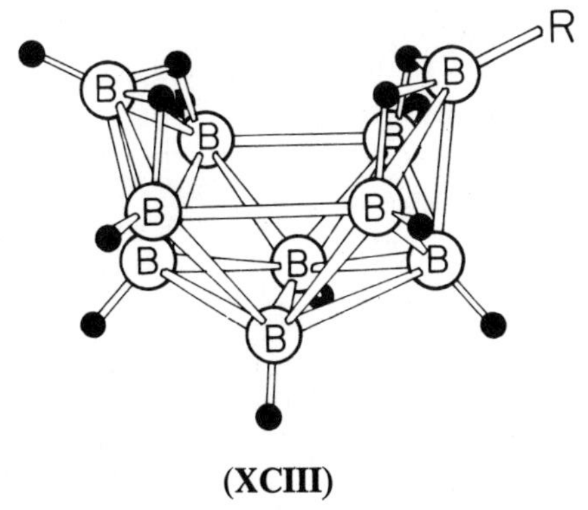

(XCIII)

Fig. 6-2. Ball and stick model of 6-alkyldecaborane(14).

reagent, decaboranylmagnesium iodide, which in turn reacts with a variety of compounds to form decaborane derivatives (*573*, *641*, *721*, *1709*).

$$B_{10}H_{14} \xrightarrow{C_2H_5MgI} \underset{10\%}{6\text{-}EtB_{10}H_{13}} + \underset{90\%}{B_{10}H_{13}MgI} \quad (468)$$

$$B_{10}H_{13}MgI \xrightarrow{Me_2SO_4} 5\text{-}MeB_{10}H_{13} + 6\text{-}MeB_{10}H_{13} \quad (469a)$$

$$B_{10}H_{13}MgI \xrightarrow{Et_3O^+BF_4^-} 5\text{-}EtB_{10}H_{13} \quad (469b)$$

$$B_{10}H_{13}MgI \xrightarrow{BuF} BuB_{10}H_{13} \quad (469c)$$

$$B_{10}H_{13}MgI \xrightarrow{PhCH_2Cl} 6\text{-}PhCH_2B_{10}H_{13} \quad (469d)$$

$$B_{10}H_{13}MgI \xrightarrow{CH_2{=}CH{-}CH_2Br} CH_2{=}CH{-}CH_2B_{10}H_{13} \quad (469e)$$

In a similar fashion, the $B_{10}H_{13}^-$ ion, prepared from sodium hydride and decaborane, reacts with methyl and ethyl sulfate (*136*) and benzyl chloride (*336*, *1523*) to give the same substituted decaboranes as are obtained from the decaboranyl Grignard reagent.

Pyrolysis of an ethylene–diborane mixture has yielded a complex product that contains, in addition to alkylated pentaboranes (Section 6-3), a mixture of mono- (probably 5-substituted), di-, tri-, and tetraethyldecaboranes (*137*); and the elevated temperature reaction between $B_{10}H_{14}$ and EtBr yields 2-$EtB_{10}H_{13}$ from a mixture that also includes diethyldecaborane (*583*). Derivatives of decaborane are listed in Table 6-5.

TABLE 6-5

Decaborane(14) Derivatives

Compound	Reference	Compound	Reference
2-$MeB_{10}H_{13}$	*1900*	$CH_2{=}CHCH_2B_{10}H_{13}$	*721*
5-$MeB_{10}H_{13}$	*136, 573*	6-$PhCH_2B_{10}H_{13}$	*136, 573, 1523, 1708, 1709*
6-$MeB_{10}H_{13}$	*136, 573, 575, 733a, 1732a*	6-$PhB_{10}H_{13}$	*733a*
1-$EtB_{10}H_{13}$	*1563*	1,2-$Me_2B_{10}H_{12}$	*1900*
2-$EtB_{10}H_{13}$	*583*	2,4-$Me_2B_{10}H_{12}$	*1900*
5-$EtB_{10}H_{13}$	*137*	6,5(or 8)-$Me_2B_{10}H_{12}$	*575*
6-$EtB_{10}H_{13}$	*136, 573, 575*	6,9-$Me_2B_{10}H_{12}$	*575*
?-$EtB_{10}H_{13}$	*1583*	1,2-$Et_2B_{10}H_{12}$	*135*
$BuB_{10}H_{13}$	*641*	2,4-$Et_2B_{10}H_{12}$	*135*
Amyl$B_{10}H_{13}$	*641*	$(PhCH_2)_2B_{10}H_{12}$	*1523*
Hexyl$B_{10}H_{13}$	*641*	1,2,3-$Me_3B_{10}H_{11}$	*1900*
Heptyl$B_{10}H_{13}$	*641*	1,2,4-$Me_3B_{10}H_{11}$	*1900*
Octyl$B_{10}H_{13}$	*641*	1,2,3,4-$Me_3B_{10}H_{10}$	*1900*
Cyclohexyl$B_{10}H_{13}$	*641*	1,2,3,5(or 8)-$Me_4B_{10}H_{10}$	*1900*

Reactions and Properties

The alkyldecaboranes, like decaborane, can be titrated as monoprotic acids in acetonitrile (*575*, *1900*). In the presence of cineole, the four bridge hydrogens of 6-benzyldecaborane exchange rapidly with deuterium oxide. This is followed by an exchange of three other hydrogens (probably 8-, 9-, and 10-positions) and eventually the remaining six terminal hydrogens (*1523*).

Mono-, di-, tri-, and tetraethyldecaboranes react with alkyl cyanides and other ligand molecules (*805*) to give materials that are most likely related to bis(acetonitrile)decaborane. In contrast to the B—N bonding in the latter compound, there is a possibility that carbon–boron bonds are formed from the reaction of cyanide ion with decaborane (*23*, *899*).

$$B_{10}H_{14} \xrightarrow{NaCN} Na_2B_{10}H_{13}CN \qquad (470)$$

$$B_{10}H_{14} \xrightarrow[Me_2S]{NaCN} NaB_{10}H_{12}CN \cdot Me_2S \qquad (471)$$

A second-order rate with an activation energy of 39.5 $\pm$ 1 kcal mole^{-1} is observed for the decomposition of 2-ethyldecaborane at about 220°C. The initial products from this pyrolysis are decaborane and diethyldecaboranes (*582*). The infrared (*107*) and mass spectra (*1583*) of ethyldecaborane have been reported; this same compound has an ionization potential of 9.0 $\pm$ 0.5 eV, which is lower than that of the parent, $B_{10}H_{14}$, 11.0 $\pm$ 0.5 eV (*1156*).

An X-ray structural study on 1-ethyldecaborane substantiates the position of attachment as well as indicating that substantially no change in the B_{10} framework has occurred upon substitution. The observed B—C (1.59 $\pm$ 0.01Å) and C—C (1.55 Å) bond distances are nearly the values expected for relatively unperturbed single bonds (*1563*).

Although no ^{11}B nmr chemical shift data are available for the various alkylated decaboranes, a comparison of published spectra with that of decaborane usually reveals the position of attachment in a rather direct fashion (*135*, *137*, *573*, *575*, *1523*, *1709*, *1900*). Unfortunately, the resonance lines of the 1,3- and 6,9-positions overlap. However, it has generally been observed that 6-alkyldecaboranes (**XCIII**) can be differentiated from 1-alkyldecaboranes simply by observing changes in the resonance lines of the apical 2- and 4-positions of B—H. The high field doublet (B—H coupling) assigned to the chemically equivalent 2,4-positions in decaborane remains as a single doublet structure in the 1-substituted compound, whereas in the 6-substituted compound the 2- and the 4-positions, no longer equivalent, are observed as two sets of closely spaced doublets (*572*, *573*, *1709*).

Carborane formation from decaborane and acetylenes and from $B_{10}H_{13}CN^{2-}$ or $B_{10}H_{12}CN^{-}$ has been reviewed elsewhere (*696a*).

6-7. Organic Derivatives of Large Closed-Cage Boron Hydrides

Neutral carbonyl derivatives of the polyhedral $B_{10}H_{10}^{2-}$ and $B_{12}H_{12}^{2-}$ ions can be prepared from the zwitterion diazonium salts (*895*, *902*). If the

$$B_{10}H_{10}^{2-} \xrightarrow[\text{2. NaBH}_4]{\text{1. HONO}} 1,10\text{-}(N_2)_2B_{10}H_8 \xrightarrow[120\text{–}140^\circ C]{CO} 1,10\text{-}(OC)_2B_{10}H_8 \qquad (472)$$

reaction, Eq. 472, is run in the presence of cyclohexane, $C_6H_{11}B_{10}H_7(CO)_2$ and $(C_6H_{11})_2B_{10}H_6(CO)_2$ are formed, suggesting a free radical mechanism (*895*, *902*). Although the carbonylation of 1,10-$(N_2)_2B_{10}H_8$ proceeds easily at 140°C, the higher temperature required for the reaction of the chlorinated derivative, 1,10-$(N_2)_2B_{10}Cl_8$, with CO is consistent with electron withdrawal by chlorine atoms, thus strengthening the boron–nitrogen bond against thermal dissociation (*895*). Whereas the $(OC)_2B_{10}H_8$ from the bisdiazonium salt is apically substituted, the monocarbonylation of 1-$B_{10}H_9SMe_2^-$ or 2-$B_{10}H_9NMe_3^-$ with oxalyl chloride gives the equatorially substituted products 1,6-$Me_2SB_{10}H_8CO$ and 2,4- and 2,7(8)-$Me_3NB_{10}H_8CO$, respectively (*792*, *795*). Both 1,12- and 1,7-$(OC)_2B_{12}H_{10}$, as well as $B_{12}H_{11}CO^-$, are products from the carbonylation of hydrated $(H_3O)_2B_{12}H_{12}$ (*900*).

The carbonyl derivatives of $B_{10}H_{10}^{2-}$ and $B_{12}H_{12}^{2-}$ react with water, alcohols, and amines to form carboxylic acids, esters, and amides, respectively (*896*, *898*, *900*) in a chemistry that parallels that of BH_3CO (Section 4-4, *Reactions*).

$$B_{10}Cl_8(CH_2OH)_2^{2-} \xleftarrow{NaBH_4} B_{10}Cl_8(CO)_2$$

$$B_{10}H_8(CO)_2 \xrightarrow[\text{2. heat}]{\text{1. Cl}_2\text{, H}_2\text{O}} B_{10}Cl_8(CO)_2$$

$$B_{10}H_8(CO)_2 \xrightarrow{CH_3OH} (H^+)_2B_{10}H_8(CO_2CH_3)_2^{2-}$$

$$B_{10}H_8(CO)_2 \xrightarrow{LiAlH_4} B_{10}H_8(CH_3)_2^{2-}$$

$$B_{10}H_8(CO)_2 \xrightarrow{NaN_3} B_{10}H_8(NCO)_2^{2-}$$

$$B_{10}H_8(CO)_2 \xrightarrow[H_2NOSO_3H]{H_2O} B_{10}H_8(NH_3)_2$$

$$B_{10}H_8(CO)_2 \xrightarrow{NH_3} (NH_4^+)_2B_{10}H_8(CONH_2)_2^{2-} \xrightarrow{Cs^+} Cs_2B_{10}H_8(CONH_2)_2$$

$$Cs_2B_{10}H_8(CN)_2 \xleftarrow{300^\circ C} Cs_2B_{10}H_8(CONH_2)_2 \qquad (473)$$

In water, the polyborane carbonyls are in equilibrium with carboxylic acid derivatives; acidity function studies indicate carboxyl proton ionization constants of about 10^{-9} (*898, 902*).

$$B_{12}H_{10}(CO)_2 \underset{}{\overset{H_2O}{\rightleftharpoons}} H^+B_{12}H_{10}COOH \cdot CO^- \underset{}{\overset{H_2O}{\rightleftharpoons}} (H^+)_2B_{12}H_{10}(COOH)_2^{2-} \xrightarrow{OH^-} B_{12}H_{10}(COO)_2^{4-} \quad (474)$$

Thermodynamic and pK_a data for the proton ionization of 1,12-$B_{12}H_{10}$-$(CO_2H)_2^{2-}$ indicate that the ionic charges on the B_{12} cage are localized in the vicinity of the carboxyl group and that the second proton ionization is largely unaffected by the increased negative charge resulting from ionization of the first proton (*734*).

Equatorial attack predominates in electrophilic substitution reactions of $B_{10}H_{10}^{2-}$, resulting, for instance, in the preparation of the 2-benzoyl derivative (**XCIV**) upon acylation with benzoyl chloride (*745a*). The $B_{10}H_9^{2-}$

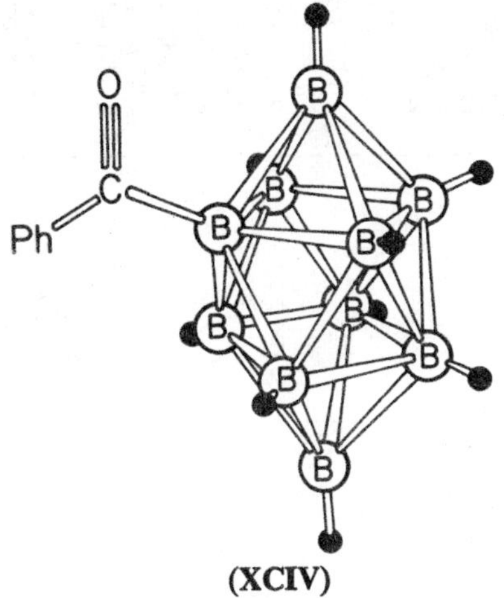

(XCIV)

moiety is a strong electron donating group, sufficiently so that 2-$PhCOB_{10}H_9^{2-}$ readily forms the protonated species 2-$PhC(OH)B_{10}H_9^-$. The polar nature of the ketone group in 2-$PhCOB_{10}H_9^{2-}$ is also shown by its low infrared stretching frequency, 1570 cm^{-1} (*901*). The ketocarbonyl group is even more basic when flanked by two B_{10} cages, and the BC(OH)B unit might be described as containing a stabilized carbonium ion (*898*). Chlorination of the B_{10} cage reduces the basicity of the carbonyl group in $B_{10}Cl_9COB_{10}Cl_8$-CO_2H^{4-} to the point that tetrabasic salts of this ion can be isolated (*898, 901*). The 1,10-$(NC)_2B_{10}H_8^{2-}$ ion rearranges about 350°C to a predominantly equatorially substituted isomeric mixture (*791*).

TABLE 6-6

Cage Polyboron Compounds

Compound	Reference	Compound	Reference
$2,4-(Me_3N)(CO)B_{10}H_8$	*792*	$1,10-(CH_3CO_2CH_2)_2B_{10}Cl_8^{2-}$	*896*
$2,7-(Me_3N)(CO)B_{10}H_8$	*792, 795*	$1,10-(Ph_3PCH_2)B_{10}Cl_8$	*896*
$1,6-(Me_2S)(CO)B_{10}H_8$	*792*	$1,10-(BrCH_2)_2B_{10}Cl_8^{2-}$	*896*
$Me_3NB_{10}H_8C(OH)B_{10}H_9^{2-}$	*898*	$1,10-(ICH_2)_2B_{10}Cl_8^{2-}$	*896*
$Me_2SB_{10}H_8C(OH)B_{10}H_9^{2-}$	*898*	$1,10-(NCCH_2)_2B_{10}Cl_8^{2-}$	*896*
$1,6-(Me_2S)(CO_2H)B_{10}H_8^{-}$	*745a, 792*	$1,10-(HO_2CCH_2)_2B_{10}Cl_8^{2-}$	*896*
$PhCOB_{10}H_9^{2-}$	*897, 901*	$1,10-(MeSOCH_2CH_2)_2B_{10}Cl_8^{2-}$	*896*
$PhCOB_{10}Cl_9^{2-}$	*897, 901*	$1,10-(Me_2NC_6H_4CO)_2B_{10}Cl_8^{2-}$	*898, 902*
$PhCOB_{10}Br_9^{2-}$	*901*	$(C_6H_{11})(CO)_2B_{10}H_7$	*895, 902*
$C_7H_6B_{10}H_9^{-}$	*737–739*	$(C_6H_{11})(HO_2C)_2B_{10}H_7^{2-}$	*900*
$1,10-Me_2B_{10}H_8^{2-}$	*900*	$(C_6H_{11})_2(CO)_2B_{10}H_6$	*895, 902*
$(C_8H_9)_2B_{10}H_9^{2-}$	*901*	$(C_6H_{11})_2(HO_2C)_2B_{10}H_6^{2-}$	*900*
$H_2NCONHN{=}C(Ph)B_{10}H_9^{2-}$	*897*	$2-PhB_{11}H_{10}S$	*790*
$H_2NCONHN{=}C(Ph)B_{10}Cl_9^{2-}$	*897*	$Ph_2B_{11}H_9S$	*790*
$1,10-(CH_2OH)_2B_{10}Cl_8^{2-}$	*895*	$B_{12}H_{11}CO^{-}$	*900*
$1,10-(CN)_2B_{10}H_8^{2-}$	*791, 902*	$B_{12}H_{11}CO_2H^{2-}$	*900*
$1,10-(CN)B_{10}H_8CO^{-}$	*900*	$B_{12}Br_{11}CO_2H^{2-}$	*900*
$1,10-NCB_{10}H_8CO_2H^{2-}$	*900*	$Me_3NB_{12}H_{10}CO$	*792*
$1,10-(NH_3CH_2)_2B_{10}Cl_8$	*896*	$PrB_{12}H_{11}^{2-}$	*901*
$1,10-(CO)_2B_{10}H_8$	*895, 902*	$C_7H_6B_{12}H_{11}^{-}$	*737, 739*
$1,10-(CO)_2B_{10}Cl_8$	*895, 900, 902*	$1,12-B_{12}H_{10}(CO)_2$	*900, 902*
$1,10-(CO)_2B_{10}Br_8$	*900, 902*	$B_{12}H_{10}(CO)CO_2H^{-}$	*902*
$1,10-(CO)_2B_{10}I_8$	*900*	$B_{12}H_{10}(CO_2H)_2^{2-}$	*745a, 900, 902*
$Me_2SB_{10}H_8C(OH)B_{10}H_8CO_2H^{2-}$	*898*	$B_{12}Br_{10}(CO_2H)_2^{2-}$	*900*
$B_{10}H_9COB_{10}Cl_8CO_2H^{4-}$	*898*	$1,12-B_{12}I_{10}(CO_2H)_2^{2-}$	*900*
$B_{10}H_9C(OH)B_{10}H_8CO_2H^{3-}$	*898*	$B_{12}H_{10}(CO_2Me)_2^{2-}$	*902*
$(Me_2S)_2B_{10}H_7C(OH)B_{10}Cl_8CO_2H^{-}$	*898*	$B_{12}H_{10}(CO_2Et)_2^{2-}$	*745a*
$(HO_2C)_2B_{10}H_8^{2-}$	*900*	$1,7-(Me_2NC_6H_4CO)_2B_{12}H_{10}$	*898*
$1,10-(H_2NCO)B_{10}H_8^{2-}$	*791, 902*	$1,12-(CN)B_{12}H_{10}CO_2H^{2-}$	*900*
$1,10-(CO_2CH_3)_2B_{10}H_8^{2-}$	*902*	$B_{12}H_{10}(CONR_2)_2^{2-}$	*745a*
$1,10-(HO_2C)_2B_{10}Cl_8^{2-}$	*900*	$B_{10}H_{13}CN^{2-}$	*899*
$1,10-(HO_2C)_2B_{10}Br_8^{2-}$	*900*	$B_{10}H_{12}CN \cdot (CH_3)_2S^{-}$	*899*
$1,10-(HO_2C)B_{10}I_8^{2-}$	*900*	$6-PhCH_2B_{10}H_{12}^{-}$	*1707*

Reaction of 2 moles of tropenylium ion with either $B_{10}H_{10}^{2-}$ or $B_{12}H_{12}^{2-}$ yields tropenyliumyl-*closo*-nonahydrodecaborate(1 −) (**XCV**) and tropenylium-*closo*-undecahydrododecaborate(1 −) (**XCVI**) ions, respectively, and cycloheptatriene. Spectroscopic investigations of these very stable and highly

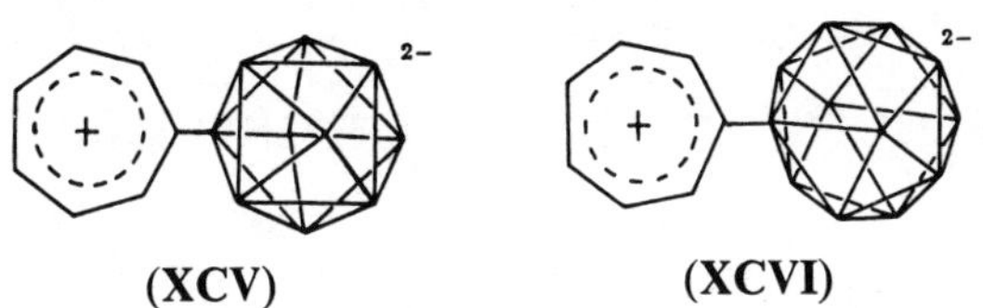

(**XCV**) (**XCVI**)

colored anions show that in both ions a cationic ring is attached to the boron cage with a C—B bond and that significant cage-to-ring electron donation occurs in both ground and excited states (*737*, *739*). A list of organic derivatives of cage polyboron compounds and related systems is presented in Table 6-6.

Chapter 7

Supplementary Chemistry

7-1. Cyclic Boron–Carbon Systems

A variety of general methods described above (Section 3-1), such as hydroboration, transmetallation, boron–boron exchange, cyclization by loss of H_2, modified Friedel–Crafts reactions (e.g., Eqs. 475–479), and a unique pathway shown in Eq. 480, have been employed for the synthesis of cyclic organoboranes.

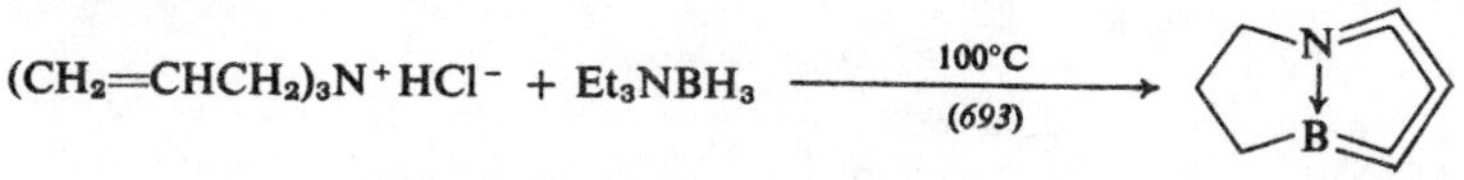

$$(CH_2{=}CHCH_2)_3N^+HCl^- + Et_3NBH_3 \xrightarrow[(693)]{100°C} \quad (475)$$

$$\text{(o-LiC}_6\text{H}_4\text{CH}_2\text{)}_2 \xrightarrow[\text{2. N-Br-succinimide; 3. NaOCH}_3\ (1833)]{\text{1. B(OBu)}_3} \text{B–OH} \quad (476)$$

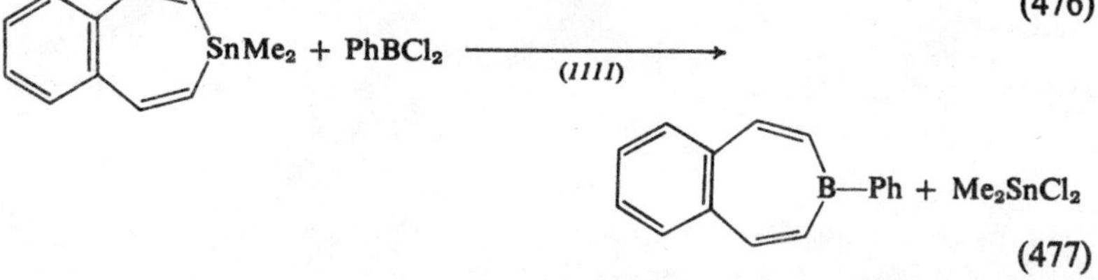

$$\text{SnMe}_2 + PhBCl_2 \xrightarrow{(1111)} \text{B—Ph} + Me_2SnCl_2 \quad (477)$$

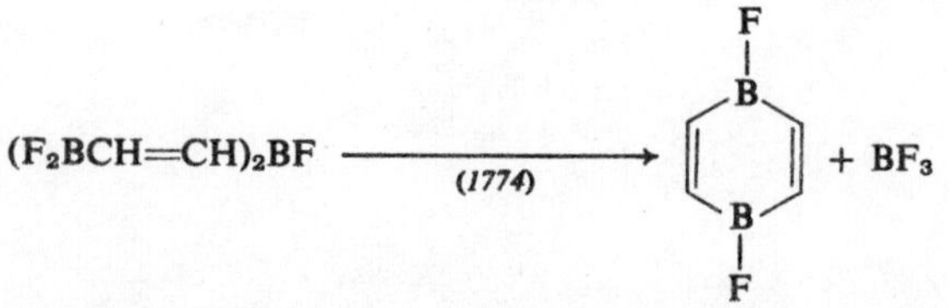

$$(F_2BCH{=}CH)_2BF \xrightarrow{(1774)} \text{F–B(CH=CH)}_2\text{B–F} + BF_3 \quad (478)$$

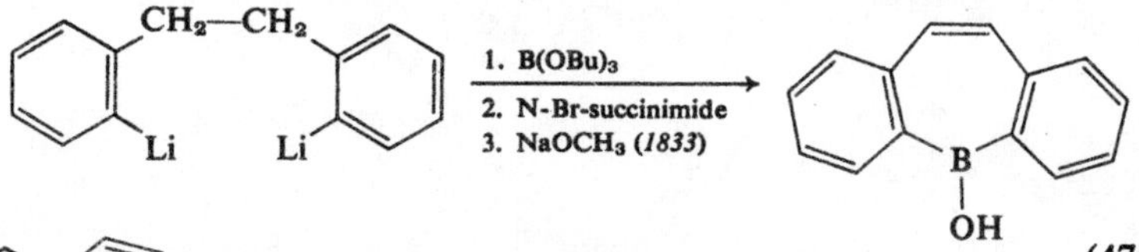

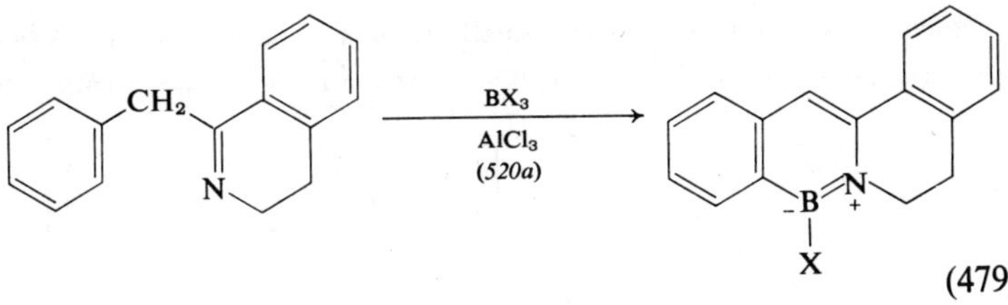

$$\text{PhCH}_2\text{-(3,4-dihydroisoquinoline)} \xrightarrow[\text{(520a)}]{\text{BX}_3,\ \text{AlCl}_3} \text{(B–X, N}^+\text{ fused heterocycle)} \tag{479}$$

$$H_2B[N(CH_3)_3]_2{}^+Cl^- \xrightarrow[(1367)]{\text{NaH}} \text{cyclo-}(H_2B\text{–}NMe_2\text{–}CH_2)_2 \tag{480}$$

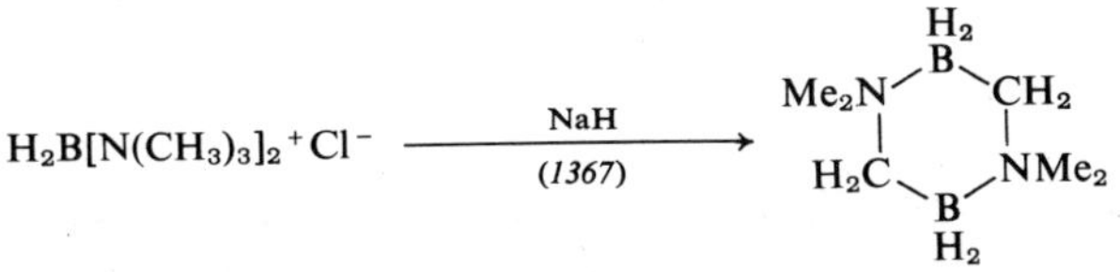

Among the three cyclic organoboranes, **(XCVII)**, **(XCVIII)**, and **(XCIX)**, the six-membered ring is found to be the most thermodynamically stable (*261, 360a, 974, 979, 1418, 1617, 1631, 1691*). This conclusion is reached from the results of the following reactions and equilibria that involve ring enlargement or contraction processes (probably via a hydroboration–dehydroboration sequence, Sections 3-1, 3-2) favoring the formation of the boracyclohexane species.

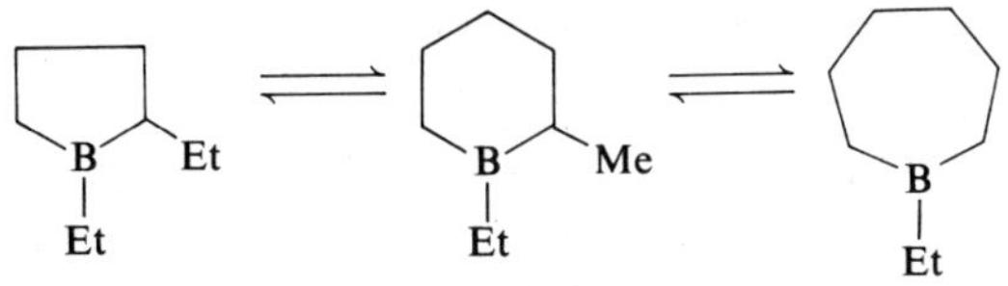

(481)

1-Et-2-Et-borolane	⇌	1-Et-2-Me-borinane	⇌	1-Et-borepane
8%		89%		3%
(XCVII)		**(XCVIII)**		**(XCIX)**

$$\text{B (tricyclic, 5/7/7 rings)} \xrightarrow{200^\circ} \text{B (perhydro-9b-boraphenalene)} \tag{482}$$

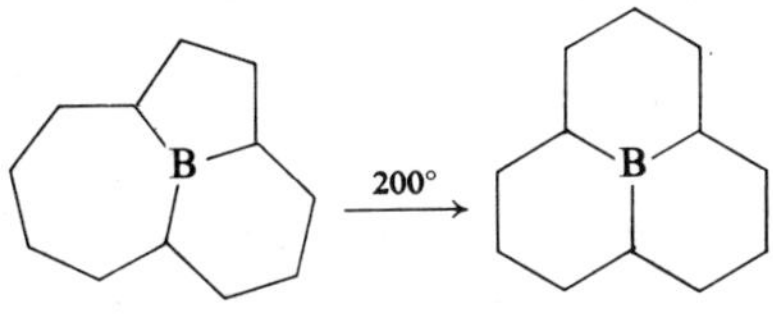

Dehydrogenation of compounds containing the heterocyclic ring (**C**) to the aromatic counterpart can be accomplished with the use of a palladium charcoal catalyst (*463, 512*).

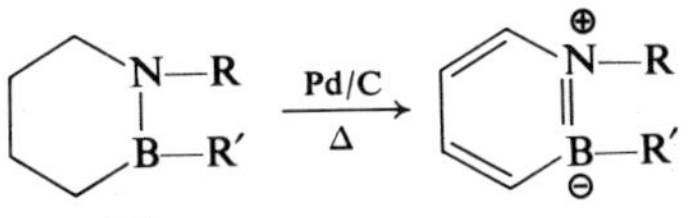

$$\underset{\textbf{(C)}}{\text{cyclo-(N–R)(B–R')}} \xrightarrow[\Delta]{\text{Pd/C}} \text{aromatic } \overset{\oplus}{N}\text{–R}=\overset{\ominus}{B}\text{–R'} \tag{483}$$

Derivatives of **(CI)** with two-coordinate boron have evaded synthesis (*1834*); however, both the anionic analogs **(CII)** (*50*) and **(CIII)** (*1835*) with three-

(CI) 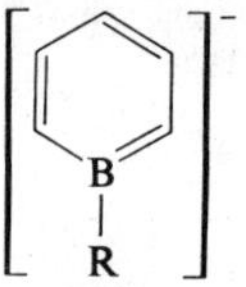 **(CII)** 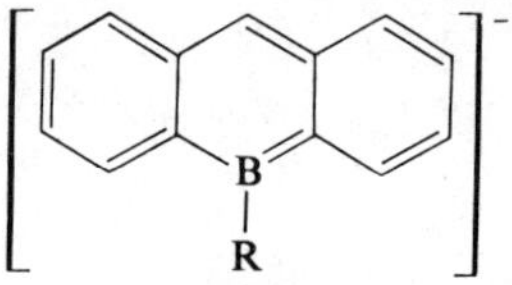 **(CIII)**

coordinate boron appear to possess a higher stability in that both have been successfully prepared.

The uv and nmr spectra of **(CIV)** (R = OH) closely resembles that of **(CV)**, suggesting that it is aromatic (*58*). Further, the olefinic pmr downfield

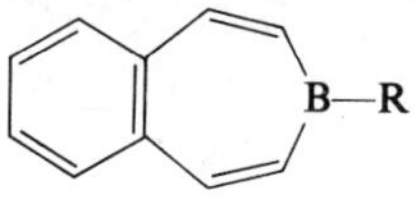

(CIV) 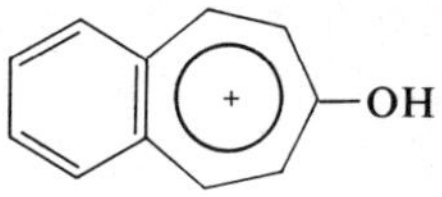 **(CV)**

shift of **(CIV)** (R = Ph), as well as the uv change upon forming a dimethylamine complex, is consistent with boron p_z orbital participation in the π electronic system (*1111*). Two 4-π electron systems, **(CVI)** (*579*) and **(CVII)** (*1774*), are derivatives of the carboranes C_4BH_5 [predicted to have a

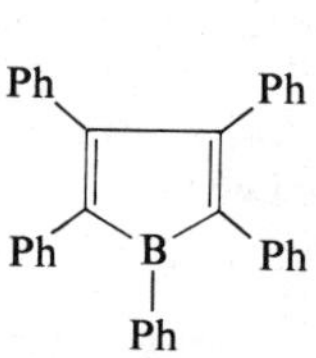

(CVI) 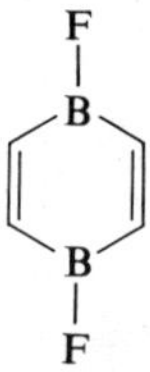 **(CVII)**

tetragonal pyramidal structure (*1120a*)] and $C_4B_2H_6$ (pentagonal pyramidal with a boron in the apex) (*706*, *1506*). Apparently, back bonding of either phenyl in **(CVI)** or fluorine in **(CVII)** inhibits the utilization of the potential vacant boron p_z orbitals for the three-center bonds demanded in the cage carborane structures.

It is suggested that the boracyclopropene **(CVIII)**, if it could be synthesized (*578*), would possess a marked dipole moment because of the substantial contribution of structure **(CIX)** (*1839*).

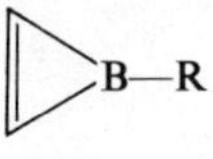

(CVIII) 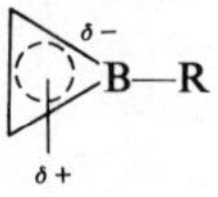 **(CIX)**

The extent of aromaticity in heteroaromatic compounds such as (**CX**)–(**CXVI**) has been the object of considerable study. Observations which

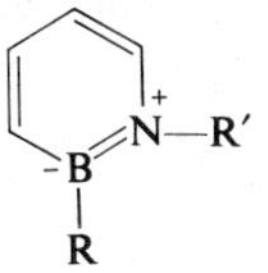

(**CX**) 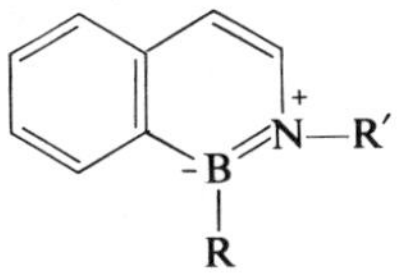 (**CXI**) 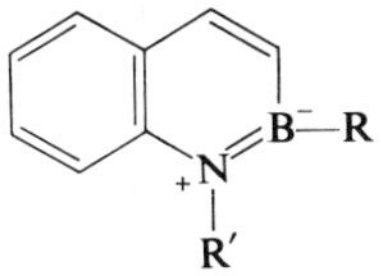 (**CXII**)

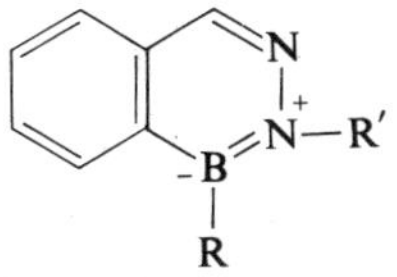

(**CXIII**) 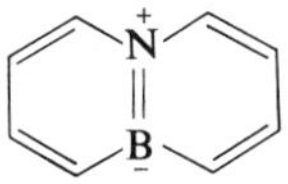 (**CXIV**)

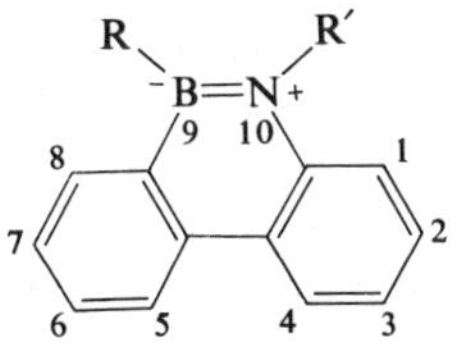

(**CXV**) 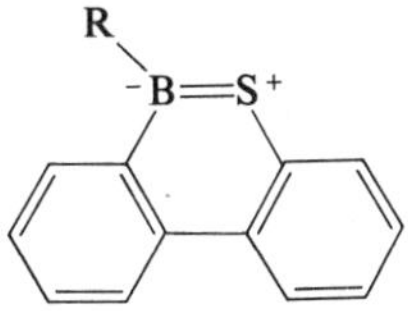 (**CXVI**)

support a rather substantial degree of π-delocalization in some of these compounds include: (a) the uv spectroscopic similarities of (**CXII**) to that of naphthalene (*523*) and of (**CXV**) to phenanthrene (*536*); (b) the high hydrolytic stability of (**CX**) (*463, 542*), (**CXII**) (*522*), (**CXIII**) (*526*), and (**CXV**) (*533*) in both acid and alkali, which is equated with highly resonance stabilized boron–nitrogen linkages; (c) behavior of (**CXII**) (R = OH) (*518, 522, 531*) and (**CXIII**) (R = OH) (*526*) as protic $>$B—OH $\xrightarrow{OH^-}$ $>$B—O$^-$ + H_2O rather than Lewis acids $>$BOH $\xrightarrow{OH^-}$ $>$B(OH)$_2^-$; (d) infrared evidence for a very acidic NH proton in (**CXIII**) (R′ = H) as would be expected if there is any significant contribution of the dipolar resonance structure (*526*); and (e) photoelectron spectra indicate a close similarity between (**CXIV**) and naphthalene (*511*).

There is evidence for less aromatic stabilization in (**CXVII**) than in (**CXI**), which may indicate that greater back-coordinative interaction occurs when heteroatoms are adjacent (*1147*). Also, a comparison of (**CXVI**) and its properties with the corresponding nitrogen and oxygen analogs shows that the sulfur compound is by far the least aromatic (*509*).

Electrophilic substitution (i.e., chlorination and nitration) of (**CXV**) occurs predominantly at the 6- and 8- positions, in agreement with the predictions of charge distribution based on molecular orbital theory (*535*). Boron, being less electronegative than carbon in the comparison compound, phenanthrene, should selectively activate positions of opposite parity to itself (i.e., 2,- 4-, 6-, and 8-positions); however, the effect of the heteroatom pair BN activates the 6- and 8-positions to a greater extent than the 2- and 4-positions (*535*).

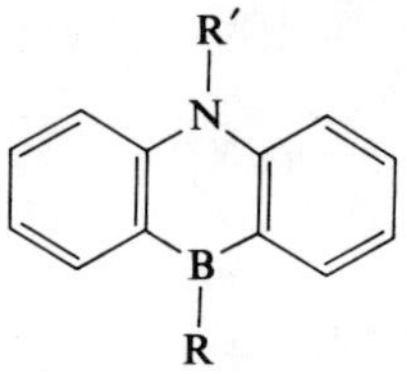

R′ N B R

(CXVII)

Similar agreement between the position of electrophilic substitution and the calculated π-electron densities has been observed for (**CXIII**) (*546*).

Total charge distributions have also been calculated for (**CX**), 1,3-dibora-2,4-diazarobenzene, and 2,1-borazaronaphthalene (**CXII**) as well as for a number of known and hypothetical organoborazo systems (*528*, *544*, *812*, *882*).

7-2. Influence of the Organic Group on the Properties of the Attached Boron-Containing Moiety

A substantial portion of the title topic is to be found in portions of other sections in the text (e.g., Sections 3-2 and 4-2, *Synthesis and Stability*, and Chapter 2), and only those studies not conveniently covered earlier are included here.

The acidity trends to be found in organoboronic acids parallel those of carboxylic acids, although the boron compounds are considerably weaker than analogs of the organic acids by ca. 4–6 pK_a units (*119*, *327*, *1926*); acid strength decreases in the order: $PhB(OH)_2 > PhCH_2B(OH)_2 > n\text{-}BuB(OH)_2$; and 2-thiophene—$B(OH)_2$ > 3-thiophene—$B(OH)_2$ > $PhB(OH)_2$. Acid strengths of various ortho-substituted phenylboronic acids have been measured and the decrease in acidity caused by groups such as methyl or chloro is attributed to F-strain in the ion $o\text{-}XC_6H_4B(OH)_3^-$ (*1210*).

Carbon–boron π bonding appears to decrease the Lewis acidity of $CH_2{=}CHBF_2$ (*449*) and this is supported by F-19 nmr studies (*448*). Bulky boron-attached organic groups stabilize the borazine ring toward hydrolysis (*654*, *1402*, *1671*), but a *B*-phenyl group is found to decrease the hydrolytic stability of borazine derivatives (*180*, *654*). The latter effect is attributed to reduced electron density on boron by an aromatic ring that is perpendicular to the borazine plane, tending to make inductive effects operate more strongly than resonance effects (*103, 1713*).

The difunctional molecule $Me_2BCH_2PMe_2$ exists as a dimer in the vapor phase, but both Me_2BCH_2SMe and $Me_2BCH_2AsMe_2$ are only weakly associated (*1592*). In the nitrogen series, $H_2BCH_2NMe_2$ can exist as a cyclic dimer (*1366, 1368, 1369*), whereas $Me_2BCH_2NH_2$ exists as both a monomer and polymer in equilibrium (*1644*). The acid–base character of the monomer suggests a metastable three-membered ring structure (*1644*). The monomeric nature of compounds of the type Ar_2BNR_2, Ar_2BPR_2, and Ar_2BAsR_2 contrasts with the dimeric Ph_2BNH_2 (*420, 422*). Other monomer–dimer equilibria involving the systems R'_2NBRX (X may be halogen or R″) (*723, 1468, 1489*) and $R'_2C{=}NBRX$ (X = hydrogen, halogen, R″S, R″) (*330, 335, 380, 382, 427, 555, 556, 753, 862, 1221–1225, 1229, 1287, 1554*) also depend upon the substitution pattern and how this affects steric and electronic factors. And, of course, all of the above equilibria are highly dependent on the physical state (gas, liquid, or solid) and solution concentrations.

The effect of a *B*-attached vinyl group on an adjacent B—N bond order, as reflected in observed infrared spectroscopic shifts, appears to be very small (*626, 1439a, 1440*). The decrease of B—N bond order for aminoboranes with a *B*-methyl substituent may be attributed to a hyperconjugative effect (*392*).

7-3. Influence of Boron Substitution on the Properties of the Attached Organic Group

The double bond of *B*-vinylboronates behaves in a reasonably normal manner toward bromine addition and also forms the expected adducts from dichlorocarbene and from Diels–Alder reagents (*395, 426a, 426b, 1193, 1194, 1246, 1536, 1920*).

$$CH_2{=}CHB(OR)_2 \xrightarrow{Br_2} CH_2BrCHBrB(OR)_2$$

Polar addition of hydrogen iodide to vinylboronates yield both the 1-iodo and 2-iodo derivatives (*1188*); and the mode of HBr addition varies according

$$CH_2{=}CHB(OR)_2 \xrightarrow{HI} \underset{\substack{|\\ I}}{CH_3CH}B(OR)_2 + ICH_2CH_2B(OR)_2$$

to the electronic factors associated with attached groups (*426b, 1178*). Bromine uptake by tris(4-butenyl)boron can take place without significant rupture of the C—B bond (*1143*) (see Section 3-2, *Halogenation*).

$$[CH_2{=}CH(CH_2)_2]_3B \xrightarrow{3\,Br_2} (CH_2BrCHBrCH_2CH_2)_3B \qquad (484)$$

Catalytic hydrogenation of both vinyl and acetylenic boron compounds is accomplished (*1105*, *1729*, *1923*) with little difficulty.

$$Me_2C{=}CHB(OH)_2 \xrightarrow[(1105)]{H_2,\ Pt} Me_2CHCH_2B(OH)_2 \tag{485}$$

$$Ph_2BC{\equiv}CR\cdot pyr \xrightarrow[(1729)]{H_2,\ Pt} Ph_2BCH{=}CHR\cdot pyr \xrightarrow{H_2,\ Pt} \text{sat'd cpd} \tag{486}$$

The triple bond of acetylene boronates may undergo typical addition reactions with RSH, Cl_3BBr, and Br_2 (*1184*) and with Diels–Alder reagents (*1923*); in some instances, difficulties are encountered (*1185*) and in those cases radical catalyzed reactions (*1170*, *1181*, *1185*, *1191*) have proved more useful for addition to both acetylenic and ethylenic boron compounds.

$$HC{\equiv}CB(OR)_2 \xrightarrow[(1181,\ 1185)]{Br_2,\ h\nu} BrCH{=}CBrB(OR)_2 \tag{487}$$

$$HO_2C(CH_2)_2SH + CH_2{=}CHB(OR)_2 \xrightarrow[(1191)]{h\nu} HO_2C(CH_2)_2SCH_2CH_2B(OR)_2 \tag{488}$$

Radical-catalyzed addition of Cl_3CBr to ethyleneboronic esters may involve an intermediate $Cl_3CCH_2\dot{C}HB(OR)_2$ stabilized by carbon–boron π bonding (*1170*).

Radical attack of halogen atoms at an α-H site of an alkylborane is a sensitive function of the structure of the organic group bonded to the boron atom and the nature of other substituents bonded to the boron atom, as well as the nature of the attacking radical (*1545a*, Section 3-2).

Nitrile oxides add to the β-carbon of dibutylethynylboronate to give the cyclic compound (**CXVIII**) (*121*).

$$Ar C{\equiv}\overset{+}{N}{-}\overset{-}{O} + HC{\equiv}CB(OR)_2 \longrightarrow \text{3-Ar-isoxazol-5-yl-}B(OH)_2 \ \textbf{(CXVIII)} \tag{489}$$

The $4n$ π electron system of pentaphenylborole (**CXIX**) (*154*, *579*, *700*) is highly reactive, undergoing a Diels–Alder reaction to give (**CXX**) (*579*).

$$\textbf{(CXIX)}\ (\text{pentaphenylborole}) + PhC{\equiv}CPh \longrightarrow \textbf{(CXX)}\ (\text{7-phenyl-7-borabicyclo[2.2.1]heptadiene, hexaphenyl}) \tag{490}$$

Dehydrobromination of $Cl_3CCH_2CHBrB(OR)_2$ is accomplished with anhydrous tributylamine (*1184*); and the light-catalyzed dehydrohalogenation reaction of the substituted vinylboron compound (**CXXI**) leads to a mixture of elimination products (*1606*).

$$\underset{\textbf{(CXXI)}}{(Cl)(H)C{=}C(H)(BCl_2)} \xrightarrow{h\nu} HC{\equiv}CBCl_2 + HC{\equiv}CH + H_2C{=}CHBCl_2 + HCl + BCl_3 \quad (491)$$

Both elimination and nucleophilic substitution have been experienced with the action of nucleophiles on 2-bromoethaneboronates. The latter route is

$$BrCH_2CH_2B(OR)_2 + Z^- \longrightarrow Br^- + C_2H_4 + ZB(OR)_2$$
$$\searrow ZCH_2CH_2B(OR)_2 + Br^- \quad (492)$$

favored for weak bases such as iodide, whereas stronger bases such as thiocyanate ion encourage the elimination pathway to predominate. The iodide displacement reaction is four to five times slower than a similar displacement using allyl bromide, but a number of times faster than the reaction using alkyl bromides (*1179*).

Treatment of Me_2BCH_2Cl with nucleophiles may give other α-substituted methylboranes (*1592*, *1644*), but a competing reaction is an alkyl rearrange-

$$Me_2BCH_2Cl \xrightarrow[(1592)]{MeSH} Me_2BCH_2SMe$$

$$Me_2BCH_2Cl \xrightarrow{N_3^-} Me_2BCH_2N_3 \xrightarrow[(1644)]{[H]} Me_2BCH_2NH_2$$

ment to give *B*-substituted methylethylboranes (Section 2-2). Displacement of bromide ion from α-bromoalkaneboronic esters by nucleophiles (I^-, RO^-, RS^-) is assisted by the neighboring boron atom (*1183*, *1188*); alkoxide attacks first at the boron to form the tetracovalent boron anion, which subsequently rearranges with expulsion of bromide ion (*1183*). Solvolysis of

$$Cl_3CCH_2CH(Br)(OBu)_2 \xrightarrow{OR^-} Cl_3CH_2CH(Br)\overset{\ominus}{B}(OR)(OBu)_2 \longrightarrow Cl_3CCH_2CH(OR)B(OBu)_2 + Br^- \quad (493)$$

dibutyl-2-bromopropane-2-boronate in aqueous ethanol yields the 2-hydroxy derivative in a process with a high degree of carbonium ion character and with significant acceleration by the boron atom (*1188*).

$$Me_2\overset{Br^{\delta-}}{C^{\delta+}}\text{—}B(OR)_2 \cdots OH_2$$

α-Iodomethaneboronate undergoes halide displacement with a number of nucleophiles (*1177*).

$$(RO)_2BCH_2I \xrightarrow[\text{aq}]{HCO_3^-} (HO)_2BCH_2OH \quad (494)$$

Evidence for the generation of a boron-stabilized carbanion $>B—CH_2^-$, has come from the incorporation of deuterium on the α-carbon of an organoborane after an initial reaction with a sterically hindered base is followed by the addition of deuterium oxide (*1594a*).

Both the $(HO)_2B—$ and $Cl_2B—$ groups are meta directors toward nitration and halogenation of the phenyl ring (*24, 743, 748, 1437, 1664, 1726, 1794*). A common side reaction is cleavage of the C—B bond (*1009, 1011, 1664, 1725*), but the use of reduced temperatures (*24, 743, 1664*), and the presence of a carboxy group para to the boron group (*1725, 1792*), serve to reduce the extent of this cleavage.

$$C_6H_5B(OH)_2 \xrightarrow[-20°C]{HNO_3\text{ (fuming)}} m\text{-}O_2NC_6H_4B(OH)_2 \quad (495)$$

When nitration of benzene boronic acid is carried out in acetic anhydride, the major product is a result of ortho substitution (*743, 1664*). The enhanced ortho reactivity in this case is ascribed to a favorable interaction between entering substituent and the donor complex (**CXXII**) (*743*).

$$C_6H_5\overset{\ominus}{B}(OH)_2\leftarrow\overset{\oplus}{O}Ac_2$$

(CXXII)

Theory suggests (*535*) that the 8-position in 10,9-borazarophenanthrene (**CXV**) should be the most reactive toward electrophilic substitution, followed closely by the 6-position; and this is supported experimentally (*534*).

Photolysis of *cis*-$Cl_2BCH{=}CHBCl_2$ rearranges the compound to the trans isomer (*446*).

7-4. Analysis of Organoboron Compounds

Analytical determinations of both carbon and boron in organoboron compounds are obviously much dependent upon the effectiveness with which the C—B bond is broken. A number of methods have been developed, each

TABLE 7-1

Degradative Reagents for the Quantitative Analysis of Organoboron Compounds

Reagent(s)	Reference
RCO_2H	*461, 1163* (Section 3-2, *Protodeboronation*)
F_3CCO_2H	*1570, 1743*
H_2O_2, NaOH	*106, 867, 1323, 1570, 1719* (Section 3-2, *Action of Peroxides*)
F_3CCO_2H, H_2O_2	*1570*
NaOH	*1570* (Section 3-2)
H_2SO_4	*1570*
Alkaline potassium persulfate	*574*
Ammoniacal silver ion	*77, 866, 1719*
H_2SO_4, Se, $CuSO_4$, K_2SO_4	*1608*
Me_3NO	*500, 967, 968, 1976*
O_2, heat	*47, 435* (Section 3-2, *Autoxidation*)

good for certain organoboron compounds. It is not the intention here to expound on the virtues of each reagent, but only to mention that the procedures are usually based on oxidative, hydrolytic, or protodeboronation (Section 3-2) routes. Common reagents used to cleave the C—B bond for analytical purposes are listed in Table 7-1. Once the boron has been converted to boric acid (or borate), it is then treated with mannitol and titrated with standard base; the carbon and hydrogen are determined in a conventional manner.

7-5. Uses and Applications of Organoboron Compounds

Organoboron compounds are now widely applied as intermediates in organic synthesis (Sections 3-1, 3-2); and BPh_4^- is a recognized standard analytical reagent for the selective precipitation of K^+ and other ions of the heavier alkali metals [see A. J. Barnard (1955) and A. J. Barnard and H. Buechl (1959), Supplementary Sources of Information]. Tetraphenylborate ion has also been used for obtaining alkaloid derivatives (*687*). Studies pertinent to biochemical and/or medicinal chemistry in which organoboron compounds have been utilized include: the inactivation of erythrocyte membrane acetylcholinesterase by tetraphenylborate (*793*); phenylboron compounds as protective groups in nucleosides (*319*); effects of arylboric acids on microorganisms and enzymes (*1795*); localization of boron in rat brain following administration of tolylboronic acid (*389*); dissociation of thyroglobin by the tetraphenylborate ion (*1198*); use of *n*-butylboronate derivatives of prostaglandins (*1508*) and corticosteroid boronates (*177a*) for gas–liquid chromatographic separation, and of organoborates of corticosteroids in yielding characteristic mass spectroscopic ions (*178*); plant physicochemical studies utilizing arylboric acids (*1800*); phenylboronic esters of carbohydrates as easily isolatable derivatives (*1682, 1796*); use of trialkylboranes in effecting the graft copolymerization of methylmethacrylate onto hemoglobin (*909*); aqueous cyanohydridoborate reduction of the rhodopsin chromophore (*588b*); action of tetraphenylborate ion on the electron flow in photosystem II of isolated chloroplasts (*830a*); and binding-catalysis relationship in α-chymotrypsin action as revealed from a reversible inhibition study using phenylalkylboronic acids (*34a*).

A considerable number of studies have dealt with: the use of trialkylboranes as catalysts for olefin polymerization and copolymerization (*51–55, 89, 174, 610, 631, 632, 635, 851, 852, 855, 856, 914–917, 1524, 1960*); also, polymerization in the presence of oxygen (*630, 633, 735, 1408, 1760, 1761, 1871*) and peroxide (*18, 431, 1458*) cocatalysts; vinyl polymerization initiated either by a trialkylborane–pyridine system (*907, 908*) or by Et_3BNH_3 (*1457*); pyridine and quinoline polymerization in a reaction with trimethylboron (*1713a*); polymerization of vinyl compounds with binary mixtures of trialkylborons and metal halides (*634*); alkali metal salt of triarylboron compounds as an initiator for the polymerization of olefins (*1407*); polymerization of diazoalkanes (*88*) and adiponitrile (*1362*); and tributylborane-initiated graft copolymerization of methylmethacrylate onto silk (*911*) and onto cotton (*910*). Other polymerization studies involving organoboron compounds include: copolymers from *p*-vinylbenzeneboronic esters (*1098*) or from *B*-trivinyl- or *B*-triallylborazine (*1556*) with styrene and diallylmaleate; free radical polymerization of *p*-vinylphenylboronic acid (*1557*); copolymerization of bis-(alkylamino)alkyl- or arylboron compounds with organic diisocyanates (*924*); and oxidative polymerization of triallylborane (*1786*). Trace amounts of Et_3B eliminate the induction period for $Al(BH_4)_3$–ethylene–O_2 explosions (*177*).

During the 1950s, a considerable effort was expended in the search for organoboranes and organopolyboranes that could serve as efficient and easy-to-handle high-energy propellants (*374, 375, 1159, 1160, 1863*). Another application includes the use of 8-quinolineboronic acid as a polyfunctional catalyst for the hydrolysis of chloroalcohols (*1095, 1101*).

Supplementary Sources of Information

Because it is impractical to include a substantial amount of background material in this monograph a number of reviews covering various aspects of boron chemistry are listed below for the interested reader.

R. M. Adams, Organoboron compounds. *Advan. Chem. Ser.* **23**, 87 (1960).

R. M. Adams, ed., "Boron, Metallo-Boron Compounds and Boranes." Wiley (Interscience), New York, 1964.

I. B. Atkinson and B. R. Currell, Boron-nitrogen polymers. *Inorg. Macromol. Rev.* **1**, 203 (1971).

B. Aylett, ed., *MTP Int. Rev. Sci., Inorg. Chem. Ser. 1* **4** (1972).

A. J. Barnard, Sodium tetraphenylboron, 1949–55. A comprehensive bibliography. *Chem. Anal.* **44**, 104 (1955).

A. J. Barnard and H. Buechl, Sodium tetraphenylboron, 1958. A bibliography. *Chem. Anal.* **48**, 44 (1959).

S. H. Bauer, Boron hydrides and related compounds. *Encycl. Chem. Tech., Suppl.* **1**, 103 (1957); also **2**, 593–600 (1948).

R. P. Bell and H. J. Emeléus, The boron hydrides and related compounds. *Quart. Rev., Chem. Soc.* **2**, 132 (1948).

"Borax to Boranes," Advan. Chem. Ser. No. 32. Amer. Chem. Soc., Washington, D.C., 1961.

D. C. Bradley, The stereochemistry of some elements of group III. *Progr. Stereochem.* **3**, 1 (1962).

R. J. Brotherton and H. Steinberg "Progress in Boron Chemistry," Vol. 3. Pergamon, Oxford, 1970.

H. C. Brown, Organoboranes, *in* "Organometallic Chemistry" (H. Zeiss, ed.), Amer. Chem. Soc. Monogr. Ser., Chapter 4, p. 150. Van Nostrand-Reinhold, Princeton, New Jersey, 1960.

H. C. Brown, Hydroboration—a powerful synthetic tool. *Tetrahedron* **12**, 117 (1961).

H. C. Brown, "Hydroboration." Benjamin, New York, 1962.

H. C. Brown, Organoborane—carbon monoxide reactions. A new versatile approach to the synthesis of carbon structure. *Accounts Chem. Res.* **2**, 65 (1969).

H. C. Brown, Versatile organoboranes. *Chem. Brit.* **7**, 458 (1971).

H. C. Brown, "Boranes in Organic Chemistry." Cornell Univ. Press, Ithaca, New York, 1972.

H. C. Brown and E. Negishi, Cyclic hydroboration of dienes. Simple convenient route to heterocyclic organoboranes. *Pure Appl. Chem.* **29**, 527 (1972).

H. C. Brown and M. M. Rogic, Organoboranes as alkylating and arylating agents. *Organometal Chem. Syn.* **1**, 305 (1972).

A. B. Burg, Chemical consequences of the borine group, BH_3. *Rec. Chem. Progr.* **15**, 159 (1954).

A. B. Burg, Chemical behavior and bonding of boron hydride derivatives. *Angew. Chem.* **72**, 183 (1960).

J. Casanova, *in* "Isonitrile Chemistry" (I. Ugi, ed.), p. 109. Academic Press, New York, 1971.

H. C. Clark, Perfluoroalkyl derivatives of the elements. *Advan. Fluorine Chem.* **3**, 19 (1963).

G. E. Coates, M. L. H. Green, and K. Wade, "Organo-Metallic Compounds," 3rd ed., Vol. 1, Chapter 3, p. 177. Methuen, London, 1967.

T. D. Coyle and J. J. Ritter, Organometallic aspects of diboron chemistry. *Advan. Organometal. Chem.* **10**, 237 (1972).

G. M. L. Cragg, Recent developments in the use of organoboranes in organic synthesis. *J. Chem. Educ.* **46**, 794 (1969).

R. H. Cragg and M. F. Lappert, Organic boron–sulfur compounds. *Organometal. Chem. Rev.* **1**, 43 (1966).

A. G. Davies, "Organic Peroxides." Butterworth, London, 1961.

A. G. Davies, Organoperoxyboranes. *Progr. Boron Chem.* **1**, 264 (1964).

A. Davison *et al.*, eds., New results in boron chemistry. "Topics in Current Chemistry." Vol. 15, Part 2. Springer-Verlag, Berlin and New York, 1970.

D. D. DeFord, C. A. Lucchesi, and J. M. Thoburn, "Analytical Chemistry of Boron." Northwestern University, Evanston, Illinois, 1953.

M. J. S. Dewar, Heteroaromatic boron compounds. *Advan. Chem. Ser.* **42**, 227 (1964).

G. R. Eaton and W. N. Lipscomb, "NMR Studies of Boron Hydrides and Related Compounds." Benjamin, New York, 1969.

A. Finch, J. B. Leach, and J. H. Morris, Boron–nitrogen ring systems. *Organometal. Chem. Rev., Sect. A* **4**, 1 (1969).

H. Flaschka and A. J. Barnard, Tetraphenylboron as an analytical reagent. *Advan. Anal. Chem. Instrum.* **1**, 1 (1961).

W. Gerrard, "The Organic Chemistry of Boron." Academic Press, New York, 1961.

W. Gerrard, *SCI* (*Soc. Chem. Ind., London*) *Monogr.* **13**, 328 (1961).

W. Gerrard and M. F. Lappert, Reactions of boron trichloride with organic compounds *Chem. Rev.* **58**, 1081 (1958).

Gmelins "Handbuch der Anorganische Chemie," Bor. Vol. 13. Verlag Chemie, GMBH, Weinheim (1954); also Springer-Verlag, Berlin and New York, 1974.

J. Goubeau, *FIAT Rev. Ger. Sci., 1939–1946, Inorg. Chem., Part I* (1948).

R. F. Gould, ed., "Boron-Nitrogen Chemistry," *Advan. Chem. Ser.* No. 42. Amer. Chem. Soc., Washington, D.C., 1964.

M. Grassberger, "Organic Boron Compounds." Verlag Chemie, Weinheim, 1971.

R. N. Grimes, "Carboranes," Academic Press, New York, 1970.

M. F. Hawthorne, Decaborane-14 and its derivatives. *Advan. Inorg. Chem. Radiochem.* **5**, 307 (1963).

M. F. Hawthorne, Polyhedral boranes, carboranes and carbametallic boron hydride derivatives. *Endeavour* **25**, 146 (1966).

M. F. Hawthorne, The chemistry of the polyhedral species derived from transition metals and carboranes. *Accounts Chem. Res.* **1**, 281 (1968).

H. G. Heal, Recent studies in boron chemistry. *Roy. Inst. Chem., Lect., Monogr., Rep.* No. 1 (1960).

R. T. Holzmann, ed., "Production of the Boranes and Related Research." Academic Press, New York, 1967.

H. D. Kaesz and F. G. A. Stone, *in* "Organometallic Chemistry" (H. Zeiss, ed.),

Amer. Chem. Soc. Monogr., Chapter 3, p. 88. Van Nostrand-Reinhold, Princeton, New Jersey, 1960.
R. Köster, Heterocyclic organoboranes. *Advan. Organometal. Chem.* **2**, 257 (1964).
R. Köster, Organoboron heterocycles. *Progr. Boron Chem.* **1**, 289 (1964).
R. Köster, Redistribution reactions of organoboranes and organoalanes. *Ann. N.Y. Acad. Sci.* **159**, 73 (1969).
R. Köster and M. A. Grassberger, Structures and syntheses of carboranes. *Angew. Chem., Int. Ed. Engl.* **6**, 218 (1967).
R. Köster, G. Benedikt, W. Larbig, K. Reinert, and G. Rotermund, Umwandlunger boroganischer Verbindungen in der Hitze. *Angew. Chem.* **75**, 1079 (1963).
M. F. Lappert, Organic compounds of boron. *Chem. Rev.* **56**, 959 (1956).
M. F. Lappert, Polymers containing boron and nitrogen. *In* "Developments in Inorganic Polymer Chemistry," Chapter 2, p. 20. Elsevier, Amsterdam, 1962.
M. F. Lappert, Boron–carbon compounds, *in* "The Chemistry of Boron and its Compounds" (E. L. Muetterties, ed.), Chapter 8, p. 443. Wiley, New York, 1967.
M. F. Lappert, ed., *Inorg. Chem. Ser. 1 MTP Int. Rev. Sci.* **1** (1972).
M. F. Lappert and H. Pyszora, Pseudohalides of group IIIB and IVB elements. *Advan. Inorg. Chem. Radiochem.* **9**, 133 (1966).
W. N. Lipscomb, Structure and reactions of the boron hydrides. *J. Inorg. Nucl. Chem.* **11**, 1 (1959).
W. N. Lipscomb, "Boron Hydrides." Benjamin, New York, 1963.
J. C. Lockhart, Redistribution and exchange reactions in Groups IIB–VIIB. *Chem. Rev.* **65**, 131 (1965).
M. L. Maddox, S. L. Stafford, and H. D. Kaesz, Applications of NMR to the study of organometallic compounds. *Advan. Organometal. Chem.* **3**, 1 (1965).
P. M. Maitlis, Heterocyclic organic boron compounds. *Chem. Rev.* **62**, 223 (1962).
D. R. Martin, Coordination compounds of boron bromide and boron iodide. *Chem. Rev.* **42**, 581 (1948).
A. G. Massey, The halides of boron. *Advan. Inorg. Chem. Radiochem.* **10**, 1 (1967).
D. S. Matteson, Organofunctional boronic esters. *Organometal. Chem. Rev.* **1**, 1 (1966).
A. Meller, The chemistry of iminoboranes. *Top. Curr. Chem.* **26**, 37 (1972).
B. M. Mikhailov, Borazole and its derivatives. *Usp. Khim.* **29**, 972 (1960).
B. M. Mikhailov, The chemistry of diborane. *Usp. Khim.* **31**, 417 (1962).
B. M. Mikhailov, Allyl boron compounds. *Organometal. Chem. Rev., Sect. A* **8**, 1 (1972).
K. Moedritzer, The redistribution reaction. *Organometal. React.* **2**, 1 (1971).
E. L. Muetteries, ed., "The Chemistry of Boron and its Compounds." Wiley, New York, 1967.
E. L. Muetterties and W. H. Knoth, "Polyhedral Boranes." Dekker, New York, 1968.
E. L. Muetterties and W. D. Phillips, The use of nuclear magnetic resonance in inorganic chemistry. *Advan. Inorg. Chem. Radiochem.* **4**, 231 (1962).
K. Niedenzu, Recent developments in the chemistry of aminoboranes. *Angew. Chem.* **76**, 165 (1964).
K. Niedenzu, Synthesis of organohaloboranes. *Organometal. Chem. Rev.* **1**, 305 (1966).
K. Niedenzu and J. W. Dawson, "Boron–Nitrogen Compounds." Springer-Verlag, Berlin and New York, 1965.
H. Nöth, Anorganische Reaktionen der Alkaliboranate. *Angew. Chem.* **73**, 371 (1961).
T. Onak, Carboranes and organosubstituted boron hydrides. *Advan. Organometal. Chem.* **3**, 263 (1965).
P. I. Paetzold, Preparation, properties and decomposition of boron azides. *Fortschr, Chem. Forsch.* **8**, 437 (1967).
R. W. Parry and M. K. Walter, The boron hydrides. *Prep. Inorg. React.* **5**, 45 (1968).

E. G. Rochow, D. T. Hurd, and R. N. Lewis, "The Chemistry of Organometallic Compounds." Wiley, New York, 1957.

W. H. Schechter, C. B. Jackson, and R. M. Adams, "Boron Hydrides and Related Compounds," 2nd ed. Callery Chemical Co., Callery, Pennsylvania, 1954.

E. Schenker, Use of complex borohydrides and diboranes in organic chemistry. *Angew. Chem.* **73**, 81 (1961).

H. I. Schlesinger and H. C. Brown, in collaboration with B. Abraham, A. C. Bond, N. Davidson, A. E. Finholt, J. R. Gilbreath, H. Hoekstra, L. Horvitz, E. K. Hyde, J. J. Katz, J. Knight, R. A. Lad, D. L. Mayfield, L. Rapp, D. M. Ritter, A. M. Schwartz, I. Sheft, L. D. Tuck, and A. O. Walker, New developments in the chemistry of diborane and the borohydrides. I. General summary. *J. Amer. Chem. Soc.* **75**, 186 (1953).

H. I. Schlesinger and A. B. Burg, Recent developments in the chemistry of the boron hydrides. *Chem. Rev.* **31**, 1 (1942).

G. Schmid, Metal–boron compounds—problems and perspectives. *Angew. Chem., Int. Ed. Engl.* **9**, 819 (1970).

F. R. Scholer and L. J. Todd, Polyhedral boranes and heteroatom boranes. *Prep. Inorg. React.* **7**, 1 (1971).

I. Shapiro, The boron–carbon–hydrogen system. *Talanta* **11**, 211 (1964).

J. C. Sheldon and B. C. Smith, The borazoles. *Quart. Rev., Chem. Soc.* **14**, 200 (1960).

O. P. Shtov, S. L. Ioffe, V. A. Tartakovski, and S. S. Novikov, Cationic boron complexes. *Usp. Khim.* **39**, 905 (1970).

B. Seigel and J. L. Mack, The boron hydrides. *J. Chem. Educ.* **34**, 314 (1957).

H. A. Skinner, The strengths of metal-to-carbon bonds. *Advan. Organometal. Chem.* **2**, 49 (1964).

W. Sneeden, The mass spectra of some borazoles, *Advan. Mass Spectrom.* **2**, 456 (1963).

A. H. Soloway, *Progr. Boron Chem.* **1**, 203 (1964).

V. I. Stanko, A. Yu. Chapovskii, V. A. Brattsev, and L. I. Zakharkin, The chemistry of decaborane and its derivatives. *Russ. Chem. Rev.* **34**, 424 (1965).

H. Steinberg and R. J. Brotherton, "Organoboron Chemistry," Vol. 2. Wiley (Interscience), New York, 1966.

H. Steinberg and A. L. McCloskey, eds., "Progress in Boron Chemistry," Vol. 1. Macmillan, New York, 1964.

A. Stock, "Hydrides of Boron and Silicon." Cornell Univ. Press, Ithaca, New York, 1933.

F. G. A. Stone, Chemistry of the boron hydrides. *Quart. Rev., Chem. Soc.* **9**, 174 (1955).

F. G. A. Stone, Stability relationships among analogous molecular addition compounds of group III elements. *Chem. Rev.* **58**, 101 (1958).

F. G. A. Stone, Chemical reactivity of the boron hydrides and related compounds. *Advan. Inorg. Chem. Radiochem.* **2**, 279 (1960).

W. Tochtermann, Structur and Reactionweise Organischer at-Komplexe. *Angew. Chem.* **18**, 355 (1966).

L. J. Todd, Transition metal—carborane complexes. *Advan. Organometal. Chem.* **8**, 87 (1970).

P. M. Treichel and F. G. A. Stone, Fluorocarbon derivatives of metals. *Advan. Organometal. Chem.* **1**, 43 (1964).

S. Trofimenko, Polypyrazolylborates. *Accounts Chem. Res.* **4**, 17 (1971).

C. Walling, The role of heteroatoms in oxidation. *Advan. Chem. Ser.* **75**, 166 (1968).

E. Wiberg, New results in preparative hydride research. *Angew. Chem.* **65**, 16 (1953)

E. Wiberg and B. Bolz, *FIAT Rev., Inorg. Chem.* **1**, 238 (1948).

G. Wittig, Komplexbildung and Reactivät in der Metallorganischen Chemie. *Angew. Chem.* **70**, 65 (1958).

V. A. Zamyatina and N. I. Bekasova, Polymeric compounds of boron. *Usp. Khim.* **30**, 48 (1961).

V. A. Zamyatina and N. I. Bekasova, Polymeric boron compounds. *Russ. Chem. Rev.* **33**, 524 (1964).

G. Zweifel and H. C. Brown, Hydration of olefins, dienes and acetylenes via hydroboration. *Org. React.* **13**, 1 (1963).

Bibliography

1. E. W. Abel, D. A. Armitage, R. P. Bush, and G. R. Willey, *J. Chem. Soc., London* p. 62 (1965).
2. E. W. Abel, S. H. Dandegaonker, W. Gerrard, and M. F. Lappert, *J. Chem. Soc., London* p. 4697 (1956).
3. E. W. Abel, W. Gerrard, and M. F. Lappert, *J. Chem. Soc., London* p. 112 (1957).
4. E. W. Abel, W. Gerrard, and M. F. Lappert, *J. Chem. Soc., London* p. 1451 (1958).
5. E. W. Abel, W. Gerrard, and M. F. Lappert, *J. Chem. Soc., London* p. 3833 (1957).
6. E. W. Abel, W. Gerrard, and M. F. Lappert, *J. Chem. Soc., London* p. 5051 (1957).
7. E. W. Abel, W. Gerrard, and M. F. Lappert, *Chem. Ind.* (*London*) p. 158 (1958).
8. E. W. Abel, W. Gerrard, M. F. Lappert, and R. Shafferman, *J. Chem. Soc., London* p. 2895 (1958).
9. E. W. Abel, N. Giles, D. J. Walker, and J. N. Wingfield, *J. Chem. Soc., A* p. 1991 (1971).
10. E. W. Abel and A. Singh, *J. Chem. Soc., London* p. 690 (1959).
11. E. W. Abel, D. J. Walker, and J. N. Wingfield, *Inorg. Nucl. Chem. Lett.* **5**, 139 (1969).
12. P. Abley and J. Halpern, *J. Chem. Soc., D* p. 1238 (1971).
13. M. H. Abraham and A. G. Davies, *Chem. Ind.* (*London*) p. 1622 (1957).
14. M. H. Abraham and A. G. Davies, *J. Chem. Soc., London* p. 429 (1959).
15. M. H. Abraham, J. H. N. Garland, J. A. Hill, and L. F. Larkworthy, *Chem. Ind.* (*London*) p. 1615 (1962).
16. M. Abufhele, C. Andersen, E. A. Lissi, and E. Sanhueza, *J. Organometal. Chem.* **42**, 19 (1972).
17. E. Abuin, J. Grotewold, E. A. Lissi, and M. C. Vara, *J. Chem. Soc., B* p. 1044 (1968).
18. E. Abuin, E. A. Lissi, and A. Yanez, *J. Polym. Sci., Part B* **8**, 515 (1970).

18a. R. M. Adams, *J. Chem. Soc.* **4**, 95 (1964).

19. R. M. Adams, *Inorg. Chem.* **7**, 1945 (1968).
20. R. M. Adams and K. A. Jensen, *Pure Appl. Chem.* **30**, 681 (1972).
21. R. M. Adams and F. D. Poholsky, *Inorg. Chem.* **2**, 640 (1963).
22. J. L. Adcock and J. J. Lagowski, *Inorg. Nucl. Chem. Lett.* **7**, 473 (1971).
23. V. D. Aftandilian, H. C. Miller, and E. L. Muetterties, *J. Amer. Chem. Soc.* **83**, 2471 (1961).
24. A. D. Ainley and F. Challenger, *J. Chem. Soc., London* p. 2171 (1930).

24a. A. A. Akhrem, I. S. Levina, Yu. A. Titov, Yu. N. Bubnov, and B. M. Mikhailov, *Izv. Akad. Nauk. SSSR, Ser. Khim.* p. 1639 (1972).

25. K. J. Alford, E. O. Bishop, P. R. Carey, and J. D. Smith, *J. Chem. Soc., A* p. 2574 (1971).
26. A. Allerhand, J. D. Odom, and R. E. Moll, *J. Chem. Phys.* **50**, 5037B (1969).
27. P. G. Allies and P. B. Brindley, *Chem. Ind. (London)* p. 319 (1967).
28. P. G. Allies and P. B. Brindley, *J. Chem. Soc., B* p. 1126 (1969).
29. E. L. Allred, C. L. Anderson, and R. L. Smith, *Tetrahedron Lett.* p. 951 (1966).
30. E. R. Altwicker, *Diss. Abstr.* **22**, 3389 (1962).
31. E. R. Altwicker, G. E. Ryschkewitsch, A. B. Garrett, and H. H. Sisler, *Inorg. Chem.* **3**, 454 (1964).
32. E. Amberger and R. Römer, *Z. Anorg. Allg. Chem.* **345**, 1 (1966).
33. H. H. Anderson, *J. Chem. Eng. Data* **8**, 576 (1963).
34. T. T. Ang and B. A. Dunell, *J. Chem. Soc., Faraday Trans. II* **68**, 1331 (1972).
34a. V. K. Antonov, T. V. Ivanina, A. G. Ivanova, I. V. Berezin, A. V. Levashov, and K. Martinek, *FEBS (Fed. Eur. Biochem. Soc.) Lett.* **20**, 37 (1972).
34b. K. Anzenhofer, *Mol. Phys.* **11**, 495 (1966).
35. R. Appel and F. Vogt, *Chem. Ber.* **95**, 2225 (1962).
36. J. F. Archer, D. R. Boyd, W. R. Jackson, M. F. Grundon, and W. A. Khan, *J. Chem. Soc., C* p. 2560 (1971).
37. D. R. Armstrong and P. G. Perkins, *Theor. Chim. Acta* **5**, 215 (1966).
38. D. R. Armstrong and P. G. Perkins, *Theor. Chim. Acta* **5**, 222 (1966).
39. D. R. Armstrong and P. G. Perkins, *Theor. Chim. Acta* **4**, 69 (1966).
40. D. R. Armstrong and P. G. Perkins, *J. Chem. Soc., A* p. 1026 (1966).
41. D. R. Armstrong and P. G. Perkins, *J. Chem. Soc., A* p. 123 (1967).
42. D. R. Armstrong and P. G. Perkins, *J. Chem. Soc., A* p. 1044 (1969).
43. D. R. Armstrong and P. G. Perkins, *Theor. Chim. Acta* **4**, 352 (1966).
44. D. R. Armstrong and P. G. Perkins, *Theor. Chim. Acta* **5**, 11 (1966).
45. S. Arnott and S. C. Abrahams, *Acta Crystallogr.* **11**, 449 (1958).
46. P. M. Aronovich, V. S. Bogdanov, and B. M. Mikhailov, *Izv. Akad. Nauk. SSSR, Ser. Khim.* p. 1682 (1970).
47. P. Arthur, R. Annino, and W. P. Donahoo, *Anal. Chem.* **29**, 1852 (1957).
48. E. C. Ashby, *J. Amer. Chem. Soc.* **81**, 4791 (1959).
49. E. C. Ashby and W. E. Foster, *J. Org. Chem.* **29**, 3225 (1964).
50. A. J. Ashe and P. Shu, *J. Amer. Chem. Soc.* **93**, 1804 (1971).
51. N. Ashikari, *Bull. Chem. Soc. Jap.* **31**, 540 (1958).
52. N. Ashikari, *J. Polym. Sci.* **28**, 250 (1958).
53. N. Ashikari, *Bull. Chem. Soc. Jap.* **31**, 229 (1958).
54. N. Ashikari, *Bull. Chem. Soc. Jap.* **32**, 1056 (1959).
55. N. Ashikari and A. Nishimura, *J. Polym. Sci.* **31**, 249 (1960).
55a. I. B. Atkinson, D. C. Blundell, and D. B. Clapp, *J. Inorg. Nucl. Chem.* **34**, 3037 (1972).
56. D. W. Aubrey, M. F. Lappert, and H. Pyszora, *J. Chem. Soc., London* p. 1931 (1961).
57. R. W. Auten and C. A. Kraus, *J. Amer. Chem. Soc.* **74**, 3398 (1952).
58. G. Axelrad and D. Halpern, *Chem. Commun.* p. 291 (1971).
58a. D. E. Axelson, A. J. Oliver, and C. E. Holloway, *Org. Magn. Resonance* **5**, 255 (1973).
59. B. J. Aylett and J. Emsley, *J. Chem. Soc., A* p. 652 (1967).
60. B. J. Aylett and L. K. Peterson, *J. Chem. Soc., London* p. 4043 (1965).
61. H. T. Baechle and H. J. Becher, *Spectrochim. Acta* **21**, 579 (1965).
62. H. T. Baechle, H. J. Becher, H. Beyer, W. S. Brey, Jr., J. W. Dawson, M. E. Fuller, and K. Niedenzu, *Inorg. Chem.* **2**, 1065 (1963).

63. C. S. L. Baker, *J. Organometal. Chem.* **19**, 287 (1969).
64. A. T. Balaban, A. Arsene, I. Bally, A. Barabas, M. Paraschiv, M. Roman, and E. Romas, *Rev. Roum. Chim.* **15**, 635 (1970).
65. A. T. Balaban and Z. Simon, *Rev. Roum. Chim.* **9**, 99 (1964).
66. I. Bally and A. T. Balaban, *Rev. Roum. Chim.* **16**, 739 (1971).
67. C. H. Bamford and D. M. Newitt, *J. Chem. Soc., London* p. 695 (1946).
68. W. R. Bamford and S. Fordham, *SCI* (*Soc. Chem. Ind., London*) *Monogr.* **13**, 320 (1961).
69. L. Banford and G. E. Coates, *J. Chem. Soc., London* p. 3564 (1964).
70. P. A. Barfield, M. F. Lappert, and J. Lee, *Proc. Chem. Soc., London* p. 421 (1961).
71. P. A. Barfield, M. F. Lappert, and J. Lee, *J. Chem. Soc., A* p. 554 (1968).
72. P. A. Barfield, M. F. Lappert, and J. Lee, *Trans. Faraday Soc.* **64**, 2571 (1968).
73. D. Barnes, W. G. Henderson, E. F. Mooney, and P. C. Uden, *J. Inorg. Nucl. Chem.* **33**, 2799 (1971).
74. R. L. Barnes and T. Wartik, *J. Amer. Chem. Soc.* **85**, 360 (1963).
75. L. S. Bartell and B. L. Carroll, *J. Chem. Phys.* **42**, 3076 (1965).
76. L. S. Bartell, B. L. Carroll, and J. P. Guillory, *Tetrahedron Lett.* **13**, 705 (1964).
77. B. Bartocha, F. E. Brinckman, H. D. Kaesz, and F. G. A. Stone, *Proc. Chem. Soc., London* p. 116 (1958).
78. B. Bartocha, C. M. Douglas, and M. Y. Gray, *Z. Naturforsch. B* **14**, 809 (1959).
79. B. Bartocha, W. A. G. Graham, and F. G. A. Stone, *J. Inorg. Nucl. Chem.* **6**, 119 (1958).
80. L. Barton and G. T. Bohn, *J. Chem. Soc. ,D* p. 77 (1971).
81. L. Barton, C. Perrin, and R. F. Porter, *Inorg. Chem.* **5**, 1446 (1966).
82. S. H. Bauer, *J. Amer. Chem. Soc.* **59**, 1804 (1937).
83. S. H. Bauer, *J. Amer. Chem. Soc.* **78**, 5775 (1956).
84. R. Bauer, *Z. Naturforsch. B* **16**, 557 (1961).
85. R. Bauer, *Z. Naturforsch. B* **16**, 839 (1961).
85a. S. H. Bauer, *Advan. Chem. Ser.* **32**, 88 (1961).
86. S. H. Bauer and J. Y. Beach, *J. Amer. Chem. Soc.* **63**, 1394 (1941).
87. S. H. Bauer and J. M. Hastings, *J. Amer. Chem. Soc.* **64**, 2686 (1942).
88. C. E. H. Bawn, A. Ledwith, and P. Matthies, *J. Polym. Sci.* **34**, 93 (1959).
89. C. E. H. Bawn, D. Margerison, and N. M. Richardson, *Proc. Chem. Soc., London* p. 397 (1959).
90. H. C. Beachell and D. W. Beistel, *Inorg. Chem.* **3**, 1028 (1964).
91. H. C. Beachell and K. R. Lange, *J. Phys. Chem.* **60**, 307 (1956).
92. O. T. Beachley, Jr., *Inorg. Chem.* **8**, 2665 (1969).
93. F. R. Bean and J. R. Johnson, *J. Amer. Chem. Soc.* **54**, 4415 (1932).
94. H. J. Becher, *Z. Anorg. Chem.* **270**, 273 (1952).
95. H. Becher, *Z. Anorg. Chem.* **271**, 243 (1953).
96. H. J. Becher, *Z. Anorg. Chem.* **288**, 235 (1956).
97. H. J. Becher, *Z. Anorg. Chem.* **289**, 262 (1957).
98. H. J. Becher, *Z. Anorg. Chem.* **291**, 151 (1957).
99. H. J. Becher, *Spectrochim. Acta* **19**, 575 (1963).
100. W. E. Becker and E. C. Ashby, *Inorg. Chem.* **4**, 1816 (1965).
101. H. J. Becher and H. T. Baechle, *Chem. Ber.* **98**, 2159 (1965).
102. H. J. Becher and H. T. Baechle, *Advan. Chem. Ser.* **42**, 71 (1964).
103. H. J. Becher and S. Frick, *Z. Phys. Chem.* (*Frankfurt am Main*) [N.S.] **12**, 241 (1957).
104. H. J. Becher and S. Frick, *Z. Anorg. Allg. Chem.* **295**, 83 (1958).
105. H. J. Becher and J. Goubeau, *Z. Anorg. Allg. Chem.* **268**, 133 (1952).

106. R. Belcher, D. Gibbons, and A. Sykes, *Mikrochim. Acta* **40**, 75 (1952).
106a. T. N. Bell and A. E. Platt, *Int. J. Chem. Kinet.* **3**, 307 (1971).
107. L. J. Bellamy, W. Gerrard, M. F. Lappert, and R. L. Willams, *J. Chem. Soc., London* p. 2412 (1958).
108. H. Bellut, C. D. Miller, and R. Köster, *Syn. Inorg. Metal.-Org. Chem.* **1**, 83 (1971).
109. J. E. Bennett and H. A. Skinner, *J. Chem. Soc., London* p. 2472 (1961).
110. H. E. Bent and M. Dorfman, *J. Amer. Chem. Soc.* **54**, 2132 (1932).
111. H. E. Bent and M. Dorfman, *J. Amer. Chem. Soc.* **57**, 1259 (1935).
112. H. E. Bent and M. Dorfman, *J. Amer. Chem. Soc.* **57**, 1924 (1935).
113. J. R. Berschied, Jr. and K. F. Purcell, *Inorg. Chem.* **9**, 624 (1970).
114. J. R. Berschied, Jr. and K. F. Purcell, *Inorg. Chem.* **11**, 930 (1972).
114a. I. Bertini, P. Dapporto, G. Fallani, and L. Sacconi, *Inorg. Chem.* **10**, 1703 (1971).
115. E. Bessler and J. Goubeau, *Z. Anorg. Allg. Chem.* **352**, 67 (1967).
116. G. W. Bethke and W. M. Kent, *J. Chem. Phys.* **26**, 1118 (1957).
117. G. W. Bethke and W. M. Kent, *J. Chem. Phys.* **27**, 978 (1957).
118. B. Bettman and G. E. K. Branch, *J. Amer. Chem. Soc.* **56**, 1616 (1934).
119. B. Bettman, G. E. K. Branch, and D. L. Yabroff, *J. Amer. Chem. Soc.* **56**, 1865 (1934).
120. H. Beyer, J. W. Dawson, H. Jenne, and K. Niedenzu, *J. Chem. Soc., London* p. 2115 (1964).
121. G. Bianchi, A. Cogoli, and P. Grunanger, *J. Organometal. Chem.* **6**, 598 (1966).
122. M. J. Biallas, *Inorg. Chem.* **10**, 1320 (1971).
123. D. B. Bigley and D. W. Payling, *Chem. Commun.* p. 938 (1968).
124. D. B. Bigley and D. W. Payling, *J. Chem. Soc., B* p. 1811 (1970).
125. D. B. Bigley and D. W. Payling, *J. Inorg. Nucl. Chem.* **33**, 1157 (1971).
126. P. Binger, *Tetrahedron Lett.* p. 2675 (1966).
127. P. Binger, *Angew. Chem.* **79**, 57 (1967); *Angew. Chem., Int. Ed. Engl.* **6**, 84 (1967).
128. P. Binger, *Angew. Chem., Int. Ed. Engl.* **7**, 286 (1968).
129. P. Binger and R. Köster, *Tetrahedron Lett.* p. 156 (1961).
130. P. Binger and R. Köster, *Angew. Chem.* **74**, 652 (1962).
131. P. Binger and R. Köster, *Tetrahedron Lett.* p. 1901 (1965).
131a. P. Binger and R. Köster, *Synthesis* p. 309 (1973).
132. J. M. Birchall, R. N. Haszeldine, and J. F. Marsh, *Chem. Ind. (London)* p. 1080 (1961).
133. J. R. Blackborow and J. C. Lockhart, *J. Chem. Soc., A* p. 1343 (1971).
134. E. J. Blau and B. W. Mulligan, *J. Chem. Phys.* **26**, 1085 (1957).
135. N. J. Blay, I. Dunstan, and R. L. Williams, *J. Chem. Soc., London* p. 430 (1960).
136. N. J. Blay, R. J. Pace, and R. L. Williams, *J. Chem. Soc., London* p. 3416 (1962).
137. N. J. Blay, J. Williams, and R. L. Williams, *J. Chem. Soc., London* p. 424 (1960).
138. T. L. Blundell and H. M. Powell, *Acta Crystallogr., Sect. B* **27**, 2304 (1971).
139. H. Bock and W. Fuss, *Chem. Ber.* **104**, 1687 (1971).
140. V. S. Bogdanov, P. M. Aronovich, A. D. Naumov, and B. M. Mikhailov, *Zh. Obshch. Khim.* **41**, 1063 (1971).
140a. V. S. Bogdanov, Yu. N. Bubnov, M. N. Bochkareva, and B. M. Mikhailov, *Dokl. Akad. Nauk. SSSR* **201**, 605 (1971).
141. V. S. Bogdanov, A. V. Kessenikh, and V. V. Negrebetsky, *J. Magn. Resonance* **5**, 145 (1971).
141a. V. S. Bogdanov, A. V. Kessenikh, and V. V. Negrebetsky, *Izv. Akad. Nauk. SSSR, Ser. Khim.* p. 1363 (1971).
141b. V. S. Bogdanov, A. V. Kessenikh, V. V. Negrebetsky, and A. Ya. Shchteinshneider, *Zh. Strukt. Khim.* **13**, 226 (1972).

141c. V. S. Bogdanov, V. G. Kiselev, A. D. Naumov, L. S. Vasil'ev, V. P. Dmitrikov, V. A. Dorokhov, and B. M. Mikhailov, *Zh. Obshch. Khim.* **42**, 1547 (1972).
142. V. S. Bogdanov, V. F. Pozdnev, Yu. N. Bubnov, and B. M. Mikhailov, *Dokl. Akad. Nauk. SSSR* **193**, 586 (1970).
143. H. Bohme and E. Boll, *Z. Anorg. Chem.* **291**, 160 (1957).
144. R. B. Booth and C. A. Kraus, *J. Amer. Chem. Soc.* **74**, 1415 (1952).
145. J. L. Boone and G. W. Willcockson, *Inorg. Chem.* **5**, 311 (1966).
146. R. F. Borch and H. D. Durst, *J. Amer. Chem. Soc.* **91**, 3996 (1969).
147. R. J. Bose and M. D. Peters, *Can. J. Chem.* **49**, 1766 (1971).
148. M. Bossa and F. Maraschini, *J. Chem. Soc., A* p. 1416 (1970).
149. R. E. Bowen and C. R. Phillips, *J. Inorg. Nucl. Chem.* **34**, 382 (1972).
150. R. A. Bowie and O. C. Musgrave, *Proc. Chem. Soc., London* p. 15 (1964).
151. G. E. K. Branch, D. L. Yabroff, and B. Bettman, *J. Amer. Chem. Soc.* **56**, 937 (1934).
152. L. M. Braun, R. A. Braun, H. R. Crissman, M. Opperman, and R. M. Adams, *J. Org. Chem.* **36**, 2388 (1971).
153. R. A. Braun, D. C. Brown, and R. M. Adams, *J. Amer. Chem. Soc.* **93**, 2823 (1971).
154. E. H. Braye and W. Hubel, *J. Amer. Chem. Soc.* **83**, 4406 (1961).
155. G. L. Brennan and R. Schaeffer, *J. Inorg. Nucl. Chem.* **20**, 205 (1961).
156. S. Bresadola, F. Rossetto, and G. Puosi, *Tetrahedron Lett.* p. 4775 (1965).
157. E. Brehm, A. Haag, and G. Hesse, *Justus Liebigs Ann. Chem.* **737**, 80 (1970).
158. E. Brehm, A. Haag, G. Hesse, and A. Witte, *Justus Liebigs Ann. Chem.* **737**, 70 (1970).
159. G. L. Brennan, G. H. Dahl, and R. Schaeffer, *J. Amer. Chem. Soc.* **82**, 6248 (1960).
160. S. Bresadola, G. Carraro, C. Pecile, and A. Turco, *Tetrahedron Lett.* p. 3185 (1964).
161. S. W. Breuer, *J. Chem. Soc., Chem. Commun.* p. 671 (1972).
162. E. Breuer and H. C. Brown, *J. Amer. Chem. Soc.* **91**, 4164 (1969).
163. S. W. Breuer and F. A. Broster, *J. Organometal. Chem.* **35**, 5C (1972).
163a. S. W. Breuer and F. A. Broster, *Tetrahedron Lett.* p. 2193 (1972).
164. S. W. Breuer, M. J. Leatham, and F. G. Thorpe, *J. Chem. Soc., D* p. 1475 (1971).
165. W. S. Brey, Jr., M. E. Fuller, G. E. Ryschkewitsch, and A. S. Marshall, *Advan. Chem. Ser.* **42**, 100 (1964).
166. V. T. Brice, H. D. Johnson, and S. G. Shore, *J. Chem. Soc., Chem. Commun.* p. 1128 (1972).
167. V. T. Brice and S. G. Shore, *Inorg. Chem.* **12**, 309 (1973).
168. F. E. Brinckman and F. G. A. Stone, *Chem. Ind. (London)* p. 254 (1959).
169. F. E. Brinckman and F. G. A. Stone, *J. Amer. Chem. Soc.* **82**, p. 6218 (1960).
170. P. B. Brindley and J. C. Dodgson, *Chem. Commun.* p. 202 (1972).
171. P. B. Brindley, W. Gerrard, and M. F. Lappert, *J. Chem. Soc., London* p. 2956 (1955).
172. P. B. Brindley, W. Gerrard, and M. F. Lappert, *J. Chem. Soc., London* p. 824 (1956).
173. P. B. Brindley, W. Gerrard, and M. F. Lappert, *J. Chem. Soc., London* p. 1540 (1956).
174. P. B. Brindley and R. G. Pearson, *J. Polym. Sci., Part B* **6**, 831 (1968).
175. R. S. Brokaw, E. J. Badin, and R. N. Pease, J. *Amer. Chem. Soc.* **70**, 1921 (1948).
176. R. S. Brokaw and R. N. Pease, *J. Amer. Chem. Soc.* **72**, 3237 (1950).
177. R. S. Brokaw and R. N. Pease, *J. Amer. Chem. Soc.* **72**, 5263 (1950).
177a. C. J. W. Brooks and D. J. Harvey, *J. Chromatogr.* **54**, 193 (1971).
178. C. J. W. Brooks, B. S. Middleditch, and D. J. Harvey, *Org. Mass Spectro.* **5**, 1429 (1971).
179. R. J. Brotherton, H. M. Manasevit, and A. L. McCloskey, *Inorg. Chem.* **1**, 749 (1962).
180. R. J. Brotherton and A. L. McCloskey, *Advan. Chem. Ser.* **42**, 131 (1964).

181. D. A. Brown, S. F. A. Kettle, J. McKenna, and J. M. McKenna, *Chem. Commun.* p. 667 (1967).
182. D. A. Brown and C. G. McCormack, *Theor. Chim. Acta* **6**, 350 (1966).
183. H. C. Brown, *J. Amer. Chem. Soc.* **67**, 374 (1945).
184. H. C. Brown, *J. Chem. Soc., London* p. 1248 (1956).
184a. H. C. Brown, "Hydroboration." Benjamin, New York, 1962.
184b. H. C. Brown, "Boranes in Organic Chemistry." Cornell Univ. Press, Ithaca, New York, 1972.
185. H. C. Brown, N. R. Ayyangar, and G. Zweifel, *J. Amer. Chem. Soc.* **84**, 4341 (1962).
186. H. C. Brown, N. R. Ayyangar, and G. Zweifel, *J. Amer. Chem. Soc.* **86**, 397 (1964).
187. H. C. Brown, N. R. Ayyangar, and G. Zweifel, *J. Amer. Chem. Soc.* **86**, 1071 (1964).
188. H. C. Brown and G. K. Barbaras, *J. Amer. Chem. Soc.* **69**, 1137 (1947).
189. H. C. Brown and G. K. Barbaras, *J. Amer. Chem. Soc.* **75**, 6 (1953).
190. H. C. Brown, H. Bartholomay, and M. D. Taylor, *J. Amer. Chem. Soc.* **66**, 435 (1944).
191. H. C. Brown and M. V. Bhatt, *J. Amer. Chem. Soc.* **82**, 2074 (1960).
192. H. C. Brown and M. V. Bhatt, *J. Amer. Chem. Soc.* **88**, 1440 (1966).
193. H. C. Brown, M. V. Bhatt, T. Munekata, and G. Zweifel, *J. Amer. Chem. Soc.* **89**, 567 (1967).
194. H. C. Brown and D. B. Bigley, *J. Amer. Chem. Soc.* **83**, 486 (1961).
195. H. C. Brown and D. B. Bigley, *J. Amer. Chem. Soc.* **83**, 3166 (1961).
196. H. C. Brown, D. B. Bigley, S. K. Arora, and N. M. Yoon, *J. Amer. Chem. Soc.* **92**, 7161 (1970).
196a. H. C. Brown and B. A. Carlson, *J. Org. Chem.* **38**, 2422 (1973).
197. H. C. Brown, B. A. Carlson, and R. H. Prager, *J. Amer. Chem. Soc.* **93**, 2070 (1971).
198. H. C. Brown and R. A. Coleman, *J. Amer. Chem. Soc.* **91**, 4606 (1969).
199. H. C. Brown, R. A. Coleman, and M. W. Rathke, *J. Amer. Chem. Soc.* **90**, 499 (1968).
200. H. C. Brown and O. J. Cope, *J. Amer. Chem. Soc.* **86**, 1801 (1964).
201. H. C. Brown and W. C. Dickason, *J. Amer. Chem. Soc.* **91**, 1226 (1969).
202. H. C. Brown and W. C. Dickason, *J. Amer. Chem. Soc.* **92**, 709 (1970).
203. H. C. Brown and V. H. Dodson, *J. Amer. Chem. Soc.* **79**, 2302 (1957).
204. H. C. Brown and C. P. Garg, *J. Amer. Chem. Soc.* **83**, 2951 (1961).
205. H. C. Brown and M. Gerstein, *J. Amer. Chem. Soc.* **72**, 2923 (1950).
206. H. C. Brown and D. Gintis, *J. Amer. Chem. Soc.* **78**, 5378 (1956).
207. H. C. Brown, D. Gintis, and L. Domash, *J. Amer. Chem. Soc.* **78**, 5387 (1956).
208. H. C. Brown and S. K. Gupta, *J. Amer. Chem. Soc.* **92**, 6983 (1970).
209. H. C. Brown and S. K. Gupta, *J. Amer. Chem. Soc.* **93**, 1816 (1971).
210. H. C. Brown and S. K. Gupta, *J. Amer. Chem. Soc.* **93**, 1818 (1971).
211. H. C. Brown and S. K. Gupta, *J. Amer. Chem. Soc.* **93**, 2802 (1971).
212. H. C. Brown and S. K. Gupta, *J. Amer. Chem. Soc.* **93**, 4062 (1971).
213. H. C. Brown and S. K. Gupta, *J. Organometal. Chem.* **32**, 1c (1971).
214. H. C. Brown and S. K. Gupta, *J. Amer. Chem. Soc.* **94**, 4370 (1972).
214a. H. C. Brown, T. Hamaoka, and N. Ravindran, *J. Chem. Soc. Amer.* **95**, 5786 (1973).
215. H. C. Brown, N. C. Hebert, and C. H. Snyder, *J. Amer. Chem. Soc.* **83**, 1001 (1961).
216. H. C. Brown, P. Heim, and N. M. Yoon, *J. Org. Chem.* **37**, 2942 (1972).
217. H. C. Brown, W. R. Heydkamp, E. Breuer, and W. S. Murphy, *J. Amer. Chem. Soc.* **86**, 3565 (1964).
218. H. C. Brown and R. B. Johannesen, *J. Amer. Chem. Soc.* **75**, 16 (1953).
219. H. C. Brown and G. W. Kabalka, *J. Amer. Chem. Soc.* **92**, 712 (1970).

220. H. C. Brown and G. W. Kabalka, *J. Amer. Chem. Soc.* **92**, 714 (1970).
221. H. C. Brown, G. W. Kabalka, and M. W. Rathke, *J. Amer. Chem. Soc.* **89**, 4530 (1967).
222. H. C. Brown, G. W. Kabalka, M. W. Rathke, and M. M. Rogic, *J. Amer. Chem. Soc.* **90**, 4165 (1968).
223. H. C. Brown and J. H. Kawakami, *J. Amer. Chem. Soc.* **92**, 1990 (1970).
224. H. C. Brown and K. A. Keblys, *J. Amer. Chem. Soc.* **86**, 1791 (1964).
225. H. C. Brown and K. A. Keblys, *J. Amer. Chem. Soc.* **86**, 1795 (1964).
226. H. C. Brown and G. J. Klender, *Inorg. Chem.* **1**, 204 (1962).
227. H. C. Brown and E. F. Knights, *J. Amer. Chem. Soc.* **90**, 4439 (1968).
228. H. C. Brown, E. F. Knights, and R. A. Coleman, *J. Amer. Chem. Soc.* **91**, 2144 (1969).
229. H. C. Brown and W. Korytnyk, *J. Amer. Chem. Soc.* **82**, 3866 (1960).
230. H. C. Brown and S. Krishnamurthy, *J. Amer. Chem. Soc.* **94**, 7159 (1972).
231. H. C. Brown and S. Krishnamurthy, *Chem. Commun.* p. 868 (1972).
231a. H. C. Brown and S. Krishnamurthy, *J. Amer. Chem. Soc.* **95**, 1669 (1973).
232. H. C. Brown and C. F. Lane, *J. Amer. Chem. Soc.* **92**, 6660 (1970).
233. H. C. Brown and C. F. Lane, *Chem. Commun.* p. 521 (1971).
233a. H. C. Brown and C. F. Lane, *Synthesis* p. 303 (1972).
234. H. C. Brown and A. B. Levy, *J. Organometal. Chem.* **44**, 233 (1972).
235. H. C. Brown and M. M. Midland, *J. Amer. Chem. Soc.* **93**, 4078 (1971).
236. H. C. Brown and M. M. Midland, *J. Amer. Chem. Soc.* **93**, 3291 (1971).
237. H. C. Brown and M. M. Midland, *Chem. Commun.* p. 699 (1971).
238. H. C. Brown and M. M. Midland, *Angew. Chem.* **84**, 702 (1972).
239. H. C. Brown, M. M. Midland, and G. W. Kabalka, *J. Amer. Chem. Soc.* **93**, 1024 (1971).
240. H. C. Brown, M. M. Midland, and A. B. Levy, *J. Amer. Chem. Soc.* **94**, 2114 (1972).
241. H. C. Brown, M. M. Midland, and A. B. Levy, *J. Amer. Chem. Soc.* **94**, 3662 (1972).
241a. H. C. Brown, M. M. Midland, and A. B. Levy, *J. Amer. Chem. Soc.* **95**, 2394 (1973).
242. H. C. Brown and A. W. Moerikofer, *J. Amer. Chem. Soc.* **83**, 3417 (1961).
243. H. C. Brown and A. W. Moerikofer, *J. Amer. Chem. Soc.* **84**, 1478 (1962).
244. H. C. Brown and A. W. Moerikofer, *J. Amer. Chem. Soc.* **85**, 2063 (1963).
245. H. C. Brown and K. J. Murray, *J. Amer. Chem. Soc.* **81**, 4108 (1959).
246. H. C. Brown and K. J. Murray, *J. Org. Chem.* **26**, 631 (1961).
247. H. C. Brown, K. J. Murray, H. Muller, and G. Zweifel, *J. Amer. Chem. Soc.* **88**, 1443 (1966).
248. H. C. Brown, K. J. Murray, L. J. Murray, J. A. Snover, and G. Zweifel, *J. Amer. Chem. Soc.* **82**, 4233 (1960).
249. H. C. Brown and H. Nambu, *J. Amer. Chem. Soc.* **92**, 1761 (1970).
250. H. C. Brown, H. Nambu, and M. M. Rogic, *J. Amer. Chem. Soc.* **91**, 6852 (1969).
251. H. C. Brown, H. Nambu, and M. M. Rogic, *J. Amer. Chem. Soc.* **91**, 6854 (1969).
252. H. C. Brown, H. Nambu, and M. M. Rogic, *J. Amer. Chem. Soc.* **91**, 6855 (1969).
253. H. C. Brown and E. Negishi, *J. Amer. Chem. Soc.* **89**, 5285 (1967).
254. H. C. Brown and E. Negishi, *J. Amer. Chem. Soc.* **89**, 5477 (1967).
255. H. C. Brown and E. Negishi, *J. Amer. Chem. Soc.* **89**, 5478 (1967).
256. H. C. Brown and E. Negishi, *Chem. Commun.* p. 594 (1968).
257. H. C. Brown and E. Negishi, *J. Amer. Chem. Soc.* **91**, 1224 (1969).
258. H. C. Brown and E. Negishi, *J. Amer. Chem. Soc.* **93**, 3777 (1971).

259. H. C. Brown and E. Negishi, *J. Amer. Chem. Soc.* **93**, 6682 (1971).
260. H. C. Brown and E. Negishi, *J. Organometal. Chem.* **28**, C1 (1971).
261. H. C. Brown and E. Negishi, *J. Organometal. Chem.* **26**, C67 (1971).
262. H. C. Brown and E. Negishi, *J. Amer. Chem. Soc.* **94**, 3567 (1972).
263. H. C. Brown, E. Negishi, and P. L. Burke, *J. Amer. Chem. Soc.* **92**, 6649 (1970).
264. H. C. Brown, E. Negishi, and P. L. Burke, *J. Amer. Chem. Soc.* **93**, 3400 (1971).
265. H. C. Brown, E. Negishi, and P. L. Burke, *J. Amer. Chem. Soc.* **94**, 3561 (1972).
266. H. C. Brown, E. Negishi, and S. K. Gupta, *J. Amer. Chem. Soc.* **92**, 2460 (1970).
267. H. C. Brown, E. Negishi, and S. K. Gupta, *J. Amer. Chem. Soc.* **92**, 6648 (1970).
268. H. C. Brown, E. Negishi, and J. J. Katz, *J. Amer. Chem. Soc.* **94**, 5893 (1972).
269. H. C. Brown and C. D. Pfaffenberger, *J. Amer. Chem. Soc.* **89**, 5475 (1967).
270. H. C. Brown and M. W. Rathke, *J. Amer. Chem. Soc.* **89**, 2737 (1967).
271. H. C. Brown and M. W. Rathke, *J. Amer. Chem. Soc.* **89**, 2738 (1967).
272. H. C. Brown and M. W. Rathke, *J. Amer. Chem. Soc.* **89**, 4528 (1967).
273. H. C. Brown, M. W. Rathke, G. W. Kabalka, and M. M. Rogic, *J. Amer. Chem. Soc.* **90**, 4166 (1968).
274. H. C. Brown, M. W. Rathke, and M. M. Rogic, *J. Amer. Chem. Soc.* **90**, 5038 (1968).
275. H. C. Brown and N. Ravindran, *J. Amer. Chem. Soc.* **94**, 2112 (1972).
275a. H. C. Brown and N. Ravindran, *J. Org. Chem.* **38**, 182 (1973).
275b. H. C. Brown and N. Ravindran, *J. Org. Chem.* **38**, 1617 (1973).
275c. H. C. Brown and N. Ravindran, *J. Amer. Chem. Soc.* **95**, 2396 (1973).
276. H. C. Brown and S. P. Rhodes, *J. Amer. Chem. Soc.* **91**, 2149 (1969).
277. H. C. Brown and S. P. Rhodes, *J. Amer. Chem. Soc.* **91**, 4306 (1969).
278. H. C. Brown and M. M. Rogic, *J. Amer. Chem. Soc.* **91**, 2146 (1969).
279. H. C. Brown and M. M. Rogic, *J. Amer. Chem. Soc.* **91**, 4304 (1969).
280. H. C. Brown, M. M. Rogic, H. Nambu, and M. W. Rathke, *J. Amer. Chem. Soc.* **91**, 2147 (1969).
281. H. C. Brown, M. M. Rogic, and M. W. Rathke, *J. Amer. Chem. Soc.* **90**, 6218 (1968).
282. H. C. Brown, M. M. Rogic, M. W. Rathke, and G. W. Kabalka, *J. Amer. Chem. Soc.* **89**, 5709 (1967).
283. H. C. Brown, M. M. Rogic, M. W. Rathke, and G. W. Kabalka, *J. Amer. Chem. Soc.* **90**, 818 (1968).
284. H. C. Brown, M. M. Rogic, M. W. Rathke, and G. W. Kabalka, *J. Amer. Chem. Soc.* **90**, 1911 (1968).
285. H. C. Brown, M. M. Rogic, M. W. Rathke, and G. W. Kabalka, *J. Amer. Chem. Soc.* **91**, 2150 (1969).
286. H. C. Brown, H. I. Schlesinger, and S. Z. Cardon, *J. Amer. Chem. Soc.* **64**, 325 (1942).
287. H. C. Brown, H. I. Schlesinger, I. Sheft, and D. M. Ritter, *J. Amer. Chem. Soc.* **75**, 192 (1953).
288. H. C. Brown and R. L. Sharp, *J. Amer. Chem. Soc.* **88**, 5851 (1966).
289. H. C. Brown and R. L. Sharp, *J. Amer. Chem. Soc.* **90**, 2915 (1968).
290. H. C. Brown, K. P. Singh, and B. J. Garner, *J. Organometal. Chem.* **1**, 2 (1963).
291. H. C. Brown and C. H. Snyder, *J. Amer. Chem. Soc.* **83**, 1002 (1961).
292. H. C. Brown and B. C. Subba Rao, *J. Amer. Chem. Soc.* **78**, 2582 (1956).
293. H. C. Brown and B. C. Subba Rao, *J. Amer. Chem. Soc.* **78**, 5694 (1956).
294. H. C. Brown and B. C. Subba Rao, *J. Org. Chem.* **22**, 1136 (1957).
295. H. C. Brown and B. C. Subba Rao, *J. Org. Chem.* **22**, 1137 (1957).
296. H. C. Brown and B. C. Subba Rao, *J. Amer. Chem. Soc.* **81**, 6423 (1959).

297. H. C. Brown and B. C. Subba Rao, *J. Amer. Chem. Soc.* **81**, 6428 (1959).
298. H. C. Brown and B. C. Subba Rao, *J. Amer. Chem. Soc.* **81**, 6434 (1959).
299. H. C. Brown and S. Sujishi, *J. Amer. Chem. Soc.* **70**, 2793 (1948).
300. H. C. Brown and S. Sujishi, *J. Amer. Chem. Soc.* **70**, 2878 (1948).
301. H. C. Brown and M. D. Taylor, *J. Amer. Chem. Soc.* **69**, 1332 (1947).
302. H. C. Brown, M. D. Taylor, and M. Gerstein, *J. Amer. Chem. Soc.* **66**, 431 (1944).
303. H. C. Brown, M. D. Taylor, and S. Sujishi, *J. Amer. Chem. Soc.* **73**, 2464 (1951).
304. H. C. Brown and A. Tsukamoto, *J. Amer. Chem. Soc.* **82**, 746 (1960).
305. H. C. Brown, A. Tsukamoto, and D. B. Bigley, *J. Amer. Chem. Soc.* **82**, 4703 (1960).
306. H. C. Brown and M. K. Unni, *J. Amer. Chem. Soc.* **90**, 2902 (1968).
307. H. C. Brown and V. Varma, *J. Amer. Chem. Soc.* **88**, 2871 (1966).
308. H. C. Brown, C. Verbrugge, and C. H. Snyder, *J. Amer. Chem. Soc.* **83**, 1001 (1961).
309. H. C. Brown and Y. Yamamoto, *J. Amer. Chem. Soc.* **93**, 2796 (1971).
310. H. C. Brown and Y. Yamamoto, *J. Chem. Soc., D* p. 1535 (1971).
311. H. C. Brown and Y. Yamamoto, *Chem. Commun.* p. 71 (1972).
311a. H. C. Brown, Y. Yamamoto, and C. F. Lane, *Synthesis* p. 304 (1972).
311b. H. C. Brown and Y. Yamamoto, *Synthesis* p. 699 (1972).
312. H. C. Brown and G. Zweifel, *J. Amer. Chem. Soc.* **81**, 247 (1959).
313. H. C. Brown and G. Zweifel, *J. Amer. Chem. Soc.* **81**, 1815 (1959).
314. H. C. Brown and G. Zweifel, *J. Amer. Chem. Soc.* **81**, 4106 (1959).
315. H. C. Brown and G. Zweifel, *J. Amer. Chem. Soc.* **81**, 5832 (1959).
316. H. C. Brown and G. Zweifel, *J. Amer. Chem. Soc.* **82**, 1504 (1960).
317. H. C. Brown and G. Zweifel, *J. Amer. Chem. Soc.* **82**, 3222 (1960).
318. H. C. Brown and G. Zweifel, *J. Amer. Chem. Soc.* **82**, 3223 (1960).
319. H. C. Brown and G. Zweifel, *J. Amer. Chem. Soc.* **82**, 4708 (1960).
320. H. C. Brown and G. Zweifel, *J. Amer. Chem. Soc.* **83**, 486 (1961).
321. H. C. Brown and G. Zweifel, *J. Amer. Chem. Soc.* **83**, 1241 (1961).
322. H. C. Brown and G. Zweifel, *J. Amer. Chem. Soc.* **83**, 2544 (1961).
323. H. C. Brown and G. Zweifel, *J. Amer. Chem. Soc.* **83**, 3934 (1961).
324. H. C. Brown and G. Zweifel, *J. Amer. Chem. Soc.* **88**, 1433 (1966).
325. H. C. Brown and G. Zweifel, *J. Amer. Chem. Soc.* **89**, 561 (1967).
325a. M. P. Brown, A. K. Holliday, and G. M. Way, *J. Chem. Soc., Chem. Commun.* p. 850 (1972).
326. R. D. Brown, A. S. Buchanan, and A. A. Humffray, *Aust. J. Chem.* **18**, 1521 (1965).
327. R. D. Brown, A. S. Buchanan, and A. A. Humffray, *Aust. J. Chem.* **18**, 1527 (1965).
328. R. L. Bruce and A. A. Humffray, *Aust. J. Chem.* **24**, 1085 (1971).
329. W. Brüser and K. H. Thiele, *Z. Anorg. Allg. Chem.* **349**, 310 (1967).
330. Y. N. Bubnov, V. S. Bogdanov, and B. M. Mikhailov, *Zh. Obshch. Khim.* **38**, 260 (1968).
331. Y. N. Bubnov, V. S. Bogdanov, and B. M. Mikhailov, *Izv. Akad. Nauk. SSSR, Ser. Khim.* p. 2416 (1970).
331a. Y. N. Bubnov, V. S. Bogdanov, I. P. Yakovlev, and B. M. Mikhailov, *Zh. Obshch. Khim.* **42**, 1313 (1972).
332. Y. N. Bubnov, S. I. Frolov, V. G. Kiselev, V. S. Bogdanov, and B. M. Mikhailov, *Organometal. Chem. Syn.* **1**, 37 (1970).
333. Y. N. Bubnov, S. I. Frolov, V. G. Kiselev, and B. M. Mikhailov, *Zh. Obshch. Khim.* **40**, 1316 (1970).

334. Y. N. Bubnov, S. I. Frolov, V. G. Kiselev, and B. M. Mikhailov, *Zh. Obshch. Khim.* **40**, 1311 (1970).
334a. Y. N. Bubnov, M. S. Grigoryan, and B. M. Mikhailov, *Zh. Obshch. Khim.* **42**, 1738 (1972).
334b. Yu. N. Bubnov, S. A. Korobeinikova, G. V. Isagoulyants, and B. M. Mikhailov, *Izv. Akad. Nauk SSSR, Ser. Khim.* p. 2023 (1970).
335. Y. N. Bubnov and B. M. Mikhailov, *Izv. Akad. Nauk SSSR, Ser. Khim.* p. 472 (1967).
335a. Yu. N. Bubnov and B. M. Mikhailov, *Izv. Akad. Nauk SSSR, Ser. Khim.* p. 2156 (1970).
336. Y. N. Bubnov, O. A. Nesmeyanova, T. Yu. Rudashevskaya, B. M. Mikhailov, and B. A. Kazansky, *Tetrahedron Lett.* p. 2153 (1971).
337. A. S. Buchanan and F. Creutzberg, *Aust. J. Chem.* **15**, 744 (1962).
338. R. Bucourt and R. Joly, *C. R. Acad. Sci.* **254**, 1655 (1962).
339. J. D. Buhler and H. C. Brown, *J. Organometal. Chem.* **40**, 265 (1972).
340. Z. J. Bujwid, W. Gerrard, and M. F. Lappert, *Chem. Ind.* (*London*) p. 1091 (1959).
341. G. J. Bullen and N. H. Clark, *J. Chem. Soc., A* p. 992 (1970).
342. G. J. Bullen and K. Wade, *Chem. Commun.* p. 1122 (1971).
343. V. L. Buls, O. L. Davis, and R. I. Thomas, *J. Amer. Chem. Soc.* **79**, 337 (1957).
343a. E. J. Bulten and J. G. Noltes, *J. Organometal. Chem.* **29**, 409 (1971).
344. J. E. Burch, W. Gerrard, M. Goldstein, E. F. Mooney, and H. A. Willis, *Spectrochim. Acta* **18**, 1403 (1962).
345. J. E. Burch, W. Gerrard, M. Goldstein, E. F. Mooney, D. E. Pratt, and H. A. Willis, *Spectrochim. Acta* **19**, 889 (1963).
346. J. E. Burch, W. Gerrard, M. Howarth, and E. F. Mooney, *J. Chem. Soc., London* p. 4916 (1960).
347. J. E. Burch, W. Gerrard, and E. F. Mooney, *J. Chem. Soc., London* p. 2200 (1962).
348. A. B. Burg, *J. Amer. Chem. Soc.* **62**, 2228 (1940).
349. A. B. Burg, *J. Amer. Chem. Soc.* **74**, 3482 (1952).
350. A. B. Burg and J. Banus, *J. Amer. Chem. Soc.* **76**, 3903 (1954).
351. A. B. Burg and J. L. Boone, *J. Amer. Chem. Soc.* **78**, 1521 (1956).
352. A. B. Burg and G. W. Campbell, *J. Amer. Chem. Soc.* **74**, 3744 (1952).
353. A. B. Burg and F. M. Graber, *J. Amer. Chem. Soc.* **78**, 1523 (1956).
354. A. B. Burg and A. A. Green, *J. Amer. Chem. Soc.* **65**, 1838 (1943).
355. A. B. Burg and J. S. Sandhu, *J. Amer. Chem. Soc.* **87**, 3787 (1965).
356. A. B. Burg and H. I. Schlesinger, *J. Amer. Chem. Soc.* **59**, 780 (1937).
357. A. B. Burg and J. R. Spielman, *J. Amer. Chem. Soc.* **81**, 3479 (1959).
358. A. B. Burg and J. R. Spielman, *J. Amer. Chem. Soc.* **83**, 2667 (1961).
359. A. B. Burg and R. I. Wagner, *J. Amer. Chem. Soc.* **75**, 3872 (1953).
360. A. B. Burg and R. I. Wagner, *J. Amer. Chem. Soc.* **76**, 3307 (1954).
360a. P. L. Burke, E. Negishi, and H. C. Brown, *J. Amer. Chem. Soc.* **95**, 3654 (1973).
361. F. K. Butcher, W. Gerrard, M. Howarth, E. F. Mooney, and H. A. Willis, *Spectrochim. Acta* **19**, 905 (1963).
362. F. K. Butcher, W. Gerrard, M. Howarth, E. F. Mooney, and H. A. Willis, *Spectrochim. Acta* **20**, 79 (1964).
363. D. N. Butler and A. H. Soloway, *J. Amer. Chem. Soc.* **86**, 2961 (1964).
364. G. B. Butler and G. L. Statton, *J. Amer. Chem. Soc.* **86**, 5045 (1964).
365. G. B. Butler and G. L. Statton, *J. Amer. Chem. Soc.* **86**, 518 (1964).
366. G. B. Butler, G. L. Statton, and W. S. Brey, Jr., *J. Org. Chem.* **30**, 4194 (1965).
367. L. Caglioti, G. Cainelli, G. Maina, and A. Selva, *Gazz. Chim. Ital.* **92**, 309 (1962).

367a. G. Cainelli, G. Dal Bello, and G. Zubiani, *Tetrahedron Lett.* p. 3429 (1965); p. 4315 (1966).
368. J. L. Calderon, F. A. Cotton, and A. Shaver, *J. Organometal. Chem.* **38**, 105 (1972).
369. G. W. Campbell, *J. Amer. Chem. Soc.* **79**, 4023 (1957).
370. G. W. Campbell, *Advan. Chem. Ser.* **32**, 195 (1961).
371. J. H. Carpenter, W. J. Jones, R. W. Jotham, and L. H. Long, *Chem. Commun.* p. 881 (1968).
372. J. H. Carpenter, W. J. Jones, R. W. Jotham, and L. H. Long, *Spectrochim. Acta, Part A* **26**, 1199 (1970).
373. J. H. Carpenter, W. J. Jones, R. W. Jotham, and L. H. Long, *Spectrochim. Acta, Part A* **27**, 1721 (1971).
374. R. A. Carpenter, *Ind. Eng. Chem.* **49**, 42A (1957).
375. R. A. Carpenter, *ARS J.* **29**, 8 (1959).
376. N. E. Carrazzoni, M. A. Molinari, and G. J. Videla, *An. Ass. Quim. Argent.* **49**, 46 (1961).
377. B. L. Carroll and L. S. Bartell, *Inorg. Chem.* **7**, 219 (1968).
378. J. C. Carter and R. W. Parry, *J. Amer. Chem. Soc.* **87**, 2354 (1965).
379. J. Casanova and H. R. Kiefer, *J. Org. Chem.* **34**, 2579 (1969).
380. J. Casanova, H. R. Kiefer, D. Kuwada, and A. H. Boulton, *Tetrahedron Lett.* p. 703 (1965).
381. J. Casanova, H. R. Kiefer, and R. E. Williams, *Org. Prep. Proced.* **1**, 57 (1969).
382. J. Casanova and R. E. Schuster, *Tetrahedron Lett.* p. 405 (1964).
383. F. F. Caserio, J. J. Cavallo, and R. I. Wagner, *J. Org. Chem.* **26**, 2157 (1961).
384. E. Caspi and K. R. Varma, *J. Org. Chem.* **33**, 2181 (1968).
385. F. P. Cassaretto, J. J. McLafferty, and C. E. Moore, *Anal. Chim. Acta* **32**, 376 (1965).
386. R. B. Castle and D. S. Matteson, *J. Organometal. Chem.* **20**, 19 (1969).
387. J. C. Catlin and H. R. Snyder, *J. Org. Chem.* **34**, 1660 (1969).
388. J. C. Catlin and H. R. Snyder, *J. Org. Chem.* **34**, 1664 (1969).
389. F. Caujolle, D. Caujolle, and G. Dousset, *C. R. Soc. Biol.* **164**, 1142 (1970).
390. P. Ceron, A. Finch, J. Frey, J. Kerrigan, T. Parsons, G. Urry, and H. I. Schlesinger, *J. Amer. Chem. Soc.* **81**, 6368 (1959).
391. M. Chaigneau, *C. R. Acad. Sci.* **239**, 1220 (1954).
392. M. R. Chakrabarty, C. C. Thompson, and W. S. Brey, Jr., *Inorg. Chem.* **6**, 518 (1967).
393. F. Challenger and O. V. Richards, *J. Chem. Soc., London* p. 405 (1934).
394. C. Chambers, A. K. Holliday, and S. M. Walker, *Proc. Chem. Soc., London* p. 286 (1964).
395. C. Chambers and A. K. Holliday, *J. Chem. Soc., London* p. 3459 (1965).
396. R. D. Chambers and T. Chivers, *J. Chem. Soc., London* p. 3933 (1965).
397. R. D. Chambers and T. Chivers, *Proc. Chem. Soc., London* p. 208 (1963).
398. R. D. Chambers, H. C. Clark, L. W. Reeves, and C. J. Willis, *Can. J. Chem.* **39**, 258 (1961).
399. R. D. Chambers, H. C. Clark, and C. J. Willis, *J. Amer. Chem. Soc.* **82**, 5298 (1960).
400. R. D. Chambers, H. C. Clark, and C. J. Willis, *Proc. Chem. Soc., London* p. 114 (1960).
401. N. G. S. Champion, R. Foster, and R. K. Mackie, *J. Chem. Soc., London* p. 5060 (1961).
402. J. Chatt, *J. Chem. Soc., London* p. 3340 (1949).

403. C. C. S. Cheung and R. A. Beaudet, *J. Mol. Spectrosc.* **36**, 337 (1970).
404. S. S. Chissick, M. J. S. Dewar, and P. M. Maitlis, *Tetrahedron Lett.* p. 8 (1960).
405. S. S. Chissick, M. J. S. Dewar, and P. M. Maitlis, *J. Amer. Chem. Soc.* **83**, 2708 (1961).
406. T. Chivers, *Chem. Commun.* p. 157 (1967).
407. G. N. Chremos, H. Weidmann, and H. K. Zimmerman, *J. Org. Chem.* **26**, 1683 (1961).
408. P. M. Christopher, *J. Chem. Eng. Data* **5**, 568 (1960).
409. P. M. Christopher and T. J. Tully, *J. Amer. Chem. Soc.* **80**, 6516 (1958).
410. T. L. Chu, *J. Amer. Chem. Soc.* **75**, 1730 (1953).
411. T. L. Chu and T. J. Weismann, *J. Phys. Chem.* **60**, 1020 (1956).
412. T. L. Chu and T. J. Weismann, *J. Amer. Chem. Soc.* **78**, 23 (1956).
413. T. L. Chu and T. J. Weismann, *J. Amer. Chem. Soc.* **78**, 3610 (1956).
414. G. F. Clark and A. K. Holliday, *J. Organometal. Chem.* **2**, 100 (1964).
415. G. M. Clark, K. G. Hancock, and G. Zweifel, *J. Amer. Chem. Soc.* **93**, 1308 (1971).
416. S. L. Clark, J. R. Jones, and H. Stange, *Advan. Chem. Ser.* **32**, 228 (1961).
416a. W. R. Clayton, D. T. Saturnino, P. W. R. Corfield, and S. G. Shore, *J. Chem. Soc., Chem. Commun.* p. 377 (1973).
417. C. G. Clear and G. E. K. Branch, *J. Org. Chem.* **2**, 522 (1938).
418. G. E. Coates and B. R. Francis, *J. Chem. Soc., A* p. 1308 (1971).
419. G. E. Coates and J. G. Livingstone, *J. Chem. Soc., London* p. 5053 (1961).
420. G. E. Coates and J. G. Livingstone, *J. Chem. Soc., London* p. 1000 (1961).
421. G. E. Coates, *J. Chem. Soc., London* p. 3481 (1950).
422. G. E. Coates and J. G. Livingstone, *J. Chem. Soc., London* p. 4909 (1961).
423. G. E. Coates and R. A. Whitcombe, *J. Chem. Soc., London* p. 3351 (1956).
424. A. T. Cocks and K. W. Egger, *J. Chem. Soc., A* p. 3606 (1971).
425. K. P. Coffin and S. H. Bauer, *J. Phys. Chem.* **59**, 193 (1955).
426. E. A. Cohen and R. A. Beaudet, *J. Chem. Phys.* **48**, 1220 (1968).
426a. G. Coindard and J. Braun, *Bull. Soc. Chim. Fr.* p. 817 (1972).
426b. G. Coindard, J. Braun, and P. Cadiot, *Bull. Soc. Chim. Fr.* p. 811 (1972).
427. M. R. Collier, M. F. Lappert, R. Snaith, and K. Wade, *J. Chem. Soc., Dalton Trans.* p. 370 (1972).
428. R. D. Compton, H. Kohl, and J. J. Lagowski, *Inorg. Chem.* **6**, 2265 (1967).
429. R. D. Compton and J. J. Lagowski, *Inorg. Chem.* **7**, 1234 (1968).
430. C. Cone, M. J. S. Dewar, R. Golden, F. Maseles, and P. Rona, *J. Chem. Soc., D* p. 1522 (1971).
431. L. Contreras, J. Grotewold, E. A. Lissi, and R. Rozas, *J. Polym. Sci., Part A* **7**, 2341 (1969).
432. W. L. Cook and K. Niedenzu, *Syn. Inorg. Metal.-Org. Chem.* **2**, 267 (1972).
433. A. S. Coolidge and H. E. Bent, *J. Amer. Chem. Soc.* **58**, 505 (1936).
434. J. H. Cooper and R. E. Powell, *J. Amer. Chem. Soc.* **85**, 1590 (1963).
435. M. Corner, *Analyst* **84**, 41 (1959).
436. S. Corsano, *Atti Accad. Naz. Lincei, Cl. Sci. Fis., Mat. Natur., Rend* [8] **34**, 430 (1963).
437. Council of the American Chemical Society, *Inorg. Chem.* **7**, 1945 (1968).
438. I. G. C. Coutts and O. C. Musgrave, *J. Chem. Soc., C* p. 2225 (1970).
439. R. D. Cowan, *J. Chem. Phys.* **17**, 218 (1949).
440. R. D. Cowan, *J. Chem. Phys.* **18**, 1101 (1950).
441. W. P. Cowie, A. H. Jackson, and O. C. Musgrave, *Chem. Ind.* (*London*) p. 1248 (1959).

442. A. H. Cowley and T. A. Furtsch, *J. Amer. Chem. Soc.* **91**, 39 (1969).
443. A. H. Cowley and T. A. Furtsch, *J. Mol. Spectrosc.* **34**, 175 (1970).
444. A. H. Cowley and J. L. Mills, *J. Amer. Chem. Soc.* **91**, 2911 (1969).
445. T. D. Coyle and J. J. Ritter, *J. Organometal. Chem.* **12**, 269 (1968).
446. T. D. Coyle and J. J. Ritter, *J. Amer. Chem. Soc.* **89**, 5739 (1967).
447. T. D. Coyle, S. L. Stafford, and F. G. A. Stone, *Spectrochim. Acta* **17**, 968 (1961).
448. T. D. Coyle, S. L. Stafford, and F. G. A. Stone, *J. Chem. Soc., London* p. 3103 (1961).
449. T. D. Coyle and F. G. A. Stone, *J. Amer. Chem. Soc.* **82**, 6223 (1960).
450. T. D. Coyle and F. G. A. Stone, *J. Amer. Chem. Soc.* **83**, 4138 (1961).
451. R. H. Cragg, *J. Chem. Soc., A* p. 2962 (1968).
451a. R. H. Cragg, *Chem. Commun.* p. 832 (1969).
452. R. H. Cragg, D. A. Gallagher, J. P. N. Husband, G. Lawson, and J. F. J. Todd, *Chem. Commun.* p. 1562 (1970).
453. R. H. Cragg and J. P. N. Husband, *Inorg. Nucl. Chem. Lett.* **6**, 773 (1970).
454. R. H. Cragg and M. F. Lappert, *Advan. Chem. Ser.* **42**, 220 (1964).
455. R. H. Cragg, M. F. Lappert, and B. P. Tilley, *J. Chem. Soc., London* p. 2108 (1964).
456. R. H. Cragg, M. F. Lappert, and B. P. Tilley, *J. Chem. Soc., A* p. 947 (1967).
457. R. H. Cragg, G. Lawson, and J. F. J. Todd, *J. Chem. Soc., Dalton Trans.* p. 878 (1972).
458. R. H. Cragg and J. E. J. Todd, *Chem. Commun.* p. 386 (1970).
459. R. H. Cragg, J. F. J. Todd, R. B. Turner, and A. F. Weston, *Chem. Commun.* p. 206 (1972).
460. R. H. Cragg, J. F. J. Todd, and A. F. Weston, *Org. Mass Spectrom.* **6**, 1077 (1972).
461. J. Crighton, A. K. Holliday, A. G. Massey, and N. R. Thompson, *Chem. Ind. (London)* p. 347 (1960).
462. J. Cueilleron and P. Guillot, *Bull. Soc. Chim. Fr.* p. 2044 (1960).
463. G. C. Culling, M. J. S. Dewar, and P. A. Marr, *J. Amer. Chem. Soc.* **86**, 1125 (1964).
464. C. S. Cundy and H. Nöth, *J. Organometal. Chem.* **30**, 135 (1971).
465. C. Curran, P. A. McCusker, and H. S. Makowski, *J. Amer. Chem. Soc.* **79**, 5188 (1957).
466. B. R. Currell, W. Gerrard, and M. Khodabocus, *J. Organometal. Chem.* **8**, 411 (1967).
467. R. Damico, *J. Org. Chem.* **29**, 1971 (1964).
468. S. H. Dandegaonker, W. Gerrard, and M. F. Lappert, *J. Chem. Soc., London* p. 2893 (1957).
469. S. H. Dandegaonker, W. Gerrard, and M. F. Lappert, *J. Chem. Soc., London* p. 2872 (1957).
470. S. H. Dandegaonker, W. Gerrard, and M. F. Lappert, *J. Chem. Soc., London* p. 2076 (1959).
471. S. D. Darling, O. N. Devgan, and R. E. Cosgrove, *J. Amer. Chem. Soc.* **92**, 696 (1970).
472. T. P. Das, *J. Chem. Phys.* **27**, 1 (1957).
473. M. K. Das, P. G. Harrison, and J. J. Zuckerman, *Inorg. Chem.* **10**, 1092 (1971)
474. M. K. Das and J. J. Zuckerman, *Inorg. Chem.* **10**, 1028 (1971).
475. J. M. Davidson and C. M. French, *J. Chem. Soc., London* p. 114 (1958).
476. J. M. Davidson and C. M. French, *J. Chem. Soc., London* p. 3364 (1962).

477. J. M. Davidson and C. M. French, *Chem. Ind.* (*London*) p. 750 (1959).
478. J. M. Davidson and C. M. French, *J. Chem. Soc., London* p. 191 (1960).
479. A. G. Davies, K. G. Foot, B. P. Roberts, and J. C. Scaiano, *J. Organometal. Chem.* **31**, 1C (1971).
480. A. G. Davies, D. Griller, and B. P. Roberts, *J. Chem. Soc., B* p. 1823 (1971).
481. A. G. Davies, D. Griller, B. P. Roberts, and J. C. Scaiano, *Chem. Commun.* p. 196 (1971).
482. A. G. Davies, D. Griller, B. P. Roberts, and R. Tudor, *Chem. Commun.* p. 640 (1970).
483. A. G. Davies and D. G. Hare, *J. Chem. Soc., London* p. 438 (1959).
484. A. G. Davies, D. G. Hare, and L. F. Larkworthy, *Chem. Ind.* (*London*) p. 1519 (1959).
485. A. G. Davies, D. G. Hare, and O. R. Khan, *J. Chem. Soc., London* p. 1125 (1963).
486. A. G. Davies, D. G. Hare, O. R. Khan, and J. Sikora, *Proc. Chem. Soc., London* p. 172 (1961).
487. A. G. Davies, D. G. Hare, and R. F. M. White, *Chem. Ind.* (*London*) p. 1315 (1959).
488. A. G. Davies, D. G. Hare, and R. F. M. White, *Chem. Ind.* (*London*) p. 566 (1960).
489. A. G. Davies, D. G. Hare, and R. F. M. White, *J. Chem. Soc., London* p. 1040 (1960).
490. A. G. Davies, D. G. Hare, and R. F. M. White, *J. Chem. Soc., London* p. 341 (1961).
491. A. G. Davies, S. C. W. Hook, and B. P. Roberts, *J. Organometal. Chem.* **23**, C11 (1970).
492. A. G. Davies, S. C. W. Hook, and B. P. Roberts, *J. Organometal. Chem.* **22**, 37C (1970).
493. A. G. Davies, K. U. Ingold, B. P. Roberts, and R. Tudor, *J. Chem. Soc., B* p. 698 (1971).
494. A. G. Davies, T. Maki, and B. P. Roberts, *J. Chem. Soc., Perkin Trans. II*, p. 744 (1972).
495. A. G. Davies and R. B. Moodie, *J. Chem. Soc., London* p. 2372 (1958).
496. A. G. Davies and B. P. Roberts, *Chem. Commun.* p. 699 (1969).
497. A. G. Davies and B. P. Roberts, *J. Chem. Soc., B* p. 17 (1967).
498. A. G. Davies and B. P. Roberts, *J. Chem. Soc., B* p. 1830 (1971).
499. A. G. Davies and B. P. Roberts, *Chem. Commun.* p. 298 (1966).
500. A. G. Davies and B. P. Roberts, *J. Chem. Soc., C* p. 1474 (1968).
501. A. G. Davies and B. P. Roberts, *J. Chem. Soc., B* p. 311 (1969).
502. A. G. Davies and B. P. Roberts, *Nature* (*London*) *Phys. Sci.* **229**, 221 (1971).
503. A. G. Davies, B. P. Roberts, and J. C. Scaiano, *J. Chem. Soc., B* p. 2171 (1971).
504. A. G. Davies, B. P. Roberts, and J. C. Scaiano, *J. Chem. Soc., Perkin Trans. II*, p. 803 (1972).
505. A. G. Davies, B. P. Roberts, and R. Tudor, *J. Organometal Chem.* **31**, 137 (1971).
506. A. G. Davies and J. C. Scaiano, *J. Chem. Soc., Perkin Trans. II*, p. 2234 (1972).
507. A. G. Davies and R. Tudor, *J. Chem. Soc., B* p. 1815 (1970).
508. D. W. Davies, *Trans. Faraday Soc.* **56**, 1713 (1960).
509. F. A. Davies and M. J. S. Dewar, *J. Amer. Chem. Soc.* **90**, 3511 (1968).
510. F. A. Davies, M. J. S. Dewar, and R. Jones, *J. Amer. Chem. Soc.* **90**, 706 (1968).
511. F. A. Davies, M. J. S. Dewar, R. Jones, and S. D. Worley, *J. Amer. Chem. Soc.* **91**, 2094 (1969).
512. K. M. Davies, M. J. S. Dewar, and P. Rona, *J. Amer. Chem. Soc.* **89**, 6294 (1967).

513. R. E. Davis, R. E. Kenson, C. L. Kibby, and H. H. Lloyd, *Chem. Commun.* p. 593 (1965).
513a. J. Dazord and H. Mongeot, *Bull. Soc. Chim. Fr.* p. 950 (1972).
513b. K. Deckelman and H. Werner, *Helv. Chim. Acta* **54**, 2189 (1971).
514. W. R. Deever, E. R. Lory, and D. M. Ritter, *Inorg. Chem.* **8**, 1263 (1969).
515. W. R. Deever and D. M. Ritter, *Inorg. Chem.* **8**, 2461 (1969).
516. H. G. Dehmelt, *Z. Phys.* **133**, 528 (1952).
517. H. G. Dehmelt, *Z. Phys.* **134**, 642 (1953).
518. J. E. DeMoor and G. P. Van der Kelen, *J. Organometal. Chem.* **6**, 235 (1966).
519. J. E. DeMoor and G. P. Van der Kelen, *J. Organometal. Chem.* **9**, 23 (1967).
519a. E. R. De-Staricco, C. Riddle, and W. L. Jolly, *J. Inorg. Nucl. Chem.* **35**, 297 (1973).
519b. D. Devaprabhakara and P. D. Gardner, *J. Amer. Chem. Soc.* **85**, 1458 (1963).
520. V. Devarajan and S. J. Cyvin, *J. Mol. Struct.* **9**, 265 (1971).
520a. M. J. S. Dewar, *Advan. Chem. Ser.* **42**, 227 (1964).
521. M. J. S. Dewar and R. Dietz, *J. Chem. Soc., London* p. 1344 (1960).
522. M. J. S. Dewar and R. Dietz, *Tetrahedron* **15**, 26 (1961).
523. M. J. S. Dewar and R. Dietz, *J. Chem. Soc., London* p. 2728 (1959).
524. M. J. S. Dewar, R. Dietz, V. P. Kubba, and A. R. Lepley, *J. Amer. Chem. Soc.* **83**, 1754 (1961).
525. M. J. S. Dewar and R. C. Dougherty, *Tetrahedron Lett.* p. 907 (1964).
526. M. J. S. Dewar and R. C. Dougherty, *J. Amer. Chem. Soc.* **86**, 433 (1964).
527. M. J. S. Dewar, R. C. Dougherty, and E. B. Fleischer, *J. Amer. Chem. Soc.* **84**, 4882 (1962).
528. M. J. S. Dewar, G. J. Gleicher, and B. P. Robinson, *J. Amer. Chem. Soc.* **86**, 5698 (1964).
529. M. J. S. Dewar, R. Golden, and P. A. Spauninger, *J. Amer. Chem. Soc.* **93**, 3298 (1971).
530. M. J. S. Dewar and P. J. Grisdale, *J. Org. Chem.* **28**, 1759 (1963).
531. M. J. S. Dewar and R. Jones, *J. Amer. Chem. Soc.* **89**, 2408 (1967); Correction: *ibid.* p. 4251.
532. M. J. S. Dewar, C. Kaneko, and M. K. Bhattacharjee, *J. Amer. Chem. Soc.* **84**, 4884 (1962).
533. M. J. S. Dewar and V. P. Kubba, *J. Amer. Chem. Soc.* **83**, 1757 (1961).
534. M. J. S. Dewar and V. P. Kubba, *J. Org. Chem.* **25**, 1722 (1960).
535. M. J. S. Dewar and V. P. Kubba, *Tetrahedron* **7**, 213 (1959).
536. M. J. S. Dewar, V. P. Kubba, and R. Pettit, *J. Chem. Soc., London* p. 3073 (1958).
537. M. J. S. Dewar, V. P. Kubba, and R. Pettit, *J. Chem. Soc., London* p. 3076 (1958).
538. M. J. S. Dewar and A. R. Lepley, *J. Chem. Amer. Soc.* **83**, 4560 (1961).
539. M. J. S. Dewar and P. M. Maitlis, *Chem. Ind. (London)* p. 1626 (1960).
540. M. J. S. Dewar and P. M. Maitlis, *Tetrahedron* **15**, 35 (1961).
541. M. J. S. Dewar and P. M. Maitlis, *J. Amer. Chem. Soc.* **83**, 187 (1961).
542. M. J. S. Dewar and P. A. Marr, *J. Amer. Chem. Soc.* **84**, 3782 (1962).
543. M. J. S. Dewar and W. H. Poesche, *J. Amer. Chem. Soc.* **85**, 2253 (1963).
544. M. J. S. Dewar and H. Rogers, *J. Amer. Chem. Soc.* **84**, 395 (1962).
545. M. J. S. Dewar and P. Rona, *J. Amer. Chem. Soc.* **91**, 2259 (1969).
546. M. J. S. Dewar and J. L. von Rosenberg, *J. Amer. Chem. Soc.* **88**, 358 (1966).
547. E. J. DeWitt, *J. Org. Chem.* **26**, 4156 (1961).
547a. U. E. Diner, M. Worsley, J. W. Lown, and J. A. Forsythe, *Tetrahedron Lett.* p. 3145 (1972).

548. J. F. Ditter, F. J. Gerhart, and R. E. Williams, *Advan. Chem. Ser.* **72**, p. 191 (1968).
549. J. F. Ditter, J. R. Spielman, and R. E. Williams, *Inorg. Chem.* **5**, 118 (1966).
550. M. DiVaira and A. B. Orlandini, *J. Chem. Soc., Dalton Trans.* p. 1704 (1972).
551. J. Dobson and R. Schaeffer, *Inorg. Chem.* **9**, 2183 (1970).
552. J. E. Dobson, P. M. Tucker, R. Schaeffer, and F. G. A. Stone, *Chem. Commun.* p. 452 (1968).
553. D. Dollimore and L. H. Long, *J. Chem. Soc., London* p. 3906 (1953).
554. V. A. Dorokhov, O. G. Boldyreva, and B. M. Mikhailov, *Zh. Obshch. Khim.* **40**, 1528 (1970).
554a. V. A. Dorokhov, O. G. Boldyreva, and B. M. Mikhailov, *Izv. Akad. Nauk SSSR, Ser. Khim.* p. 191 (1971).
554b. V. A. Dorokhov, O. G. Boldyreva, V. S. Bogdanov, and B. M. Mikhailov, *Zh. Obshch. Khim.* **42**, 1558 (1972).
555. V. A. Dorokhov and M. F. Lappert, *Chem. Commun.* p. 250 (1968).
556. V. A. Dorokhov and M. F. Lappert, *J. Chem. Soc., A* p. 433 (1969).
556a. V. A. Dorokhov, L. I. Lavrinovich, and B. M. Mikhailov, *Dokl. Akad. Nauk SSSR* **195**, 1100 (1970).
556b. V. A. Dorokhov, L. I. Lavrinovich, I. P. Yakovlev, and B. M. Mikhailov, *Zh. Obshch. Khim.* **41**, 2501 (1971).
557. V. A. Dorokhov and B. M. Mikhailov, *Izv. Akad. Nauk SSSR, Ser. Khim.* p. 1804 (1970).
558. V. A. Dorokhov and B. M. Mikhailov, *Izv. Akad. Nauk SSSR, Ser. Khim.* p. 364 (1966).
559. V. A. Dorokhov and B. M. Mikhailov, *Dokl. Akad. Nauk SSSR* **187**, 1300 (1969).
559a. V. A. Dorokhov and B. M. Mikhailov, *Zh. Obshch. Khim.* **41**, 2702 (1971).
559b. V. A. Dorokhov and B. M. Mikhailov, *Izv. Akad. Nauk SSSR, Ser. Khim.* p. 1895 (1972).
560. A. Dornow and D. Wille, *Chem. Ber.* **98**, 1505 (1965).
561. J. C. Doty, B. Babb, P. J. Grisdale, M. Glogowski, and J. L. R. Williams, *J. Organometal. Chem.* **38**, 229 (1972).
562. J. C. Doty, P. J. Grisdale, T. R. Evans, and J. L. R. Williams, *J. Organometal. Chem.* **32**, 35C (1971).
563. J. E. Douglass, *J. Org. Chem.* **26**, 1312 (1961).
564. J. E. Douglass, G. R. Roehrig, and O. Ma, *J. Organometal. Chem.* **8**, 421 (1967).
565. J. E. Douglass, *J. Amer. Chem. Soc.* **84**, 121 (1962).
566. J. E. Douglass, *J. Amer. Chem. Soc.* **86**, 5431 (1964).
567. P. M. Draper, T. H. Chan, and D. N. Harpp, *Tetrahedron Lett.* p. 1687 (1970).
568. G. Drefahl, H. H. Höerhold, and K. D. Hofmann, *J. Prakt. Chem.* [4] **37**, 97 (1968).
569. G. Drefahl and E. Keil, *J. Prakt. Chem.* [4] **6**, 80 (1958).
570. R. Dulou and Y. Chrétien-Bessière, *Bull. Soc. Chim. Fr.* p. 1362 (1959).
571. E. Dumont, *Kunstst.-Plast.* (*Solothurn*) **6**, 30 (1959).
572. L. A. Duncanson, W. Gerrard, M. F. Lappert, H. Pyszora, and R. Shafferman, *J. Chem. Soc., London* p. 3652 (1958).
573. I. Dunstan, N. J. Blay, and R. L. Williams, *J. Chem. Soc., London* p. 5016 (1960).
574. I. Dunstan and J. V. Griffiths, *Anal. Chem.* **33**, 1598 (1961).
575. I. Dunstan, R. L. Williams, and N. J. Blay, *J. Chem. Soc., London* p. 5012 (1960).
576. A. S. Dworkin and E. R. Van Artsdalen, *J. Amer. Chem. Soc.* **76**, 4316 (1954).
576a. G. R. Eaton, *Inorg. Nucl. Chem. Lett.* **8**, 643 (1972).

577. C. A. Eggers and S. F. A. Kettle, *Inorg. Chem.* **6**, 160 (1967).
578. J. J. Eisch and L. J. Gonsior, *J. Organometal. Chem.* **8**, 53 (1967).
579. J. J. Eisch, N. K. Hota, and S. Kozima, *J. Amer. Chem. Soc.* **91**, 4575 (1969).
580. P. D. Ellis, J. D. Odom, D. W. Lowman, and A. D. Cardin, *J. Amer. Chem. Soc.* **93**, 6704 (1971).
581. G. F. El-Fayoumy, M. A. Wahab, and M. I. Roushdy, *J. Chem. U.A.R.* **12**, 371 (1969).
582. F. W. Emery, P. L. Harold, and A. J. Owen, *J. Chem. Soc., London* p. 426 (1964).
583. F. W. Emery, P. L. Harold, and A. J. Owen, *J. Chem. Soc., London* p. 4931 (1964).
584. M. V. Encina and E. A. Lissi, *J. Organometal. Chem.* **29**, 21 (1971).
585. W. D. English, A. L. McCloskey, and H. Steinberg, *J. Amer. Chem. Soc.* **83**, 2122 (1961).
586. C. E. Erickson and F. C. Gunderloy, *J. Org. Chem.* **24**, 1161 (1959).
587. I. D. Eubanks and J. J. Lagowski, *J. Amer. Chem. Soc.* **88**, 2425 (1966).
588. E. C. Evers, W. O. Freitag, W. A. Kriner, and A. G. MacDiarmid, *J. Amer. Chem, Soc.* **81**, 5106 (1959).
588a. O. Exner and V. Jehlička, *Collect. Czech. Chem. Commun.* **37**, 2169 (1972).
588b. R. S. Fager, P. Sejnowski and E. W. Abramson, *Biochem. Biophys. Res. Commun.* **47**, 1244 (1972).
589. J. N. G. Faulks, N. N. Greenwood, and J. H. Morris, *J. Inorg. Nucl. Chem.* **29**. 329 (1967).
590. E. Favre and M. Gaudemar, *C.R. Acad. Sci., Ser. C* **272**, 111 (1971).
591. E. M. Fedneva and I. V. Kryakova, *Russ. J. Inorg. Chem.* **13**, 772 (1968).
592. R. Fellous and R. Luft, *Tetrahedron Lett.* p. 1505 (1970).
593. R. Fellous, R. Luft, and A. Puill, *Tetrahedron Lett.* p. 1509 (1970).
593a. R. Fellous, R. Luft, and A. Puill, *Bull. Soc. Chim. Fr.* p. 1801 (1972).
594. T. P. Fehlner, *J. Amer. Chem. Soc.* **93**, 6366 (1971).
595. T. P. Fehlner and W. S. Koski, *J. Amer. Chem. Soc.* **87**, 409 (1965).
596. T. P. Fehlner and G. W. Mappes, *J. Phys. Chem.* **73**, 873 (1969).
597. W. Fenzyl and R. Köster, *Angew. Chem.* **83**, 807 (1971); *Angew. Chem., Int. Ed. Engl.* **10**, 750 (1971).
598. S. A. Fieldhouse and I. R. Peat, *J. Phys. Chem.* **73**, 275 (1969).
599. B. Figgis and R. L. Williams, *Spectrochim. Acta* **15**, 331 (1959).
600. A. Finch and P. J. Gardner, *Trans. Faraday Soc.* **62**, 3314 (1966).
601. A. Finch, P. J. Gardner, J. C. Lockhart, and E. J. Pearn, *J. Chem. Soc., London* p. 1428 (1962).
602. A. Finch, P. J. Gardner, P. M. McNamara, and G. R. Wellum, *J. Chem. Soc., (A) London* p. 3339 (1970).
603. A. Finch, P. J. Gardner, E. J. Pearn, and G. B. Watts, *Trans. Faraday Soc.* **63**, 1880 (1967).
604. A. Finch, P. J. Gardner, and G. B. Watts, *Chem. Commun.* p. 1054 (1967).
605. A. Finch, P. J. Hendra, and E. J. Pearn, *Spectrochim. Acta* **18**, 51 (1962).
606. A. Finch and J. C. Lockhart, *Chem. Ind. (London)* p. 497 (1964).
607. A. Finch and E. J. Pearn, *Tetrahedron* **20**, 173 (1964).
608. W. Fink, *Chem. Ber.* **96**, 1071 (1963).
608a. R. H. Fish, *J. Amer. Chem. Soc.* **90**, 4435 (1968).
608b. R. H. Fish, *J. Org. Chem.* **38**, 158 (1973).
609. K. G. Foot and B. P. Roberts, *J. Chem. Soc., C* p. 3475 (1971).
610. J. W. L. Fordham and C. L. Sturm, *J. Polym. Sci.* **33**, 503 (1958).
611. R. Foster, *Nature (London)* **195**, 490 (1962).

612. D. L. Fowler and C. A. Kraus, *J. Amer. Chem. Soc.* **62**, 1143 (1940).
613. D. L. Fowler and C. A. Kraus, *J. Amer. Chem. Soc.* **62**, 2237 (1940).
614. W. B. Fox and T. Wartick, *J. Amer. Chem. Soc.* **83**, 498 (1961).
615. J. Francois and E. Franta, *C. R. Acad. Sci., Ser. C* **274**, 1237 (1972).
616. E. Frankland, *Ann. Chem. Pharm.* **124**, 129 (1862).
617. E. Frankland, *J. Chem. Soc., London* **15**, 363 (1862).
618. E. Frankland, *Proc. Roy. Soc., London* **12**, 123 (1863).
619. E. Frankland and B. F. Duppa, *Proc. Roy. Soc., London* **10**, 568 (1859).
620. E. Frankland and B. F. Duppa, *Ann. Chem. Pharm.* **115**, 319 (1860).
621. R. C. Freidlina, A. N. Nesmejanow, and K. A. Kozeschkow, *Chem. Ber.* **68**, 565 (1935).
622. F. A. French and R. S. Rasmussen, *J. Chem. Phys.* **14**, 389 (1946).
623. E. I. Frenkin, A. A. Prokhorova, Ya. M. Paushkin, and A. V. Topchiev, *Izv. Akad. Nauk SSSR, Otd. Khim. Nauk* p. 1507 (1960).
624. F. W. Frey, Jr., P. Kobetz, G. C. Robinson, and T. O. Sistrunk, *J. Org. Chem.* **26**, 2950 (1961).
625. L. B. Friedman and W. N. Lipscomb, *Inorg. Chem.* **5**, 1752 (1966).
626. P. Fritz, K. Niedenzu, and J. W. Dawson, *Inorg. Chem.* **3**, 626 (1964).
627. P. Fritz, K. Niedenzu, and J. W. Dawson, *Inorg. Chem.* **4**, 886 (1965).
628. S. I. Frolov, Y. N. Bubnov, and B. M. Mikhailov, *Izv. Akad. Nauk SSSR, Ser. Khim.* p. 1996 (1969).
629. Y.-C. Fu and G. R. Hill, *J. Amer. Chem. Soc.* **84**, 353 (1962).
630. J. Furukawa and T. Tsuruta, *J. Polym. Sci.* **28**, 227 (1958).
631. J. Furukawa, T. Tsuruta, T. Imada, and H. Fukutani, *Makromol. Chem.* **31**, 122 (1959).
632. J. Furukawa, T. Tsuruta, and S. Ione, *Polym. Sci.* **26**, 234 (1957).
633. J. Furukawa, T. Tsuruta, A. Onishi, T. Sagusa, T. Fueno, H. Fujutani, and T. Imada, *Kogyo Kagaku Zasshi* **61**, 728 (1958).
634. J. Furukawa, T. Tsuruta, T. Sagusa, A. Onishi, A. Kawasaki, T. Fueno, S.-T. Chen, N. Yamamoto, and T. Matsumoto, *Kogyo Kagaku Zasshi* **61**, 1046 (1958).
635. J. Furukawa, T. Tsuruta, and S. Shiotani, *J. Polym. Sci.* **40**, 237 (1959).
636. D. F. Gaines, *J. Amer. Chem. Soc.* **91**, 6503 (1969).
637. D. F. Gaines and T. V. Iorns, *J. Amer. Chem. Soc.* **92**, 4571 (1970).
638. H. J. Galbraith, *J. Chem. Phys.* **22**, 1461 (1954).
639. G. L. Gal'chenko, R. M. Varushchenko, Yu. N. Bubnov, and B. M. Mikhailov, *Zh. Obshch. Khim.* **32**, 2405 (1962).
640. G. L. Gal'chenko, R. M. Varuschenko, Yu. N. Bubnov, and B. M. Mikhailov, *Zh. Obshch. Khim.* **32**, 284 (1962).
640a. G. L. Gal'chenko and R. M. Varushchenko, *Zh. Fiz. Khim.* **37**, 2513 (1963).
641. J. Callaghan and B. Siegel, *J. Amer. Chem. Soc.* **81**, 504 (1959).
641a. F. Gallais, J. P. Laurent, and G. Jugie, *J. Chim. Phys. Physiochim. Biol.* **67**, 934 (1970).
642. P. S. Ganguli, L. P. Gordon, and H. A. McGee, *J. Chem. Phys.* **53**, 782 (1970).
643. P. S. Ganguli and H. A. McGee, *J. Chem. Phys.* **50**, 4658 (1969).
644. M. E. Garabedian and S. W. Benson, *J. Amer. Chem. Soc.* **86**, 176 (1964).
644a. P. J. Garegg and K. Lindstrom, *Acta Chem. Scand.* **25**, 1559 (1971).
645. W. Geilmann and W. Gebauhr, *Z. Anal. Chem.* **139**, 161 (1953).
646. I. Geisler and H. Nöth, *J. Chem. Soc., D* p. 775 (1969).
647. S. Geller, *J. Chem. Phys.* **32**, 1569 (1960).
648. T. A. George and M. F. Lappert, *Chem. Commun.* p. 463 (1966).

649. W. Gerrard, M. Howarth, E. F. Mooney, and D. E. Pratt, *J. Chem. Soc., London* p. 1582 (1963).
650. W. Gerrard, M. F. Lappert, and R. Shafferman, *J. Chem. Soc., London* p. 3828 (1957).
651. W. Gerrard, M. F. Lappert, and R. Shafferman, *Chem. Ind.* (*London*) p. 722 (1958).
652. W. Gerrard, M. F. Lappert, and R. Shafferman, *J. Chem. Soc., London* p. 3648 (1958).
653. W. Gerrard and E. F. Mooney, *J. Chem. Soc., London* p. 4028 (1960).
654. W. Gerrard, E. F. Mooney, and D. E. Pratt, *J. Appl. Chem.* **13**, 127 (1963).
655. W. Gerrard, E. F. Mooney, and R. G. Rees, *J. Chem. Soc., London* p. 740 (1964).
656. W. Gerrard, E. F. Mooney, and H. A. Willis, *J. Chem. Soc., London* p. 3153 (1961).
657. D. H. Geske, *J. Phys. Chem.* **63**, 1062 (1959).
658. D. H. Geske, *J. Phys. Chem.* **66**, 1743 (1962).
659. P. Geymayer and E. G. Rochow, *Monatsh. Chem.* **97**, 429 (1966).
660. P. Geymayer and E. G. Rochow, *Monatsh. Chem.* **97**, 437 (1966).
661. H. Gilman and L. O. Moore, *J. Amer. Chem. Soc.* **80**, 3609 (1958).
661a. M. Giambiagi, M. S. Giambiagi, and R. Carbo, *J. Chim. Phys. Physicochim. Biol.* **69**, 1298 (1972).
662. H. Gilman, L. Santucci, D. R. Swayampati, and R. O. Ranck, *J. Amer. Chem. Soc.* **79**, 3077 (1957).
663. H. Gilman, D. R. Swayampati, and R. O. Ranck, *J. Amer. Chem. Soc.* **80**, 1355 (1958).
664. H. Gilman and C. C. Vernon, *J. Amer. Chem. Soc.* **48**, 1063 (1926).
665. E. Gipstein, P. R. Kippur, M. A. Higgins, and B. F. Clark, *J. Org. Chem.* **26**, 943 (1961).
666. E. Gipstein, P. R. Kippur, M. A. Higgins, and B. F. Clark, *J. Org. Chem.* **26**, 2947 (1961).
667. D. Giraud, J. Soulié, and P. Cadiot, *C. R. Acad. Sci.* **254**, 319 (1962).
668. T. A. Giudici and A. L. Fluharty, *J. Org. Chem.* **32**, 2043 (1967).
669. F. Glockling and R. G. Strafford, *J. Chem. Soc., A* p. 1761 (1971).
670. C. D. Good and D. M. Ritter, *J. Amer. Chem. Soc.* **84**, 1162 (1962).
671. W. Gordy, H. Ring, and A. B. Burg, *Phys. Rev.* **74**, 1191 (1948).
672. W. Gordy, H. Ring, and A. B. Burg, *Phys. Rev.* **75**, 208 (1949).
673. W. Gordy, H. Ring, and A. B. Burg, *Phys. Rev.* **78**, 512 (1950).
674. J. Goubeau, *Advan. Chem. Ser.* **42**, 87 (1964).
675. J. Goubeau and J. H. Becher, *Z. Anorg. Chem.* **268**, 1 (1952).
676. J. Goubeau and R. Epple, *Chem. Ber.* **90**, 171 (1957).
677. J. Goubeau, R. Epple, D. Ulmschneider, and H. Lehmann, *Angew. Chem.* **67**, 710 (1955).
678. J. Goubeau and J. W. Ewers, *Z. Anorg. Chem.* **304**, 230 (1960).
679. J. Goubeau and J. W. Ewers, *Z. Phys. Chem.* (*Frankfurt am Main*) [N.S.] **25**, 276 (1960).
680. J. Goubeau and D. Hummel, *Z. Phys. Chem.* (*Frankfurt am Main*) [N.S.] **20**, 15 (1959).
681. J. Goubeau and H. Grabner, *Chem. Ber.* **93**, 1379 (1960).
682. J. Goubeau and H. Keller, *Z. Anorg. Chem.* **267**, 1 (1951).
683. J. Goubeau and H. Keller, *Z. Anorg. Chem.* **272**, 303 (1953).
684. J. Goubeau and H. Lehmann, *Z. Anorg. Chem.* **322**, 224 (1963).

685. J. Goubeau and K. H. Rohwedder, *Justus Liebigs Ann. Chem.* **604**, 168 (1957).
686. J. Goubeau and A. Zappel, *Z. Anorg. Chem.* **279**, 39 (1955).
687. M. Gracza-Lukácz, G. Szász, L. Buda, and A. Végh, *Z. Anal. Chem.* **253**, 126 (1971).
688. W. A. G. Graham and F. G. A. Stone, *J. Inorg. Nucl. Chem.* **3**, 164 (1956).
689. W. A. G. Graham and F. G. A. Stone, *Chem. Ind. (London)* p. 1096 (1957).
690. M. A. Grassberger, E. G. Hoffmann, G. Schomburg, and R. Köster, *J. Amer. Chem. Soc.* **90**, 56 (1968).
691. N. N. Greenwood and J. H. Morris, *Proc. Chem. Soc., London* p. 25 (1960).
692. N. N. Greenwood and J. H. Morris, *J. Chem. Soc., London* p. 2922 (1960).
693. N. N. Greenwood, J. H. Morris, and J. C. Wright, *J. Chem. Soc., London* p. 4753 (1964).
694. N. N. Greenwood and K. Wade, *J. Chem. Soc., London* p. 1130 (1960).
695. N. N. Greenwood and J. C. Wright, *J. Chem. Soc., London* p. 448 (1965).
696. R. G. Griffin and H. Van-Willigen, *J. Chem. Phys.* **57**, 86 (1972).
696a. R. N. Grimes, "Carboranes." Academic Press, New York, 1970.
697. W. Grimme, K. Reinert, and R. Köster, *Tetrahedron Lett.* p. 624 (1961).
698. P. J. Grisdale, B. E. Babb, J. C. Doty, T. H. Regan, D. P. Maier, and J. L. R. Williams, *J. Organometal. Chem.* **14**, 63 (1968).
699. P. J. Grisdale, M. E. Glogowski, and J. L. R. Williams, *J. Org. Chem.* **36**, 3821 (1971).
700. P. J. Grisdale and J. L. R. Williams, *J. Organometal. Chem.* **22**, C19 (1970).
701. P. J. Grisdale and J. L. R. Williams, *J. Org. Chem.* **34**, 1675 (1969).
702. P. J. Grisdale, J. L. R. Williams, M. E. Glogowski, and B. E. Babb, *J. Org. Chem.* **36**, 544 (1971).
703. S. Gronowitz and A. Bugge, *Acta Chem. Scand.* **19**, 1271 (1965).
703a. S. Gronowitz and A. Maltesson, *Chem. Scripta* **2**, 79 (1972).
703b. S. Gronowitz and V. Michael, *Ark. Kemi* **32**, 283 (1971).
704. S. Gronowitz and J. Namtvedt, *Tetrahedron Lett.* p. 2967 (1966).
705. S. Gronowitz and J. Namtvedt, *Acta Chem. Scand.* **21**, 2151 (1967).
706. E. Groszek, J. B. Leach, G. T. F. Wong, C. Ungermann, and T. P. Onak, *Inorg. Chem.* **10**, 2770 (1971).
707. S. J. Groszos and S. F. Stafiej, *J. Amer. Chem. Soc.* **80**, 1357 (1958).
708. J. Grotewold, J. Hernandez, and E. A. Lissi, *J. Chem. Soc., B* p. 182 (1971).
709. J. Grotewold and E. A. Lissi, *Chem. Commun.* p. 21 (1965).
710. J. Grotewold and E. A. Lissi, *J. Chem. Soc., B* p. 264 (1968).
711. J. Grotewold and E. A. Lissi, *Chem. Commun.* p. 1367 (1968).
712. J. Grotewold, E. A. Lissi, and J. C. Scaiano, *J. Organometal. Chem.* **19**, 431 (1969).
713. J. Grotewold, E. A. Lissi, and J. C. Scaiano, *J. Chem. Soc., B* p. 475 (1969).
714. J. Grotewold, E. A. Lissi, and J. C. Scaiano, *J. Chem. Soc., B* p. 1187 (1971).
715. J. Grotewold, E. A. Lissi, and A. E. Villa, *J. Chem. Soc., A* p. 1034 (1966).
716. J. Grotewold, E. A. Lissi, and A. E. Villa, *J. Chem. Soc., A* p. 1038 (1966).
717. D. Groves, W. Rhine, and G. D. Stucky, *J. Amer. Chem. Soc.* **93**, 1553 (1971).
718. O. Grummitt, *J. Amer. Chem. Soc.* **64**, 1811 (1942).
719. H. Grundke and P. I. Paetzold, *Chem. Ber.* **104**, 1136 (1971).
719a. H. Grundke and P. I. Paetzold, *Angew. Chem.* **83**, 447 (1971).
720. M. F. Grundon, W. A. Khan, D. R. Boyd, and W. R. Jackson, *J. Chem. Soc., C* p. 2557 (1971).
721. E. Gryszkiewicz-Trochimowski, J. Maurel, and O. Gryszkiewicz-Trochimowski, *Bull. Soc. Chim. Fr.* p. 1953 (1959).

722. M. F. Guest, I. H. Hillier, and V. R. Saunders, *J. Organometal. Chem.* **44**, p. 59 (1972).
723. F. C. Gunderloy and C. E. Erickson, *Inorg. Chem.* **1**, 349 (1962).
724. V. Gutmann, A. Meller, and E. Schaschel, *Monatsh. Chem.* **95**, 1188 (1964).
725. V. Gutmann, A. Meller, and E. Schaschel, *J. Organometal. Chem.* **2**, 287 (1964).
726. V. Gutmann, A. Meller, and R. Schlegel, *Monatsh. Chem.* **94**, 1071 (1963).
727. V. Gutmann, A. Meller, and R. Schlegel, *Monatsh. Chem.* **94**, 733 (1963).
728. V. Gutmann, A. Meller, and R. Schlegel, *Monatsh. Chem.* **95**, 314 (1964).
729. J. Guy and M. Chaigneau, *Bull. Soc. Chim. Fr.* p. 257 (1956).
729a. W. D. Haack and W. Rüdorff, *Angew. Chem.* **71**, 464 (1959).
730. A. Haag and G. Hesse, *Justus Liebigs Ann. Chem.* **751**, 95 (1971).
731. R. J. Haines and A. L. du Preez, *J. Amer. Chem. Soc.* **93**, 2820 (1971).
732. K. G. Hancock and D. A. Dickinson, *J. Chem. Soc., Chem. Commun.* p. 962B (1972).
732a. K. G. Hancock and D. A. Dickinson, *J. Amer. Chem. Soc.* **95**, 280 (1973).
733. K. G. Hancock and A. K. Uriarte, *J. Amer. Chem. Soc.* **92**, 6374 (1970).
733a. F. Hanousek, B. Stibr, S. Heřmánek, J. Plešek, A. Vítek, and F. Haruda, *Collect. Czech. Chem. Commun.* **37**, 3001 (1972).
734. L. D. Hansen, J. A. Partridge, R. M. Izatt, and J. J. Christensen, *Inorg. Chem.* **5**, 569 (1966).
735. R. L. Hansen, *J. Polym. Sci., Part A* **2**, 4215 (1964).
736. R. L. Hansen and R. R. Hamann, *J. Phys. Chem.* **67**, 2868 (1963).
737. A. B. Harmon and K. M. Harmon, *J. Amer. Chem. Soc.* **88**, 4093 (1966).
738. K. M. Harmon, A. B. Harmon, and A. A. MacDonald, *J. Amer. Chem. Soc.* **86**, 5036 (1964).
739. K. M. Harmon, A. B. Harmon, and A. A. MacDonald, *J. Amer. Chem. Soc.* **91**, 323 (1969).
740. J. J. Harris, *J. Org. Chem.* **26**, 2155 (1961).
741. B. C. Harrison, I. J. Solomon, R. D. Hites, and M. J. Klein, *J. Inorg. Nucl. Chem.* **14**, 195 (1960).
742. H. Hartmann and K. H. Birr, *Z. Anorg. Chem.* **299**, 174 (1959).
743. D. R. Harvey and R. O. C. Norman, *J. Chem. Soc., London* p. 3822 (1962).
744. E. A. Haseley, Ph.D. Dissertation, Ohio State University, Columbus (1956).
745. E. A. Haseley, A. B. Garrett, and H. H. Sisler, *J. Phys. Chem.* **60**, 1136 (1956).
745a. F. Haslinger, A. H. Soloway, and D. N. Butler, *J. Med. Chem.* **9**, 581 (1966).
746. A. Hassner and B. H. Braun, *J. Org. Chem.* **28**, 261 (1963).
746a. W. Haubold and R. Schaeffer, *Chem. Ber.* **104**, 513 (1971).
747. J. Havir, *Collect. Czech. Chem. Commun.* **26**, 1775 (1961).
748. R. T. Hawkins, W. J. Lennarz, and H. R. Snyder, *J. Amer. Chem. Soc.* **82**, 3053 (1960).
749. R. T. Hawkins and H. R. Snyder, *J. Amer. Chem. Soc.* **82**, 3863 (1960).
750. D. T. Haworth and L. F. Hohnstedt, *J. Amer. Chem. Soc.* **82**, 3860 (1960).
751. M. F. Hawthorne, *J. Amer. Chem. Soc.* **80**, 3480 (1958).
752. M. F. Hawthorne, *J. Org. Chem.* **23**, 1788 (1959).
753. M. F. Hawthorne, *Tetrahedron* **17**, 117 (1962).
754. M. F. Hawthorne, *J. Org. Chem.* **22**, 1001 (1957).
755. M. F. Hawthorne, *J. Org. Chem.* **23**, 1788 (1958).
756. M. F. Hawthorne, *Chem. Ind.* (*London*) p. 1242 (1957).
757. M. F. Hawthorne, *J. Amer. Chem. Soc.* **80**, 4291 (1958).
758. M. F. Hawthorne, *J. Amer. Chem. Soc.* **80**, 4293 (1958).

759. M. F. Hawthorne, *J. Amer. Chem. Soc.* **81**, 5836 (1959).
760. M. F. Hawthorne, *J. Amer. Chem. Soc.* **82**, 748 (1960).
761. M. F. Hawthorne, *J. Amer. Chem. Soc.* **82**, 1886 (1960).
762. M. F. Hawthorne, *J. Amer. Chem. Soc.* **83**, 367 (1961).
763. M. F. Hawthorne, J. *Amer. Chem. Soc.* **83**, 831 (1961).
764. M. F. Hawthorne, *J. Amer. Chem. Soc.* **83**, 833 (1961).
765. M. F. Hawthorne, *J. Amer. Chem. Soc.* **83**, 1345 (1961).
766. M. F. Hawthorne, *J. Amer. Chem. Soc.* **83**, 2541 (1961).
767. M. F. Hawthorne, *J. Amer. Chem. Soc.* **83**, 2671 (1961).
768. M. F. Hawthorne, *Tetrahedron* **17**, 117 (1962).
769. M. F. Hawthorne and W. L. Budde, *J. Amer. Chem. Soc.* **86**, 5337 (1964).
770. M. F. Hawthorne, W. L. Budde, and D. Walmsley, *J. Amer. Chem. Soc.* **86**, 5337 (1964).
771. M. F. Hawthorne and J. A. Dupont, *J. Amer. Chem. Soc.* **80**, 5830 (1958).
772. M. F. Hawthorne and E. S. Lewis, *J. Amer. Chem. Soc.* **80**, 4296 (1958).
773. M. F. Hawthorne and M. Reintjes, *J. Amer. Chem. Soc.* **86**, 951 (1964).
774. M. F. Hawthorne and M. Reintjes, *J. Amer. Chem. Soc.* **86**, 5016 (1964).
775. M. F. Hawthorne and M. Reintjes, *J. Amer. Chem. Soc.* **87**, 4585 (1965).
776. C. W. Heitsch, *Inorg. Chem.* **4**, 1019 (1965).
777. C. W. Heitsch and J. G. Verkade, *Inorg. Chem.* **1**, 392 (1962).
778. R. Hemming and D. G. Johnston, *J. Chem. Soc., London* p. 466 (1964).
779. D. Henneberg, H. Damen, and R. Köster, *Justus Liebigs Ann. Chem.* **640**, 52 (1961).
780. G. F. Hennion, P. A. McCusker, E. C. Ashby, and A. J. Rutkowski, *J. Amer. Chem. Soc.* **79**, 5194 (1957).
781. G. F. Hennion, P. A. McCusker, E. C. Ashby, and A. J. Rutkowski, *J. Amer. Chem. Soc.* **79**, 5190 (1957).
782. G. F. Hennion, P. A. McCusker, and J. V. Marra, *J. Amer. Chem. Soc.* **80**, 3481 (1958).
783. G. F. Hennion, P. A. McCusker, and J. V. Marra, *J. Amer. Chem. Soc.* **81**, 1768 (1959).
784. G. F. Hennion, P. A. McCusker, and A. J. Rutkowski, *J. Amer. Chem. Soc.* **80**, 617 (1958).
784a. G. E. Herberich and G. Greiss, *Chem. Ber.* **105**, 3413 (1972).
785. G. E. Herberich, G. Greiss, and H. F. Heil, *Angew. Chem., Int. Ed. Engl.* **9**, 805 (1970).
786. G. E. Herberich, G. Greiss, H. F. Heil, and J. Mueller, *J. Chem. Soc., D* p. 1328 (1971).
787. G. E. Herberich and H. Muller, *Angew. Chem.* **83**, 1020 (1971).
788. O. Herstad, G. A. Pressley, and F. E. Stafford, *J. Phys. Chem.* **74**, 874 (1970).
789. W. R. Hertler, *Inorg. Chem.* **3**, 1195 (1964).
790. W. R. Hertler, F. Klanberg, and E. L. Muetterties, *Inorg. Chem.* **6**, 1696 (1967).
791. W. R. Hertler, W. H. Knoth, and E. L. Muetterties, *J. Amer. Chem. Soc* .**86**, 5434 (1964).
792. W. R. Hertler., W. H. Knoth, and E. L. Muetterties, *Inorg. Chem.* **4**, 288 (1965)
793. F. Herz, E. Kaplan, and I. G. Luna, *Experientia* **27**, 1260 (1971).
794. J. E. Herz and L. A. Marquez, *J. Chem. Soc., C* p. 3504 (1971).
795. W. R. Hertler, *J. Amer. Chem. Soc.* **86**, 2949 (1964).
796. J. E. Herz and L. A. Marquez, *J. Chem. Soc., C* p. 2243 (1969).
797. H. Hess, *Acta Crystallogr., Sect. B* **25**, 2334 (1969).

798. G. Hesse and A. Haag, *Tetrahedron Lett.* p. 1123 (1965).
799. G. Hesse and H. Witte, *Angew. Chem.* **75**, 791 (1963).
800. G. Hesse and H. Witte, *Justus Liebigs Ann. Chem.* **687**, 1 (1965).
801. G. Hesse, H. Witte, and G. Bittner, *Justus Liebigs Ann. Chem.* **687**, 9 (1965).
802. G. Hesse, H. Witte, and W. Gulden, *Tetrahedron Lett.* p. 2707 (1966).
803. B. Hessett, J. H. Morris, and P. G. Perkins, *J. Chem. Soc., A* p. 2466 (1971).
804. R. Heyes and J. C. Lockhart, *J. Chem. Soc., A* p. 326 (1968).
805. T. L. Heying and C. Naar-Colin, *Inorg. Chem.* **3**, 282 (1964).
806. T. L. Heying and H. D. Smith, Jr., *Advan. Chem. Ser.* **42**, 201 (1964).
807. M. E. D. Hillman, *J. Amer. Chem. Soc.* **84**, 4715 (1962).
808. M. E. D. Hillman, *J. Amer. Chem. Soc.* **85**, 982 (1963).
809. M. E. D. Hillman, *J. Amer. Chem. Soc.* **85**, 1626 (1963).
810. H. Hock and F. Ernst, *Chem. Ber.* **92**, 2716 (1959).
811. B. D. Hoewe, L. J. Malone, and R. M. Manley, *Inorg. Chem.* **10**, 930 (1971).
811a. K. F. Hoffmann and U. Engelhardt, *Z. Anorg. Allg. Chem.* **389**, 97 (1972).
812. R. Hoffmann, *Advan. Chem. Ser.* **42**, 78 (1964).
813. R. Hoffmann, *J. Chem. Phys.* **40**, 2474 (1964).
814. H. K. Hofmeister and J. R. Van Wazer, *J. Inorg. Nucl. Chem.* **26**, 1209 (1964).
815. L. F. Hohnstedt, J. P. Brennan, and K. A. Reynard, *J. Chem. Soc., A* p. 2455 (1970).
816. A. K. Holliday and W. Jeffers, *J. Inorg. Nucl. Chem.* **6**, 134 (1958).
817. A. K. Holliday and G. N. Jessop, *J. Chem. Soc., A* p. 889 (1967).
818. A. K. Holliday and G. N. Jessop, *J. Organometal. Chem.* **10**, 291 (1967).
819. A. K. Holliday and A. G. Massey, *J. Chem. Soc., London* p. 43 (1960).
820. A. K. Holliday and A. G. Massey, *J. Chem. Soc., London* p. 2075 (1960).
821. A. K. Holliday and A. G. Massey, *Chem. Rev.* **62**, 303 (1962).
822. A. K. Holliday and R. P. Ottley, *Chem. Commun.* p. 336 (1969).
823. A. K. Holliday and R. P. Ottley, *J. Chem. Soc., A* p. 886 (1971).
824. A. K. Holliday, W. Reade, R. A. W. Johnstone, and A. F. Neville, *J. Chem. Soc., D* p. 51 (1971).
825. A. K. Holliday and F. B. Taylor, *J. Chem. Soc., London* p. 2731 (1964).
826. A. K. Holliday and N. R. Thompson, *J. Chem. Soc., London* p. 2695 (1960).
827. R. E. Hollins and F. E. Stafford, *Inorg. Chem.* **9**, 877 (1970).
828. R. R. Holmes and R. P. Carter, Jr., *Inorg. Chem.* **2**, 1146 (1963).
829. R. R. Holmes and R. P. Wagner, *J. Amer. Chem. Soc.* **84**, 357 (1962).
830. H. E. Holmquist and R. E. Benson, *J. Amer. Chem. Soc.* **84**, 4720 (1962).
830a. P. H. Homann, *Biochim. Biophys. Acta* **256**, 336 (1972).
831. J. B. Honeycutt, Jr. and J. M. Riddle, *J. Amer. Chem. Soc.* **81**, 2593 (1959).
832. J. B. Honeycutt, Jr. and J. M. Riddle, *J. Amer. Chem. Soc.* **82**, 3051 (1960).
833. J. B. Honeycutt, Jr. and J. M. Riddle, *J. Amer. Chem. Soc.* **83**, 369 (1961).
833a. J. Hooz, J. N. Bridson, J. G. Calzada, H. C. Brown, M. M. Midland, and A. B. Levy, *J. Org. Chem.* **38**, 2574 (1973).
834. J. Hooz and D. M. Gunn, *Tetrahedron Lett.* p. 3455 (1969).
835. J. Hooz and D. M. Gunn, *J. Amer. Chem. Soc.* **91**, 6195 (1969).
836. J. Hooz and D. M. Gunn, *Chem. Commun.* p. 139 (1969).
837. J. Hooz, D. M. Gunn, and H. Kono, *Can. J. Chem.* **49**, 2371 (1971).
837a. J. Hooz and R. B. Layton, *Can. J. Chem.* **50**, 1105 (1972).
838. J. Hooz and S. Linke, *J. Amer. Chem. Soc.* **90**, 5936 (1968).
839. J. Hooz and S. Linke, *J. Amer. Chem. Soc.* **90**, 6891 (1968).
840. J. Hooz and G. F. Morrison, *Can. J. Chem.* **48**, 868 (1970).

841. W. V. Hough, L. J. Edwards, and A. F. Stang, *J. Amer. Chem. Soc.* **85**, 831 (1963).
842. A. J. Hubert, *J. Chem. Soc., London* p. 6669 (1965).
843. A. J. Hubert, *J. Chem. Soc., London* p. 6679 (1965).
844. A. J. Hubert and J. Dale, *J. Chem. Soc., London* p. 4091 (1963).
845. R. Huisgen, I. Ugi, I. Zeigler, and H. Huber, *Tetrahedron* **15**, 44 (1961).
846. A. A. Humffray and L. F. G. Williams, *Chem. Commun.* p. 616 (1965).
846a. A. A. Humffray and L. F. G. Williams, *Electrochim. Acta* **17**, 1157 (1972).
847. D. T. Hurd, *J. Org. Chem.* **13**, 711 (1948).
848. D. T. Hurd, *J. Amer. Chem. Soc.* **70**, 2053 (1948).
848a. R. O. Hutchins and D. Kandasamy, *J. Amer. Chem. Soc.* **95**, 6131 (1973).
849. R. O. Hutchins, B. E. Maryanoff, and C. A. Milewski, *J. Amer. Chem. Soc.* **93**, 1793 (1971).
850. R. O. Hutchins, B. E. Maryanoff, and C. A. Milewski, *J. Chem. Soc., D* p. 1097 (1971).
850a. G. Huttner and B. Krieg, *Chem. Ber.* **105**, 3437 (1972).
850b. G. Huttner, B. Krieg, and W. Gartzke, *Chem. Ber.* **105**, 3424 (1972).
851. F. Ide and Y. Takayama, *Kogyo Kagaku Zasshi* **63**, 529 (1960).
852. F. Ide and Y. Takayama, *Kogyo Kagaku Zasshi* **63**, 533 (1960).
853. M. Inatome and L. P. Kuhn, *Advan. Chem. Ser.* **42**, 183 (1964).
854. K. U. Ingold, *Chem. Commun.* p. 911 (1969).
855. M. Ishii, *Kogyo Kagaku Zasshi* **65**, 458 (1962).
856. M. Ishii and H. Suyama, *Sen-i Gakkaishi* **18**, 22 (1962).
857. K. Ito, H. Watanabe, and M. Kubo, *J. Chem. Phys.* **34**, 1043 (1961).
858. J. Iyoda and I. Shiihara, *Bull. Chem. Soc. Jap.* **32**, 304 (1959).
859. H. Jäger and G. Hesse, *Chem. Ber.* **95**, 345 (1962).
860. R. Jefferson, M. F. Lappert, B. Prokai, and B. P. Tilley, *J. Chem. Soc., A* p. 1584 (1966).
861. H. Jenne and K. Niedenzu, *Inorg. Chem.* **3**, 68 (1964).
862. J. R. Jennings, I. Pattison, C. Summerford, K. Wade, and B. K. Wyatt, *Chem. Commun.* p. 250 (1968).
862a. R. E. Jensen and P. S. Roiger, *Anal. Chem.* **44**, 846 (1972).
863. H. D. Johnson, V. T. Brice, G. L. Brubaker, and S. G. Shore, *J. Amer. Chem. Soc.* **94**, 6711 (1972).
863a. H. D. Johnson, V. T. Brice, and S. G. Shore, *Inorg. Chem.* **12**, 689 (1973).
864. W. H. Johnson, M. V. Kilday, and E. J. Prosen, *J. Res. Nat. Bur. Stand., Sect. A* **65**, 215 (1961).
865. J. R. Johnson, H. R. Snyder, and M. G. Van Campen, Jr., *J. Amer. Chem. Soc.* **60**, 115 (1938).
866. J. R. Johnson, M. G. Van Campen, Jr., and O. Grummitt, *J. Amer. Chem. Soc.* **60**, 111 (1938).
867. J. R. Johnson and M. G. Van Campen, Jr., *J. Amer. Chem. Soc.* **60**, 121 (1938).
868. C. Jolicoeur, N. D. The, and A. Cabana, *Can. J. Chem.* **49**, 2008 (1971).
869. P. R. Jones, *J. Org. Chem.* **37**, 1886 (1972).
869a. W. J. Jones and N. Sheppard, *Proc. Roy. Soc., Ser. A* **304**, 135 (1968).
870. F. Joy and M. F. Lappert, *Proc. Chem. Soc., London* p. 353 (1960).
871. F. Joy, M. F. Lappert, and B. Prokai, *J. Organometal. Chem.* **5**, 506 (1966).
872. G. Jugie and J. P. Laurent, *Bull. Soc. Chim. Fr.* p. 2010 (1968).
873. G. Jugie, J. P. Pouyanne, and J. P. Laurent, *C. R. Acad. Sci.* **268**, 1377 (1969).
874. P. Jutzi, *J. Organometal. Chem.* **19**, 1 (1969).
874a. P. Jutzi, *Angew. Chem.* **84**, 28 (1972).

875. G. W. Kabalka, *J. Organometal. Chem.* **33**, 25C (1971).
876. G. W. Kabalka, H. C. Brown, A. Suzuki, S. Honma, A. Arase, and M. Itoh, *J. Amer. Chem. Soc.* **92**, 710 (1970).
876a. L. I. Kachaeva, S. F. Zhil'tsov, O. N. Druzhkov, L. F. Kudryavtsev, and G. G. Petukhov, *Zh. Obshch. Khim.* **42**, 1029 (1972).
877. H. D. Kaesz, S. L. Stafford, and F. G. A. Stone, *J. Amer. Chem. Soc.* **81**, 6336 (1959).
878. H. D. Kaesz and F. G. A. Stone, *J. Amer. Chem. Soc.* **82**, 6213 (1960).
879. A. Kaldor and R. F. Porter, *J. Amer. Chem. Soc.* **93**, 2140 (1971).
879a. V. N. Kapshtal, *Russ. J. Phys. Chem.* **40**, 508 (1966).
879b. V. N. Kapshtal and L. M. Sverdlov, *Russ. J. Phys. Chem.* **39**, 1169 (1965).
879c. V. N. Kapshtal and L. M. Sverdlov, *Russ. J. Phys. Chem.* **40**, 1514 (1966).
879d. E. I. Karpeiskaya, Yu. N. Kukushkin, and V. A. Trofimov, *Zh. Neorg. Khim.* **17**, 1484 (1972).
880. S. Kato, M. Wada, and Y. Tsuzuki, *Bull. Chem. Soc. Jap.* **36**, 868 (1963).
881. J. J. Kaufman and J. R. Hamann, *Advan. Chem. Ser.* **42**, 95 (1964).
882. J. J. Kaufman and J. R. Hamann, *Advan. Chem. Ser.* **42**, 273 (1964).
882a. R. N. Keller and E. M. V. Wall, *Advan. Chem. Ser.* **32**, 221 (1961).
883. E. Khotinsky and M. Melamed, *Chem. Ber.* **42**, 3090 (1909).
884. K. Khol'tzanfel and K. Richter, *Zh. Obshch. Khim.* **32**, 1358 (1962).
885. R. W. Kirk, D. L. Smith, W. Airey, and P. L. Timms, *J. Chem. Soc., Dalton Trans.* p. 1392 (1972).
886. K. Kiss-Eroess, L. Erdey, and I. Buzas, *Talanta* **17**, 1209 (1970).
887. W. Kitching, K. V. G. Das, and C. J. Moore, *J. Organometal. Chem.* **22**, p. 399 (1970).
888. A. J. Klanica and J. P. Faust, *Inorg. Chem.* **7**, 1037 (1968).
889. A. J. Klanica, J. P. Faust, and C. S. King, *Inorg. Chem.* **6**, 840 (1967).
890. R. Klein, A. D. Bliss, L. J. Schoen, and H. G. Nadeau, *J. Amer. Chem. Soc.* **83**, 4131 (1961).
891. J. Klein, E. Dunkelblum, and M. A. Wolff, *J. Organometal. Chem.* **7**, 377 (1967).
892. E. F. Knights and H. C. Brown, *J. Amer. Chem. Soc.* **90**, 5280 (1968).
893. E. F. Knights and H. C. Brown, *J. Amer. Chem. Soc.* **90**, 5281 (1968).
894. E. F. Knights and H. C. Brown, *J. Amer. Chem. Soc.* **90**, 5283 (1968).
895. W. H. Knoth, *J. Amer. Chem. Soc.* **88**, 935 (1966).
896. W. H. Knoth, *J. Amer. Chem. Soc.* **89**, 4850 (1967).
897. W. H. Knoth, H. C. Miller, D. C. England, G. W. Parshall, and E. L. Muetterties, *J. Amer. Chem. Soc.* **84**, 1056 (1962).
898. W. H. Knoth, N. E. Miller, and W. R. Hertler, *Inorg. Chem.* **6**, 1977 (1967).
899. W. H. Knoth and E. L. Muetterties, *J. Inorg. Nucl. Chem.* **20**, 66 (1961).
900. W. H. Knoth, J. C. Sauer, J. H. Balthis, H. C. Miller, and E. L. Muetterties, *J. Amer. Chem. Soc.* **89**, 4842 (1967).
901. W. H. Knoth, J. C. Sauer, D. C. England, W. R. Hertler, and E. L. Muetterties, *J. Amer. Chem. Soc.* **86**, 3973 (1964).
902. W. H. Knoth, J. C. Sauer, H. C. Miller, and E. L. Muetterties, *J. Amer. Chem. Soc.* **86**, 115 (1964).
903. G. Kobrich and H. R. Merkle, *Angew. Chem.* **79**, 50 (1967).
904. G. Kobrich and H. R. Merkle, *Chem. Ber.* **100**, 3371 (1967).
905. G. Kodama, *J. Amer. Chem. Soc.* **94**, 5907 (1972).
906. K. A. Koehler, R. C. Jackson, and G. E. Lienhard, *J. Org. Chem.* **37**, 2232 (1972).
907. K. Kojima, T. Habu, S. Iwabuchi, and M. Yoshikuni, *Nippon Kagaku Kaishi* p. 2165 (1972).

908. K. Kojima, Y. Iwata, M. Nagayama, and S. Iwabuchi, *J. Polym. Sci., Part B* **8**, 541 (1970).
909. K. Kojima, S. Iwabuchi, K. Kojima, and N. Tarumi, *Bull. Chem. Soc. Jap.* **44**, 1891 (1971).
910. K. Kojima, S. Iwabuchi, K. Murakami, K. Kojima, and F. Ichikawa, *J. Appl. Polym. Sci.* **16**, 1139 (1972).
911. K. Kojima, T. Suzuki, S. Iwabuchi, and N. Tarumi, *Nippon Kagaku Kaishi* p. 1943 (1972).
912. A. A. Koksharova, G. G. Petukhov, and S. F. Zhil'tsov, *Zh. Obshch. Khim.* **40**, p. 2446 (1970).
912a. A. A. Koksharova, G. G. Petukhov, and S. F. Zhil'tsov, *Zh. Obshch. Khim.* **40**, p. 2449 (1970).
913. G. S. Kolesnikov, S. L. Davydova, M. A. Yampol'skay, and N. V. Klimentova, *Izv. Akad. Nauk SSSR, Otd. Khim. Nauk* p. 841 (1962).
914. G. S. Kolesnikov and L. S. Fedorova, *Izv. Akad. Nauk SSSR, Otd. Khim. Nauk* p. 236 (1957).
915. G. S. Kolesnikov and L. S. Fedorova, *Izv. Akad. Nauk. SSSR, Otd. Khim. Nauk* p. 906 (1958).
916. G. S. Kolesnikov and N. V. Klimentova, *Izv. Akad. Nauk SSSR, Otd. Khim. Nauk* p. 652 (1957).
917. G. S. Kolesnikov, A. P. Suprun, and T. A. Soboleva, *Vysokomol. Soedin.* **1**, 627 (1957).
918. J. Kollonitsch, *Nature* **189**, 1005 (1961).
919. I. I. Kolodkina, A. S. Guseva, E. A. Ivanova, L. S. Varshavskara, and A. M. Yurkevich, *Zh. Obshch. Khim.* **40**, 2489 (1970).
920. W. König and W. Scharrnbeck, *J. Prakt. Chem.* **128**, 153 (1930).
921. S. Korcek, G. B. Watts, and K. U. Ingold, *J. Chem. Soc., Perkin Trans. II*, p. 242 (1972).
922. V. V. Korshak, N. I. Bekasova, L. M. Chursina, and V. A. Zamyatina, *Izv. Akad. Nauk SSSR, Ser. Khim.* p. 1645 (1963).
922a. V. V. Korshak, A. I. Solomatina, N. I. Bekasova, and V. A. Zamyatina, *Izv. Akad. Nauk SSSR, Ser. Khim.* p. 1856 (1963).
923. V. V. Korshak, V. A. Zamyatina, and N. I. Bekasova, *Izv. Akad. Nauk SSSR, Ser. Khim.* p. 1648 (1963).
924. V. V. Korshak, V. A. Zamyatina, N. I. Bekasova, and Ma Zhui-Zhan, *Vysokomolekul. Soedin.* **3**, 521 (1961).
925. V. V. Korshak, V. A. Zamyatina, and R. M. Oganesyan, *Bull. Acad. Sci. USSR, Div. Chem. Sci.* p. 1580 (1962).
926. R. Köster, *Angew. Chem.* **69**, 94 (1957).
927. R. Köster, *Angew. Chem.* **69**, 94 (1957).
928. R. Köster, *Angew. Chem.* **68**, 383 (1956).
929. R. Köster, *Angew. Chem.* **69**, 684 (1957).
930. R. Köster, *Oester. Chem.-Ztg.* **57**, 136 (1956).
931. R. Köster, *Angew. Chem.* **70**, 413 (1958).
932. R. Köster, *Justus Liebigs Ann. Chem.* **618**, 31 (1958).
933. R. Köster, *Angew. Chem.* **70**, 371 (1958).
934. R. Köster, *Angew. Chem.* **70**, 743 (1958).
935. R. Köster, *Angew. Chem.* **71**, 31 (1959).
936. R. Köster, *Angew. Chem.* **71**, 520 (1959).
937. R. Köster, *Angew. Chem.* **72**, 626 (1960).
938. R. Köster, *Angew. Chem.* **73**, 66 (1961).

939. R. Köster, *Angew. Chem., Int. Ed. Engl.* **3**, 174 (1964).
940. R. Köster, K.-L. Amen, H. Bellut, and W. Fenzyl, *Angew. Chem., Int. Ed. Engl.* **10**, 748 (1971).
941. R. Köster, H. Bellut, G. Benedikt, and E. Ziegler, *Justus Liebigs Ann. Chem.* **724**, 34 (1969).
942. R. Köster, H. Bellut, and S. Hattori, *Justus Liebigs Ann. Chem.* **720**, 1 (1969).
943. R. Köster, H. Bellut, S. Hattori, and L. Weber, *Justus Liebigs Ann. Chem.* **720**, 32 (1968).
944. R. Köster, H. Bellut, and E. Ziegler, *Angew. Chem.* **79**, 241 (1967); *Angew. Chem., Int. Ed. Engl.* **6**, 255 (1967).
945. R. Köster and G. Benedikt, *Angew. Chem.* **75**, 346 (1963).
946. R. Köster and G. Benedikt, *Angew. Chem.* **76**, 650 (1964).
947. R. Köster and G. Benedikt, *Angew. Chem.* **75**, 419 (1963).
948. R. Köster and G. Benedikt, *Angew. Chem.* **74**, 589 (1962); *Angew. Chem., Int. Ed. Engl.* **1**, 507 (1962).
949. R. Köster, G. Benedikt, and M. A. Grassberger, *Justus Liebigs Ann. Chem.* **719**, 187 (1968).
950. R. Köster, G. Benedikt, and H. W. Schrotter, *Angew. Chem.* **76**, 649 (1964).
951. R. Köster and G. Bruno, *Justus Liebigs Ann. Chem.* **629**, 89 (1960).
952. R. Köster, G. Bruno, and P. Binger, *Justus Liebigs Ann. Chem.* **644**, 1 (1961).
953. R. Köster and W. Fenzyl, *Angew. Chem.* **80**, 756 (1968).
954. R. Köster and M. A. Grassberger, *Angew. Chem.* **77**, 457 (1965); *Angew. Chem., Int. Ed. Engl.* **4**, 139 (1965).
955. R. Köster and M. A. Grassberger, *Angew. Chem.* **78**, 590 (1966).
956. R. Köster and M. A. Grassberger, *Justus Liebigs Ann. Chem.* **719**, 169 (1969).
957. R. Köster, M. A. Grassberger, E. G. Hoffmann, and G. W. Rotermund, *Tetrahedron Lett.* p. 905 (1966).
958. R. Köster and G. Griasnow, *Angew. Chem.* **73**, 171 (1961).
959. R. Köster, G. Griasnow, W. Larbig, and P. Binger, *Justus Liebigs Ann. Chem.* **672**, 1 (1964).
960. R. Köster, S. Hattori, and Y. Morita, *Angew. Chem.* **77**, 719B (1965).
961. R. Köster, J. H. Horstschäfer, and P. Binger, *Angew. Chem., Int. Ed. Engl.* **5**, 730 (1966).
962. R. Köster, H. J. Horstschäfer, and P. Binger, *Justus Liebigs Ann. Chem.* **717**, 1 (1968).
963. R. Köster and K. Iwasaki, *Advan. Chem. Ser.* **42**, 148 (1964).
964. R. Köster, K. Iwasaki, S. Hattori, and Y. Morita, *Justus Liebigs Ann. Chem.* **720**, 23 (1969).
965. R. Köster, W. Larbig, and G. W. Rotermund, *Justus Liebigs Ann. Chem.* **682**, 21 (1965).
966. R. Köster and W. Larbig, *Angew. Chem.* **73**, 620 (1961).
967. R. Köster and Y. Morita, *Justus Liebigs Ann. Chem.* **704**, 70 (1967).
968. R. Köster and Y. Morita, *Angew. Chem.* **78**, 589 (1966).
969. R. Köster and Y. Morita, *Angew. Chem.* **77**, 589 (1965).
970. R. Köster and K. Reinert, *Angew. Chem.* **71**, 521 (1959).
971. R. Köster, K. Reinert, and K. H. Müller, *Angew. Chem.* **72**, 78 (1960).
972. R. Köster and Γ. Rickborn, *J. Amer. Chem. Soc.* **89**, 2782 (1967).
973. R. Köster and G. W. Rotermund, *Tetrahedron Lett.* p. 777 (1965).
974. R. Köster and G. W. Rotermund, *Angew. Chem.* **72**, 563 (1960).
975. R. Köster and G. W. Rotermund, *Tetrahedron Lett.* p. 1667 (1964).
976. R. Köster and G. W. Rotermund, *Justus Liebigs Ann. Chem.* **689**, 40 (1965).

977. R. Köster and G. W. Rotermund, *Angew. Chem.* **72**, 138 (1960).
978. R. Köster and G. W. Rotermund, *Angew. Chem.* **74**, 252 (1962).
979. R. Köster and G. Schomburg, *Angew. Chem.* **72**, 567 (1960).
980. R. Köster and K. Ziegler, *Angew. Chem.* **69**, 94 (1957).
981. J. C. Kotz and W. J. Painter, *J. Organometal. Chem.* **32**, 231 (1971).
982. A. A. Koksharova, G. G. Petukhov, and S. F. Zhil'tsov, *Zh. Obshch. Khim.* **40**, 2449 (1970).
983. J. C. Kotz and E. W. Post, *J. Amer. Chem. Soc.* **90**, 4503 (1968).
984. J. C. Kotz and E. W. Post, *Inorg. Chem.* **9**, 1661 (1970).
985. A. I. Kovredov and L. I. Zakharkin, *Izv. Akad. Nauk SSSR, Ser. Khim.* p. 50 (1964).
986. V. P. Kozitskii, *Izv. Akad. Nauk SSSR, Ser. Khim.* p. 8 (1972).
987. V. P. Kozitskii, *Izv. Akad. Nauk SSSR, Ser. Khim.* p. 12 (1972).
988. E. Krause, *Chem. Ber.* **57**, 813 (1924).
989. E. Krause, *Chem. Ber.* **57**, 216 (1924).
990. E. Krause and P. Dittmar, *Chem. Ber.* **63**, 2347 (1930).
991. C. A. Kraus and W. W. Hawes, *J. Amer. Chem. Soc.* **55**, 2776 (1933).
992. E. Krause and R. Nitsche, *Chem. Ber.* **54**, 2784 (1921).
993. D. Krause and R. Nitsche, *Chem. Ber.* **55**, 1261 (1922).
994. E. Krause and P. Nobbe, *Chem. Ber.* **64**, 2112 (1931).
995. E. Krause and H. Polack, *Chem. Ber.* **59**, 777 (1926).
996. E. Krause and H. Polack, *Chem. Ber.* **61**, 271 (1928).
997. E. Krause and P. Nobbe, *Chem. Ber.* **63**, 934 (1930).
998. M. M. Kreevoy and J. E. C. Hutchins, *J. Amer. Chem. Soc.* **91**, 4329 (1969).
998a. P. Krohmer and J. Goubeau, *Chem. Ber.* **104**, 1347 (1971).
999. U. Krüerke, *Z. Naturforsch. B* **11**, 364 (1956).
1000. U. Krüerke, *Z. Naturforsch. B* **11**, 676 (1956).
1001. U. Krüerke, *Z. Naturforsch. B* **11**, 606 (1956).
1002. P. J. Krusic and J. K. Kochi, *J. Amer. Chem. Soc.* **91**, 3942 (1969).
1003. W. Kuchen and R. D. Brinckman, *Z. Anorg. Allg. Chem.* **325**, 225 (1963).
1004. L. P. Kuhn and M. Inatome, *J. Amer. Chem. Soc.* **85**, 1206 (1963).
1005. H. G. Kuivila, *J. Amer. Chem. Soc.* **76**, 870 (1954).
1006. H. G. Kuivila, *J. Amer. Chem. Soc.* **77**, 4014 (1955).
1007. H. G. Kuivila and A. G. Armour, *J. Amer. Chem. Soc.* **79**, 5659 (1957).
1008. H. G. Kuivila and L. E. Benjamin, *J. Amer. Chem. Soc.* **77**, 3823 (1955).
1009. H. G. Kuivila, L. E. Benjamin, C. J. Murphy, A. D. Price, and J. H. Polevy, *J. Org. Chem.* **27**, 825 (1962).
1010. H. G. Kuivila and E. K. Easterbrook, *J. Amer. Chem. Soc.* **73**, 4629 (1951).
1011. H. G. Kuivila and A. R. Henrickson, *J. Amer. Chem. Soc.* **74**, 5068 (1952).
1012. H. G. Kuivila, A. H. Keough, and E. J. Soboczenski, *J. Org. Chem.* **19**, 780 (1954).
1013. H. G. Kuivila and T. C. Muller, *J. Amer. Chem. Soc.* **84**, 377 (1962).
1014. H. G. Kuivila and K. V. Nahabedian, *Chem. Ind.* (*London*) p. 1120 (1959).
1015. H. G. Kuivila and K. V. Nahabedian, *J. Amer. Chem. Soc.* **83**, 2159 (1961).
1016. H. G. Kuivila and K. V. Nahabedian, *J. Amer. Chem. Soc.* **83**, 2164 (1961).
1017. H. G. Kuivila, J. F. Reuwer, and J. A. Mangravite, *J. Amer. Chem. Soc.* **86**, 2666 (1964).
1018. H. G. Kuivila, J. F. Reuwer, and J. A. Mangravite, *Can. J. Chem.* **41**, 3081 (1963).
1019. H. G. Kuivila and E. J. Soboczenski, *J. Amer. Chem. Soc.* **76**, 2675 (1959).
1020. H. G. Kuivila and R. A. Wiles, *J. Amer. Chem. Soc.* **77**, 4830 (1955).
1021. H. G. Kuivila and R. M. Williams, *J. Amer. Chem. Soc.* **76**, 2679 (1954).

1022. M. Kumada, N. Imaki, and K. Yamamoto, *J. Organometal. Chem.* **6**, 490 (1966).
1023. R. M. Kuznesof, F. E. Stafford, and D. F. Shriver, *J. Phys. Chem.* **71**, 1939 (1967).
1023a. J. F. Labarre and C. Leibovici, *J. Chim. Phys. Physicochim. Biol.* **69**, 404 (1972).
1024. W. J. Lafferty and J. J. Ritter, *Chem. Commun.* p. 909 (1969).
1025. W. J. Lafferty and J. J. Ritter, *J. Mol. Spectrosc.* **38**, 181 (1971).
1026. J. J. Lagowski and P. G. Thompson, *Proc. Chem. Soc., London* p. 301 (1959).
1027. F. J. Lalor, T. Paxson, and M. F. Hawthorne, *J. Amer. Chem. Soc.* **93**, 3156 (1971).
1027a. L. Lambert, C. Pepin, and A. Cabana, *J. Mol. Spectrosc.* **44**, 578 (1972).
1028. L. Lamneck and S. Kaye, *Nat. Adv. Comm. Aeronaut., Res. Memo.* **E58 E12** (1958).
1029. H. Landesman, U.S. Patent 2,977,392 (1961).
1030. H. Landesman, U.S. Patent 2,983,761 (1961).
1031. H. Landesman, U.S. Patent 2,964,568 (1960).
1032. H. Landesman and R. E. Williams, *J. Amer. Chem. Soc.* **83**, 2663 (1961).
1033. C. F. Lane, *J. Organometal. Chem.* **31**, 421 (1971).
1034. C. F. Lane and H. C. Brown, *J. Amer. Chem. Soc.* **92**, 7212 (1970).
1035. C. F. Lane and H. C. Brown, *J. Organometal. Chem.* **26**, 51c (1971).
1036. C. F. Lane and H. C. Brown, *J. Amer. Chem. Soc.* **93**, 1025 (1971).
1037. C. F. Lane and H. C. Brown, *J. Organometal. Chem.* **34**, 29c (1972).
1038. I. I. Lapkin and G. A. Yuzhakova, *Zh. Obshch. Khim.* **32**, 1967 (1962).
1038a. I. I. Lapkin, G. A. Yuzhakova, and T. Y. Sergienko, *Zh. Obshch. Khim.* **42**, 401 (1972).
1039. M. F. Lappert, *Angew. Chem.* **69**, 684 (1957).
1040. M. F. Lappert, *Angew. Chem.* **72**, 36 (1960).
1041. M. F. Lappert, *J. Chem. Soc., London* p. 542 (1962).
1042. M. F. Lappert and M. K. Majumdar, *Advan. Chem. Ser.* **42**, 208 (1964).
1043. M. F. Lappert and M. K. Majumdar, *J. Organometal. Chem.* **6**, 316 (1966).
1044. M. F. Lappert, M. K. Majumdar, and B. P. Tilley, *J. Chem. Soc., A* p. 1590 (1966).
1044a. M. F. Lappert, J. B. Pedley, P. N. K. Riley, and A. Tweedale, *Chem. Commun.* p. 788 (1966).
1045. M. F. Lappert and B. Prokai, *J. Chem. Soc., London* p. 4223 (1963).
1046. M. F. Lappert and B. Prokai, *J, Organometal. Chem.* **1**, 384 (1964).
1047. M. F. Lappert and B. Prokai, *J. Chem. Soc., A* p. 129 (1967).
1048. M. F. Lappert and H. Pyszora, *Proc. Chem. Soc., London* p. 350 (1960).
1049. M. F. Lappert, H. Pyszora, and M. Rieber, *J. Chem. Soc., London* p. 4256 (1965).
1050. M. D. La Prades and C. E. Nordman, *Inorg. Chem.* **8**, 1669 (1969).
1051. L. Lardicci, T. P. Giacomelli, and L. DeBernardi, *J. Organometal. Chem.* **39**, 245 (1972).
1052. R. C. Larock and H. C. Brown, *J. Amer. Chem. Soc.* **92**, 2467 (1970).
1053. R. C. Larock and H. C. Brown, *J. Organometal. Chem.* **26**, 35 (1971).
1054. R. C. Larock and H. C. Brown, *J. Organometal. Chem.* **36**, 1 (1972).
1055. A. W. Laubengayer and W. F. Gilliam, *J. Amer. Chem. Soc.* **63**, 477 (1941).
1056. J.-L. Laurent and J. P. Laurent, *Bull. Soc. Chim. Fr.* p. 3565 (1968).
1057. J. P. Laurent, *C. R. Acad. Sci.* **253**, 1812 (1961).
1058. J. P. Laurent, *C. R. Acad. Sci.* **256**, 3283 (1963).
1059. J. P. Laurent, *Bull. Soc. Chim. Fr.* p. 558 (1963).
1060. J. P. Laurent, *C. R. Acad. Sci.* **258**, 1233 (1964).

1061. J. P. Laurent, *C. R. Acad. Sci.* **258:** 1481, (1964).
1062. J. P. Laurent, G. Jugie, and G. Commenges, *J. Inorg. Nucl. Chem.* **31**, 1353 (1969).
1063. J. P. Laurent and M. Pasdelorys, *C. R. Acad. Sci.* **256,** 133 (1963).
1064. J. P. Laurent, J. P. Tuchagues, and F. Gallais, *C. R. Acad. Sci.* **267,** 789 (1968).
1065. R. W. Law and J. L. Margrave, *J. Chem. Phys.* **25**, 1086 (1956).
1065a. S. O. Lawesson, *Ark. Kemi.* **10,** 171 (1956).
1065b. S. O. Lawesson, *Ark. Kemi* **11,** 387 (1957).
1065c. S. O. Lawesson, *Acta Chem. Scand.* **11,** 1075 (1957).
1065d. E. A. Laws, R. M. Stevens, and W. N. Lipscomb, *J. Amer. Chem. Soc.* **94,** 4467 (1972).
1066. J. B. Leach and J. H. Morris, *J. Organometal. Chem.* **13**, 313 (1968).
1067. J. B. Leach and T. P. Onak, *J. Magn. Resonance* **4,** 30 (1971).
1068. J. B. Leach, C. B. Ungermann, and T. P. Onak, *J. Magn. Resonance* **6,** 74 (1972).
1069. D. A. Lee, *J. Inorg. Nucl. Chem.* **34,** 2895 (1972).
1070. A. J. Leffler, *Inorg. Chem.* **3,** 145 (1964).
1071. J. E. Leffler, E. Dolan, and T. Tanigaki, *J. Amer. Chem. Soc.* **87,** 927 (1965).
1072. J. E. Leffler and B. G. Ramsey, *Proc. Chem. Soc., London* p. 117 (1961).
1073. J. E. Leffler and L. J. Todd, *Chem. Ind. (London)* p, 512 (1961).
1074. J. E. Leffler, G. B. Watts, T. Tanigaki, E. Dolan, and D. S. Miller, *J. Amer. Chem. Soc.* **92,** 6825 (1970).
1075. H. Lehmann, J. Goubeau, R. Epple, and D. Ulmschneider, *Angew. Chem.* **67,** 710 (1955).
1076. W. J. Lehmann, *J. Mol. Spectrosc.* **7,** 1 (1961).
1077. W. J. Lehmann and J. F. Ditter, *J. Chem. Phys.* **31,** 549 (1959).
1078. W. J. Lehmann and I. Shapiro, *Spectrochim. Acta* **17,** 396 (1961).
1079. W. J. Lehmann, C. O. Wilson, J. F. Ditter, and I. Shapiro, *Advan. Chem. Ser.* **32,** 139 (1961).
1080. W. J. Lehmann, C. O. Wilson, and I. Shapiro, *J. Chem. Phys.* **28,** 777 (1958).
1081. W. J. Lehmann, C. O. Wilson, and I. Shapiro, *J. Chem. Phys.* **28,** 781 (1958).
1082. W. J. Lehmann, C. O. Wilson, and I. Shapiro, *J. Inorg. Nucl. Chem.* **11,** 91 (1959).
1083. W. J. Lehmann, C. O. Wilson, and I. Shapiro, *J. Chem. Phys.* **32,** 1088 (1960).
1084. W. J. Lehmann, C. O. Wilson, and I. Shapiro, *J. Chem. Phys.* **31,** 1071 (1959).
1085. W. J. Lehmann, C. O. Wilson, and I. Shapiro, *J. Chem. Phys.* **32,** 1786 (1960).
1086. W. J. Lehmann, C. O. Wilson, and I. Shapiro, *J. Chem. Phys.* **33,** 590 (1960).
1087. W. J. Lehmann, C. O. Wilson, and I. Shapiro, *J. Chem. Phys.* **34,** 476 (1961).
1088. W. J. Lehmann, C. O. Wilson, and I. Shapiro, *J. Chem. Phys.* **34,** 783 (1961).
1089. W. J. Lehmann, C. O. Wilson, and I. Shapiro, *J. Inorg. Nucl. Chem.* **21,** 25 (1961).
1090. B. Lengyel and B. Csakvari, *Z. Anorg. Chem.* **322,** 103 (1963).
1091. W. J. Lennarz and H. R. Snyder, *J. Amer. Chem. Soc.* **82,** 2169 (1960).
1092. W. J. Lennarz and H. R. Snyder, *J. Amer. Chem. Soc.* **82,** 2172 (1960).
1093. R. L. Letsinger, *Advan. Chem. Ser.* **42,** 1 (1964).
1094. R. L. Letsinger and S. H. Dandegaonker, *J. Amer. Chem. Soc.* **81,** 498 (1959).
1095. R. L. Letsinger, S. Dandegaonker, W. J. Vullo, and J. D. Morrison, *J. Amer. Chem. Soc.* **85,** 2223 (1963).
1096. R. L. Letsinger, T. E. Feare, T. J. Savereide, and J. R. Nazy, *J. Org. Chem.* **26,** 1271 (1961).
1097. R. L. Letsinger and S. B. Hamilton, *J. Amer. Chem. Soc.* **80,** 5411 (1958).
1098. R. L. Letsinger and S. B. Hamilton, *J. Amer. Chem. Soc.* **81,** 3009 (1959).
1099. R. L. Letsinger and S. B. Hamilton, *J. Org. Chem.* **25,** 592 (1960).
1100. R. L. Letsinger and D. B. Maclean, *J. Amer. Chem. Soc.* **85,** 2230 (1963).

1101. R. L. Letsinger and J. D. Morrison, *J. Amer. Chem. Soc.* **85**, 2227 (1963).
1102. R. L. Letsinger and J. R. Nazy, *J. Org. Chem.* **23**, 914 (1958).
1103. R. L. Letsinger and J. R. Nazy, *J. Amer. Chem. Soc.* **81**, 3013 (1959).
1104. R. L. Letsinger and N. Remes, *J. Amer. Chem. Soc.* **77**, 2489 (1955).
1105. R. L. Letsinger and I. H. Skoog, *J. Org. Chem.* **18**, 895 (1953).
1106. R. L. Letsinger and I. H. Skoog, *J. Amer. Chem. Soc.* **76**, 4174 (1954).
1107. R. L. Letsinger and I. H. Skoog, *J. Amer. Chem. Soc.* **77**, 2491 (1955).
1108. R. L. Letsinger and I. H. Skoog, *J. Amer. Chem. Soc.* **77**, 5176 (1955).
1109. R. L. Letsinger, I. H. Skoog, and N. Remes, *J. Amer. Chem. Soc.* **76**, 4047 (1954).
1110. R. L. Letsinger and A. J. Wysocki, *J. Org. Chem.* **28**, 3199 (1963).
1111. A. J. Leusink, W. Drenth, J. G. Noltes, and G. J. M. van der Kerk, *Tetrahedron Lett.* p. 1263 (1967).
1112. K. A. Levison and P. G. Perkins, *Theor. Chim. Acta* **17**, 1 (1970).
1112a. A. B. Levy and H. C. Brown, *J. Amer. Chem. Soc.* **95**, 4067 (1973).
1113. H. A. Levy and L. O. Brockway, *J. Amer. Chem. Soc.* **59**, 2085 (1937).
1114. L. A. Levy and L. Fishbein, *Tetrahedron Lett.* p. 3773 (1969).
1115. E. S. Lewis and R. H. Grinstein, *J. Amer. Chem. Soc.* **84**, 1158 (1962).
1116. D. R. Lide, Jr., *J. Chem. Phys.* **32**, 1570 (1960).
1117. D. R. Lide, Jr., R. W. Taft, and P. Love, *J. Chem. Phys.* **31**, 561 (1959).
1118. H. H. Lindner and T. Onak, *J. Amer. Chem. Soc.* **88**, 1886 (1966).
1119. H. H. Lindner and T. Onak, *J. Amer. Chem. Soc.* **88**, 1890 (1966).
1120. S. J. Lippard and P. S. Welcker, *Inorg. Chem.* **11**, 6 (1972).
1120a. W. N. Lipscomb, "Boron Hydrides." Benjamin, New York, 1963.
1121. W. N. Lipscomb, *Science* **153**, 373 (1966).
1122. E. A. Lissi and E. Sanhueza, *J. Organometal. Chem.* **26**, C59 (1971).
1123. E. A. Lissi and E. Sanhueza, *J. Organometal. Chem.* **32**, 285 (1971).
1124. D. R. Lloyd and N. Lynaugh, *J. Chem. Soc., D* p. 1545 (1970).
1125. D. R. Lloyd and N. Lynaugh, *J. Chem. Soc., D* p. 125 (1971).
1126. D. R. Lloyd and N. Lynaugh, *J. Chem. Soc., Faraday Trans.* **2**, 947 (1972).
1127. J. E. Lloyd and K. Wade, *J. Chem. Soc., London* p. 1649 (1964).
1128. J. C. Lockhart, *J. Chem. Soc., London* p. 1197 (1962).
1129. J. C. Lockhart, *J. Chem. Soc., A* p. 1552 (1966).
1130. J. C. Lockhart, *Spectrochim. Acta, Part A* **24**, 1205 (1968).
1131. J. C. Lockhart, *J. Chem. Soc., A* p. 809 (1966).
1131a. B. Lockman and T. Onak, *J. Org. Chem.* **38**, 2552 (1973).
1132. T. J. Logan, *J. Org. Chem.* **26**, 3657 (1961).
1133. T. J. Logan and T. J. Flautt, *J. Amer. Chem. Soc.* **82**, 3446 (1960).
1134. L. H. Long and D. Dollimore, *J. Chem. Soc., London* p. 3902 (1953).
1135. L. H. Long and R. W. Norrish, *Phil. Trans. Roy. Soc. London, Ser. A* **241**, 587 (1949).
1136. L. H. Long and A. C. Sanhueza, *Chem. Ind.* (*London*) p. 588 (1961).
1137. L. H. Long and M. G. H. Wallbridge, *Chem. Ind.* (*London*) p. 295 (1959).
1138. L. H. Long and M. G. H. Wallbridge, *J. Chem. Soc., London* p. 2181 (1963).
1139. L. H. Long and M. G. H. Wallbridge, *J. Chem. Soc., London* p. 2513 (1965).
1140. P. Love, *J. Chem. Phys.* **39**, 3044 (1963).
1141. P. Love, R. W. Taft, Jr., and T. Wartik, *Tetrahedron* **5**, 116 (1959).
1141a. D. W. Lowman, P. D. Ellis, and J. D. Odom, *Inorg. Chem.* **12**, 681 (1973).
1142. C. A. Lutz and D. M. Ritter, *Can. J. Chem.* **41**, 1344 (1963).
1143. R. E. Lyle, E. J. Dewitt, and I. C. Pattison, *J. Org. Chem.* **21**, 61 (1956).
1144. L. Lynds and D. R. Stern, *J. Amer. Chem. Soc.* **81**, 5006 (1959).

1145. A. R. Lyons and M. C. R. Symons, *J. Chem. Soc., Faraday Trans.*, Section II p. 502 (1972).
1146. R. G. Maguire, I. J. Solomon, and M. J. Klein, *Inorg. Chem.* **2**, 1133 (1963).
1147. P. M. Maitlis, *J. Chem. Soc., London* p. 425 (1961).
1148. P. M. Maitlis, *J. Chem. Soc., London* p. 3149 (1961).
1148a. G. I. Makin, V. P. Maslennikov, and V. A. Shushunov, *Zh. Obshch. Khim.* **42**, 834 (1972).
1149. S. C. Malhotra, *Inorg. Chem.* **3**, 862 (1963).
1150. L. J. Malone, *Inorg. Chem.* **7**, 1039 (1968).
1151. L. J. Malone and M. R. Manley, *Inorg. Chem.* **6**, 2260 (1967).
1152. L. J. Malone and R. W. Parry, *Inorg. Chem.* **6**, 817 (1967).
1153. G. W. Mappes and T. P. Fehlner, *J. Amer. Chem. Soc.* **92**, 1562 (1970).
1154. G. W. Mappes, S. A. Fridmann, and T. P. Fehlner, *J. Phys. Chem.* **74**, 3307 (1970).
1155. M. Margoshes, F. Fillwalk, V. A. Fassel, and R. E. Rundle, *J. Chem. Phys.* **22**, 381 (1954).
1156. J. L. Margrave, *J. Chem. Phys.* **32**, 1889 (1960).
1157. G. Marr, R. E. Moore, and B. W. Rockett, *J. Chem. Soc., C* p. 24 (1968).
1158. J. A. Marshall, *Synthesis* p. 229 (1971).
1159. D. R. Martin, *Advan. Chem. Ser.* **32**, 1 (1961).
1160. D. R. Martin, *J. Chem. Educ.* **36**, 208 (1959).
1161. D. Marynick and T. P. Onak, *J. Chem. Soc., A* p. 1797 (1969).
1162. P. G. Maslov, *Zh. Obshch. Khim.* **33**, 1054 (1963).
1163. A. G. Massey, *J. Chem. Soc., London* p. 5264 (1960).
1164. A. G. Massey and A. J. Park, *J. Organometal. Chem.* **2**, 245 (1964).
1165. A. G. Massey and A. J. Park, *J. Organometal. Chem.* **5**, 218 (1966).
1166. A. G. Massey and A. J. Park, *J. Organometal. Chem.* **2**, 461 (1964).
1167. A. G. Massey, A. J. Park, and F. G. A. Stone, *Proc. Chem. Soc., London* p. 212 (1963).
1168. A. G. Massey, E. W. Randall, and D. Shaw, *Spectrochim. Acta* **20**, 379 (1964).
1168a. A. G. Massey, E. W. Randall, and D. Shaw, *Spectrochim. Acta* **21**, 263 (1965).
1169. Y. Matsui and R. C. Taylor, *J. Amer. Chem. Soc.* **90**, 1363 (1968).
1170. D. S. Matteson, *J. Amer. Chem. Soc.* **82**, 4228 (1960).
1171. D. S. Matteson, *J. Org. Chem.* **27**, 275 (1962).
1172. D. S. Matteson, *J. Org. Chem.* **27**, 3712 (1962).
1173. D. S. Matteson, *J. Org. Chem.* **27**, 4293 (1962).
1174. D. S. Matteson, *J. Org. Chem.* **29**, 3399 (1964).
1175. D. S. Matteson and P. G. Allies, *J. Amer. Chem. Soc.* **92**, 1801 (1970); *J. Organometal Chem.* **54**, 35 (1973).
1176. D. S. Matteson and R. A. Bowie, *J. Amer. Chem. Soc.* **87**, 2587 (1965).
1177. D. S. Matteson and T. Cheng, *J. Organometal. Chem.* **6**, 100 (1966).
1177a. D. S. Matteson, L. A. Hagelee, and R. J. Willsek, *J. Amer. Chem. Soc.* **95**, 5096 (1973).
1178. D. S. Matteson and J. D. Liedtke, *Chem. Ind.* (*London*) p. 1241 (1963).
1179. D. S. Matteson and J. D. Liedtke, *J. Org. Chem.* **28**, 1924 (1963).
1180. D. S. Matteson and J. D. Liedtke, *J. Amer. Chem. Soc.* **87**, 1526 (1965).
1181. D. S. Matteson and R. W. H. Mah, *J. Org. Chem.* **28**, 2171 (1963).
1182. D. S. Matteson and R. W. H. Mah, *J. Org. Chem.* **28**, 2174 (1963).
1183. D. S. Matteson and R. W. H. Mah, *J. Amer. Chem. Soc.* **85**, 2599 (1963).
1183a. D. S. Matteson and P. K. Mattschei, *Inorg. Chem.* **12**, 2472 (1973).
1184. D. S. Matteson and K. Peacock, *J. Amer. Chem. Soc.* **82**, 5759 (1960).

1185. D. S. Matteson and K. Peacock, *J. Org. Chem.* **28**, 369 (1963).
1186. D. S. Matteson and K. Peacock, *J. Organometal. Chem.* **2**, 190 (1964).
1187. D. S. Matteson and K. Peacock, *J. Organometal. Chem.* **2**, 192 (1964).
1188. D. S. Matteson and G. D. Schaumberg, *J. Org. Chem.* **31**, 726 (1966).
1189. D. S. Matteson and J. G. Shdo, *J. Amer. Chem. Soc.* **85**, 2684 (1963).
1190. D. S. Matteson and J. G. Shdo, *J. Org. Chem.* **29**, 2742 (1964).
1191. D. S. Matteson, A. H. Soloway, D. W. Tomlinson, J. D. Campbell, and G. A. Nixon, *J. Med. Chem.* **7**, 640 (1964).
1192. D. S. Matteson and M. L. Talbot, *J. Amer. Chem. Soc.* **89**, 1119 (1967).
1193. D. S. Matteson and M. L. Talbot, *J. Amer. Chem. Soc.* **89**, 1123 (1967).
1194. D. S. Matteson and J. O. Waldbillig, *J. Org. Chem.* **28**, 366 (1963).
1195. D. S. Matteson and J. O. Waldbillig, *J. Amer. Chem. Soc.* **85**, 1019 (1963).
1196. D. S. Matteson and J. O. Waldbillig, *J. Amer. Chem. Soc.* **86**, 3778 (1964).
1197. D. S. Matteson, J. O. Waldbillig, and S. W. Peterson, *J. Amer. Chem. Soc.* **86**, 3781 (1964).
1198. J. Mauchamp, *Biochim. Biophys. Acta* **251**, 281 (1971).
1199. E. Mayer, *Monatsh. Chem.* **102**, 940 (1971).
1200. R. E. McCoy and S. H. Bauer, *J. Amer. Chem. Soc.* **78**, 2061 (1956).
1201. P. A. McCusker, E. C. Ashby, and H. S. Makowski, *J. Amer. Chem. Soc.* **79**, 5179 (1957).
1202. P. A. McCusker, E. C. Ashby, and H. S. Makowski, *J. Amer. Chem. Soc.* **79**, 5182 (1957).
1203. P. A. McCusker and J. H. Bright, *J. Org. Chem.* **29**, 2093 (1964).
1204. P. A. McCusker and L. J. Glunz, *J. Amer. Chem. Soc.* **77**, 4253 (1955).
1205. P. A. McCusker, G. F. Hennion, and E. C. Ashby, *J. Amer. Chem. Soc.* **79**, 5192 (1957).
1206. P. A. McCusker and H. S. Makouski, *J. Amer. Chem. Soc.* **79**, 5185 (1957).
1207. P. A. McCusker, J. V. Marra, and G. F. Hennion, *J. Amer. Chem. Soc.* **83**, 1924 (1961).
1208. P. A. McCusker, P. L. Pennartz, and R. C. Pilger, *J. Amer. Chem. Soc.* **84**, 4362 (1962).
1209. P. A. McCusker, F. M. Rossi, J. H. Bright, and G. F. Hennion. *J. Org. Chem.* **28**, 2889 (1963).
1210. D. H. McDaniel and H. C. Brown, *J. Amer. Chem. Soc.* **77**, 3756 (1955).
1211. I. R. McKinley and H. Weigel, *J. Chem. Soc., Chem. Commun.* p. 1051 (1972).
1212. D. E. McLaughlin, M. Tamres, S. Searles, Jr., and F. Block, *J. Inorg. Nucl. Chem.* **18**, 118 (1961).
1213. J. C. McMullen and N. E. Miller, *Inorg. Chem.* **9**, 2291 (1970).
1214. H. Meerwein, G. Hinz, H. Majert, and H. Sonke, *J. Prakt. Chem.* **147**, 226 (1937).
1215. H. Meerwein and H. Sonke, *J. Prakt. Chem.* **147**, 251 (1937).
1215a. B. Meissner and H. A. Staab, *Justus Liebigs Ann. Chem.* **753**, 92 (1971).
1216. I. Mehrotra and D. Devaprabhakara, *J. Organometal. Chem.* **33**, 287 (1971).
1217. L. A. Melcher, J. L. Adcock, and J. J. Lagowski, *Inorg. Chem.* **11**, 1247 (1972).
1218. A. Meller, *Monatsh. Chem.* **94**, 183 (1963).
1219. A. Meller and H. Egger, *Monatsh. Chem.* **97**, 790 (1966).
1220. A. Meller and H. Marecek, *Monatsh. Chem.* **99**, 1666 (1968).
1221. A. Meller and W. Maringgele, *Monatsh. Chem.* **99**, 1909 (1968).
1222. A. Meller and W. Maringgele, *Monatsh. Chem.* **99**, 2504 (1968).
1223. A. Meller and W. Maringgele, *Monatsh. Chem.* **101**, 387 (1970).

1224. A. Meller and W. Maringgele, *Monatsh. Chem.* **101**, 753 (1970).
1225. A. Meller and W. Maringgele, *Monatsh. Chem.* **102**, 118 (1971).
1226. A. Meller and W. Maringgele, *Monatsh. Chem.* **102**, 121 (1971).
1227. A. Meller and A. Ossko, *Monatsh. Chem.* **99**, 1217 (1968).
1228. A. Meller and A. Ossko, *Monatsh. Chem.* **100**, 1187 (1969).
1229. A. Meller and A. Ossko, *Monatsh. Chem.* **102**, 131 (1971).
1230. A. Meller, M. Wechsberg, and V. Gutmann, *Monatsh. Chem.* **96**, 388 (1965).
1231. A. Meller, M. Wechsberg, and V. Gutmann, *Monatsh. Chem.* **97**, 1163 (1966).
1232. A. Meller, M. Wojnowsko, and H. Marecek, *Monatsh. Chem.* **100**, 175 (1968).
1233. E. K. Mellon and J. J. Lagowski, *Nature* (*London*) **199**, 997 (1963).
1234. A. Michaelis, *Chem. Ber.* **27**, 244 (1894).
1235. A. Michaelis and P. Becker, *Chem. Ber.* **15**, 180 (1882).
1236. A. Michaelis and P. Becker, *Chem. Ber.* **13**, 58 (1880).
1237. A. Michaelis, G. Thevénot, E. Richter, and E. Hillringhaus, *Justus Liebigs Ann. Chem.* **315**, 19 (1901).
1238. M. M. Midland and H. C. Brown, *J. Amer. Chem. Soc.* **93,** 1506 (1971).
1238a. M. M. Midland and H. C. Brown, *J. Amer. Chem. Soc.* **95**, 4069 (1973).
1239. L. Miginiac and J. Blais, *J. Organometal. Chem.* **29**, 349 (1971).
1240. B. M. Mikhailov, A. A. Akhhazaryan, and L. S. Vasil'ev, *Dokl. Akad. Nauk SSSR* **136**, 828 (1961).
1241. B. M. Mikhailov and P. M. Aronovich, *Izv. Akad. Nauk SSSR, Otd. Khim. Nauk* p. 946 (1955).
1242. B. M. Mikhailov and P. M. Aronovich, *Izv. Akad. Nauk SSSR, Otd. Khim. Nauk* p. 322 (1956).
1243. B. M. Mikhailov and P. M. Aronovich, *Izv. Akad. Nauk SSSR, Otd. Khim. Nauk* p. 1123 (1957).
1244. B. M. Mikhailov and P. M. Aronovich, *Zh. Obshch. Khim.* **29**, 1254 (1959).
1245. B. M. Mikhailov and P. M. Aronovich, *Zh. Obshch. Khim.* **29**, 1257 (1959).
1246. B. M. Mikhailov and P. M. Aronovich, *Izv. Akad. Nauk SSSR, Otd. Khim. Nauk* p. 927 (1961).
1247. B. M. Mikhailov and P. M. Aronovich, *Izv. Akad. Nauk SSSR, Otd. Khim. Nauk* p. 1233 (1963).
1248. B. M. Mikhailov, P. M. Aronovich, and V. G. Kiselov, *Izv. Akad. Nauk SSSR, Ser. Khim.* p. 146 (1968).
1249. B. M. Mikhailov and T. K. Baryshnikova, *Zh. Obshch. Khim.* **41**, 1303 (1971).
1249a. B. M. Mikhailov, T. K. Baryshnikova, and V. S. Bogdanov, *Dokl. Akad. Nauk SSSR* **202**, 358 (1972).
1250. B. M. Mikhailov and A. V. Bazhenova, *Izv. Akad. Nauk SSSR, Otd. Khim. Nauk* p. 76 (1959).
1251. B. M. Mikhailov and A. Ya. Bezmenov, *Izv. Akad. Nauk SSSR, Ser. Khim.* p. 931 (1965).
1252. B. M. Mikhailov, A. Ya. Bezmenov, L. S. Vasil'ev, and V. G. Kiselev, *Dokl. Akad. Nauk SSSR* **155**, 141 (1964).
1253. B. M. Mikhailov and A. N. Blokhina, *Zh. Obshch. Khim.* **30**, 3615 (1960).
1254. B. M. Mikhailov and A. N. Blokhina, *Izv. Akad. Nauk SSSR, Otd. Khim. Nauk* p. 1373 (1962).
1255. B. M. Mikhailov, V. S. Bogdanov, G. V. Logodzinskaya, and V. F. Pozdnev, *Izv. Akad. Nauk SSSR, Ser. Khim.* p. 386 (1966).
1255a. B. M. Mikhailov, B. I. Bryantsev, and T. K. Kozminskaya, *Dokl. Akad. Nauk SSSR* **203**, 837 (1972).

1256. B. M. Mikhailov and Yu. N. Bubnov, *Dokl. Akad. Nauk SSSR* **127**, 571 (1959).
1257. B. M. Mikhailov and Yu. N. Bubnov, *Izv. Akad Nauk SSSR, Otd. Khim. Nauk* p. 172 (1959).
1258. B. M. Mikhailov and Yu. N. Bubnov, *Zh. Obshch. Khim.* **29**, 1648 (1959).
1259. B. M. Mikhailov and Yu. N. Bubnov, *Izv. Akad. Nauk SSSR, Otd. Khim. Nauk* p. 368 (1960).
1260. B. M. Mikhailov and Yu. N. Bubnov, *Izv. Akad. Nauk SSSR, Otd. Khim. Nauk* p. 370 (1960).
1261. B. M. Mikhailov and Yu. N. Bubnov, *Izv. Akad. Nauk SSSR, Otd. Khim. Nauk* p. 1872 (1960).
1262. B. M. Mikhailov and Yu. N. Bubnov, *Izv. Akad. Nauk SSSR, Otd. Khim. Nauk* p. 1883 (1960).
1263. B. M. Mikhailov and Yu. N. Bubnov, *Zh. Obshch. Khim.* **31**, 160 (1961).
1264. B. M. Mikhailov and Yu. N. Bubnov, *Zh. Obshch. Khim.* **31**, 577 (1961).
1265. B. M. Mikhailov and Yu. N. Bubnov, *Zh. Obshch. Khim.* **32**, 1969 (1962).
1266. B. M. Mikhailov and Yu. N. Bubnov, *Izv. Akad. Nauk SSSR, Ser. Khim.* p. 2248 (1964).
1267. B. M. Mikhailov and Yu. N. Bubnov, *Izv. Akad. Nauk SSSR, Ser. Khim.* p. 1874 (1964).
1268. B. M. Mikhailov and Yu. N. Bubnov, *Izv. Akad. Nauk SSSR, Ser. Khim.* p. 1310 (1965).
1269. B. M. Mikhailov and Yu. N. Bubnov, *Zh. Obshch. Khim.* **41**, 2039 (1971).
1269a. B. M. Mikhailov and Yu. N. Bubnov, *Tetrahedron Lett.* p. 2127 (1971).
1270. B. M. Mikhailov, Yu. N. Bubnov, and S. I. Frolov, *Izv. Akad. Nauk SSSR, Ser. Khim.* p. 2290 (1967).
1271. B. M. Mikhailov, Yu. N. Bubnov, S. I. Frolov, and S. A. Korobeinikova, *Izv. Akad. Nauk SSSR, Ser. Khim.* p. 1631 (1969).
1271a. B. M. Mikhailov, Yu. N. Bubnov, and M. S. Grigoryan, *Izv. Akad. Nauk SSSR, Ser. Khim.* p. 1842B (1971).
1272. B. M. Mikhailov, Yu N. Bubnov, and V. G. Kiselev, *Zh. Obshch. Khim.* **36,** p. 62 (1966).
1273. B. M. Mikhailov, Yu. N. Bubnov, and S. A. Korobeinikova, *Izv. Akad. Nauk SSSR, Ser. Khim.* p. 2465 (1969).
1273a. B. M. Mikhailov, Yu. N. Bubnov, and S. A. Korobeinikova, *Izv. Akad. Nauk SSSR, Ser. Khim.* p. 2631 (1970).
1274. B. M. Mikhailov, Yu. N. Bubnov, S. A. Korobeinikova, and V. S. Bogdanov, *Zh. Obshch. Khim.* **40**, 1321 (1970).
1275. B. M. Mikhailov, Y. B. Bubnov, S. A. Korobeinikova, and S. I. Frolov, *Izv. Akad. Nauk SSSR, Ser. Khim.* p. 1923 (1968).
1275a. B. M. Mikhailov, Yu. N. Bubnov, S. A. Korobeinikova, and S. I. Frolov, *J. Organometal. Chem.* **27**, 165 (1971).
1275b. B. M. Mikhailov, Yu. N. Bubnov, O. A. Nesmeyanova, V. G. Kiselev, T. Yu. Rudashevskaya, and B. A. Kazansky, *Tetrahedron Lett.* p. 4627 (1972).
1276. B. M. Mikhailov and K. L. Cherkasova, *Zh. Obshch. Khim.* **41**, 2578 (1971).
1276a. B. M. Mikhailov and K. L. Cherkasova, *Izv. Akad. Nauk SSSR, Ser. Khim.* p. 1244 (1971).
1276b. B. M. Mikhailov and K. L. Cherkasova, *Zh. Obshch. Khim.* **42**, 138 (1972).
1276c. B. M. Mikhailov and K. L. Cherkasova, *Zh. Obshch. Khim.* **42**, 1744 (1972).
1277. B. M. Mikhailov and V. A. Dorokhov, *Dokl. Akad. Nauk SSSR*, **130**, 782 (1960).
1278. B. M. Mikhailov and V. A. Dorokhov, *Dokl. Akad. Nauk SSSR* **133**, 119 (1960).
1279. B. M. Mikhailov and V. A. Dorokhov, *Zh. Obshch. Khim.* **31**, 4020 (1961).

1280. B. M. Mikhailov and V. A. Dorokhov, *Dokl. Akad. Nauk SSSR* **136**, 356 (1961).
1281. B. M. Mikhailov and V. A. Dorokhov, *Izv. Akad. Nauk SSSR, Otd. Khim. Nauk* p. 623 (1962).
1282. B. M. Mikhailov and V. A. Dorokhov, *Izv. Akad. Nauk SSSR, Otd. Khim. Nauk* p. 1213 (1962).
1282a. B. M. Mikhailov and V. A. Dorokhov, *Izv. Akad. Nauk SSSR, Ser. Khim.* p. 1446 (1970).
1282b. B. M. Mikhailov and V. A. Dorokhov, *Izv. Akad. Nauk SSSR, Ser. Khim.* p. 201 (1971).
1283. B. M. Mikhailov, V. A. Dorokhov, V. S. Bogdanov, I. P. Yakovlev, and A. D. Naumov, *Dokl. Akad. Nauk SSSR*, **194**, 595 (1970).
1283a. B. M. Mikhailov, V. A. Dorokhov, and L. I. Lavrinovich, *Zh. Obshch. Khim.* **42**, 1553 (1972).
1284. B. M. Mikhailov, V. A. Dorokhov, and N. V. Mostovoi, *Izv. Akad. Nauk SSSR, Ser. Khim.* p. 199 (1I64).
1285. B. M. Mikhailov, V. A. Dorokhov, and N. V. Mostovoi, *Izv. Akad. Nauk SSSR, Ser. Khim.* p. 201 (1964).
1286. B. M. Mikhailov, V. A. Dorokhov, and N. V. Mostovoi, *Dokl. Akad. Nauk SSSR* **166**, 1114 (1966).
1286a. B. M. Mikhailov, V. A. Dorokhov, N. V. Mostovoi, O. G. Boldyreva, and M. N. Bochkareva, *Zh. Obshch. Khim.* **40**, 1817 (1970).
1286b. B. M. Mikhailov, V. A. Dorokhov, and V. I. Seredenko, *Dokl. Akad. Nauk SSSR* **199**, 1328 (1971).
1287. B. M. Mikhailov, V. A. Dorokhov, and I. P. Yakolev, *Izv. Akad. Nauk SSSR, Ser. Khim.* p. 332 (1966).
1288. B. M. Mikhailov and N. S. Fedotov, *Izv. Akad. Nauk SSSR, Otd. Khim. Nauk* p. 375 (1956).
1289. B. M. Mikhailov and N. S. Fedotov, *Izv. Akad. Nauk SSSR, Otd. Khim. Nauk* p. 1511 (1956).
1290. B. M. Mikhailov and N. S. Fedotov, *Izv. Akad. Nauk SSSR, Otd. Khim. Nauk* p. 857 (1958).
1291. B. M. Mikhailov and N. S. Fedotov, *Izv. Akad. Nauk SSSR, Otd. Khim. Nauk* p. 1482 (1959).
1292. B. M. Mikhailov and N. S. Fedotov, *Zh. Obshch. Khim.* **29**, 2244 (1959).
1293. B. M. Mikhailov and N. S. Fedotov, *Izv. Akad. Nauk SSSR, Otd. Khim. Nauk* p. 1590 (1960).
1294. B. M. Mikhailov and N. S. Fedotov, *Izv. Akad. Nauk SSSR, Otd. Khim. Nauk* p. 1913 (1961).
1295. B. M. Mikhailov and N. S. Fedotov, *Zh. Obshch. Khim.* **32**, 93 (1962).
1296. B. M. Mikhailov, N. S. Fedotov, T. A. Shchegoleva, and V. D. Sheludyakov, *Dokl. Akad. Nauk SSSR*, **145**, 340 (1962).
1297. B. M. Mikhailov and A. F. Galkin, *Izv. Akad. Nauk SSSR, Otd. Khim. Nauk* p. 619 (1962).
1298. B. M. Mikhailov and A. F. Galkin, *Izv. Akad. Nauk SSSR, Otd. Khim. Nauk* p. 641 (1963).
1299. B. M. Mikhailov and T. V. Kostroma, *Izv. Akad. Nauk SSSR, Otd. Khim. Nauk* p. 376 (1956).
1300. B. M. Mikhailov and T. V. Kostroma, *Izv. Akad. Nauk SSSR, Otd. Khim. Nauk* p. 1125 (1957).
1301. B. M. Mikhailov, T. V. Kostroma, and N. S. Fedotov, *Izv. Akad. Nauk SSSR, Otd. Khim. Nauk* p. 589 (1957).

1302. B. M. Mikhailov and T. K. Kozminskaya, *Dokl. Akad. Nauk SSSR* **121**, 656 (1958).
1303. B. M. Mikhailov and T. K. Kozminskaya, *Izv. Akad. Nauk SSSR, Otd. Khim. Nauk* p. 80 (1959).
1304. B. M. Mikhailov and T. K. Kozminskaya, *Izv. Akad. Nauk SSSR, Otd. Khim. Nauk* p. 2247 (1960).
1305. B. M. Mikhailov and T. K. Kozminskaya, *Zh. Obshch. Khim.* **30**, 3619 (1960).
1306. B. M. Mikhailov and T. K. Kozminskaya, *Izv. Akad. Nauk SSSR, Ser. Khim.* p. 256 (1962).
1307. B. M. Mikhailov and T. K. Kozminskaya, *Izv. Akad. Nauk SSSR, Ser. Khim.* p. 1703 (1963).
1308. B. M. Mikhailov and T. K. Kozminskaya, *Izv. Akad. Nauk SSSR, Ser. Khim.* p. 439 (1965).
1309. B. M. Mikhailov, T. K. Kozminskaya, and A. Ya. Bezmenov, *Izv. Akad. Nauk SSSR, Ser. Khim.* p. 355 (1965).
1310. B. M. Mikhailov, T. K. Kozminskaya, A. N. Blockhina, and T. A. Shchegoleva, *Izv. Akad. Nauk SSSR, Otcd. Khim. Nauk* p. 692 (1956).
1311. B. M. Mikhailov, T. K. Kozminskaya, N. S. Fedotov, and V. A. Dorokhov, *Dokl. Akad. Nauk SSSR*, **127**, 1023 (1959).
1312. B. M. Mikhailov and M. E. Kuimova, *Zh. Obshch. Khim.* **41**, 1714 (1971).
1313. B. M. Mikhailov, M. E. Kuimova, and E. A. Shagova, *Izv. Akad. Nauk SSSR, Ser. Khim.* p. 548 (1968).
1314. B. M. Mikhailov, M. E. Kuimova, and E. A. Shagova, *Dokl. Akad. Nauk SSSR* **179**, 1344 (1968).
1315. B. M. Mikhailov and M. E. Nikolaeva, *Izv. Akad. Nauk SSSR, Ser. Khim.* p. 1368 (1963).
1316. B. M. Mikhailov and V. F. Pozdnev, *Izv. Akad. Nauk SSSR, Otd. Khim. Nauk* p. 1475 (1962).
1317. B. M. Mikhailov and V. F. Pozdnev, *Izv. Akad. Nauk SSSR, Otd. Khim. Nauk* p. 1698 (1962).
1318. B. M. Mikhailov and V. F. Pozdnev, *Izv. Akad. Nauk SSSR, Otd. Khim. Nauk* p. 1861 (1962).
1319. B. M. Mikhailov and V. F. Pozdnev, *Dokl. Akad. Nauk SSSR*, **151**, 340 (1963).
1320. B. M. Mikhailov, V. F. Pozdnev, and V. G. Kiselev, *Dokl. Akad. Nauk SSSR* **151**, 577 (1963).
1321. B. M. Mikhailov and E. N. Safonova, *Izv. Akad. Nauk SSSR, Ser. Khim.* p. 1487 (1965).
1322. B. M. Mikhailov and T. A. Shchegoleva, *Izv. Akad. Nauk SSSR, Otd. Khim. Nauk* p. 1124 (1955).
1323. B. M. Mikhailov and T. A. Shchegoleva, *Dokl. Akad. Nauk SSSR* **108**, 481 (1956).
1324. B. M. Mikhailov and T. A. Shchegoleva, *Izv. Akad. Nauk SSSR, Otd. Khim. Nauk* p. 508 (1956).
1325. B. M. Mikhailov and T. A. Shchegoleva, *Izv. Akad. Nauk SSSR, Otd. Khim. Nauk* p. 1080 (1957).
1326. B. M. Mikhailov and T. A. Shchegoleva, *Izv. Akad. Nauk SSSR, Otd. Khim. Nauk* p. 357 (1959).
1327. B. M. Mikhailov and T. A. Shchegoleva, *Izv. Akad. Nauk SSSR, Otd. Khim. Nauk* p. 546 (1959).
1328. B. M. Mikhailov and T. A. Shchegoleva, *Zh. Obshch. Khim.* **29**, 3130 (1959).

1329. B. M. Mikhailov and T. A. Shchegoleva, *Zh. Obshch. Khim.* **29**, 3443 (1959).
1330. B. M. Mikhailov and T. A. Shchegoleva, *Izv. Akad. Nauk SSSR, Otd. Khim. Nauk* p. 1142 (1961).
1330a. B. M. Mikhailov and T. A. Shchegoleva, *Izv. Akad. Nauk SSSR, Ser. Khim.* p. 661 (1971).
1331. B. M. Mikhailov, T. A. Shchegoleva, and A. N. Blokhina, *Izv. Akad. Nauk SSSR, Otd. Khim. Nauk* p. 1307 (1960).
1332. B. M. Mikhailov, T. A. Shchegoleva, and Yu. N. Bubnov, *Izv. Akad. Nauk SSSR, Otd. Khim. Nauk* p. 413 (1962).
1332a. B. M. Mikhailov, T. A. Shchegoleva, and E. M. Shashkova, *Izv. Akad. Nauk SSSR, Otd. Khim. Nauk* p. 916 (1961).
1333. B. M. Mikhailov, T. A. Schhegoleva, V. D. Sheludyakov, and A. N. Blokhina, *Izv. Akad. Nauk SSSR, Otd. Khim. Nauk* p. 646 (1963).
1333a. B. M. Mikhailov and V. N. Smirnov, *Dokl. Akad. Nauk SSSR* **193**, p. 1311 (1970).
1333b. B. M. Mikhailov and V. N. Smirnov, *Izv. Akad. Nauk SSSR, Ser. Khim.* p. 1672 (1972).
1333c. B. M. Mikhailov, V. N. Smirnov, and E. P. Prokof'ev, *Dokl. Akad. Nauk SSSR* **206**, 125 (1972).
1333d. B. M. Mikhailov, V. N. Smirnov, and O. D. Ryazanova, *Dokl. Akad. Nauk SSSR* **204**, 612 (1972).
1334. B. M. Mikhailov and G. S. Ter-Sarkisyan, *Izv. Akad. Nauk SSSR, Ser. Khim.* p. 380 (1966).
1335. B. M. Mikhailov and G. S. Ter-Sarkisyan, and N. A. Nikolaeva, *Izv. Akad. Nauk SSSR, Ser, Khim.* p. 541 (1968).
1336. B. M. Mikhailov, G. S. Ter-Sarkisyan, and N. A. Nikolaeva, *Zh. Obshch. Khim.* **41**, 1721 (1971).
1337. B. M. Mikhailov, F. B. Tutorskaya, *Dokl. Akad. Nauk SSSR* **123**, 479 (1958).
1338. B. M. Mikhailov and F. B. Tutorskaya, *Izv. Akad. Nauk SSSR, Otd. Khim. Nauk* p. 1127 (1959).
1339. B. M. Mikhailov and F. B. Tutorskaya, *Izv. Akad. Nauk SSSR, Otd. Khim. Nauk* p. 1865 (1959).
1340. B. M. Mikhailov and F. B. Tutorskaya, *Izv. Akad. Nauk SSSR, Otd. Khim. Nauk* p. 2068 (1960).
1341. B. M. Mikhailov and F. B. Tutorskaya, *Izv. Akad. Nauk SSSR, Otd. Khim. Nauk* p. 1158 (1961).
1342. B. M. Mikhailov and F. B. Tutorskaya, *Zh. Obshch. Khim.* **32**, 833 (1962).
1343. B. M. Mikhailov and L. S. Vasil'ev, *Izv. Akad. Nauk SSSR, Otd. Khim. Nauk* p. 530 (1961).
1344. B. M. Mikhailov and L. S. Vasil'ev, *Dokl. Akad. Nauk SSSR* **139**, 385 (1961).
1345. B. M. Mikhailov and L. S. Vasil'ev, *Izv. Akad. Nauk SSSR, Otd. Khim. Nauk* p. 628 (1962).
1346. B. M. Mikhailov and L. S. Vasil'ev, *Izv. Akad. Nauk SSSR, Otd. Khim. Nauk* p. 827 (1962).
1347. B. M. Mikhailov and L. S. Vasil'ev, *Izv. Akad. Nauk SSSR, Otd. Khim. Nauk* p. 1756 (1962).
1348. B. M. Mikhailov and L. S. Vasil'ev, *Zh. Obshch. Khim.* **35**, 925 (1965).
1349. B. M. Mikhailov and L. S. Vasil'ev, *Zh. Obshch. Khim.* **35**, 1073 (1965).
1350. B. M. Mikhailov, L. S. Vasil'ev, and A. Ya. Bezmenov, *Izv. Akad. Nauk SSSR, Ser. Khim.* p. 712 (1965).

1351. B. M. Mikhailov, L. S. Vasil'ev, and V. P. Dimitrikov, *Izv. Akad. Nauk SSSR, Ser. Khim.* p. 2164 (1968).
1352. B. M. Mikhailov, L. S. Vasil'ev, and E. N. Safonova, *Dokl. Akad. Nauk SSSR* **147**, 630 (1962).
1352a. B. M. Mikhailov and V. A. Vaver, *Dokl. Akad. Nauk SSSR* **102**, 531 (1955).
1353. B. M. Mikhailov and V. A. Vaver, *Izv. Akad. Nauk SSSR, Otd. Khim. Nauk* p. 451 (1956).
1354. B. M. Mikhailov and V. A. Vaver, *Dokl. Akad. Nauk SSSR* **109**, 94 (1957).
1355. B. M. Mikhailov and V. A. Vaver, *Izv. Akad. Nauk SSSR, Otd. Khim Nauk* p. 812 (1957).
1356. B. M. Mikhailov and V. A. Vaver, *Izv. Akad. Nauk SSSR, Otd. Khim Nauk* p. 989 (1957).
1357. B. M. Mikhailov and V. A. Vaver, *Izv. Akad. Nauk SSSR, Otd. Khim Nauk* p. 419 (1958).
1358. B. M. Mikhailov and V. A. Vaver, *Zh. Obshch. Khim.* **29**, 2248 (1959).
1359. B. M. Mikhailov and V. A. Vaver, *Izv. Akad. Nauk SSSR, Otd. Khim. Nauk* p. 852 (1960).
1360. B. M. Mikhailov and V. A. Vaver, *Zh. Obshch. Khim.* **31**, 574 (1961).
1361. B. M. Mikhailov, V. A. Vaver, and Yu. N. Bubnov, *Dokl. Akad. Nauk SSSR* **126**, 575 (1959).
1362. L. N. Mikhailova and B. I. Tikhomirov, *Vestn. Leningrad. Univ., Fiz. Khim.* p. 119 (1971).
1363. F. M. Miller and D. M. Ritter, *Inorg. Chem.* **9**, 1284 (1970).
1364. J. J. Miller and F. A. Johnson, *J. Amer. Chem. Soc.* **90**, 218 (1968).
1365. J. J. Miller and F. A. Johnson, *Inorg. Chem.* **9**, 69 (1970).
1366. N. E. Miller, *J. Amer. Chem. Soc.* **88**, 4284 (1966).
1367. N. E. Miller and E. L. Muetterties, *Inorg. Chem.* **3**, 1196 (1964).
1368. N. E. Miller, M. D. Murphy, and D. L. Reznicek, *Inorg. Chem.* **5**, 1832 (1966).
1369. N. E. Miller and D. L. Reznicek, *Inorg. Chem.* **8**, 275 (1969).
1370. F. J. Millero, *J. Phys. Chem.* **75**, 280 (1971).
1371. J. E. Milks, G. W. Kennerly, and J. H. Polevy, *J. Amer. Chem. Soc.* **84**, 2529 (1962).
1372. H. Minato, J. C. Ware, and T. G. Traylor, *J. Amer. Chem. Soc.* **85**, 3024 (1963).
1373. Y. Minoura and H. Ikeda, *J. Appl. Polym. Sci.* **15**, 2219 (1971).
1374. S. B. Mirviss, *J. Amer. Chem. Soc.* **83**, 3051 (1961).
1375. S. B. Mirviss, *J. Org. Chem.* **32**, 1713 (1967).
1376. N. Miyamoto, S. Isiyama, K. Utimoto, and H. Nozaki, *Tetrahedron Lett.* p. 4597 (1971).
1377. N. Miyaura, M. Itoh, A. Suzuki, H. C. Brown, M. M. Midland, and P. Jacob, *J. Amer. Chem. Soc.* **94**, 6549 (1972).
1378. C. W. Moeller and W. K. Wilmarth, *J. Amer. Chem. Soc.* **81**, 2638 (1959).
1379. P. C. Moews, Jr. and A. W. Laubengayer, *Inorg. Chem.* **2**, 1072 (1963).
1380. P. C. Moews, Jr. and R. W. Parry, *Inorg. Chem.* **5**, 1552 (1966).
1381. L. G. Mogel and A. M. Yurkevich, *Zh. Obshch. Khim.* **40**, 708 (1970).
1382. R. K. Mohanty, T. S. Sarma, S. Subramanian, and J. C. Ahluwalia, *Trans. Faraday Soc..* **67**, 305 (1971).
1382a. H. Möhrle, B. Gusowski, and R. Feil, *Tetrahedron* **27**, 221 (1971).
1383. M. A. Molinari and P. A. McCusker, *J. Org. Chem.* **29**, 2094 (1964).
1384. E. F. Mooney, *Spectrochim. Acta* **18**, 1355 (1962).
1385. E. F. Mooney and M. A. Qaseem, *J. Inorg. Nucl. Chem.* **30**, 1439 (1968).

1386. E. F. Mooney and M. A. Qaseem, *J. Inorg. Nucl. Chem.* **30.**, 1638 (1968).
1387. E. F. Mooney and M. A. Qaseem, *Spectrochim. Acta, Part A* **24**, 969 (1968).
1388. E. F. Mooney and P. H. Winson, *Chem. Commun.* p. 341 (1967).
1389. C. E. Moore, F. P. Cassaretto, H. Posvic, and J. J. McLafferty, *Anal. Chim. Acta* **35**, 1 (1966).
1389a. E. B. Moore and W. N. Lipscomb, *Acta Crystallogr.* **9**, 668 (1956).
1389b. W. R. Moore, H. W. Anderson, and S. D. Clark, *J. Amer. Chem. Soc.* **95**, 835 (1973).
1390. N. V. Mostovoi, V. A. Dorokhov, and B. M. Mikhailov, *Izv. Akad. Nauk SSSR Ser. Khim.* p. 90 (1966).
1391. M. Mousseron and P. Chamayou, *Bull. Soc. Chim. Fr.* **3**, 403 (1962).
1392. E. L. Muetterties, *J. Amer. Chem. Soc.* **80**, 4526 (1958).
1393. E. L. Muetterties, *J. Amer. Chem. Soc.* **81**, 2597 (1959).
1394. E. L. Muetterties, *J. Amer. Chem. Soc.* **82**, 4163 (1960).
1395. E. L. Muetterties and F. N. Tebbe, *Inorg. Chem.* **7**, 2663 (1968).
1395a. T. Mukaiyama, K. Inomata, and M. Muraki, *J. Amer. Chem. Soc.* **95**, 967 (1973).
1395b. T. Mukaiyama, S. Yamamoto, and M. Shiono, *Bull. Chem. Soc. Jap.* **45**, 2244 (1972).
1396. R. S. Mulliken, *Chem. Rev.* **41**, 207 (1947).
1397. C. B. Murphy and R. E. Enrione, *Int. J. Mass Spectrom. Ion Phys.* **5**, 157 (1970).
1398. C. B. Murphy and R. E. Enrione, *Int. J. Mass Spectrom. Ion Phys.* **7**, 327 (1971).
1399. O. C. Musgrave, *Chem. Ind.* (*London*) p. 1152 (1957).
1400. O. C. Musgrave and T. O. Park, *Chem. Ind.* (*London*) p. 1552 (1955).
1401. K. A. Muszkat and B. Kirson, *Isr. J. Chem.* **2**, 63 (1964).
1402. K. Nagasawa, *Inorg. Chem.* **5**, 442 (1966).
1403. K. Nagasawa, T. Yoshizaki, and H. Watanabe, *Inorg. Chem.* **4**, 275 (1965).
1404. K. V. Nahabedian and H. G. Kuivila, *J. Amer. Chem. Soc.* **83**, 2167 (1961).
1405. F. C. Nahm, E. F. Rothergy, and K. Niedenzu, *J. Organometal. Chem.* **35**, 9 (1972).
1406. R. Nakane, T. Watanabe, and T. Oyama, *Bull. Chem. Soc. Jap.* **37**, 381 (1964).
1407. Y. Nakayama, J. Furukawa, T. Tsuruta, and J. Yamuchi, *Kogyo Kagaku Zasshi* **63**, 1477 (1960).
1408. Y. Nakayama, T. Tsuruta, and J. Furukawa, *Makromol. Chem.* **40**, 79 (1960).
1409. M. H. Nam, A. J. Russo, and R. F. Nystrom, *Chem. Ind.* (*London*) p. 1876 (1963).
1410. H. Nambu and H. C. Brown, *Organometal. Chem. Syn.* **1**, 95 (1970/1971).
1411. H. Nambu and H. C. Brown, *J. Amer. Chem. Soc.* **92**, 5790 (1970).
1412. J. Namtvedt and S. Gronowitz, *Acta Chem. Scand.* **20**, 1448 (1966).
1413. J. Namtvedt and S. Gronowitz, *Acta Chem. Scand.* **22**, 1373 (1968).
1414. M. Naruse, K. Utimoto, and H. Nozaki, *Tetrahedron Lett.* p 1605 (1972).
1415. R. E. Naylor, Jr. and E. B. Wilson, Jr., *J. Chem. Phys.* **26**, 1057 (1957).
1416. E. Negishi and H. C. Brown, *Synthesis* p. 196 (1972).
1417. E. Negishi and H. C. Brown, *Synthesis* p. 197 (1972).
1418. E. Negishi, P. L. Burke, and H. C. Brown, *J. Amer. Chem. Soc.* **94**, 7431 (1972).
1419. E. Negishi, J. J. Katz, and H. C. Brown, *J. Amer. Chem. Soc.* **94**, 4025 (1972).
1419a. E. Negishi, J. J. Katz, and H. C. Brown, *Synthesis* p. 555 (1972).
1419b. A. N. Nesmeyanov, N. E. Kolobova, Yu. V. Makarov, and K. N. Anisimov, *Izv. Akad. Nauk SSSR, Ser. Khim.* p. 1992 (1969).

1420. A. N. Nesmeyanov and V. A. Sazonova, *Izv. Akad. Nauk SSSR, Ser. Khim.* p. 187 (1955).

1420a. A. N. Nesmeyanov, V. A. Sazonova, V. A. Blinova, and N. N. Sedova, *Izv. Akad. Nauk SSSR, Ser. Khim.* p. 2583 (1971).

1421. A. N. Nesmeyanov, V. A. Sazonova, and V. N. Drozd, *Izv. Akad. Nauk SSSR, Otd. Khim. Nauk* p. 1389 (1957).

1422. A. N. Nesmeyanov, V. A. Sazonova, and V. N. Drozd, *Izv. Akad. Nauk SSSR, Otd. Khim. Nauk* p. 163 (1959).

1423. A. N. Nesmeyanov, V. A. Sazonova, and V. N. Drozd, *Dokl. Akad. Nauk SSSR* **126**, 1004 (1959).

1424. A. N. Nesmeyanov, W. A. Sazonova, and V. N. Drozd, *Chem. Ber.* **93**, 2717 (1960).

1425. A. N. Nesmeyanov, V. A. Sazonova, A. V. Gerasimenko, and V. G. Medvedera, *Izv. Akad. Nauk SSSR, Otd. Khim. Nauk* p. 2073 (1962).

1426. A. N. Nesmeyanov, V. A. Sazonova, G. S. Liberman, and L. I. Emelyanova, *Izv. Akad. Nauk SSSR, Otd. Khim. Nauk* p. 48 (1955).

1426a. A. N. Nesmeyanov, V. A. Sazonova, and N. N. Sedova, *Dokl. Akad. Nauk SSSR* **194**, 825 (1970).

1427. R. Neu, *Chem. Ber.* **88**, 1761 (1955).

1428. R. Neu, *Naturwissenschaften* **46**, 262 (1959).

1428a. R. Neu, *Arch. Pharm.* (*Weinheim*) **292**, 437 (1959).

1428b. R. Neu, *Arch. Pharm.* (*Weinheim*) **294**, 7 (1961).

1429. R. Neu, *Tetrahedron Lett.* p. 917 (1962).

1430. H. C. Newsom, W. D. English, A. L. McCloskey, and W. G. Woods, *J. Amer. Chem. Soc.* **83**, 4134 (1961).

1431. H. C. Newsom, W. G. Woods, and A. L. McCloskey, *Inorg. Chem.* **2**, 36 (1963).

1432. K. Niedenzu, H. Beyer, and J. W. Dawson, *Inorg. Chem.* **1**, 738 (1962).

1433. K. Niedenzu, H. Beyer, J. W. Dawson, and H. Jenne, *Chem. Ber.* **96**, 2653 (1963).

1434. K. Niedenzu and J. W. Dawson, *J. Amer. Chem. Soc.* **81**, 5553 (1959).

1435. K. Niedenzu and J. W. Dawson, *Angew. Chem.* **71**, 651 (1959).

1436. K. Niedenzu and J. W. Dawson, *J. Amer. Chem. Soc.* **82**, 4223 (1960).

1437. K. Niedenzu and J. W. Dawson, *J. Org. Chem.* **26**, 1671 (1961).

1438. K. Niedenzu, J. W. Dawson, P. Fritz, and H. Jenne, *Chem. Ber.* **98**, 3050 (1965).

1439. K. Niedenzu, J. W. Dawson, P. W. Fritz, and W. Weber, *Chem. Ber.* **100**, 1898 (1967).

1439a. K. Niedenzu, J. W. Dawson, G. A. Neece, W. Sawodny, D. R. Squire, and W. Weber, *Inorg. Chem.* **5**, 2161 (1966).

1440. K. Niedenzu, P. Fritz, and J. W. Dawson, *Inorg. Chem.* **3**, 778 (1964).

1441. K. Niedenzu, P. Fritz, and J. W. Dawson, *Inorg. Chem.* **3**, 1077 (1964).

1442. K. Niedenzu, P. Fritz, and H. Jenne, *Angew. Chem.* **76**, 535A (1964).

1443. K. Niedenzu, P. Fritz, and W. Weber, *Z. Naturforsch. B* **22**, 225 (1967).

1444. K. Niedenzu, D. H. Harrelson, W. George, and J. W. Dawson, *J. Org. Chem.* **26**, 3037 (1961).

1445. K. Niedenzu, C. D. Miller, and F. C. Nahm, *Tetrahedron Lett.* p. 2441 (1970).

1446. K. Niedenzu and W. Sawodny, *Z. Anorg. Allg. Chem.* **344**, 179 (1966).

1447. D. R. Nielsen and W. E. McEwen, *J. Amer. Chem. Soc.* **79**, 3081 (1957).

1448. D. R. Nielsen, W. E. McEwen, and C. A. Vanderwerf, *Chem. Ind.* (*London*) p. 1069 (1957).

1449. A. N. Nikitina, V. A. Petukhov, A. F. Galkin, N. S. Fedotov, Y. B. Bubnov, and P. M. Aronovich, *Opt. Spectrosc.* **16**, 528 (1964).

1450. A. N. Nikitina, V. A. Vaver, N. S. Fedotov, and B. M. Mikhailov, *Opt. Spektr.* 7, 644 (1959).
1451. J. F. Nobis, L. F. Moormeier, and R. E. Robinson, *Advan. Chem. Ser.* **23**, 63 (1959).
1452. D. Nölle and H. Nöth, *Angew. Chem.* **83**, 112 (1971); *Angew. Chem., Int. Ed. Engl.* **10**, 126 (1971).
1453. M. J. Nolte, G. Gafner, and L. M. Haines, *Chem. Commun.* p. 1406 (1969).
1453a. "Nomenclature of Inorganic Boron Compounds," Inform. Bull. No. 8. Int. Union Pure Appl. Chem., Oxford, England, 1970.
1454. A. D. Norman and R. Schaeffer, *J. Amer. Chem. Soc.* **88**, 1143 (1966).
1455. H. Normant and J. Braun, *C. R. Acad. Sci.* **248**, 828 (1959).
1456. H. Normant and J. Braun, *Izv. Akad. Nauk SSSR, Otd. Khim. Nauk* p. 1397 (1959).
1457. K. Noro and H. Kawazura, *J. Polym. Sci.* **45**, 264 (1960).
1458. K. Noro, H. Kawazura, and E. Uemura, *Kogyo Kagaku Zasshi* **65**, 973 (1962).
1459. H. Nöth, *Z. Naturforsch. B* **16**, 470 (1961).
1460. H. Nöth, *Z. Naturforsch. B* **16**, 471 (1961).
1461. H. Nöth, *Z. Naturforsch. B* **16**, 618 (1961).
1462. H. Nöth, *Chem. Ber.* **104**, 558 (1971).
1463. H. Nöth and G. Abeler, *Chem. Ber.* **101**, 969 (1968).
1464. H. Nöth, H. Beyer, and H. J. Vetter, *Chem. Ber.* **97**, 110 (1964).
1465. H. Nöth and P. Fritz, *Angew. Chem.* **73**, 408 (1961).
1466. H. Nöth and P. Fritz, *Z. Anorg. Chem.* **322**, 297 (1963).
1467. H. Nöth and P. Fritz, *Z. Anorg. Chem.* **324**, 129 (1963).
1468. H. Nöth and P. Fritz, *Z. Anorg. Chem.* **324**, 270 (1963).
1469. H. Nöth, P. Fritz, H. K. Hermannsdorfer, W. Meister, H. Schick, and G. Schmidt, *Angew. Chem., Int. Ed. Engl.* **3**, 148 (1964).
1470. H. Nöth, S. Lukas, and P. Schweizer, *Chem. Ber.* **98**, 962 (1965).
1471. H. Nöth and G. Mikulaschek, *Chem. Ber.* **97**, 202 (1964).
1472. H. Nöth and G. Mikulaschek, *Chem. Ber.* **97**, 709 (1964).
1473. H. Nöth and W. Petz, *J. Organometal. Chem.* **20**, 35 (1969).
1474. H. Nöth and W. Regnet, *Z. Naturforsch. B* **18**, 1138 (1963).
1475. H. Nöth and W. Regnet, *Advan. Chem. Ser.* **42**, 166 (1964).
1476. H. Nöth and W. Regnet, *Z. Anorg. Allg. Chem.* **352**, 1 (1967).
1477. H. Nöth and W. Regnet, *Chem. Ber.* **102**, 167 (1969).
1478. H. Nöth and W. Regnet, *Chem. Ber.* **102**, 2241 (1969).
1479. H. Nöth, W. Regnet, H. Rihl, and R. Standfest, *Chem. Ber.* **104**, 722 (1971).
1480. H. Nöth, H. Schaefer, and G. Schmid, *Angew. Chem.* **81**, 530A (1969); *Angew Chem., Int. Ed. Engl.* **8**, 515 (1969).
1481. H. Nöth, H. Schaefer, and G. Schmid, *Z. Naturforsch. B* **26**, 497 (1971).
1482. H. Nöth and G. Schmid, *Angew. Chem.* **75**, 861 (1963).
1483. H. Nöth and G. Schmid, *Z. Anorg. Chem.* **345**, 69 (1966).
1484. H. Nöth and W. Schrägle, *Z. Natursforsch. B* **16**, 473 (1961).
1485. H. Nöth and W. Schrägle, *Chem. Ber.* **97**, 2374 (1964).
1486. H. Nöth and M. J. Sprague, *J. Organometal. Chem.* **23**, 323 (1970).
1487. H. Nöth and W. Storch, *Syn. Inorg. Metal.-Org. Chem.* **1**, 197 (1971).
1488. H. Nöth and H. Vahrenkamp, *Chem. Ber.* **99**, 1049 (1966).
1489. H. Nöth and H. Vahrenkamp, *Chem. Ber.* **100**, 3353 (1967).
1490. H. Nöth and H. Vahrenkamp, *J. Organometal. Chem.* **11**, 399 (1968).
1491. H. Nöth and H. Vahrenkamp, *J. Organometal. Chem.* **12**, 23 (1968).

1492. M. Nussim, Y. Mazur, and F. Sondheimer, *J. Org. Chem.* **29**, 1120 (1964).
1493. R. J. O'Brien and G. A. Ozin, *J. Chem. Soc., A* p. 1136 (1971).
1494. O. Ohashi, Y. Kurita, T. Totani, H. Watanabe, T. Nakagawa, and M. Kubo, *Bull. Chem. Soc. Jap.* **35**, 1317 (1962).
1494a. K. Ohkubo, H. Shimada, and M. Okada, *Bull. Chem. Soc. Jap.* **44**, 2025 (1971).
1495. Y. Okamoto and A. J. Gordon, *Chem. Ind.* (*London*) p. 528 (1963).
1496. T. P. Onak, *J. Amer. Chem. Soc.* **83**, 2584 (1961).
1497. T. P. Onak, *Advan. Organometal. Chem.* **3**, 263 (1965).
1498. T. P. Onak, R. Drake, and G. Dunks, *J. Amer. Chem. Soc.* **87**, 2505 (1965).
1499. T. P. Onak, R. P. Drake, and I. W. Searcy, *Chem. Ind.* (*London*) p. 1865 (1964).
1500. T. P. Onak, G. B. Dunks, I. W. Searcy, and J. Spielman, *Inorg. Chem.* **6**, 1465 (1967).
1501. T. P. Onak, L. B. Friedman, J. A. Hartsuck, and W. N. Lipscomb, *J. Amer. Chem. Soc.* **88**, 3439 (1966).
1502. T. P. Onak and F. J. Gerhart, *Inorg. Chem.* **1**, 742 (1962).
1502a. T. Onak, F. J. Gerhart, and R. E. Williams, *J. Amer. Chem. Soc.* **85**, 1754 (1963).
1503. T. P. Onak, H. Landesman, R. E. Williams, and I. Shapiro, *J. Phys. Chem.* **63** 1533 (1959).
1504. T. P. Onak, P. Mattschei, and E. Groszek, *J. Chem. Soc., A* p. 1990 (1969).
1505. T. P. Onak and R. E. Williams, *Inorg. Chem.* **1**, 106 (1962).
1506. T. P. Onak and G. T. F. Wong, *J. Amer. Chem. Soc.* **92**, 5226 (1970).
1507. M. M. Otto, *J. Amer. Chem. Soc.* **57**, 1476 (1935).
1507a. E. Pace, *Mem. Atti. Accad. Naz. Lincei.* **10**, 193 (1929).
1508. C. Pace-Asciak and L. S. Wolfe, *J. Chromatogr.* **56**, 129 (1971).
1509. P. I. Paetzold, *Angew. Chem., Int. Ed. Engl.* **1**, 515 (1962).
1510. P. I. Paetzold, *Z. Anorg. Chem.* **326**, 53 (1963).
1511. P. I. Paetzold, *Z. Anorg. Chem.* **326**, 58 (1963).
1512. P. I. Paetzold, *Z. Anorg. Chem.* **326**, 64 (1963).
1513. P. I. Paetzold and P. P. Habereder, *J. Organometal. Chem.* **7**, 61 (1967).
1514. P. I. Paetzold, P. P. Habereder, and R. Müllbauer, *J. Organometal. Chem.* **7**, 45 (1967).
1515. P. I. Paetzold, P. P. Habereder, and R. Müllbauer, *J. Organometal. Chem.* **7**, 51 (1967).
1516. P. I. Paetzold and H. Hansen, *Z. Anorg. Chem.* **345**, 79 (1966).
1517. P. I. Paetzold and W. Simpson, *Angew. Chem.* **78**, 825 (1966).
1518. P. I. Paetzold and H. G. Smolka, *Chem. Ber.* **103**, 289 (1970).
1519. P. I. Paetzold, G. Stohr, H. Maisch, and H. Lenz, *Chem. Ber.* **101**, 2881 (1968).
1520. M. Pailer and W. Fenzl, *Monatsh. Chem.* **92**, 1294 (1961).
1521. M. Pailer and H. Huemer, *Monatsh. Chem.* **95**, 373 (1964).
1522. R. T. Paine and R. W. Parry, *Inorg. Chem.* **11**, 268 (1972).
1523. R. J. F. Palchak, J. H. Norman, and R. E. Williams, *J. Amer. Chem. Soc.* **83**, 3380 (1961).
1524. I. M. Panaiotov, *C. R. Acad. Bulg. Sci.* **14**, 147 (1961).
1525. R. W. Parry, C. E. Nordman, J. C. Carter, and G. Terhaar, *Advan. Chem. Ser.* **42**, 302 (1964).
1526. T. D. Parsons, E. D. Baker, A. B. Burg, and G. L. Juvinall, *J. Amer. Chem. Soc.* **83**, 250 (1961).
1527. T. D. Parsons and D. M. Ritter, *J. Amer. Chem. Soc.* **76**, 1710 (1954).
1528. T. D. Parsons, J. M. Self, and L. H. Schaad, *J. Amer. Chem. Soc.* **89**, 3446 (1967).

1529. T. D. Parsons, M. B. Silverman, and D. M. Ritter, *J. Amer. Chem. Soc.* **79**, 5091 (1957).
1530. L. Parts and J. T. Miller, Jr., *Inorg. Chem.* **3**, 1483 (1964).
1531. D. J. Pasto, *J. Amer. Chem. Soc.* **86**, 3039 (1964).
1532. D. J. Pasto, S. K. Arora, and J. Chow, *Tetrahedron* **25**, 1571 (1969).
1533. D. J. Pasto and P. Balasubramaniyan, *J. Amer. Chem. Soc.* **89**, 295 (1967).
1534. D. J. Pasto, V. Balasubramaniyan, and P. W. Wojtkowski, *Inorg. Chem.* **8**, 594 (1969).
1535. D. J. Pasto, J. Chow, and S. K. Arora, *Tetrahedron Lett.* p. 723 (1967).
1536. D. J. Pasto, J. Chow, and S. K. Arora, *Tetrahedron* **25**, 1557 (1969).
1537. D. J. Pasto and C. C. Cumbo, *J. Amer. Chem. Soc.* **86**, 4343 (1964).
1538. D. J. Pasto, C. C. Cumbo, and P. Balasubramaniyan, *J. Amer. Chem. Soc.* **88**, 2187 (1966).
1539. D. J. Pasto and J. Hickman, *J. Amer. Chem. Soc.* **89**, 5608 (1967).
1540. D. J. Pasto and J. Hickman, *J. Amer. Chem. Soc.* **90**, 4445 (1968).
1541. D. J. Pasto, J. Hickman, and T. Cheng, *J. Amer. Chem. Soc.* **90**, 6259 (1968).
1542. D. J. Pasto and S. Z. Kang, *J. Amer. Chem. Soc.* **90**, 3797 (1968).
1543. D. J. Pasto and F. M. Klein, *J. Org. Chem.* **33**, 1468 (1968).
1544. D. J. Pasto, B. Lepeska, and V. Balasubramaniyan, *J. Amer. Chem. Soc.* **94**, 6090 (1972).
1545. D. J. Pasto, B. Lepeska, and T. C. Cheng, *J. Amer. Chem. Soc.* **94**, 6083 (1972).
1545a. D. J. Pasto and K. McReynolds, *Tetrahedron Lett.* p. 801 (1971).
1546. D. J. Pasto and J. L. Miesel, *J. Amer. Chem. Soc.* **84**, 4991 (1962).
1547. D. J. Pasto and J. L. Miesel, *J. Amer. Chem. Soc.* **85**, 2118 (1963).
1548. D. J. Pasto and R. Snyder, *J. Org. Chem.* **31**, 2773 (1966).
1549. D. J. Pasto and R. Snyder, *J. Org. Chem.* **31**, 2777 (1966).
1550. D. J. Pasto and P. W. Wojtkowski, *Tetrahedron Lett.* p. 215 (1970).
1551. D. J. Pasto and P. W. Wojtkowski, *J. Organometal. Chem.* **34**, 251 (1972).
1552. D. J. Pasto and P. W. Wojtkowski, *J. Org. Chem.* **36**, 1790 (1971).
1553. W. G. Paterson and M. Onyszchuk, *Can. J. Chem.* **39**, 2324 (1961).
1554. I. Pattison and K. Wade, *J. Chem. Soc., A* p. 1098 (1967).
1555. I. Pattison and K. Wade, *J. Chem. Soc., A* p. 2618 (1968).
1556. J. Pellon, W. G. Deichert, and W. M. Thomas, *J. Polym. Sci.* **55**, 153 (1961).
1557. J. Pellon, L. H. Schwind, M. J. Guinard, and W. M. Thomas, *J. Polym. Sci.* **55** 161 (1961).
1558. A. Pelter, M. G. Hutchings, and K. Smith, *Chem. Commun.* p. 1529 (1970).
1559. A. Pelter, M. G. Hutchings, and K. Smith, *J. Chem. Soc., D* p. 1048A (1971).
1560. A. Pelter, M. G. Hutchings, and K. Smith, *J. Chem. Soc., D* p. 1048B (1971).
1561. A. Pelter and D. N. Sharrocks, *J. Chem. Soc., Chem. Commun.* p. 566 (1972).
1562. P. G. Perkins and D. H. Wall, *J. Chem. Soc., A* p. 1207 (1966).
1563. A. Perloff, *Acta Crystallogr.* **17**, 332 (1964).
1564. J. C. Perrine and R. N. Keller, *J. Amer. Chem. Soc.* **80**, 1823 (1958).
1565. A. A. Petrov, V. S. Zavgorodnii, and V. A. Korner, *Zh. Obshch. Khim.* **32**, 1349 (1962).
1566. R. C. Petry and F. H. Verhoek, *J. Amer. Chem. Soc.* **78**, 6416 (1956).
1567. C. S. G. Phillips, P. Powell, J. A. Semlyen, and P. L. Timms, *Z. Anal. Chem.* **197**, 202 (1963).
1568. J. R. Phillips and F. G. A. Stone, *J. Chem. Soc., London* p. 94 (1962).
1569. W. D. Phillips, H. C. Miller, and E. L. Muetterties, *J. Amer. Chem. Soc.* **81**, 4496 (1959).

1569a. C. G. Pierpont and R. Eisenberg, *Inorg. Chem.* **12**, 199 (1973).
1570. R. H. Pierson, *Anal. Chem.* **34**, 1642 (1962).
1571. K. S. Pitzer and H. S. Gutowsky, *J. Amer. Chem. Soc.* **68**, 2204 (1946).
1572. J. Ploquin and L. Sparfel, *C. R. Acad. Sci., D* **268**, 1820 (1969).
1573. H. E. Podall, H. E. Petree, and J. R. Zietz, *J. Org. Chem.* **24**, 1222 (1959).
1574. A. E. Pope and H. A. Skinner, *J. Chem. Soc., London* p. 3704 (1963).
1575. T. P. Povlock and W. T. Lippincott, *J. Amer. Chem. Soc.* **80**, 5409 (1958).
1576. P. Powell, J. A. Semlyen, R. E. Blofeld, and C. S. G. Phillips, *J. Chem. Soc., London* p. 280 (1964).
1577. W. C. Price, *J. Chem. Phys.* **17**, 1044 (1949).
1578. R. Prinz and H. Werner, *Angew. Chem.* **79**, 63 (1967); *Angew. Chem., Int. Ed. Engl.* **6**, 91 (1967).
1579. A. A. Prokhorova and Ya. M. Paushkin, *Dokl. Akad. Nauk SSSR* **135**, 84 (1960).
1580. K. F. Purcell, *Inorg. Chem.* **11**, 891 (1972).
1581. J. M. Purser and B. F. Spielvogel, *Inorg. Chem.* **7**, 2156 (1968).
1582. J. M. Purser and B. F. Spielvogel, *Chem. Commun.* p. 386 (1968).
1582a. K. V. Puzitskii, S. D. Pirozhkov, K. G. Ryabova, I. V. Pastukhova, and Ya. T. Eidus, *Izv. Akad. Nauk SSSR, Ser. Khim.* p. 1998 (1972).
1583. A. Quayle, *J. Appl. Chem.* **9**, 395 (1959).
1583a. H. Raber and W. Likussar, *Mikrochim. Acta* p. 92 (1972).
1584. F. L. Ramp, E. J. DeWitt, and L. E. Trapasso, *J. Org. Chem.* **27**, 4368 (1964).
1585. B. G. Ramsay, *J. Phys. Chem.* **70**, 611 (1966).
1586. B. G. Ramsay, *J. Phys. Chem.* **70**, 4097 (1966).
1587. B. Ramsey, M. Ashraf El-Bayoumi, and M. Kasha, *J. Chem. Phys.* **35**, 1502 (1961).
1588. B. G. Ramsey and N. K. Das, *J. Amer. Chem. Soc.* **91**, 6191 (1969).
1589. B. G. Ramsey and N. K. Das, *J. Amer. Chem. Soc.* **94**, 4227 (1972).
1590. B. G. Ramsey and H. Ito, *Theor. Chim. Acta* **17**, 188 (1970).
1591. B. G. Ramsey and J. E. Leffler, *J. Phys. Chem.* **67**, 2242 (1963).
1592. J. Rathke and R. Schaeffer, *Inorg. Chem.* **11**, 1150 (1972).
1593. M. W. Rathke and H. C. Brown, *J. Amer. Chem. Soc.* **88**, 2606 (1966).
1594. M. W. Rathke and H. C. Brown, *J. Amer. Chem. Soc.* **89**, 2740 (1967).
1594a. M. W. Rathke and R. Kow, *J. Amer. Chem. Soc.* **94**, 6854 (1972).
1595. M. W. Rathke, N. Inoue, K. R. Varma, and H. C. Brown, *J. Amer. Chem. Soc.* **83**, 2870 (1966).
1596. G. A. Razuvaev and T. G. Brilkina, *Dokl. Akad. Nauk SSSR* **91**, 861 (1953).
1597. G. A. Razuvaev and T. G. Brilkina, *Zh. Obshch. Khim.* **24**, 1415 (1954).
1598. G. A. Razuvaev, A. A. Koksharova, G. G. Petukhov, and S. F. Zhiltsov, *Dokl. Akad. Nauk SSSR* **187**, 821 (1969).
1599. C. W. Rector, G. W. Schaeffer, and J. R. Platt, *J. Chem. Phys.* **17**, 460 (1949).
1600. P. R. Reed and R. W. Lovejoy, *Spectrochim. Acta, Part A* **26**, 1087 (1970).
1601. D. L. Reznicek and N. E. Miller, *Inorg. Chem.* **11**, 858 (1972).
1602. B. Rice, J. M. G. Barredo, and T. F. Young, *J. Amer. Chem. Soc.* **73**, 2306 (1951).
1603. B. Rice, T. F. Young, and J. M. González Barredo, *An. Real Soc. Espan. Fis. Quim., Ser. B* **48**, 191 (1952).
1604. B. Rickborn and S. E. Wood, *J. Amer. Chem. Soc.* **93**, 3940 (1971).
1605. J. J. Ritter and T. D. Coyle, *J. Chem. Soc., A* p. 1303 (1970).
1606. J. J. Ritter, T. D. Coyle, and J. M. Bellama, *J. Chem. Soc., D* p. 908 (1969).
1607. J. J. Ritter, T. D. Coyle, and J. M. Bellama, *J. Organometal. Chem.* **29**, 175 (1971).

1608. R. C. Rittner and R. Culmo, *Anal. Chem.* **35**, 1268 (1963).
1609. C. S. Rondestvedt, R. M. Scribner, and C. E. Wulfman, *J. Org. Chem.* **20**, 9 (1955).
1609a. B. Roques and D. Florentin, *J. Organometal. Chem.* **46**, C38 (1972).
1610. A. Rosen and M. Zeldin, *J. Organometal. Chem.* **31**, 319 (1971).
1611. J. Rosenbaum and M. C. R. Symons, *Proc. Chem. Soc., London* p. 92 (1959).
1612. J. Rosenbaum and M. C. R. Symons, *Mol. Phys.* **3**, 205 (1960).
1613. L. Rosenblum, *J. Amer. Chem. Soc.* **77**, 5016 (1955).
1614. L. Rosenblum, *J. Org. Chem.* **25**, 1652 (1960).
1615. F. M. Rossi, P. A. McCusker, and G. F. Hennion, *J. Org. Chem.* **32**, 450 (1967).
1616. F. M. Rossi, P. A. McCusker, and G. F. Hennion, *J. Org. Chem.* **32**, 1233 (1967).
1617. G. Rotermund and R. Köster, *Angew. Chem.* **74**. 329 (1962).
1618. G. W. Rotermund and R. Köster, *Justus Liebigs Ann. Chem.* **686**, 153 (1965).
1619. H. J. Roth and B. Miller, *Naturwissenschaften* **50**, 732 (1963).
1620. H. J. Roth and B. Miller, *Arch. Pharm.* (*Weinheim*) **297**, 744 (1964).
1621. E. Rothstein and R. W. Saville, *J. Chem. Soc., London* p. 2987 (1952).
1622. R. W. Rudolph, *J. Amer. Chem. Soc.* **89**, 4216 (1967).
1623. W. Rüdorff and H. Zannier, *Z. Naturforsch. B* **8**, 611 (1953).
1624. J. K. Ruff, *J. Org. Chem.* **27**, 1020 (1962).
1625. W. I. Ruigh, *Angew. Chem.* **69**, 688 (1957).
1626. W. L. Ruigh, W. R. Dunnavant, F. C. Gunderloy, N. G. Steinberg, M. Sedlak, and A. D. Olin, *Advan. Chem. Ser.* **32**, 241 (1961).
1627. G. E. Ryschkewitsch, W. S. Brey, Jr., and A. Saji, *J. Amer. Chem. Soc.* **83**, 1010 (1961).
1628. G. E. Ryschkewitsch, S. W. Harris, E. J. Mezey, H. H. Sisler, E. A. Weilmuenster, and A. B. Garrett, *Inorg. Chem.* **2**, 890 (1963).
1629. G. E. Ryschkewitsch, J. J. Harris, and H. H. Sisler, *J. Amer. Chem. Soc.* **80**, 4515 (1958).
1630. G. E. Ryschkewitsch, E. J. Mezey, E. R. Altwicker, H. H. Sisler, and A. B. Garrett, *Inorg. Chem.* **2**, 893 (1963).
1631. K. A. Saegebarth, *J. Amer. Chem. Soc.* **82**, 2081 (1960).
1632. H. K. Saha, L. J. Glicenstein, and G. Urry, *J. Organometal. Chem.* **8**, 37 (1967).
1633. B. Samuel and K. Wade, *J. Chem. Soc., A* p. 1742 (1969).
1634. R. T. Sanderson, *J. Chem. Phys.* **21**, 571 (1953).
1634a. E. I. Sandvik, *Dissertation Abstr.* **22**, 1030 (1961).
1635. L. Santucci and H. Gilman, *J. Amer. Chem. Soc.* **80**, 193 (1958).
1636. L. Santucci, L. Travoletti, and D. Montalbano, *J. Org. Chem.* **27**, 2257 (1962).
1637. L. Santucci and C. Triboulet, *J. Chem. Soc., A* p. 392 (1969).
1638. R. Sayre, *J. Chem. Eng. Data* **8**, 244 (1963).
1639. V. A. Sazonova, A. V. Gerasimenko, and N. A. Shiller, *Zh. Obshch. Khim.* **33**, 2042 (1963).
1640. V. A. Sazonova and N. Y. Konrod, *Zh. Obshch. Khim.* **26**, 1876 (1956).
1641. V. A. Sazonova and I. I. Nazarova, *Zh. Obshch. Khim.* **26**, 3440 (1956).
1642. V. A. Sazonova, E. P. Serebryakov, and L. S. Kovaleva, *Dokl. Akad. Nauk SSSR* **113**, 1295 (1957).
1643. W. Schabacter and J. Goubeau, *Z. Anorg. Chem.* **294**, 183 (1958).
1643a. R. Schaeffer, ARL Tech. Rept. No. 60-334, U.S. Air Force Contract No. 33(616)5827. Aeronautical Research Laboratory, Ohio, 1960.
1644. R. Schaeffer and L. J. Todd, *J. Amer. Chem. Soc.* **87**, 488 (1965).
1645. G. D. Schaumberg and S. Donovan, *J. Organometal. Chem.* **20**, 261 (1969).

1645a. G. P. Schiemenz, *J. Mol. Struct.* **16**, 99 (1973).
1645b. G. P. Schiemenz, *J. Organometal. Chem.* **52**, 349 (1973).
1646. H. I. Schlesinger and H. C. Brown, *J. Amer. Chem. Soc.* **62**, 3429 (1940).
1647. H. I. Schlesinger, H. C. Brown, L. Horvitz, A. C. Bond, L. D. Tuck, and A. O. Walker, *J. Amer. Chem. Soc.* **75**, 222 (1953).
1648. H. I. Schlesinger, H. C. Brown, and G. W. Schaeffer, *J. Amer. Chem. Soc.* **65**, 1786 (1943).
1649. H. I. Schlesinger, N. W. Flodin, and A. B. Burg, *J. Amer. Chem. Soc.* **61**, 1078 (1939).
1650. H. I. Schlesinger, L. Horvitz, and A. B. Burg, *J. Amer. Chem. Soc.* **58**, 407 (1936).
1651. H. I. Schlesinger, L. Horvitz, and A. B. Burg, *J. Amer. Chem. Soc.* **58**, 409 (1936).
1652. H. I. Schlesinger, D. M. Ritter, and A. B. Burg, *J. Amer. Chem. Soc.* **60**, 1296 (1938).
1653. H. I. Schlesinger, R. T. Sanderson, and A. B. Burg, *J. Amer. Chem. Soc.* **62**, 3421 (1940).
1654. H. I. Schlesinger and A. O. Walker, *J. Amer. Chem. Soc.* **57**, 621 (1935).
1655. G. Schmid, *Chem. Ber.* **102**, 191 (1969).
1655a. G. Schmid, *Chem. Ber.* **103**, 528 (1970).
1656. G. Schmid and H. Nöth, *Chem. Ber.* **101**, 2502 (1968).
1657. M. Schmidt and F. R. Rittig, *Angew. Chem., Int. Ed. Engl.* **9**, 738 (1970).
1657a. M. Schmidt and F. R. Rittig, *Chem. Ber.* **103**, 3343 (1970).
1657b. M. Schmidt and F. R. Rittig, *Z. Anorg. Allg. Chem.* **394**, 152 (1972).
1658. M. Schmidt and W. Siebert, *Chem. Ber.* **102**, 2752 (1969).
1659. M. Schmidt, W. Siebert, and E. Gast, *Z. Naturforsch. B* **22**, 557 (1967).
1660. M. Schmidt, W. Siebert, and F. R. Rittig, *Chem. Ber.* **101**, 281 (1968).
1661. G. N. Schrauzer, *Chem. Ber.* **95**, 1438 (1962).
1662. R. R. Schrock and J. A. Osborn, *Inorg. Chem.* **9**, 2339 (1970).
1663. K. N. Scott and W. S. Brey, Jr., *Inorg. Chem.* **8**, 1703 (1969).
1664. W. Seaman and J. R. Johnson, *J. Amer. Chem. Soc.* **53**, 711 (1931).
1665. G. R. Seely, J. P. Oliver, and D. M. Ritter, *Anal. Chem.* **31**, 1993 (1959).
1666. J. A. Semlyen and P. J. Flory, *J. Chem. Soc., A* p. 191 (1966).
1667. B. Serafin, H. Duda, and M. Makosza, *Rocz. Chem.* **37**, 765 (1963).
1668. B. Serafin and M. Makosza, *Tetrahedron* **19**, 821 (1963).
1669. D. Seyferth, *J. Inorg. Nucl. Chem.* **7**, 152 (1958).
1670. D. Seyferth, *J. Amer. Chem. Soc.* **81**, 1844 (1959).
1671. D. Seyferth, W. R. Freyer, and M. Takamizawa, *Inorg. Chem.* **1**, 710 (1962).
1672. D. Seyferth and S. O. Grim, *J. Amer. Chem. Soc.* **83**, 1613 (1961).
1673. D. Seyferth and H. P. Kögler, *J. Inorg. Nucl. Chem.* **15**, 99 (1960).
1674. D. Seyferth, H. P. Kögler, W. R. Freyer, M. Takamizawa, H. Yamazaki, and Y. Sato, *Advan. Chem. Ser.* **42**, 259 (1964).
1675. D. Seyferth and B. Prokai, *J. Amer. Chem. Soc.* **88**, 1834 (1966).
1675a. D. Seyferth, G. Raab, and S. O. Grim, *J. Org. Chem.* **26**, 3034 (1961).
1676. D. Seyferth, Y. Sato, and M. Takamizawa, *J. Organometal. Chem.* **2**, 367 (1964).
1677. D. Seyferth and M. Takamizawa, *Inorg. Chem.* **2**, 731 (1963).
1678. D. Seyferth and M. Takamizawa, *J. Org. Chem.* **28**, 1142 (1963).
1679. D. Seyferth and M. A. Weiner, *J. Org. Chem.* **26**, 4797 (1961).
1680. D. Seyferth and M. A. Weiner, *J. Amer. Chem. Soc* **83**, 3583 (1961).
1681. D. Seyferth, H. Yamazaki, and Y. Sato, *Inorg. Chem.* **2**, 734 (1963).
1682. F. Shafizadeh, G. D. McGinnis, and P. S. Chin, *Carbohyd. Res.* **18**, 357 (1971).
1683. I. Shapiro, C. O. Wilson, J. F. Ditter, and W. J. Lehmann, *Advan. Chem. Ser.* **32**, 127 (1961).

1684. I. Shapiro, R. E. Williams, and S. G. Gibbins, *J. Phys. Chem.* **65**, 1061 (1961).
1685. I. Shapiro, C. O. Wilson, and W. J. Lehmann, *J. Chem. Phys.* **29**, 237 (1958).
1686. J. G. Sharefkin and H. D. Banks, *J. Org. Chem.* **30**, 4313 (1965).
1687. R. K. Sharma, B. A. Shoulders, and P. D. Gardner, *Chem. Ind.* (*London*) p. 2087 (1962).
1688. A. Sharma and J. C. Joshi, *J. Quant. Spectrosc. & Radiat. Transfer* **12**, 1073 (1972).
1689. T. A. Shchegoleva and E. M. Belyavskaya, *Dokl. Akad. Nauk SSSR* **136**, 638 (1961).
1690. T. A. Shchegoleva and B. M. Mikhailov, *Izv. Akad. Nauk SSSR, Ser. Khim.* p. 714 (1965).
1691. T. A. Shchegoleva, E. M. Shashkova, V. G. Kiselev, and B. M. Mikhailov, *Zh. Obshch. Khim.* **35**, 1078 (1965).
1692. T. A. Shchegoleva, E. M. Shashkova, V. G. Kiselev, and B. M. Mikhailov, *Izv. Akad. Nauk SSSR, Ser. Khim.* p. 365 (1964).
1693. D. Sheehan, A. P. Bentz, and J. C. Petropoulos, *J. Appl. Polym. Sci.* **6**, 47 (1962).
1694. A. Shepp and S. H. Bauer, *J. Amer. Chem. Soc.* **76**, 265 (1954).
1695. F. G. Sherif and C. D. Schmulbach, *Inorg. Chem.* **5**, 322 (1966).
1696. S. G. Shore, J. L. Crist, B. Lockman, J. R. Long, and A. D. Coon, *J. Chem. Soc., Dalton Trans.* p. 1123 (1972).
1696a. P. P. Shorygin, B. V. Lopatin, and O. G. Boldyreva, *Izv. Akad. Nauk SSSR, Ser. Khim.* p. 1196 (1972).
1697. D. F. Shriver and M. J. Biallas, *J. Amer. Chem. Soc.* **89**, 1078 (1967).
1698. D. F. Shriver, D. E. Smith, and P. Smith, *J. Amer. Chem. Soc.* **86**, 5153 (1964).
1698a. V. A. Shushunov, V. P. Maslennikov, G. I. Makin, and A. V. Gorbunov, *Zh. Obshch. Khim.* **42**, 1577 (1972).
1698b. V. A. Shushunov, V. P. Maslennikov, G. I. Makin, and S. I. Kulakov, *Zh. Org. Khim.* **8**, 1409 (1972).
1699. H. Siebert, *Z. Anorg. Chem.* **268**, 13 (1952).
1699a. W. Siebert, *Angew. Chem.* **82**, 699 (1970).
1700. W. Siebert, *Chem. Ber.* **103**, 2308 (1970).
1701. W. Siebert, E. Gast, and M. Schmidt, *J. Organometal. Chem.* **23**, 329 (1970).
1702. W. Siebert and A. Ospici, *Chem. Ber.* **105**, 454 (1972).
1702a. W. Siebert, A. Ospici, and F. Reigel, *Angew. Chem., Int. Ed. Engl.* **10**, 858 (1971).
1703. W. Siebert, F. R. Rittig, and M. Schmidt, *J. Organometal. Chem.* **25**, 305 (1970).
1704. W. Siebert, F. R. Rittig, and M. Schmidt, *J. Organometal. Chem.* **22**, 511 (1970).
1705. W. Siebert, K. J. Schaper, and M. Schmidt, *J. Organometal. Chem.* **25**, 315 (1970).
1706. W. Siebert, M. Schmidt, and E. Gast, *J. Organometal. Chem.* **20**, 29 (1969).
1707. A. R. Siedle, *J. Inorg. Nucl. Chem.* **33**, 3671 (1971).
1708. B. Siegel and J. L. Mack, *J. Phys. Chem.* **63**, 1212 (1960).
1709. B. Siegel, J. L. Mack, J. U. Lowe, and J. Gallaghan, *J. Amer. Chem. Soc.* **80**, 4523 (1958).
1709a. R. J. Sime, R. P. Dodge, A. Zalkin, and D. H. Templeton, *Inorg. Chem.* **10**, 537 (1971).
1709b. E. Siska and E. Pungor, *Z. Anal. Chem.* **257**, 7 (1971).
1709c. E. Siska and E. Pungor, *Z. Anal. Chem.* **257**, 12 (1971).
1709d. H. A. Skinner, *Adv. Organometal. Chem.* **2**, 49 (1964).
1710. H. A. Skinner and T. F. S. Tees, *J. Chem. Soc., London* p. 3378 (1953).
1711. I. H. Skoog, *J. Org. Chem.* **29**, 492 (1964).

1712. B. Skowronska-Serafin and M. Makosza, *Rocz. Chem.* **35**, 359 (1961).
1713. J. H. Smalley and S. F. Stafiej, *J. Amer. Chem. Soc.* **81**, 582 (1959).
1713a. R. F. Smirnov, B. I. Tikhomirov, M. I. Bitsenko, and A. I. Yakubchik, *Vysokomol. Soedin., Ser. A.* **13**, 1618 (1971).
1714. E. B. Smith and C. A. Kraus, *J. Amer. Chem. Soc.* **73**, 2751 (1951).
1715. J. A. S. Smith and D. A. Tong, *J. Chem. Soc., A* p. 173 (1971).
1716. J. A. S. Smith and D. A. Tong, *J. Chem. Soc., A* p. 178 (1971).
1717. G. Smolinsky, *J. Org. Chem.* **26**, 4915 (1961).
1718. H. R. Snyder, M. S. Konecky, and W. J. Lennarz, *J. Amer. Chem. Soc.* **80**, 3611 (1958).
1719. H. R. Snyder, J. A. Kuck, and J. R. Johnson, *J. Amer. Chem. Soc.* **60**, 105 (1938).
1720. H. R. Snyder, A. J. Reedy, and W. J. Lennarz, *J. Amer. Chem. Soc.* **80**, 835 (1958).
1721. H. R. Snyder and C. Weaver, *J. Amer. Chem. Soc.* **70**, 232 (1948).
1722. H. R. Snyder and F. W. Wyman, *J. Amer. Chem. Soc.* **70**, 234 (1948).
1723. I. J. Solomon, M. J. Klein, and K. Hattori, *J. Amer. Chem. Soc.* **80**, 4520 (1958).
1724. I. J. Solomon, M. J. Klein, R. G. Maguire, and K. Hattori, *Inorg. Chem.* **2**, 1136 (1963).
1725. A. H. Soloway, *J. Amer. Chem. Soc.* **81**, 3017 (1959).
1726. A. H. Soloway, *J. Amer. Chem. Soc.* **82**, 2442 (1960).
1727. A. H. Soloway and P. Szabady, *J. Org. Chem.* **25**, 1683 (1960).
1728. F. Sondheimer and S. Wolfe, *Can. J. Chem.* **37**, 1870 (1959).
1729. J. Soulié and P. Cadiot, *Bull. Soc. Chim. Fr.* p. 1981 (1966).
1730. J. Soulié and A. Willemart, *C. R. Acad. Sci.* **251**, 727 (1960).
1731. J. R. Spielman and A. B. Burg, *Inorg. Chem.* **2**, 1139 (1963).
1732. B. F. Spielvogel, R. F. Bratton, and C. G. Moreland, *J. Amer. Chem. Soc.* **94**, 8597 (1972).
1732a. R. F. Sprecher and J. C. Carter, *J. Amer. Chem. Soc.* **95**, 2369 (1973).
1732b. H. A. Staab and B. Meissner, *Justus Liebigs Ann. Chem.* **753**, 80 (1971).
1733. S. L. Stafford, *Can. J. Chem.* **41**, 807 (1963).
1734. S. L. Stafford and F. G. A. Stone, *Spectrochim. Acta* **17**, 412 (1961).
1735. S. L. Stafford and F. G. A. Stone, *J. Amer. Chem. Soc.* **82**, 6238 (1960).
1736. S. J. Steck, G. A. Pressley, and F. E. Stafford, *J. Phys. Chem.* **73**, 1000 (1969).
1737. W. C. Steele, L. D.Nichols, and F. G. A. Stone, *J. Amer. Chem. Soc.* **84**, 1154 (1962).
1738. R. N. Sterlin and S. S. Dubov, *Zh. Vses. Khim. Obshchest.* **7**, 117 (1962).
1739. J. E. Stewart, *J. Res. Nat. Bur. Stand.* **56**, 337 (1956).
1740. A. Stock and F. Zeidler, *Chem. Ber.* **54**, 531 (1921).
1741. F. G. A. Stone and H. J. Emeléus, *J. Chem. Soc., London* p. 2755 (1950).
1742. F. G. A. Stone and W. A. G. Graham, *Chem. Ind.* (*London*) p. 1181 (1955).
1743. R. D. Strahm and M. F. Hawthorne, *Anal. Chem.* **32**, 530 (1960).
1744. M. W. P. Strandberg, C. S. Pearsall, and M. T. Weiss, *J. Chem. Phys.* **17**, 429 (1949).
1745. W. Strecker, *Chem. Ber.* **43**, 1131 (1910).
1746. A. Streitwieser, Jr., L. Verbit, and R. Bittman, *J. Org. Chem.* **32**, 1530 (1967).
1747. W. Strehmeier and K. H. Hümpfner, *Z. Elektrochem.* **61**, 1010 (1957).
1748. C. Summerford and K. Wade, *J. Chem. Soc., A* p. 2010 (1970).
1749. C. Summerford and K. Wade, *J. Chem. Soc., A* p. 1487 (1969).
1750. S. Sundaram and F. F. Cleveland, *J. Chem. Phys.* **32**, 166 (1960).
1751. N. Sutton and H. Schneider, *Microchem. J.* **9**, 209 (1965).

1752. A. Suzuki, A. Arase, H. Matsumoto, M. Itoh, H. C. Brown, M. M. Rogic, and M. W. Rathke, *J. Amer. Chem. Soc.* **89**, 5708 (1967).
1752a. A. Suzuki, N. Miyaura, S. Abiko, M. Itoh, H. C. Brown, J. A. Sinclair, and M. M. Midland, *J. Amer. Chem. Soc.* **94**, 3080 (1973).
1752b. A. Suzuki, M. Miyaura, and M. Itoh, *Tetrahedron* **27**, 2775 (1971).
1753. A. Suzuki, N. Miyayra, and M. Itoh, *J. Amer. Chem. Soc.* **93**, 2797 (1971).
1753a. A. Suzuki, N. Miyaura, M. Itoh, H. C. Brown, and P. Jacob, *Synthesis* p. 305 (1973).
1754. A. Suzuki, S. Nozawa, M. Harada, I. Mitsuomi, H. C. Brown, and M. M. Midland, *J. Amer. Chem. Soc.* **93**, 1508 (1971).
1755. A. Suzuki, S. Nozawa, M. Itoh, H. C. Brown, G. W. Kabalka, and G. W. Holland, *J. Amer. Chem. Soc.* **92**, 3503 (1970).
1756. A. Suzuki, S. Nozawa, M. Itoh, H. C. Brown, E. Negishi, and S. K. Gupto, *J. Chem. Soc., D* p. 1009 (1969).
1757. A. Suzuki, S. Nozawa, N. Miyaura, M. Itoh, and H. C. Brown, *Tetrahedron Lett.* p. 2955 (1969).
1758. A. Suzuki, S. Sono, M. Itoh, H. C. Brown, and M. M. Midland, *J. Amer. Chem. Soc.* **93**, 4329 (1971).
1759. E. Switkes, I. R. Epstein, J. A. Tossell, R. M. Stevens, and W. N. Lipscomb, *J. Amer. Chem. Soc.* **92**, 3837 (1970).
1760. G. Talamini and G. Vidotto, *Makromol. Chem.* **50**, 129 (1961).
1761. G. Talamini and G. Vidotto, *Makromol. Chem.* **53**, 21 (1962).
1762. J. Tanaka and J. C. Carter, *Tetrahedron Lett.* p. 329 (1965).
1763. Y. Tanigawa, I. Moritani, and S. Nishida, *J. Organometal. Chem.* **28**, 73 (1971).
1763a. Y. Tanahashi, J. Lhomme, and G. Ourisson, *Tetrahedron* **28**, 2655 (1972).
1764. R. C. Taylor, *J. Chem. Phys.* **27**, 979 (1957).
1765. R. C. Taylor, *J. Chem. Phys.* **26**, 1131 (1957).
1766. G. T. Terhaar, Sr, M. A. Fleming, and R. W. Parry, *J. Amer. Chem. Soc.* **84**, 1767 (1962).
1766a. G. Tersac and S. Boileau, *J. Chim. Phys. Physicochim. Biol.* **68**, 903 (1971).
1766b. G. S. Ter-Sarkisyan, N. A. Nikolaeva, and B. M. Mikhailov, *Izv. Akad. Nauk SSSR, Ser. Khim.* p. 876 (1970).
1766c. G. S. Ter-Sarkisyan, N. A. Nikolaeva, V. G. Kiselev, and B. M. Mikhailov, *Zh. Obshch. Khim.* **41**, 152 (1971).
1767. A. D. Tevebaugh, *Inorg. Chem.* **3**, 302 (1964).
1768. B. L. Therrell and E. K. Mellon, *Inorg. Chem.* **11**, 1137 (1972).
1769. G. Thielens, *Naturwissenschaften* **45**, 543 (1958).
1769a. P. C. Thomas and I. C. Paul, *Chem. Commun.* p. 1130 (1968).
1770. N. R. Thompson, *J. Chem. Soc., London* p. 6290 (1965).
1771. R. J. Thompson and J. C. Davis, Jr., *Inorg. Chem.* **4**, 1464 (1965).
1772. P. L. Timms, *J. Amer. Chem. Soc.* **89**, 1629 (1967).
1773. P. L. Timms, *Chem. Commun.* p. 1525 (1968).
1774. P. L. Timms, *J. Amer. Chem. Soc.* **90**, 4585 (1968).
1775. E. A. Timofeeva, G. A. Lutsenko, and N. I. Shuikin, *Izv. Akad. Nauk SSSR, Ser. Khim.* p. 1152 (1967).
1776. W. Tochtermann and B. Knickel, *Chem. Ber.* **102**, 3508 (1969).
1777. W. Tochtermann and H. D. Trautwein, *Justus Liebigs Ann. Chem.* **735**, 189 (1970).
1778. R. H. Toeniskoetter and F. R. Hall, *Inorg. Chem.* **2**, 29 (1963).
1779. R. H. Toeniskoetter and K. A. Killip, *J. Amer. Chem. Soc.* **86**, 690 (1964).

1780. B. C. Tollin, R. Schaeffer, and H. J. Svec, *J. Inorg. Nucl. Chem.* **4**, 273 (1957).
1781. A. V. Topchiev, Ya. M. Paushkin, and A. A. Prokhorova, *Dokl. Akad. Nauk SSSR* **129**, 598 (1959).
1782. A. V. Topchiev, Ya. M. Paushkin, A. A. Prokhorova, E. I. Frenkin, and M. V. Kurashev, *Dokl. Akad. Nauk SSSR* **134**, 364 (1960).
1783. A. V. Topchiev, Ya. M. Paushkin, A. A. Prokhorova, and M. V. Kurashev, *Dokl. Akad. Nauk SSSR* **128**, 110 (1959).
1784. A. V. Topchiev, A. A. Prokhorova, and M. V. Kurashev, *Dokl. Akad. Nauk SSSR* **141**, 1386 (1961).
1785. A. V. Topchiev, A. A. Prokhorova, Ya. M. Paushkin, and M. V. Kurashev, *Izv. Akad. Nauk SSSR Otd. Khim. Nauk.* p. 370 (1958).
1786. A. V. Topchiev, A. A. Prokhorova, Ya. M. Paushkin, and M. V. Kurashev, *Dokl. Akad. Nauk SSSR* **131**, 105 (1960).
1787. L. H. Toporcer, R. E. Dessy, and S. I. E. Green, *J. Amer. Chem. Soc.* **87**, 1236 (1965).
1788. L. H. Toporcer, R. E. Dessy, and S. I. E. Green, *Inorg. Chem.* **4**, 1649 (1965).
1789. K. Torssell, *Acta Chem. Scand.* **8**, 1779 (1954).
1790. K. Torssell, *Acta Chem. Scand.* **9**, 239 (1955).
1791. K. Torssell, *Acta Chem. Scand.* **9**, 242 (1955).
1792. K. Torssell, *Ark. Kemi* **10**, 473 (1957).
1793. K. Torssell, *Ark. Kemi* **10**, 507 (1957).
1794. K. Torssell, *Ark. Kemi* **10**, 513 (1957).
1795. K. Torssell, *Ark. Kemi* **10**, 529 (1957).
1796. K. Torssell, *Ark. Kemi* **10**, 541 (1957).
1797. K. Torssell, *Acta Chem. Scand.* **13**, 115 (1959).
1798. K. Torssell, *Acta Chem. Scand.* **16**, 87 (1962).
1799. K. Torssell and E. N. V. Larsson, *Acta Chem. Scand.* **11**, 404 (1957).
1800. K. Torssell, J. H. McClendon, and G. F. Somers, *Acta Chem. Scand.* **12**, 1373 (1958).
1801. K. Torssell, H. Meyer, and B. Zacharias, *Ark. Kemi* **10**, 497 (1957).
1802. P. M. Treichel, J. Benedict, and R. G. Haines, *Inorg. Syn.* **13**, 32 (1972).
1803. S. Trofimenko, *J. Amer. Chem. Soc.* **88**, 1842 (1966).
1804. S. Trofimenko, *J. Amer. Chem. Soc.* **89**, 3165 (1967).
1805. S. Trofimenko, *J. Amer. Chem. Soc.* **89**, 4948 (1967).
1806. S. Trofimenko, *J. Amer. Chem. Soc.* **89**, 7014 (1967).
1807. S. Trofimenko, *J. Amer. Chem. Soc.* **91**, 2139 (1969).
1808. S. Trofimenko, *J. Amer. Chem. Soc.* **89**, 3903 (1967).
1809. S. Trofimenko, *J. Org. Chem.* **36**, 1161 (1971).
1810. C. C. Tsai and W. E. Streib, *Tetrahedron Lett.* p. 669 (1968).
1810a. C. Tsai and W. E. Streib, *Acta Crystallogr., Part B* **26**, 835 (1970).
1811. J. P. Tuchagues and J. P. Laurent, *Bull. Soc. Chim. Fr.* p. 4160 (1967).
1812. J. P. Tuchagues, J. P. Laurent, and G. Commenges, *Bull. Soc. Chim. Fr.* p. 385 (1969).
1813. J. P. Tuchagues, J. P. Laurent, and F. Gallais, *C. R. Acad. Sci., Ser. C* **268**, 2125 (1969).
1814. P. M. Tucker and T. P. Onak, *J. Amer. Chem. Soc.* **91**, 6869 (1969).
1815. P. M. Tucker, T. P. Onak, and J. B. Leach, *Inorg. Chem.* **9**, 1430 (1970).
1816. J. J. Tufariello and M. M. Hovey, *J. Amer. Chem. Soc.* **92**, 3221 (1970).
1817. J. J. Tufariello and L. T. C. Lee, *J. Amer. Chem. Soc.* **88**, 4757 (1966).
1818. J. J. Tufariello, L. T. C. Lee, and P. Wojtkowski, *J. Amer. Chem. Soc.* **89**, 6804 (1967).

1819. J. J. Tufariello, P. Wojtkowski, and L. T. C. Lee, *Chem. Commun.* p. 505 (1967).
1820. H. S. Turner, *Chem. Ind.* (*London*) p. 1405 (1958).
1821. D. Ulmschneider and J. Goubeau, *Z. Phys. Chem.* **14**, 56 (1958).
1822. D. Ulmschneider and J. Goubeau, *Chem. Ber.* **90**, 2733 (1957).
1823. F. Umland and D. Thierig, *Angew. Chem.* **74**, 388 (1962).
1824. F. Umland and C. Schleyerbach, *Angew. Chem.* **77**, 169 (1965).
1825. F. Umland and C. Schleyerbach, *Angew. Chem.* **77**, 426 (1965).
1826. G. Urry, A. G. Garrett, and H. I. Schlesinger, *Inorg. Chem.* **2**, 395 (1963).
1827. G. Urry, J. Kerrigan, T. D. Parsons, and H. I. Schlesinger, *J. Amer. Chem. Soc.* **76**, 5299 (1954).
1827a. K. Utimoto, T. Tanaka, and H. Nozaki, *Tetrahedron Lett.* p. 1167 (1972).
1828. H. Vahrenkamp, *J. Organometal Chem.* **28**, 167 (1971).
1829. H. Vahrenkamp, *J. Organometal. Chem.* **28**, 181 (1971).
1830. L. Van Alten, G. R. Seely, J. Oliver, and D. M. Ritter, *Advan. Chem. Ser.* **32**, 107 (1961).
1831. J. T. Vandeberg, C. E. Moore, F. P. Cassaretto, and H. Posvic, *Anal. Chim. Acta* **44**, 175 (1969).
1832. J. T. Vandeberg, C. E. Moore, and F. P. Cassaretto, *Spectrochim. Acta, Part A* **27**, 501 (1971).
1833. E. E. Van Tamelen, G. Brieger, and K. G. Untch, *Tetrahedron Lett.* p. 14 (1960).
1834. R. Van Veen and F. Bickelhaupt, *J. Organometal. Chem.* **24**, 589 (1970).
1835. R. Van Veen and F. Bickelhaupt, *J. Organometal. Chem.* **30**, 51C (1971).
1835a. R. Van Veen and F. Bickelhaupt, *J. Organometal. Chem.* **47**, 33 (1973).
1836. K. R. Varma and E. Caspi, *Tetrahedron* **24**, 6365 (1968).
1836a. L. S. Vasil'ev, V. P. Dmitrikov, and B. M. Mikhailov, *Zh. Obshch. Khim.* **42**, 1015 (1972).
1836b. L. S. Vasil'ev, V. P. Dmitrikov, V. S. Bogdanov, and B. M. Mikhailov, *Zh. Obshch. Khim.* **42**, 1318 (1972).
1836c. L. S. Vasil'ev, M. M. Vartanyan, V. S. Bogdanov, V. G. Kiselev, and B. M. Mikhailov, *Zh. Obshch. Khim.* **42**, 1540 (1972).
1837. J. G. Verkade and C. W. Heitsch, *Inorg. Chem.* **2**, 512 (1963).
1838. J. G. Verkade, R. W. King, and C. W. Heitsch, *Inorg. Chem.* **3**, 884 (1964).
1839. M. E. Volpin, Y. D. Koreshkov, V. Dulova, and D. N. Kursanov, *Tetrahedron* **18**, 107 (1962).
1840. R. C. Wade, E. A. Sullivan, J. R. Berschied, and K. F. Purcell, *Inorg. Chem.* **9**, 2146 (1970).
1841. T. C. Waddington and F. Klanberg, *Z. Anorg. Chem.* **304**, 185 (1960).
1842. R. I. Wagner and J. L. Bradford, *Inorg. Chem.* **1**, 93 (1962).
1843. R. I. Wagner and J. L. Bradford, *Inorg. Chem.* **1**, 99 (1962).
1844. L. A. Wall, S. Straus, R. E. Florin, F. L. Mohler, and P. Bradt, *J. Res. Nat. Bur. Stand., Sect. A* **63**, 63 (1959).
1845. D. E. Walmsley, W. L. Budde, and M. F. Hawthorne, *J. Amer. Chem. Soc.* **93**, 3150 (1971).
1846. T. T. Wang and K. Niedenzu, *J. Organometal. Chem.* **35**, 231 (1972).
1847. J. C. Ware and T. G. Traylor, *J. Amer. Chem. Soc.* **85**, 3026 (1963).
1848. T. Wartik and R. Pearson, *J. Inorg. Nucl. Chem.* **5**, 250 (1958).
1849. T. Wartik and H. I. Schlesinger, *J. Amer. Chem. Soc.* **75**, 835 (1953).
1850. R. M. Washburn, F. A. Billig, M. Bloom, C. F. Albright, and E. Levens, *Advan. Chem. Ser.* **32**, 208 (1961).
1851. R. M. Washburn, E. Levens, C. F. Albright, F. A. Billig, and E. S. Cernak, *Advan. Chem. Ser.* **23**, 102 (1959).

1852. R. M. Washburn, E. Levens, C. F. Albright, and F. A. Billig, *Org. Syn.* **39**, 3 (1959).
1853. S. K. Wason and R. F. Porter, *Inorg. Chem.* **5**, 161 (1966).
1854. H. Watanabe, K. Ito, and M. Kubo, *J. Amer. Chem. Soc.* **82**, 3294 (1960).
1855. H. Watanabe and M. Kubo, *J. Amer. Chem. Soc.* **82**, 2428 (1960).
1856. H. Watanabe, K. Nagasawa, T. Totani, T. Yoshizaki, and T. Nakagawa, *Advan. Chem. Ser.* **42**, 116 (1964).
1857. H. Watanabe, K. Nagasawa, T. Totani, T. Yoshizaki, T. Nakagawa, O. Ohashi, and M. Kubo, *Advan. Chem. Ser.* **42**, 108 (1964).
1858. H. Watanabe, M. Narisada, T. Nakagawa, and M. Kubo, *Spectrochim. Acta* **16**, 78, (1960).
1859. H. Watanabe, T. Tatani, and T. Yoshizaki, *Inorg. Chem.* **4**, 657 (1965).
1860. W. Weber, J. W. Dawson, and K. Niedenzu, *Inorg. Chem.* **5**, 726 (1966).
1861. W. J. Wechter, *Chem. Ind.* (*London*) p. 294 (1959).
1862. H. Wiedmann and H. K. Zimmerman, Jr., *J. Phys. Chem.* **64**, 182 (1960).
1862a. F. J. Weigert and J. D. Roberts, *J. Amer. Chem. Soc.* **91**, 4940 (1969).
1862b. F. J. Weigert and J. D. Roberts, *Inorg. Chem.* **12**, 313 (1973).
1863. E. A. Weilmuenster, *Ind. Eng. Chem.* **49**, 1337 (1957).
1864. A. J. Weinheimer and W. E. Marsico, *J. Org. Chem.* **27**, 1926 (1964).
1865. T. J. Weismann and J. C. Schug, *J. Chem. Phys.* **40**, 956 (1964).
1866. H. G. Weiss, W. J. Lehmann, and I. Shapiro, *J. Amer. Chem. Soc.* **84**, 3840 (1962).
1867. M. T. Weiss, M. W. P. Strandberg, R. B. Lawrance, and C. C. Loomis, *Phys. Rev.* **78**, 202 (1950).
1868. S. I. Weissman, J. Townsend, D. E. Paul, and G. E. Pake, *J. Chem. Phys.* **21**, 2227 (1953).
1869. S. I. Weissman and H. Van-Willigen, *J. Amer. Chem. Soc.* **87**, 2285 (1965).
1870. C. N. Welch and S. G. Shore, *Inorg. Chem.* **8**, 2810 (1969).
1871. F. J. Welch, *J. Polym. Sci.* **61**, 243 (1962).
1871a. W. W. Wendlandt, *Chemist-Analyst* **47**, 38 (1958).
1872. H. Werner, R. Prinz, and E. Decklemann, *Chem. Ber.* **102**, 95 (1969).
1873. A. T. Whatley and R. N. Pease, *J. Amer. Chem. Soc.* **76**, 835 (1954).
1874. D. G. White, *J. Amer. Chem. Soc.* **85**, 3634 (1963).
1875. E. Wiberg, *Naturwissenschaften* **35**, 212 (1948).
1876. E. Wiberg, J. E. F. Evans, and H. Nöth, *Z. Naturforsch. B* **13**, 263 (1958).
1877. E. Wiberg, J. E. F. Evans, and H. Nöth, *Z. Naturforsch. B* **13**, 265 (1958).
1878. E. Wiberg and H. Hertwig, *Z. Anorg. Allg. Chem.* **257**, 138 (1948).
1879. E. Wiberg and H. Hertwig, *Z. Anorg. Chem.* **255**, 141 (1947).
1880. E. Wiberg, H. Hertwig, and A. Bolz, *Z. Anorg. Chem.* **256**, 177 (1948).
1881. E. Wiberg and U. Krüerke, *Z. Naturforsch. B* **8**, 608–610 (1953).
1882. E. Wiberg and G. Horeld, *Z. Naturforsch. B* **6**, 338 (1951).
1883. E. Wiberg and W. Ruschmann, *Chem. Ber.* **70**B, 1583 (1937).
1884. E. Wiberg and P. Strebel, *Justus Liebigs Ann. Chem.* **607**, 9 (1957).
1885. E. Wiberg and W. Sturm, *Z. Naturforsch. B* **8**, 689 (1953).
1886. E. Wiberg and W. Sturm, *Angew. Chem.* **67**, 483 (1955).
1887. E. Wiberg and W. Sturm, *Z. Naturforsch. B* **10**, 108 (1955).
1888. H. Wille and J. Goubeau, *Chem. Ber.* **105**, 2156 (1972).
1889. E. Wiberg and W. Sturm, *Z. Naturforsch. B* **10**, 112 (1955).
1890. E. Wiberg and W. Sturm, *Z. Naturforsch. B* **10**, 113 (1955).
1891. G. Wilke and P. Heimbach, *Justus Liebigs Ann. Chem.* **652**, 7 (1962).
1892. J. L. R. Williams, J. C. Doty, P. J. Grisdale, T. H. Regan, and D. G. Borden, *Chem. Commun.* p. 109 (1967).

1893. J. L. R. Williams, J. C. Doty, P. J. Grisdale, T. H. Regan, G. P. Happ, and D. P. Maier, *J. Amer. Chem. Soc.* **90**, 53 (1968).
1894. J. L. R. Williams, J. C. Doty, P. J. Grisdale, R. Searle, T. H. Regan, G. P. Happ, and D. P. Maier, *J. Amer. Chem. Soc.* **89**, 5153 (1967).
1895. J. L. R. Williams, P. J. Grisdale, and J. C. Doty, *J. Amer. Chem. Soc.* **89**, 4538 (1967).
1896. R. E. Williams, U.S. Patent 2,917,547 (1959).
1897. R. E. Williams, *Inorg. Chem.* **1**, 971 (1962).
1898. R. E. Williams, H. D. Fisher, and C. O. Wilson, *J. Chem. Phys.* **64**, 1583 (1960).
1899. R. E. Williams and F. J. Gerhart, *J. Organometal. Chem.* **10**, 168 (1967).
1900. R. L. Williams, I. Dunstan, and N. J. Blay, *J. Chem. Soc., London* p. 5006 (1960).
1901. P. F. Winternitz and A. A. Carotti, *J, Amer. Chem. Soc.* **82**, 2430 (1960).
1901a. C. O. Wilson and I. Shapiro, *Anal. Chem.* **32**, 78 (1960).
1902. H. Witte, *Tetrahedron Lett.* p. 1127 (1965).
1902a. H. Witte, P. Mischke, and G. Hesse, *Justus Liebigs Ann. Chem.* **722**, 21 (1969).
1903. G. Wittig, *Angew. Chem.* **62**, 231 (1950).
1904. G. Wittig, L. Gonsior, and H. Vogel, *Justus Liebigs Ann. Chem.* **688**, 1 (1965).
1905. G. Wittig and W. Haag, *Chem. Ber.* **88**, 1654 (1955).
1906. G. Wittig and W. Herwig, *Chem. Ber.* **88**, 962 (1955).
1907. G. Wittig and G. Keicher, *Naturwissenschaften* **34**, 216 (1947).
1908. G. Wittig, G. Keicher, A. Rückert, and P. Raff, *Justus Liebigs Ann. Chem.* **563**, 110 (1949).
1909. G. Wittig and G. Kolb, *Chem. Ber.* **93**, 1469 (1960).
1910. G. Wittig and P. Raff, *Justus Liebigs Ann. Chem.* **573**, 195 (1951).
1911. G. Wittig and P. Raff, *Z. Naturforsch. B* **6**, 225 (1951).
1912. G. Wittig, H. G. Reppe, and T. Eicher, *Justus Liebigs Ann. Chem.* **643**, 47 (1961).
1913. G. Wittig and A. Rückert, *Justus Liebigs Ann. Chem.* **566**, 101 (1950).
1914. G. Wittig and W. Stilz, *Justus Liebigs Ann. Chem.* **598**, 85 (1956).
1915. G. Wittig and D. Wittenberg, *Justus Liebigs Ann. Chem.* **606**, 1 (1957).
1916. C. J. Wolf and R. H. Toeniskoetter, *J. Phys. Chem.* **66**, 1526 (1962).
1917. S. Wolfe, M. Nussim, Y. Mazur, and F. Sondheimer, *J. Org. Chem.* **24**, 1034 (1959).
1918. J. E. Wollrab, E. A. Rinehart, P. B. Reinhart, and P. R. Reed, *J. Chem. Phys.* **55**, 1998 (1971).
1919. W. S. Wood, *Chem. Ind.* (*London*) p. 136 (1959).
1920. W. G. Woods and I. S. Bengelsdorf, *J. Org. Chem.* **31**, 2769 (1966).
1921. W. G. Woods, I. S. Bengelsdorf, and D. L. Hunter, *J. Org. Chem.* **31**, 2766 (1966).
1922. W. G. Woods and A. L. McCloskey, *Inorg. Chem.* **2**, 861 (1963).
1923. W. G. Woods and P. L. Strong, *J. Organometal. Chem.* **7**, 371 (1967).
1923a. W. G. Woods and P. L. Strong, *J. Amer. Chem. Soc.* **88**, 4667 (1966).
1924. L. A. Woodward, J. R. Hall, R. N. Dixon, and N. Sheppard, *Spectrochim. Acta* **15**, 249 (1959).
1925. D. L. Yabroff and G. E. K. Branch, *J. Amer. Chem. Soc.* **55**, 1663 (1933).
1926. D. L. Yabroff, G. E. K. Branch, and B. Bettman, *J. Amer. Chem. Soc.* **56**, 1850 (1934).
1926a. H. L. Yale, *J. Heterocycl. Chem.* **8**, 205 (1971).
1927. H. L. Yale, F. H. Bergein, F. A. Sowinski, J. Bernstein, and J. Fried, *J. Amer. Chem. Soc.* **84**, 688 (1962).
1928. Z. Yoshida, T. Ogushi, O. Manabe, and H. Hiyamra, *Tetrahedron Lett.* p. 753 (1965).

1929. D. E. Young and S. G. Shore, *J. Amer. Chem. Soc.* **91**, 3497 (1969).
1930. A. M. Yurkevich and O. N. Shevtsova, *Zh. Obshch. Khim.* **42**, 1993 (1972).
1931. L. I. Zakharkin and A. I. Kovredov, *Izv. Akad. Nauk SSSR, Otd. Khim. Nauk* p. 357 (1962).
1932. L. I. Zakharkin and A. I. Kovredov, *Izv. Akad. Nauk SSSR, Otd. Khim. Nauk* p. 2247 (1962).
1933. L. I. Zakharkin and A. I. Kovredov, *Zh. Obshch. Khim.* **32**, 1421 (1962).
1934. L. I. Zakharkin and A. I. Kovredov, *Izv. Akad. Nauk SSSR, Otd. Khim. Nauk* p. 362 (1962).
1935. L. I. Zakharkin and A. I. Kovredov, *Izv. Akad. Nauk SSSR, Ser. Khim.* p. 393 (1964).
1936. L. I. Zakharkin and A. I. Kovredov, *Izv. Akad. Nauk SSSR, Otd. Khim. Nauk* p. 1564 (1962).
1937. L. I. Zakharkin and O. Y. Okhlobystin, *Izv. Akad. Nauk SSSR, Otd. Khim. Nauk* p. 1135 (1959).
1938. L. I. Zakharkin and O. Y. Okhlobystin, *Dokl. Akad. Nauk SSSR* **116**, 236 (1957).
1939. L. I. Zakharkin and O. Y. Okhlobystin, *Zh. Obshch. Khim.* **30**, 2134 (1960).
1940. L. I. Zakharkin and O. Y. Okhlobystin, *Zh. Obshch. Khim.* **31**, 3662 (1961).
1941. L. I. Zakharkin and O. Y. Okhlobystin, *Izv. Akad. Nauk SSSR, Otd. Khim. Nauk* p. 181 (1959).
1942. L. I. Zakharkin, O. Y. Okhlobystin, and B. N. Strunin, *Izv. Akad. Nauk SSSR, Otd. Khim. Nauk* p. 2002 (1962).
1943. L. I. Zakharkin, O. Y. Okhlobystin, and B. N. Strunin, *Dokl. Akad. Nauk SSSR* **144**, 1299 (1962).
1944. L. I. Zakharkin and V. I. Stanko, *Izv. Akad. Nauk SSSR, Otd. Khim. Nauk* p. 1896 (1960).
1945. V. S. Zavgorodnii and A. A. Petrov, *Zh. Obshch. Khim.* **31**, 2433 (1961).
1946. P. Zdunneck and K. H. Thiele, *J. Organometal. Chem.* **22**, 659 (1970).
1947. M. Zeldin, *J. Inorg. Nucl. Chem.* **33**, 1179 (1971).
1948. M. Zeldin, A. R. Gatti, and T. Wartik, *J. Amer. Chem. Soc.* **89**, 4217 (1967).
1949. M. Zeldin and A. Rosen, *J. Organometal Chem.* **34**, 259 (1972).
1950. M. Zeldin and T. Wartik, *Inorg. Chem.* **4**, 1372 (1965).
1951. M. Zeldin and T. Wartik, *J. Amer. Chem. Soc.* **88**, 1336 (1966).
1952. M. J. Zetlmeisl and L. J. Malone, *Inorg. Chem.* **11**, 1245 (1972).
1953. A. F. Zhigach, E. B. Kazakova, and E. S. Krongauz, *Dokl. Akad. Nauk SSSR* **111**, 1029 (1957).
1954. A. F. Zhigach, V. N. Siryatskaya, I. S. Antonov, and S. Z. Makaeva, *Zh. Obshch. Khim.* **30**, 227 (1960).
1954a. A. F. Zhigach, R. A. Svitsyn, E. S. Sobolev, and I. V. Persianova, *Dokl. Akad. Nauk SSSR* **196**, 1349 (1971).
1954b. S. F. Zhil'tsov, A. A. Koksharova, and G. G. Petukhov, *Zh. Obshch. Khim.* **41**, 1060 (1971).
1955. K. Ziegler and H. Hoberg, *Angew. Chem.* **73**, 577 (1961).
1956. K. Ziegler and O. W. Steudel, *Justus Liebigs Ann. Chem.* **652**, 1 (1962).
1957. H. Zimmer and G. Singh, *Advan. Chem. Ser.* **42**, 17 (1964).
1958. H. K. Zimmerman, *J. Org. Chem.* **26**, 5214 (1961).
1959. H. K. Zimmerman, *Advan. Chem. Ser.* **42**, 23 (1964).
1960. N. L. Zutty and F. J. Welch, *J. Polym. Sci.* **43**, 445 (1960).
1961. N. L. Zutty and F. J. Welch, *J. Org. Chem.* **25**, 861 (1960).
1961a. Z. V. Zvonkova, *Kristallografiya* **2**, 408 (1957).
1961b. Z. V. Zvonkova and V. P. Glushkova, *Kristallografiya* **3**, 559 (1958).

1962. G. Zweifel, *J. Organometal. Chem.* **9**, 215 (1967).
1962a. G. Zweifel and H. Arzoumanian, *Tetrahedron Lett.* p. 2535 (1966).
1963. G. Zweifel and H. Arzoumanian, *J. Amer. Chem. Soc.* **89**, 291 (1967).
1964. G. Zweifel, H. Arzoumanian, and C. C. Whitney, *J. Amer. Chem. Soc.* **89**, 3652 (1967).
1965. G. Zweifel, N. R. Ayyangar, and H. C. Brown, *J. Amer. Chem. Soc.* **84**, 4342 (1962).
1966. G. Zweifel, N. R. Ayyangar, and H. C. Brown, *J. Amer. Chem. Soc.* **85**, 2072 (1963).
1967. G. Zweifel, N. R. Ayyangar, T. Munekata, and H. C. Brown, *J. Amer. Chem. Soc.* **86**, 1076 (1964).
1968. G. Zweifel and H. C. Brown, *J. Amer. Chem. Soc.* **85**, 2066 (1963).
1969. G. Zweifel and H. C. Brown, *J. Amer. Chem. Soc.* **86**, 393 (1964).
1970. G. Zweifel, G. M. Clark, and N. L. Polston, *J. Amer. Chem. Soc.* **93**, 3395 (1971).
1970a. G. Zweifel and R. P. Fisher, *Synthesis* p. 557 (1972).
1971. G. Zweifel, R. P. Fisher, J. T. Snow, and C. C. Whitney, *J. Amer. Chem. Soc.* **93**, 6309 (1971).
1972. G. Zweifel, A. Hornig, and J. T. Snow, *J. Amer. Chem. Soc.* **92**, 1427 (1970).
1973. G. Zweifel, K. Nagase, and H. C. Brown, *J. Amer. Chem. Soc.* **84**, 183 (1962).
1974. G. Zweifel, K. Nagase, and H. C. Brown, *J. Amer. Chem. Soc.* **84**, 190 (1962).
1975. G. Zweifel and N. L. Polston, *J. Amer. Chem. Soc.* **92**, 4068 (1970).
1976. G. Zweifel, N. L. Polston, and C. C. Whitney, *J. Amer. Chem. Soc.* **90**, 6243 (1968).

Author Index

Numbers in parentheses are reference numbers and indicate that an author's work is referred to although his name is not cited in the text. Numbers in italics show the page on which the complete reference is listed.

A

Abel, E. W., 24(4, 6), 49(8), 52(2, 3, 4, 5, 6, 7, 9, 10, 11), 54(5), 55(2, 5), 56(1), 62(6, 9, 11), 63(10), 65(2, 6), 77(1), 79(2, 6), 92(4, 6), 97(4, 7), 102(6), 151(8), 152(8), 153(8), *230*

Abeler, G., 55(1463), *271*

Abley, P., 144(12), *230*

Abraham, M. H., 47(14), 61(14), 97(14), 99(14), 100(13), 107(15), *230*

Abrahams, S. C., 138(45), *231*

Abramson, E. W., 224(588b), *246*

Abufhele, M., 108(16), *230*

Abuin, E., 111(17), 224(18), *230*

Adams, R. M., 1(18a, 19, 20), 25(153), 27(21), 28(153), 29(21, 152, 153), 33(21, 153), 46(21, 153), 196(152), *225, 228, 230, 234*

Adcock, J. L., 21(22), 22(1217), 77(22, 1217), 78(22, 1217), *230, 262*

Aflandilian, V. D., 83(23), 137(23), 199(23), 209(23), *230*

Ahluwalia, J. C., 138(1382), *268*

Ainley, A. D., 92(24), 94(24), 100(24), 102(24), 109(24), 110(24), 133(24), 222(24), *230*

Airey, W., 159(885), 160(885), 212(885), *254*

Akhhazaryan, A. A., 50(1240), 52(1240), 53(1240), 54(1240), 92(1240), 165(1240), 175(1240), 176(1240), 182(1240), 185(1240), *263*

Akhrem, A. A., 115(24a), *230*

Albright, C. F., 21(1850, 1851), 23(1850, 1851), 51(1850), 57(1850, 1851), 62(1850), 79(1852), 92(1850), *281 282*

Alford, K. J., 154(25), *231*

Allerhand, A., 47(26), 49(26), *231*

Allies, P. G., 79(28), 98(27, 28), 109(1175), 110(1175), *231, 261*

Allred, E. L., 113(29), *231*

Altwicker, E. R., 196(30), 197(31, 1630), 201(31, 1630), *231, 275*

Amberger, E., 57(32), *231*

Amen, K.-L., 92(940), 97(940), *256*

Andersen, C., 108(16), *230*

Anderson, C. L., 113(29), *231*

Anderson, H. H., 18(33), *231*

Anderson, H. W., 182(1389b), *269*

Ang, T. T., 148(34), *231*

Anisimov, K. N., 89(1419b), 90(1419b), *269*

Annino, R., 223(47), *231*

Antonov, I. S., 28(1954), 47(1954), 48(1954), *284*

Antonov, V. K., 224(34a), *231*

Anzenhofer, K., 4(34b), *231*

Appel, R., 51(35), 92(35), 151(35), *231*

Arase, A., 115(1752), 117(1752), 118(876), *254, 279*

Archer, J. F., 145(36), *231*

Armitage, D. A., 56(1), 77(1), *230*

Armour, A. G., 100(1007), *257*

Armstrong, D. R., 5(39), 12(37, 44), 13(38), 16(39, 41, 43, 44), 157(40), 163(42), *231*
Arnott, S., 138(45), *231*
Aronovich, P. M., 21(1241, 1242, 1244), 23(1244), 29(1248), 45(1247), 50(1245), 51(1245), 52(1241, 1242, 1244, 1245), 53(1241), 54(1244), 55(1244), 59(1246), 60(1246), 62(1242), 77(1243), 80(1247), 83(46, 140), 92(1245), 137(46), 140(1242), 219(1246), *231, 233, 263*
Arora, S. K., 31(1535, 1536), 57(1535), 61(1535), 1536), 80(1535, 1536), 81(1535), 1536), 90(1536), 102(1532), 108(1536), 113(1536), 124(1536), 182, 219(1536), *235, 273*
Arsene, A., 62(64), 86(64), 92(64), 140(64), *232*
Arthur, P., 223(47), *231*
Arzoumanian, H., 45(1962a, 1963), 104(1964), *285*
Ashboy, E. C., 21(49, 100, 781), 22(48, 781), 24(48, 780, 1205), 25(1205), 29(48), 33(48), 34(48), 47(48, 781), 48(48, 781), 49(48), 50(100), 55(1201, 1205), 56(1201), 64(1202), 65(1202), 79(48, 780, 1201), 111(48, 780, 781), 149(49), 150(49), 151(49), 173(48, 781), *231, 232, 251, 262*
Ashe, A. J., 7(50), 22(50), 67(50), 216(50), *231*
Ashikari, N., 92(54), 224(51, 52, 53, 54, 55), *231*
Ashraf El-Bayoumi, M., 13(1587), *274*
Atkinson, I. B., 78(55a), *225, 231*
Aubrey, D. W., 12(56), 59(56), 79(56), *231*
Auten, R. W., 40(57), 47(57), 109(57), *231*
Axelrad, G., 22(58), 67(58), 216(58), *231*
Axelson, D. E., 10(58a), *231*
Aylett, B. J., 148(59, 60), 225, 231
Ayyangar, N. R., 27(1966), 28(185, 186, 187, 1965, 1966, 1967), 29(186, 187), 32(185, 186, 187, 1965, 1967), 49(186), 170(186, 1966), 173(187), 177(186, 187), 181(185, 186, 1966), 182(185, 186, 187, 1965, 1967), *235, 285*

B

Babb, B. E., 13(561), 49(561), 145(698, 702), *245, 249*
Badin, E. J., 97(175), *234*
Baechle, H. T., 8(61), 12(102), 49(61, 102), 50(61, 62), 51(61, 101, 102), 56(61), *231, 232*
Baker, C. S. L., 28(63), 29(63), *232*
Baker, E. D., 40(1526), 41(1526), 125(1526), 153(1526), *272*
Balaban, A. T., 62(64), 66(85), 86(64, 66), 92(64), 140(64, 66), *232*
Balasubramaniyan, P., 15(1533), 28(1533, 1538), 29(1538), 33(1533), 153(1533), *273*
Balasubramaniyan, V., 15(1534), 31(1544), 165(1534), 166(1534, 1544), 186(1534, 1544), *273*
Bally, I., 62(64), 86(64, 66), 92(64), 140(64, 66), *232*
Balthis, J. H., 210(900), 212(900), *254*
Baly, N. J., 207(1900), 208(1900), 209(1900), *283*
Bamford, C. H., 97(67), 99(67), *232*
Bamford, W. R., 22(68), 56(68), 57(68), 62(68), 63(68), 80(68), 81(68), *232*
Banford, L., 13(69), 158(69), *232*
Banks, H. D., 106(1686), *277*
Banus, J., 49(350), 53(350), *239*
Barabous, A., 62(64), 86(64), 92(64), 140(64), *232*
Barbaras, G. K., 148(188, 189), 153(188, 189), *235*
Barfield, P. A., 8(71, 72), 9(72), 50(71), 56(70, 71), 57(70, 71), 80(71), *232*
Barnard, A. J., 223, *225, 226*
Barnes, D., 62(73), 63(73), *232*
Barnes, R. L., 22(74), 42(74), 54(74), 64(74), 92(74), *232*
Barredo, J. M. G., 12(1602), 176(1602), *274*

Bartell, L. S., 3(377), 4(75), 5(377), 6(76), 190(377), *232, 240*
Bartholomay, H., 46(190), 148(190), 153(190), *235*
Bartocha, B., 22(77, 78), 25(79), 28(79), 62(77, 79), 64(78), 97(77), 113(79), 124(79), 125(79), 151(79), 152(79), 223(77), *232*
Barton, L., 100(80), *232*
Baryshnikova, T. K., 21(1249a), 60(1249a), 66(1249), 67(1249), 74(1249), 115(1249), 116(1249), *263*
Bauer, R., 21(85), 136(84), 138(84, 85), 174(84), *232*
Bauer, S. H., 3(87), 4(86, 87, 425), 160(83, 85a), 162(82), 163(1200, 1694), 164(1200), 187(1200), *225, 232, 241, 262, 277*
Bawn, C. E. H., 224(89), *232*
Bazhenova, A. V., 21(1250), 63(1250), 65(1250), 79(1250), *263*
Beach, J. Y., 4(86), *232*
Beachell, H. C., 7(90), 46(91), 58(90), *232*
Beachley, O. T., Jr., 78(92), *232*
Bean, F. R., 21(93), 57(93), 58(93), 59(93), *232*
Beaudet, R. A., 5(403), 62(403), 203(426), 204(426), *241*
Becher, H. J., 6(103), 8(61), 11(95, 105), 12(95, 98, 99, 102, 105), 49(61, 94, 96, 97, 99, 102, 105), 50(61, 62), 51(61, 102), 53(96, 98), 54(95), 56(61), 62(98), 64(95), 77(103, 104), 148(94), 151(96), 219(103), *231, 232*
Becher, J. H., 11(675), 12(675), *248*
Becker, P., 20(1236), 22(1236), 57(1235), 62(1235), 65(1235, 1236), 79(1235), 106(1236), *263*
Beistel, D. W., 7(90), 58(90), *232*
Bekasova, N. I., 56(922), 58(923), 59(923), 77(922, 922a), 78(922), 148(922), 185(922, 923), *229, 255*
Belcher, R., 223(106), *233*
Bell, R. P., *225*
Bell, T. N., 46(106a), 108(106a), *233*
Bellama, J. M., 22(1606), 39(1607), 62(1606), 64(1606, 1607), 82(1607), 92(1606), 221(1606), *274*
Bellamy, L. J., 12(107), 209(107), *233*
Bellut, H., 24(943), 29(943), 46(942, 943), 50(942, 943, 944), 51(943), 66(943), 69(943), 84(108), 92(108, 940), 97(940), 109(941, 944), 133(941), 174(942), 185(942), *233, 256*
Belyavskaya, E. M., 29(1689), 54(1689), 63(1689), *277*
Benedict, J., 22(1802), 55(1802), 65(1802), *280*
Benedikt, G., 21(946, 947), 22(948), 24(945), 38(947), 41(946), 46(947), 58(948), 66(945), 68(947), 69(945, 946, 947), 74(949), 82(946), 109(941, 946, 949, 950), 133(941), 168(945), 176(945, 949), 178(947, 949), 180, 187(948), *227, 256*
Bengelsdorf, I. S., 21(1921), 59(1920, 1921), 60(1920), 61(1920), 88(1920), 219(1920), *283*
Benjamin, L. E., 58(1009), 59(1008), 104(1008), 222(1009), *257*
Bennett, J. E., 32(109), *233*
Benson, R. E., 83(830), *252*
Benson, S. W., 160(644), 187(644), *247*
Bent, H. E., 131(111, 112, 433), 132(110, 111, 112), *233, 241*
Bentz, A. P., 77(1693), *277*
Berezin, I. V., 224(34a), *231*
Bergein, F. H., 56(1927), 57(1927), 71(1927), *283*
Bernstein, J., 56(1927), 57(1927), 71(1927), *283*
Berschied, J. R., Jr., 10(114), 12(113, 114), 83(1840), 137(1840), 138(1840), *233, 281*
Bertini, I., 3(114a), *233*
Bessler, E., 21(115), 22(115), 49(115), 55(115), 152(115), *233*
Bethke, G. W., 12(116), 162(116, 117), *233*
Bettman, B., 5(119, 151), 57(119, 1926), 58(118, 119), 59(119, 151, 1926), 218(119, 1926), *233, 234, 283*
Beyer, H., 23(1433), 24(1432, 1433),

49(1433), 50(62, 1433), 51(120, 1432, 1433), 55(1432), 56(120, 1432), 57(1433), 115(120), 116(120), 138(1464), *231, 233, 270, 271*

Bezmenov, A. Ya., 66(1252, 1350), 68(1252), 69(1252), 72(1350), 80(1350), 92(1251), 93(1251), *263, 267*

Bhatt, M. V., 29(193), 111(191, 192 193), 119(191, 192, 193), *235*

Bhattacharjee, M. K., 72(532), 73(532), *244*

Biallas, M. J., 39(122), 66(1697), 81(122), 89(1697), 169(122), *233, 277*

Bianchi, G., 86(121), 92(121), 220(121), *233*

Bickelhauph, F., 38(1835a), 68(1834, 1835), 69(1835), 216(1834, 1835), *281*

Bigley, D. B., 28(195), 53(305), 60(305), 61(305), 92(125), 93(125), 102(123, 124), 148(305), 165(305), 169(305), 175(305), 182(194, 195, 196), *233, 235, 238*

Billig, F. A., 21(1850, 1851), 23(1850, 1851), 51(1850), 57(1850, 1851), 62(1850), 79(1852), 92(1850), *281, 282*

Binger, P., 21(959, 962), 22(129), 24(130), 25(130, 959), 28(129, 130), 959), 45(130), 46(131, 959), 47(130, 131a, 952), 48(952, 959), 49(959), 50(952), 52(131a), 66(959), 67(959), 68(128), 69(959), 75(127), 77(952), 81(126, 131), 92(952), 109(129), 113(129), 114(129, 130), 124(129), 131(952), 137(131), 138(131), 139(131), 141(131), 142(127), 143(130), 145(130), 148(962), 149(962), 150(962), 151(962), 152(962), 176(952, 961), 177(952), 178(952), 180(952), 182(130, 959), *233, 256*

Birchall, J. M., 28(132), 32(132), 46(132), 47(132), 53(132), 62(132), 113(132), 124(132), 182(132), *233*

Birr, K. H., 22(742), 50(742), 55(742), 56(742), 65(742), 81(742), *250*

Bishop, E. O., 154(25), *231*

Bitsenko, M. I., 224(1713a), *278*

Bittman, R., 31(1746), 32(1746), 182(1746), *278*

Bittner, G., 71(801), 127(801), 139(801), *252*

Blackborow, J. R., 57(133), 140(133), 152(133), *233*

Blais, J., 21(1239), 22(1239), 62(1239), *263*

Blau, E. J., 12(134), *233*

Blay, N. J., 85(575), 196(135), 199(137), 201(135), 204(137), 207(135, 573, 575), 208(135, 136, 137, 573, 575), 209(135, 139, 573, 575), *233, 245*

Blinova, V. A., 89(1420a), *270*

Bliss, A. D., 131(890), 174(890), *254*

Block, F., 148(1212), *262*

Blofeld, R. E., 21(1576), 77(1576), 78(1576), *274*

Blokhina, A. N., 28(1254), 29(1331, 1333), 31(1254), 46(1253, 1254), 50(1253), 51(1254), 52(1253, 1254), 54(1253, 1331), 56(1254), 59(1253, 1254), 63(1331), 66(1333), 67(1333), 92(1253), 114(1254), *263, 267*

Bloom, M., 21(1850), 23(1850), 51(1850), 57(1850), 62(1850), 92(1850), *281*

Blundell, D. C., 78(55a), *231*

Blundell, T. L., 3(138), *233*

Bochkareva, M. N., 48(140a), 71(1286a), 134(140a), *233, 265*

Bock, H., 49(139), 55(139), 72(139), *233*

Bogdanov, V. S., 10(141, 141a, 141b), 21(1249a), 24(331), 28(331, 331a), 46(331, 331a), 48(140a), 51(1836b, 1836c), 52(331), 53(1836b, 1836c), 55(1836c), 59(331a), 60(1249a), 61(332), 62(332), 66(1274), 67(332), 71(554b, 1283, 1836b), 74(1836c), 79(332), 83(46, 140), 85(1283), 92(331), 102(1836b), 115(330, 332, 1274), 116(332, 1274), 134(140a, 142, 1255),

137(46), 156(141c, 1836b), 219(330), *231, 233, 234, 238, 245, 263, 264, 265, 281*
Bohme, H., 146(143), 151(143), *234*
Bohn, G. T., 100(80), *232*
Boileau, S., 138(1766a), *279*
Boldyreva, O. G., 29(554), 51(1696a), 52(1696a), 53(1696a), 71(554, 554b, 1286a), 75(554a), 76(554a), *245*, 265, *277*
Boll, E., 146(143), 151(143), *234*
Bolz, B., *228*
Bond, A. C., 138(1647), *276*
Boone, J. L., 22(145), 46(351), 49(351), 84(145), 181(351), *234, 239*
Booth, R. B., 50(144), 55(144), 64(144), 91(144), 92(144), 109(144), *234*
Borch, R. F., 145(146), *234*
Borden, D. G., 144(1892), *282*
Bose, R. J., 13(147), *234*
Bossa, M., 16(148), *234*
Boulton, A. H., 56(380), 71(380), 127(380), 219(380), *240*
Bowen, R. E., 195(149), *234*
Bowie, R. A., 43(150), 61(1176), 109(1176), *234, 261*
Boyd, D. R., 138(720), 145(36, 720), *231, 249*
Bradford, J. L., 24(1843), 77(1842), 78(1842), 92(1842), *281*
Bradley, D. C., *225*
Bradt, J., 88(1844), 89(1844), *281*
Branch, G. E. K., 5(119, 151, 417), 57(119, 1925, 1926), 58(118, 119, 417), 59(119, 151, 1926), 218(119, 1926), *233, 234, 241, 283*
Bratton, R. F., 83(1732), 92(1732), *278*
Brattsev, V. A., *228*
Braun, B. H., 29(746), 33(746), *250*
Braun, J., 21(1455, 1456), 56(1455), 57(426b), 58(1455), 59(1455), 60(426b, 1455), 61(426b, 1455), 63(1455), 64(426b), 100(1455), 219(426a, 426b), *241, 271*
Braun, L. M., 29(152), 196(152), *234*
Braun, R. A., 25(153), 28(153), 29(152, 153), 33(153), 46(153), 196(152), *234*
Braye, E. H., 21(154), 66(154), 220(154), *234*
Brehm, E., 76(158), 139(157), 143(158), *234*
Brennan, G. L., 22(159), 77(159), 194(155), 195(155), *234*
Brennan, J. P., 50(815), 51(815), 55(815), *252*
Bresadola, S., 43(156), 71(160), 72(156, 160), 127(160), *234*
Breuer, E., 106(27), 133(217), 170(162), 175(162), 177(162), *234, 235*
Breuer, S. W., 21(161, 163), 22(164), 23(161), 79(163a), *234*
Brey, W. S., Jr., 8(165, 1627, 1663), 9(1663), 16(392), 21(392), 28(366), 35(366), 49(165), 50(62, 165), 51(165), 71(366), 79(1663), 83(366), 219(392), *231, 234, 239, 240, 275, 276*
Brice, V. T., 199(167), 202(167), 206(166, 863, 863a), *234, 253*
Bridson, J. N., 127(833a), *252*
Brieger, G., 21(1833), 68(1833), 214(1833), *281*
Bright, J. H., 24(1203), 79(1203), 111(1209), *262*
Brilkina, T. G., 138(1596), 139(1596, 1597), 143(1597), 144(1597), 145(1597), 146(1596, 1597), 151(1596, 1597), *274*
Brinckman, F. E., 20(169), 22(77, 168, 169), 23(169), 25(169), 53(169), 55(169), 62(77, 168, 169), 64(168, 169), 94(169), 97(77), 140(168), 152(168), 223(77), *232, 234*
Brinckman, R. D., 83(1003), *257*
Brindley, P. B., 21(171), 61(171), 62(171, 172, 173), 63(172, 173), 79(28), 91(172), 98(27, 28, 152(173), 224(174), *231, 234*
Brockway, L. O., 4(1113), *260*
Brokaw, R. S., 28(176, 177), 30(176), 43(176), 224(177), *234*
Brooks, C. J. W., 223(177a), 224(178), *234*
Broster, F. A., 21(163), 79(163a), *234*
Brotherton, R. J., 44(179), 57(179),

77(180), 80(179), 152(179), 218(180), *225, 228, 234*
Brown, D. A., 15(182), 30(181), *235*
Brown, D. C., 25(153), 28(153), 29(153), 33(153), 46(153), *234*
Brown, H. C., 6(299, 320), 9(184), 21(183, 203), 24(263, 264, 304), 25(208, 211), 26, 27(242, 243, 244, 290, 297, 306, 317, 318, 321, 893, 1966, 1968, 1969), 28(185, 186, 187, 195, 204, 224, 226, 229, 240, 242, 244, 248, 260, 261, 262, 268, 269, 275, 275a, 275b, 275c, 276, 290, 294, 296, 306, 314, 315, 317, 318, 320, 321, 322, 893, 894, 1418, 1965, 1967, 1968, 1969), 29(186, 187, 193, 200, 209, 214, 223, 224, 227, 243, 245, 248, 253, 257, 260, 264, 265, 266, 271, 288, 292, 293, 294, 296, 297, 304, 306, 314, 317, 319, 321, 323, 325, 892, 1418, 1969), 30(242, 312, 319, 322, 1969), 31(200, 224, 225, 227, 288, 289, 306, 319), 32(185, 186, 187, 1965, 1967), 33(214, 244, 268, 275, 315, 321, 323, 892, 893, 894, 1968, 1973, 1974), 38(247), 43(279), 44(1593), 45(323), 46(183, 190, 309, 1419), 47(183), 48(212, 298), 49(186, 203, 233, 243, 294, 298, 299, 320), 50(309, 311), 52(309), 53(203, 211, 226, 243, 275, 304, 305), 54(309), 55(275), 56(214), 57(214), 60(208, 209, 214, 226, 268, 275c, 305), 61(208, 209, 210, 214, 268, 304, 305), 62(209, 210, 212, 268, 1417), 64(234), 65(234), 66(259, 262, 264, 265, 266, 267, 269, 360a, 892, 1035, 1418), 67(247, 257, 260, 261, 262, 263, 265, 267, 269, 360a, 1418), 79(1593), 80(263, 264), 91(184b, 245, 246, 323), 92(203, 245, 246, 263, 266), 93(203, 245, 246, 263, 266), 95(250, 251, 283), 96(245), 97(203, 238, 239, 259, 281, 1238, 1238a, 1754), 98(237, 1238), 99(203, 237), 100(184a, 184b, 235, 293, 294, 320, 322, 1969), 101(323), 102(236, 1034), 103(233a, 309, 311a, 1034, 1035, 1036), 104(214a), 105(232, 233, 274), 106(217), 107(204), 108(219, 220, 233a, 236, 309, 310, 311a, 1034), 109(215, 291, 308, 339, 1052, 1053, 1054), 110(339, 1053, 1054), 111(191, 192, 193, 243, 295, 296, 298, 316, 324, 325, 892, 1418, 1968), 112(324, 325), 114(276, 277), 115(219, 220, 222, 238, 273, 282, 1377, 1752, 1753a, 1752), 118(219, 220, 258, 876, 1755), 117(222, 273, 282, 1377, 1756), 119(191, 192, 193, 295, 298, 1037), 120(295, 298), 122(221, 253, 254, 255, 256, 257, 267, 270, 271, 894, 1417), 123(198, 199, 201, 228, 255, 270, 271, 272, 1416, 1594), 124(285), 126(249, 250, 251, 252, 278, 279, 280, 281, 283, 284, 285, 311, 1410, 1411), 127(196a, 197, 233a, 241, 311a, 311b, 833a), 128(240, 241a, 1112a, 1758), 129(1757), 130(1419a), 131(203), 133(217, 307, 1595), 136(287, 1646), 138(202, 230, 231, 231a, 259, 287, 1647), 139(1646), 145(202, 230, 231a), 146(1752a), 148(183, 188, 189, 190, 205, 206, 207, 212, 218, 226, 268, 286, 300, 301, 302, 303, 305), 149(183, 210, 213, 226), 150(183, 299), 151(309), 153(184, 184b, 188, 189, 190, 206, 207, 210, 218, 229, 286, 299, 300, 301, 302, 303), 156(268), 165(304, 305), 166(263, 264, 265, 1418), 169(184a, 184b, 294, 297, 304, 305), 170(162, 186, 226, 242, 243, 259, 265, 296, 297, 319, 322, 1966, 1968, 1969, 1973), 171(260, 261), 172(226, 260, 892), 173(187, 226, 243, 298, 1646, 1648), 174(212), 175(162, 304, 305), 176(213, 259, 263, 264, 1648), 177(162, 186, 187, 213, 226, 243, 259, 260, 261, 263, 892, 1418, 1419, 1968), 178(1418), 179(184a, 892), 180(259, 261, 1418), 181, 182(185, 186, 187, 194, 195, 196, 216, 242, 244, 262, 269, 307, 320, 1965, 1967,

1969), 186(226, 1419), 215(261, 360a, 1418), 218(1210), 223(319), *225, 228, 229, 234, 235, 236, 237, 238, 239, 252, 254, 258, 260, 262, 263, 268, 269, 274, 276, 279, 285*
Brown, M. P., 80(325a), 135(325a), *238*
Brown, R. D., 92(327), 93(326), 104(327), 218(327), *238*
Brubaker, G. L., 206(863), *253*
Bruce, R. L., 59(328), 104(328), *238*
Brüser, W., 83(329), *238*
Bruno, G., 22(951), 46(951), 47(951, 952), 48(951, 952), 49(951), 50(952), 53(951), 77(952), 92(952), 109(951), 131(952), 132(951), 148(952), 150(951), 176(952), 177(952), 178(952), 180(952), *256*
Bryantsev, B. I., 66(1255a), 67(1255a), *263*
Bubnov, Y. N., 16(639, 640), 18(1332), 24(331), 28(331, 331a), 29(330), 46(331, 331a), 47(333, 334, 336, 1256, 1273, 1275), 48(140a), 49(335a, 1273, 1275), 50(1257, 1258, 1259, 1260, 1261, 1264, 1265, 1361), 51(1261, 1264), 52(331, 1258, 1261, 1263, 1361), 53(1263, 1264, 1267, 1361), 54(1256, 1257, 1258, 1263, 1361), 59(331, 331a), 60(628, 1256, 1267, 1275), 61(332, 1271), 62(332, 628), 63(1266), 66(333, 334, 334a, 1268, 1270, 1271a, 1273a, 1274), 67(332, 333, 334, 628, 1271), 77(1265), 79(332, 334, 628), 83(330, 335, 1256), 86(1262), 91(334b), 92(331, 335a, 336, 1259, 1260, 1261, 1262, 1263, 1264, 1265, 1266, 1271, 1273, 1361), 96(334b, 1261, 1263, 1264, 1266), 107(1257, 1258), 108(1263, 1266), 113(1269a), 114(1269a), 115(24a, 330, 332, 333, 334, 335, 336, 628, 1268, 1269, 1270, 1271, 1271a, 1273, 1273a, 1274, 1275, 1275a, 1275b), 116(332, 333, 334, 334a, 628, 1267, 1268, 1269, 1269a, 1270, 1271, 1273, 1273a, 1274, 1275, 1275a), 117(336), 121(1272, 1275), 134(140a, 142), 148(1256), 149(1260), 150(1260), 208(336), 219(330, 335), *230, 233, 234, 238, 239, 247, 264, 267, 268*
Buchanan, A. S., 92(327), 93(326), 104(327), 135(337), 218(327), *238, 239*
Bucourt, R., 22(338), 46(338), 47(338), *239*
Buda, L., 138(687), 223(687), *249*
Budde, W. L., 148(1845), 149(1845), 154(769, 770, 1845), *251, 281*
Buechl, H., 223, *225*
Bugge, A., 76(703), 88(703), *249*
Buhler, J. D., 109(339), 110(339), *239*
Bujwid, Z. J., 36(340), *239*
Bullen, G. J., 3(341, 342), *239*
Buls, V. L., 24(343), 53(343), 55(343), 62(343), 63(343), 64(343), *239*
Bulten, E. J., 140(343a), *239*
Burch, J. E., 12(344), 22(346), 56(347), 62(344), 65(346), 77(345, 347), *239*
Burg, A. B., 4(671, 672, 673), 20 (358, 1653), 22(358, 1653), 24(353, 1650, 1652), 25(348), 40(1526), 41(1526), 46(351, 359, 1653), 49(350, 351, 1651), 51(348, 359), 53(348, 350, 358, 359), 54(353, 360), 55(1651), 56(348), 62(348), 63(353), 76(1651), 77(1649, 1651, 1652), 78(1651, 1652), 79(348), 88(352), 125(1526), 148(359, 1649), 149(359, 1649), 151(354, 1649), 152(354), 153(354, 1526, 1649), 155(359, 1649), 159(356, 1731), 160(349, 356, 1731), 162(671, 672, 673), 163(349, 673), 164(1651), 165(1649), 170(1731), 175(360, 1650), 176(1649, 1650), 181(351), 183(352, 1649, 1651), 184(359, 1649, 1651), 189(352), 194(357, 1731), 195(357, 1731), 196(1731), 198(355), 199(355), 201(355), *226, 228, 239, 248, 272, 276, 278*
Burke, P. L., 24(263, 264), 28(1418), 29(264, 265, 1418), 66(264, 265, 360a, 1418), 67(263, 265, 360a, 1418), 80(263, 264), 92(263), 93(263), 111(1418), 166(263, 264,

265, 1418), 176(263, 264, 265), 177(263, 1418), 178(1418), 180(1418), 215(360a, 1418), *237, 239, 269*
Bush, R. P., 56(1), 77(1), *230*
Butcher, F. K., 12(361, 362), *239*
Butler, D. N., 27(363), 56(363), 59(363), 211(745a), 212(745a), *239, 250*
Butler, G. B., 28(364, 365, 366), 35(364, 365, 366), 71(365, 366), 72(365), 75(364), 83(366), *239*
Buzas, I., 138(886), 212(886), *254*

C

Cabana, A., 138(868), 159(1027a), *253, 258*
Cadiot, P., 21(667, 1729), 22(667, 1729), 52(667, 1729), 57(426b), 60(426b), 61(426b), 64(426b), 65(1729), 149(667, 1729), 150(667, 1729), 151(667, 1729), 219(426b), 220(1729), *241, 248, 278*
Caglioti, L., 28(367), 92(367), *239*
Cainelli, G., 21(367a), 28(367), 45(367a), 92(367), *239, 240*
Calderon, J. L., 140(368), *240*
Callaghan, J., 208(641), *247*
Calzada, J. G., 127(833a), *252*
Campbell, G. W., 88(352, 370), 183(352), 189(352, 369, 370), *239, 240*
Campbell, J. D., 56(1191), 76(1191), 220(1191), *262*
Carbo, R., 71(661a), *248*
Cardin, A. D., 10(580), 204(580), *246*
Cardon, S. Z., 148(286), 153(286), *237*
Carey, P. R., 154(25), *231*
Carlson, B. A., 127(196a, 197), *235*
Carotti, A. A., 38(1901), 66(1901), 67(1901), 111(1901), 173(1901), *283*
Carpenter, J. H., 12(371), 190(371, 372, 372), 192(373), *240*
Carpenter, R. A., 224(374, 375), *240*
Carraro, G., 71(160), 72(160), 127(160), *234*
Carrazzoni, N. E., 77(376), *240*
Carroll, B. L., 3(377), 4(75), 5(377), 6(77), 190(377), *232, 240*
Carter, J. C., 71(1762), 138(378, 1525), 161(378, 1525), 194(1525), 196(1525), 208(1732a), *240, 272, 278, 279*
Carter, R. P., Jr., 8(828), 148(828), 153(828), *252*
Casanova, J., 10(382), 20(381), 22(381), 46(381), 56(380), 71(379, 380), 127(379, 380, 382), 136(382), 139(382), 141(382), 147(382), 219(380, 382), *226, 240*
Caserio, F. F., 56(383), *240*
Caspi, E., 30(1836), 182(384, 1836), *240, 281*
Cassaretto, F. P., 10(1832), 12(1832), 21(385, 1389, 1831), 139(385, 1389, 1831), *240, 269, 281*
Castle, R. B., 40(386), 44(386), 57(386), 80(386), 82(386), 92(386), 95(386), *240*
Catlin, J. C., 46(388), 58(387), 63(388), 73(387, 388), 74(387), 75(387), *240*
Caujolle, D., 223(389), *240*
Caujolle, F., 223(389), *240*
Cavallo, J. J., 56(383), *240*
Cernak, E. S., 21(1851), 23(1851), 57(1851), *281*
Ceron, P., 20(390), 22(390), 38(390), 39(390), 80(390), 81(390), *240*
Chaigneau, M., 22(391), 46(391, 729), *240, 250*
Chakrabarty, M. R., 16(392), 21(392), 219(392), *240*
Challenger, F., 92(24), 94(24), 100(24), 102(24), 109(24, 393), 110(24), 133(24), 222(24), *230, 240*
Chamayou, P., 176(1391), 181(1391), *269*
Chambers, C., 38(394, 395), 81(395), 82(394), 92(395), 219(395), *240*
Chambers, R. D., 22(396, 397), 23(399, 400), 51(396), 54(396), 59(396, 397), 63(396, 397), 64(396, 397), 95(397), 125(396), 136(399, 400), 140(398, 399, 400), 147(399, 400), 152(397), *240*
Champion, N. G. S., 77(401), *240*

Chan, T. H., 133(567), *245*
Chapovskii, A. Yu., *228*
Chatt, J., 6(402), *240*
Chen, S.-T., 224(634), *247*
Cheng, T., 22(1177), 45(1177), 56(1177), 59(1177), 61(1177), 80(1177), 222(1177), *261*
Cheng, T. C., 31(1545), *273*
Cherkasova, K. L., 21(1276), 66(1276, 1276a, 1276c), 67(1276c), 68(1276b), 115(1276a), 116(1276a), *264*
Cheung, C. C. S., 5(403), 62(403), *241*
Chin, P. S., 62(1682), 224(1682), *276*
Chissick, S. S., 37(404), 56(405), 72(404), 75(404), *241*
Chivers, T., 21(406), 22(396, 397), 51(396), 54(396), 55(406), 59(396, 397), 60(406), 63(396, 397), 64(396, 397), 92(406), 95(397), 125(396), 152(397), *240, 241*
Chow, J., 31(1535, 1536), 57(1535), 61(1535, 1536), 80(1535, 1536), 81(1535, 1536), 90(1536), 108(1536), 113(1536), 124(1536), 219(1536), *273*
Chremos, G. N., 51(407), 52(407), *241*
Chrétien-Bessière, P., 28(570), *245*
Christensen, J. J., 211(734), *250*
Christopher, P. M., 18(408, 409), *241*
Chu, T. L., 131(413), 132(410, 411, 412, 413), *241*
Chursina, L. M., 56(922), 77(922), 78(922), 148(922), 185(922), *255*
Clapp, D. B., 78(55a), *231*
Clark, B. F., 24(665, 666), 35(665, 666), 46(665, 666), 55(665, 666), 57(665), 64(666), *248*
Clark, G. F., 24(414), 28(414), 33(414), 45(414), 80(414), 92(414), 171(414), *241*
Clark, G. M., 33(1970), 134(415), *241, 285*
Clark, H. C., 23(399, 400), 136(399, 400), 140(398, 399, 400), 147(399, 400), *226, 240*
Clark, N. H., 3(341), *239*
Clark, S. D., 182(1389b), *269*
Clark, S. L., 22(416), 46(416), 66(416), 67(416), 80(416), 92(416), *241*
Clayton, W. R., 3(416a), *241*
Clear, C. G., 5(417), 58(417), *241*
Cleveland, F. F., 10(1750), 12(1750), 159(1750), *278*
Coates, G. E., 13(69), 21(418, 422), 22(421), 23(422), 24(421), 25(421), 49(421), 51(420, 422), 52(422), 53(420, 422), 54(420), 55(422), 109(418), 110(418), 151(421), 152(419), 153(423), 158(69), 219(420, 422), *226, 232, 241*
Cocks, A. T., 111(424), 112(424), *241*
Coffin, K. P., 4(425), *241*
Cogoli, A., 86(121), 92(121), 220(121), *233*
Cohen, E. A., 203(426), 204(426), *241*
Coindard, G., 57(426b), 60(426b), 61(426b), 64(426b), 219(426a, 426b), *241*
Coleman, R. A., 123(198, 199, 228), *235, 236*
Collier, M. R., 21(427), 50(427), 51(427), 219(427), *241*
Commenges, G., 24(1812), 60(1812), 148(1062), 151(1812), 152(1812), *259, 280*
Compton, R. D., 77(429), *241*
Cone, C., 13(430), 14(430), *241*
Contreras, L., 224(431), *241*
Cook, W. L., 56(432), *241*
Coolidge, A. S., 131(433), *241*
Coon, A. D., 59(1696), 63(1696), *277*
Cooper, J. H., 92(434), 137(434), *241*
Cope, O. J., 29(200), 31(200), 181(200), *235*
Corfield, P. W. R., 3(416a), *241*
Corner, M., 223(435), *241*
Corsano, S., 32(436), 33(436), *241*
Cosgrove, R. E., 132(471), *242*
Cotton, F. A., 140(368), *240*
Coutts, I. G. C., 21(438), 80(438), 82(438), 88(438), *241*
Cowan, R. D., 10(440), 12(440), 162(439, 440), 163(440), 190(439), *241*
Cowie, W. P., 21(441), 57(441), 79(441), *241*

Cowley, A. H., 8(442), 9(443), 47(442), 62(442), 64(442), 148(444), 149(442), 153(442), 154(444), *242*
Coyle, T. D., 5(448, 449), 9(450), 15(448), 22(1606), 25(448), 38(445, 446), 39(445, 1607), 46(447), 62(447, 1606), 64(447, 1605, 1606, 1607), 80(445), 81(445, 446), 82(445, 1607), 92(1606), 113(445), 124(445), 218(448, 449), 221(1606), 222(446), *226, 242, 274*
Cragg, G. M. L., *226*
Cragg, R. H., 13(457, 458, 460), 14(452, 459), 51(455), 55(454), 56(451, 454, 455, 456), 57(451a), 63(451, 451a, 456), 79(453), 88(456), 115(455), 116(455), *226, 242*
Creutzberg, F., 135(337), *239*
Crighton, J., 91(461), 92(461), 223(461), *242*
Crissman, H. R., 29(152), 196(152), *234*
Crist, J. L., 59(1696), 63(1696), *277*
Csakvari, B., 64(1090), 65(1090), *259*
Cueilleron, J., 207(462), *242*
Culling, G. C., 71(463), 72(463), 75(463), 76(463), 215(463), 217(463), *242*
Culmo, R., 223(1608), *275*
Cumbo, C. C., 28(1538), 29(1538), 35(1537), 113(1537), 114(1537), *273*
Cundy, C. S., 90(464), *242*
Curran, C., 5(465), *242*
Currell, B. R., 51(466), 77(466), 92(466), 140(466), 141(466), 151(466), 152(466), 153(466), *225, 242*
Cyvin, S. J., 163(520), *244*

D

Dahl, G. H., 22(159), 77(159), *234*
Dal Bello, G., 21(367a), 45(367a), *240*
Dale, J., 120(844), *253*
Damen, H., 13(779), *251*
Damico, R., 8(467), 139(467), 141(467), *242*
Dandegaonker, S. H., 21(1094), 52(2), 55(2), 56(1095), 58(470), 62(469), 63(468, 469), 65(2, 468), 79(2), 87(1094), 224(1095), *230, 242, 259*
Dapporto, P., 3(114a), *233*
Darling, S. D., 132(471), *242*
Das, K. V. G., 138(887), 143(887), *254*
Das, M. K., 56(473, 474), 62(473), *242*
Das, N. K., 12(1588, 1589), 49(1589), *274*
Das, T. P., 162(472), *242*
Davidson, J. M., 21(478), 51(478), 53(478), 59(478), 70(478), 157(475, 476, 477), 158(476, 477), *242, 243*
Davies, A. G., 11(484), 21(489, 505), 47(14, 505), 48(483, 487, 489, 490, 505), 49(485), 50(492), 53(488, 489), 56(485, 489), 57(497), 61(14, 483, 487, 488, 489, 497), 79(487, 489), 92(500), 93(500), 95(500), 97(14, 485, 488, 502), 98(483, 489, 493, 497, 499, 501), 99(14), 100(13), *226, 230, 243*
Davies, D. W., 77(508), *243*
Davies, F. A., 7(510), 16(509), 37(509), 75(509), 79(509), 217(509, 511), *243*
Davies, K. M., 21(512), 29(512), 71(512), 215(512), *243*
Davis, J. C., Jr., 7(1771), 138(1771), 139(1771), *279*
Davis, O. L., 24(343), 53(343), 55(343), 62(343), 63(343), 64(343), *239*
Davis, R. E., 156(513), *244*
Davison, A., *226*
Davydova, S. I., 48(913), 53(913), 92(913), *255*
Dawson, J. W., 20(1434), 22(626, 1434, 1436, 1438), 23(1433, 1434, 1440), 24(1432, 1433, 1444), 37(1435), 49(1433, 1434, 1438, 1440), 50(62, 1433, 1434, 1436, 1438, 1440), 51(120, 1432, 1433, 1434, 1436, 1438, 1444), 55(626, 1432, 1441, 1444, 1860), 56(120, 1432, 1438, 1441, 1860), 57(626, 1432, 1436,

1439), 64(626), 77(626, 627, 1435), 79(1437, 1444), 92(1444), 116(120), 219(626, 1439a, 1440), 222(1437), *227, 231, 233, 247, 270, 282*

Dazord, J., 194(513a), *244*

DeBernardi, L., 109(1051), *258*

Deckelman, K., 89(513b), *244*

Decklemann, E., 88(1872), *282*

Deever, W. R., 8(515), 193(515), 194(514, 515), *244*

DeFord, D. D., *226*

Dehmelt, H. G., 18(516, 517), *244*

Deichert, W. G., 77(1556), 224(1556), *273*

DeMoor, J. E., 9(518), 18(519), 153(519), 217(518), *244*

Dessy, R. E., 86(1788), 87(1787, 1788), 91(1787), 92(1787), 96(1788), *280*

De-Staricco, E. R., 140(519a), *244*

Devaprabhakara, D., 32(519b), 33(519b), 93(1216), *244, 262*

Devarajan V., 163(520), *244*

Devgan, O. N., 132(471), *242*

Dewar, M. J. S., 7(510, 531), 8(545), 13(430), 14(430), 16(509, 528, 535, 538, 544, 546), 21(512), 22(543), 24(525), 26(525), 29(512, 528), 37(404, 509, 521, 530, 543), 55(529), 56(405, 529, 537), 57(522, 537), 62(537), 63(537), 71(463, 512, 542), 72(404, 463, 524, 527, 528, 530, 532, 533, 534, 535, 537, 540, 541), 73(522, 523, 524, 526, 528, 532, 543, 546), 75(404, 463, 509, 521, 539, 543), 76(463, 526), 79(509), 86(525), 109(541), 152(527), 215(463, 512, 520a), 217(463, 509, 512, 522, 523, 526, 531, 533, 535, 536, 542), 218(528, 544, 546), 22(534, 535), *226, 241, 242, 243, 244*

DeWitt, E. J., 11(1143), 21(1143), 47(1143), 48(1143), 49(1143), 53(547), 109(574), 174(1584), 219(1143), *244, 260, 274*

Dickason, W. C., 69(202), 123(201), 138(202), 145(202), *235*

Dickinson, D. A., 108(732, 732a), *250*

Dietz, R., 37(521), 57(522), 72(524), 73(522, 523, 524), 75(521), 217(522, 523), *244*

Diner, U. E., 182(547a), *244*

Ditter, J. F., 46(1683), 47(1683), 176(1683), 190(1077, 1079), 192(1683), 201(549), 204(548), *245, 259, 276*

Dittmar, P., 48(990), 49(990), 79(990), 150(990), 151(990), *257*

DiVaira, M., 3(550), 138(550), *245*

Dixon, R. N., 11(1924), 12(1924), *283*

Dmitrikov, V. P., 51(1351, 1836b), 53(1836a, 1836b), 61(1836a), 71(1351, 1836b), 74(1836a), 102(1836a, 1836b), 103(1836a), 156(141c, 1836a, 1836b), *234, 268, 281*

Dobson, J. E., 41(552), 80(552), 81(552), 82(552), 166(551), 183(551), *245*

Dodge, R. P., 3(1709a), *277*

Dodson, V. H., 21(203), 49(203), 53(203), 92(203), 93(203), 97(203), 99(203), 131(203), *235*

Dolan, E., 49(1074), 109(1071), 132(1071, 1074), *259*

Dollimore, D., 47(1134), 55(553, 1134), 108(1134), *245, 260*

Domash, L., 148(207), 153(207), *235*

Donahoo, W. P., 223(47), *231*

Donovan, S., 21(1645), 61(1645), 79(1645), *275*

Dorfman, M., 131(111, 112), 132(110, 111, 112), *233*

Dornow, A., 56(560), *245*

Dorokhov, V. A., 24(1277, 1279, 1390), 28(1284, 1285, 1286, 1390), 29(554, 1280, 1285), 35(1284), 46(1280, 1284, 1390), 47(1390), 48(1278, 1279), 50(555, 556, 556b, 558, 1280, 1284), 51(1280, 1281, 1282), 52(1277, 1282a, 1286, 1390), 53(1277, 1278), 54(1278, 1390), 56(1281, 1282), 57(1282), 60(1390), 66(1278), 71(554, 554b, 557, 1283, 1285, 1286a), 75(554a, 1286), 76(554a), 83(559, 1287), 84(599a, 1286b), 85(1283), 86(556a), 87(557), 92(556a, 556b, 559b, 1278, 1282b, 1283a, 1285),

93(1278), 96(1278), 115(555), 148(1280, 1281, 1282, 1286), 149(1281, 1282), 165(1279), 172(1277, 1279, 1281), 175(1277, 1278), 177(1277, 1278, 1279, 1281), 179(1277), 180(1277, 1279), 182(1278), 183(1278), 185(1280, 1281, 1282), 187(1277), 219(555, 556, 1287), *245, 264, 265, 269*

Doty, J. C., 13(561), 49(561), 134(1895), 144(562, 1892, 1893, 1894, 1895), 145(698), *245, 249, 282, 283*

Doughtery, R. C., 24(525), 26(525), 72(527), 73(526), 76(526), 152(527), 217(526), *244*

Douglas, C. M., 22(78,) 64(78), *232*

Douglass, J. E., 48(565), 50(563), 52(563), 53(563), 56(564), 155(565), 157(564, 565, 566), 158(564, 565, 566), 159(564), 186(565), *245*

Dousset, G., 223(389), *240*

Drake, R. P., 202(1498), 203(1499), *272*

Draper, P. M., 133(567), *245*

Drefahl, G., 21(568), 58(568), 145(569), *245*

Drenth, W., 22(1111), 67(1111), 214(1111), 216(1111), *260*

Drozd, V. N., 21(1421, 1422, 1423), 89(1423, 1424), 92(1424), 97(1423, 1424), 109(1424), 133(1424), 139(1421, 1422), 150(1422), *270*

Druzhkov, O. N., 143(876a), *254*

Dubov, S. S., 12(1738), 46(1738), 92(1738), *278*

Duda, H., 58(1667), 106(1667), *276*

Dulou, R., 28(570), *245*

Dulova, V., 216(1839), *281*

Dumont, E., 77(571), *245*

Duncanson, L. A., 52(572), 53(572), 62(572), 209(572), *245*

Dunell, B. A., 148(34), *231*

Dunkelblum, E., 29(891), 30(891), 31(891), *254*

Dunks, G. B., 196(1500), 199(1500), 201(1500), 202(1498), 204(1500), *272*

Dunnavant, W. R., 47(1626), 77(1626), *275*

Dunstan, I., 85(575), 196(135), 201(135), 207(135, 573, 575, 1900), 208(135, 573, 575, 1900), 209(135, 573, 575, 1900), 223(574), *233, 245, 283*

Dupont, J. A., 28(771), 31(771), 47(771), 55(771), 113(771), 114(771), 124(771), 152(771), *251*

Duppa, B. F., 21(620), 22(619, 620), 47(619), 59(619), 97(619), 99(620), *247*

du Preez, A. L., 90(731), 146(731), *250*

Durst, H. D., 145(146), *234*

Dworkin, A. S., 61(576), *245*

E

Easterbrook, E. K., 104(1010), *257*

Eaton, G. R., 52(576a), *226, 245*

Edwards, L. J., 199(841), *253*

Egger, H., 84(1219), *262*

Egger, K. W., 111(424), 112(424), *241*

Eggers, C. A., 59(577), 65(577), *246*

Eicher, T., 138(1912), *283*

Eidus, Ya. T., 122(1582a), *274*

Eisch, J. J., 21(579), 22(579), 66(579), 216(578, 579), 220(579), *246*

Eisenberg, R., 3(1569a), *274*

El-Fayoumy, G. F., 24(581), 111(581), *246*

Ellis, P. D., 10(580), 201(1141a), 204(580), *246, 260*

Emeléus, H. J., 28(1741), 32(1741), 48(1741), *225, 278*

Emelyanova, L. I., 21(1426), 48(1426), 49(1426), 138(1426), 139(1426), *270*

Emery, F. W., 208(583), 209(582), *246*

Emsley, J., 148(59), *231*

Encina, M. V., 108(584), *246*

Engelhardt, U., 162(811a), *252*

England, D. C., 211(901), 212(897, 901), *254*

English, W. D., 55(585), 135(1430), *246, 270*

Enrione, R. E., 204(1397, 1398), *269*

Epple, R., 70(676, 677), 91(677), 92(1075), 93(677), 135(676, 677, 1075), *248, 259*
Epstein, I. R., 197(1759), *279*
Erdey, L., 138(886), 212(886), *254*
Erickson, C. E., 22(586), 23(586), 24(723), 49(586), 57(723), 92(586), 219(723), *246, 250*
Ernst, F., 63(810), *252*
Eubanks, I. D., 49(587), 50(587), *246*
Evans, J. E. F., 24(1876), 57(1876), 136(1877), 138(1876, 1877), 148(1876), 165(1876), 174(1876), 175(1876), 177(1876), 179(1876), 180(1876), 185(1876), 187(1876), *282*
Evans, T. R., 144(562), *245*
Evers, E. C., 22(588), 47(588), 92(588), *246*
Ewers, J. W., 26(678), 140(678, 679), 148(679), 151(678), *248*
Exner, O., 52(588a), 61(588a), 63(588a), 80(588a), 82(588a), *246*

F

Fager, R. S., 224(588b), *246*
Fallani, G., 3(114a), *233*
Fassel, V. A., 162(1155), 163(1155), *261*
Faulks, J. N. G., 148(589), 152(589), 186(589), *246*
Faust, J. P., 21(889), 22(888), 77(889), 78(888, 889), *254*
Favre, E., 115(590), *246*
Feare, T. E., 21(1096), 57(1096), 58(1096), *259*
Fedneva, E. M., 77(591), *246*
Fedorova, L. S., 224(914, 915), *255*
Fedotov, N. S., 13(1449, 1450), 21(1292, 1450), 22(1301), 23(1301), 48(1301, 1449, 1450), 51(1289, 1291, 1292, 1293, 1295, 1301, 1449, 1450), 52(1292, 1301, 1449, 1450), 53(1449, 1450), 54(1288, 1292, 1301, 1450), 55(1288), 57(1450), 59(1449), 62(1290, 1449, 1450), 63(1290), 77(1450), 79(1292), 92(1290, 1293), 139(1301), 140(1291, 1293), 151(1289, 1292, 1295, 1449, 1450), 158(1296), *265, 270, 271*
Fehlner, T. P., 28(594), 30(594), 160(595, 1153, 1154), 163(595, 596), 187(594), *246, 261*
Feil, R., 52(1382a), *268*
Fellous, R., 27(592), 28(592, 593a), 181(592), 182(593), *246*
Fenzl, W., 51(597), 56(1520), 57(1520), 62(1520), 92(597, 940), 97(940), *246, 256, 272*
Fieldhouse, S. A., 62(598), *246*
Figgis, B., 196(599), 197(599), 201(599), *246*
Fillwalk, F., 162(1155), 163(1155), *261*
Finch, A., 16(603, 604), 20(390), 22(390, 603), 23(601), 38(390), 39(390), 48(603), 49(603), 51(600), 54(603), 55(603), 57(600), 61(601, 602), 62(601, 602), 63(607), 64(606), 65(603), 66(605), 80(390), 81(390), 176(605), *226, 240, 246*
Fink, W., 56(608), *246*
Fish, R. H., 29(608a, 608b), 32(608a), 33(608a), *246*
Fishbein, L., 124(1114), 131(1114), *260*
Fisher, H. D., 190(1898), *283*
Fisher, R. P., 28(1971), 92(1971), 93(1971), 104(1971), 130(1970a), *285*
Flaschka, H., *226*
Flautt, T. J., 28(1133), 46(1133), 170(1133), *260*
Fleischer, E. B., 72(527), 152(527), *244*
Fleming, M. A., 195(1766), *279*
Flodin, N. W., 77(1649), 148(1649), 149(1649), 151(1649), 153(1649), 155(1649), 165(1649), 176(1649), 183(1649), 184(1649), *276*
Florentin, D., 76(1609a), *275*
Florin, R. E., 88(1844), 89(1844), *281*
Flory, P. J., 77(1666), *276*
Fluharty, A. L., 182(668), *248*
Foot, K. G., 53(609), 92(609), 121(479, 609), *243, 246*
Fordham, J. W. L., 224(610), *246*
Fordham, S., 22(68), 56(68), 57(68), 62(68), 63(68), 80(68), 81(68), *232*

Forsythe, J. A., 182(547a), *244*
Foster, R., 77(401, 611), *240, 246*
Foster, W. E., 21(49), 149(49), 150(49), 151(49), *231*
Fowler, D. L., 140(612, 613), *247*
Fox, W. B., 40(614), 65(614), 82(614), *247*
Francis, B. R., 21(418), 109(418), 110(418), *241*
Francois, J., 138(615), *247*
Frankland, E., 21(617, 620), 22(616, 617, 618, 619, 620), 46(617, 618), 47(616, 617, 618, 619), 55(616, 617), 56(616, 617), 59(616, 617, 619), 79(617), 92(616), 97(616, 619), 99(617, 618, 620), 140(618), 148(616, 617, 618), *247*
Franta, E., 138(615), *247*
Freidlina, R. C., 109(621), *247*
Freitag, W. O., 22(588), 47(588), 92(588), *246*
French, C. M., 21(478), 51(478), 53(478), 59(478), 70(478), 157(475, 476, 477), 158(476, 477), *242, 243*
French, F. A., 5(622), *247*
Frenkin, E. I., 43(623), 53(1782), 60(1782), 65(623), 92(1782), 121(1782), *247, 280*
Frey, F. W., Jr., 143(624), *247*
Frey, J., 20(390), 22(390), 38(390), 39(390), 80(390), 81(390), *240*
Freyer, W. R., 77(1671, 1674), 78(1671), 85(1671), 218(1671), *276*
Frick, S., 6(103), 77(103, 104), 219(103), *232*
Fridmann, S. A., 160(1154), *261*
Fried, J., 56(1927), 57(1927), 71(1927), *283*
Friedman, L. B., 4(625, 1501), 199(1501), 201(1501), 203(625, 1501), *247, 272*
Fritz, P., 21(1466, 1468), 22(626, 1438), 23(1440, 1465, 1466, 1468), 44(1467), 49(1438, 1440, 1443, 1468), 50(1438, 1440, 1443, 1469), 51(1438), 55(626, 1441, 1466), 56(1438, 1441, 1442), 57(626, 1439, 1465, 1466, 1467, 1468), 60(1467), 61(1467), 64(626), 77(626, 627), 80(1465, 1467), 140(1466), 158(1466), 219(626, 1439a, 1440, 1468), *247, 270, 271*
Frolov, S. I., 47(333, 334, 1275), 49(1275), 60(628, 1275), 61(332, 1271), 62(332, 628), 66(333, 334, 1270), 67(332, 333, 334, 628, 1271), 79(332, 334, 628), 92(1271), 115(332, 333, 334, 628, 1270, 1271, 1275, 1276), 116(332, 333, 334, 628, 1270, 1271, 1275, 1276), *238, 239, 247, 264*
Fu, Y.-C., 159(629), 160(629), *247*
Fueno, T., 224(633, 634), *247*
Fukutani, H., 224(631, 633), *247*
Fuller, M. E., 8(165), 49(165), 50(62, 165), 51(165), *231, 234*
Furtsch, T. A., 8(442), 9(443), 47(442), 62(442), 64(442), 149(442), 153(442), *242*
Furukawa, J., 224(630, 631, 632, 633, 634, 635, 1407, 1408), *247, 269*
Fuss, W., 49(139), 55(139), 72(139), *233*

G

Gafner, G., 3(1453), 90(1453), 146(1453), *271*
Gaines, D. F., 173(636), 176(636), 193(636), 194(636), 201(637), 206(637), *247*
Galbraith, H. J., 159(638), *247*
Gal'chenko, G. L., 16(639, 640, 640a), *247*
Galkin, A. F., 21(1297), 78(1297, 1298), *265*
Gallaghan, J., 208(1709), 209(1709), *277*
Gallagher, D. A., 14(452), *242*
Gallais, F., 53(1064), 62(1064), 148(641a), 151(1813), 152(1813), *247, 259, 280*
Ganguli, P. S., 160(642), 163(643), *247*
Garabedian, M. E., 160(644), 187(644), *247*
Gardner, P. D. 29(1687), 32(519b), 33(519b), 66(1687), *244, 277*
Gardner, P. J., 16(603, 604), 22(603),

23(601), 48(603), 49(603), 51(600), 54(603), 55(603), 57(600), 61(601, 602), 62(601, 602), 65(603), *246*
Garegg, P. J., 51(644a), *247*
Garg, C. P., 28(204), 107(204), *235*
Garland, J. H. N., 107(15), *230*
Garner, B. J., 27(290), 28(290), 181(290), *237*
Garrett, A. B., 16(745), 197(31, 1630), 201(31, 1630), *231, 250, 275*
Garrett, A. G., 80(1826), *281*
Gartzke, W., 3(850b), 4(850b), 90(850b), *253*
Gast, E., 24(1706), 54(1701, 1706), 55(1706), 88(1659), 90(1659), 143(1706), *276, 277*
Gatti, A. R., 38(1948), 39(1948), *284*
Gaudemar, M., 115(590), *246*
Gebauhr, W., 138(645), *247*
Geilmann, W., 138(645), *247*
Geisler, I., 55(646), *247*
Geller, S., 3(647), *247*
George, T. A., 50(648), 51(648), 109(648), 110(648), *247*
George, W., 24(1444), 51(1444), 55(1444), 79(1444), 92(1444), *270*
Gerasimenko, A. V., 89(1425), 115(1639), *270, 275*
Gerhart, F. J., 194(1899), 195(1899), 196(1502), 199(1502, 1502a), 201(1502), 204(548), *272, 283*
Gerrard, W., 12(107, 344, 361, 362), 21(171, 650, 653), 22(346, 649, 655), 24(4, 6), 36(340), 49(8), 50(650, 653), 51(466, 653, 655), 52(2, 3, 4, 5, 6, 572, 651, 652), 53(572, 650, 651, 652), 54(5, 650, 655, 656), 55(2, 5, 650, 655), 56(347), 58(470), 60(655), 61(171, 655), 62(6, 171, 172, 173, 344, 469, 572, 652), 63(172, 173, 468, 469), 64(655), 65(2, 6, 346, 468, 649, 655), 77(345, 347, 466, 654, 655, 656), 79(2, 5, 6, 656), 91(172), 92(4, 6, 466, 650, 652), 97(4), 99(651), 102(6), 140(466), 141(466), 151(8, 466, 650), 152(8, 173, 466), 153(8, 466), 158(650), 209(107, 572), 218(654), *226, 230, 233, 234, 239, 242, 245, 248*
Gerstein, M., 148(205, 302), 153(302), *235, 238*
Geske, D. H., 144(657, 658), *248*
Geymayer, P., 22(660), 55(660), 57(659), 77(659), *248*
Giacomelli, T. P., 109(1051), *258*
Giambiagi, M., 71(661a), *248*
Giambiagi, M. S., 71(661a), *248*
Gibbins, S. G., 194(1684), *277*
Gibbons, D., 223(106), *233*
Giles, N., 52(9), 62(9), *230*
Gilliam, W. F., 6(1055), *258*
Gilman, H., 20(661), 21(662, 663, 664, 1635), 22(661), 57(661), 58(1635), 59(1635), 79(662), *248, 275*
Gintis, S. K., 148(206, 207), 153(206, 207), *235*
Gipstein, E., 24(665, 666), 35(665, 666), 46(665, 666), 55(665, 666), 57(665), 64(666), *248*
Giraud, D., 21(667), 22(667), 52(667), 149(667), 150(667), 151(667), *248*
Giudici, T. A., 182(668), *248*
Gleicher, G. J., 16(528), 29(528), 72(528), 73(528), 218(528), *244*
Glicenstein, L. J., 22(1632), 39(1632), 81(1632), 82(1632), *275*
Glockling, F., 13(669), *248*
Glogowski, M. E., 13(561), 49(561), 72(699), 145(702), *245, 249*
Glunz, L. J., 25(1204), 62(1204), 98(1204), *262*
Glushkova, V. P., 4(1961b), *284*
Golden, R., 13(430), 14(430), 55(529), 56(529), *241, 244*
Goldstein, M., 12(344), 62(344), 77(345), *239*
Gonsior, L. J., 86(1904), 216(578), *246, 283*
González Barredo, J. M., 17(1603), 176(1603), *274*
Good, C. D., 7(670), 8(670), 12(670), 16(670), 22(670), 46(670), 47(670), 54(670), 149(670), *248*
Gorbunov, A. V., 102(1698a), *277*
Gordon, A. J., 77(1495), *272*
Gordon, L. P., 160(642), *247*

Gordy, W., 4(671, 672, 673), 162(671, 672, 673), 163(673), *248*
Goubeau, J., 11(105, 675, 1643), 12(105, 675), 21(115), 22(115), 24(682, 1643), 26(678), 28(1888), 46(680), 49(105, 115, 681, 686), 50(680, 1821, 1822), 51(684, 1822), 54(1643), 55(115, 686, 1822), 56(680), 57(1888), 59(1822), 62(685), 64(1643), 69(998a), 70(676, 677), 71(1888), 77(674, 683, 1888), 79(682, 683, 1822), 80(998a), 91(677, 684), 92(684, 686, 1075, 1822), 93(677), 95(1822), 127(685), 135(676, 677, 1075), 140(678, 679), 148(674, 679), 151(678), 152(115), *226, 232, 233, 248, 249, 257, 259, 275, 281, 282*
Gould, R. F., *226*
Graber, F. M., 24(353), 54(353), 63(353), *239*
Grabner, H., 49(681), *248*
Graeza-Lukácz, M., 138(687), 223(687), *249*
Graham, W. A. G., 25(79), 28(79, 689), 32(1742), 47(689), 53(1742), 62(79, 1742), 113(79, 1742), 124(79, 1742), 125(79, 1742), 148(688), 151(79), 152(79, 1742), 174(689), 176(689), *232, 249, 278*
Grassberger, M. A., 5(690), 25(956), 53(956), 54(956), 55(956), 62(690, 955, 956), 64(956), 65(956), 66(949, 956), 67(956), 68(956), 69(949), 74(949), 109(949), 114(956), 176(949, 954, 957), 178(949), *226, 227, 249, 256*
Green, A. A., 151(354), 152(354), 153(354), *239*
Green, M. L. H., *226*
Green, S. I. E., 86(1788), 87(1787, 1788), 91(1787), 92(1787), 96(1788), *280*
Greenwood, N. N., 12(695), 28(693, 695), 29(692, 693), 66(695), 69(691, 692), 83(693), 148(589), 151(692), 152(589), 186(589), 214(693), *246, 249*
Greiss, G., 89(784a, 785), 90(784a, 786), 134(785, 786), *251*
Griasnow, G., 21(959), 24(958), 25(958, 959), 28(958, 959), 46(959), 48(959), 49(959), 66(958, 959), 67(959), 69(958, 959), 180(958), 182(958, 959), *256*
Griffin, R. G., 132(696), *249*
Griffiths, J. V., 223(574), *245*
Grigoryan, M. S., 66(334a, 1271a), 115(1271a), 116(334a), *239, 264*
Griller, D., 101(480), 108(480, 481, 482), *243*
Grim, S. O., 138(1672), 140(1672, 1675a), *276*
Grimes, R. N., 109(696a), 202(696a), 209(696a), *226, 249*
Grimme, W., 67(697), *249*
Grinstein, R. H., 156(1115), *260*
Grisdale, P. J., 13(561), 21(700), 37(530), 49(561), 66(700), 72(530, 699, 701), 75(701), 134(1895), 144(562, 1892, 1893, 1894, 1895), 145(698, 702), 146(700), 220(700), *244, 245, 249, 282, 283*
Gronowitz, S., 51(703b), 71(703a, 1413), 76(703, 703a, 703b, 704, 705, 1412), *249, 269*
Groszek, E., 188(706), 202(706, 1504), 216(706), *249, 272*
Groszek, S. J., 22(707), 77(707), *249*
Grotewold, J., 17(712), 98(708, 713), 99(713), 103(712), 104(712), 108(709, 710, 711, 712, 714), 111(17), 160(715, 716), 194(714), 224(431), *230, 241, 249*
Groves, D., 147(717), *249*
Grummitt, O., 21(866), 56(866), 57(866), 92(866), 94(866), 97(866), 98(866), 99(718, 866), 102(866), 109(866), 133(866), 223(866), *249, 253*
Grunanger, P., 86(121), 92(121), 220(121), *233*
Grundke, H., 46(719, 719a), 60(719), 148(719, 719a), *249*
Grundon, M. F., 138(720), 145(36, 720), *231, 249*

Gryszkiewicz-Trochimowski, E., 208(721), *249*
Gryszkiewicz-Trochimowski, O., 208(721), *249*
Guest, M. F., 15(722), *250*
Guillory, J. P., 6(76), *232*
Guillot, P., 207(462), *242*
Gulden, W., 71(802), 141(802), *252*
Gunderloy, F. C., 22(586), 23(586), 24(723), 47(1626), 49(586), 57(723), 77(1626), 92(586), 219(723), *246, 250, 275*
Gunn, D. M., 127(834, 835, 836, 837), *252*
Gupta, S. K., 25(208, 211), 29(209, 214, 266), 33(214), 48(212), 53(211), 56(214), 57(214), 60(208, 209, 214), 61(208, 209, 210, 214), 62(209), 66(266, 267), 67(267), 92(266), 93(266), 118(1756), 122(267), 148(212), 149(210, 213), 153(210), 174(212), 176(213), 177(213), *235, 237, 279*
Guseva, A. S., 62(919), 92(919), *255*
Gusowski, B., 52(1382a), *268*
Gutmann, V., 21(725, 1231), 22(724, 1230), 77(724, 725, 1230, 1231), 78(724, 725, 726, 727), 84(728), *250, 263*
Gutowsky, H. S., 6(1571), *274*
Guy, J., 46(729), *250*

H

Haack, W. D., 139(729a), *250*
Haag, A., 55(798), 76(158), 130(798), 138(730), 139(157, 730), 142(730), 143(158), *234, 250, 252*
Haag, W., 151(1905), *283*
Habereder, P. P., 50(1514), 54(1514), 55(1514), 79(1515), 129(1513, 1515), 150(1514), 151(1514), *272*
Habu, T., 224(907), *254*
Hagelee, L. A., 89(1177a), *261*
Haines, L. M., 3(1453), 90(1453), 146(1453), *271*
Haines, R. G., 22(1802), 55(1802), 65(1802), *280*
Haines, R. J., 90(731), 146(731), *250*
Hall, F. R., 22(1778), 78(1778), *279*
Hall, J. R., 11(1924), 12(1924), *283*
Halpern, D., 22(58), 67(58), 216(58), *231*
Halpern, J., 144(12), *230*
Hamann, J. R., 5(881), 16(881, 882), 218(882), *254*
Hamann, R. R., 97(736), *250*
Hamaoka, T., 104(214a), *235*
Hamilton, S. B., 55(1097), 56(1097), 57(1099), 58(1098), 62(1099), 224(1098), *259*
Hancock, K. G., 41(733), 50(733), 57(733), 108(732, 732a), 134(415), *241, 250*
Hanousek, F., 208(733a), *250*
Hansen, H., 51(1516), *272*
Hansen, L. D., 211(734), *250*
Hansen, R. L., 34(735), 97(736), *250*
Happ, G. P., 144(1893, 1894), *283*
Harada, M., 97(1754), *279*
Hare, D. G., 11(484), 21(489), 48(483, 487, 489, 490), 49(485), 53(488, 489), 56(485, 489), 61(483, 487, 488, 489, 490), 79(487, 489), 97(488), 98(483, 489), 99(483, 487, 489, 490), 127(486), *243*
Harmon, A. B., 212(737, 738, 739), 213(737, 739), *250*
Harmon, K. M., 212(737, 738, 739), 213(737, 739), *250*
Harold, P. L., 208(583), 209(582), *246*
Harpp, D. N., 133(567), *245*
Harrelson, D. H., 24(1444), 51(1444), 55(1444), 79(1444), 92(1444), *270*
Harris, J. J., 22(740, 1629), 77(1629), 78(1629), 84(740), *250, 275*
Harris, S. W., 196(1628), 197(1628), 201(1628), 202(1628), *275*
Harrison, B. C., 194(741), 195(741), *250*
Harrison, P. G., 56(473), 62(473), *242*
Hartmann, H., 22(742), 50(742), 55(742), 56(742), 65(742), 81(742), *250*
Hartsuck, J. A., 4(1501), 199(1501), 201(1501), 203(1501), *272*
Haruda, F., 208(733a), *250*
Harvey, D. J., 223(177a), 224(178), *234*
Harvey, D. R., 58(743), 222(743), *250*

Haseley, E. A., 16(745), 197(744), *250*
Haslinger, F., 211(745a), 212(745a), *250*
Hassner, A., 29(746), 33(746), *250*
Hastings, J. M., 3(87), 4(87), *232*
Haszeldine, R. N., 28(132), 32(132), 46(132), 47(132), 53(132), 62(132), 113(132), 124(132), 182(132), *233*
Hattori, K., 88(1723), 165(1723), 175(1723), 176(1723), 205(1724), 206(1724), *278*
Hattori, S., 24(943), 29(943), 38(960, 964), 46(942, 943), 50(942, 943, 964), 51(943), 66(943), 69(943), 72(964), 73(964), 77(960, 964), 174(942), 185(942), *256*
Haubold, W., 42(746a), 46(746a), 49(746a), *250*
Havir, J., 139(747), *250*
Hawes, W. W., 140(991), *257*
Hawkins, R. T., 49(748), 51(748), 53(748), 56(748), 59(748, 749), 72(749), 74(748), 79(748, 749), 222(748), *250*
Haworth, D. T., 77(750), *250*
Hawthorne, M. F., 21(754), 25(766), 28(759, 760, 761, 766, 768, 771), 29(755), 31(771), 33(752), 34(755, 759, 760, 766), 46(760, 766, 767), 47(766, 771), 48(752, 760, 766), 49(752, 755, 766), 50(775), 52(758), 53(758), 54(760), 55(759, 763, 767, 771), 56(763), 62(757), 66(760, 766), 67(766), 70(760, 766), 77(759, 764), 83(753), 86(774), 92(760, 774), 95(761), 96(774), 100(754), 113(771), 114(761, 771), 115(773, 775), 118(773, 775), 124(771), 136(751, 762), 138(751, 762), 148(756, 757, 759, 763, 1845), 149(756, 757, 758, 1027, 1845), 152(771), 154(749, 762, 764, 769, 770, 1027, 1845), 155(756, 757, 758, 759, 760, 763, 765, 772), 156(772), 168(757), 179(757), 180(758), 185(756), 186(757, 758), 219(753), 223(1743), *226, 250, 251, 258, 278, 281*
Heal, H. G., *226*
Hebert, N. C., 109(215), *235*
Heil, H. F., 89(785), 90(786), 134(785, 786), *251*
Heim, P., 182(216), *235*
Heimbach, P., 24(1891), 48(1891), 49(1891), 150(1891), *282*
Heitsch, C. W., 9(776), 148(777, 1837, 1838), *251, 281*
Hemming, R., 56(778), 57(778), 62(778), 63(778), 130(778), *251*
Henderson, W. G., 62(73), 63(73), *232*
Hendra, P. J., 66(605), 76(605), *246*
Henneberg, D., 13(779), *251*
Hennion, G. F., 21(781, 782, 783, 784, 1207, 1616), 22(781, 782, 784), 23(784), 24(780, 782, 783, 784, 1205, 1207), 25(1205), 28(1207), 47(781, 1207), 48(781, 783, 784, 1207), 49(782, 1616), 53(784), 54(1207), 55(1205), 79(780), 111(780, 781, 1209, 1615), 112(1615), 173(781), *251, 262, 275*
Henrickson, A. R., 57(1011), 58(1011), 59(1011), 104(1011), 222(1011), *257*
Herberich, G. E., 48(787), 89(784a, 785, 787), 90(784a, 786), 134(785, 786, 787), *251*
Heřmánek, S., 208(733a), *250*
Hermannsdorfer, H. K., 50(1469), *271*
Hernandez, J., 98(708), *249*
Herstad, O. 15(788), 163(788), *251*
Hertler, W. R., 210(792, 795, 898), 211(791, 898, 901), 212(790, 791, 792, 795, 898, 901), *251, 254*
Hertwig, H., 49(1878, 1879, 1880), 56(1879, 1880), 57(1879), 77(1878, 1879, 1880), 79(1879, 1880), 92(1878, 1879, 1880), 148(1878, 1879, 1880), 152(1879), 174(1879), *282*
Herwig, W., 21(1906), 49(1906), 69(1906), 81(1906), 139(1906), *283*
Herz, F., 223(793), *251*
Herz, J. E., 28(794, 796), 182(794), *251*
Hess, H., 3(797), 69(797), *251*
Hesse, G., 55(798), 71(799, 800, 801, 802), 72(799, 800, 1902a), 76(158),

127(799, 800, 801, 1902a), 130(798), 138(730), 139(157, 730, 801), 141(802), 142(730, 859), 143(158), *234, 250, 252, 253, 283*
Hessett, B., 55(803), 56(803), *252*
Heydkamp, W. R., 106(217), 133(217), *235*
Heyes, R., 62(804), *252*
Heying, T. L., 56(806), *252*
Hickman, J., 31(1540), 56(1541), 124(1539, 1540, 1541), 149(1541), 176(1541), 188(1539, 1540, 1541), *273*
Higgins, H. A., 24(665, 666), 35(665, 666), 46(665, 666), 55(665, 666), 57(665), 64(666), *248*
Hill, G. R., 159(629), 160(629), *247*
Hill, J. A., 107(15), *230*
Hillier, I. H., 15(722), *250*
Hillman, M. E. D., 60(808), 62(808), 74(807), 75(807), 79(807), 122(807), 124(808, 809), 127(807, 808, 809), 159(809), *252*
Hillringhaus, E., 51(1237), 54(1237), 58(1237), 64(1237), 65(1237), 79(1237), *263*
Hinz, G., 51(1214), 92(1214), 93(1214), 121(1214), *262*
Hites, R. D., 194(741), 195(741), *250*
Hiyamra, H., 53(1928), 107(1928), *283*
Hoberg, H., 109(1955), 140(1955), *284*
Hock, H., 63(810), *252*
Höerhold, H. H., 21(568), 58(568), *245*
Hoewe, B. D., 138(811), 161(811), *252*
Hoffmann, E. G., 5(690), 62(690), 176(957), *249, 256*
Hoffmann, K. F., 162(811a), *252*
Hoffmann, R., 16(812, 813), 218(812), *252*
Hofmann, K. D., 21(568), 58(568), *245*
Hofmeister, H. K., 24(814), *252*
Hohnstedt, L. F., 50(815), 51(815), 55(815), 77(750), *250, 252*
Holland, G. W., 115(1755), *279*
Holliday, A. K., 18(824), 20(817), 22(817, 818), 24(414, 822, 825), 28(414), 33(414), 38(394, 395, 821), 39(822, 823), 45(414), 46(817), 55(823), 62(825), 80(325a, 414, 819, 820), 81(395, 819, 823), 82(394, 817, 822, 823), 91(461), 92(395, 414, 461, 826), 109(823), 111(823), 115(819), 135(325a), 137(826), 139(826), 143(826), 171(414), 174(816), 219(395), 223(461), *238, 240, 241, 242, 252*
Hollins, R. E., 196(827), *252*
Holloway, C. E., 10(58a), *231*
Holmes, R. R., 8(828), 49(829), 53(829), 133(829), 148(828), 153(828), *252*
Holmquist, H. E., 83(830), *252*
Holzmann, R. T., *226*
Homann, P. H., 224(830a), *252*
Honeycutt, J. B., Jr., 28(833), 109(831, 832), 110(831), 136(833), 138(833), 139(833), 141(833), 145(833), 147(833), *252*
Honma, S., 118(876), *254*
Hook, S. C. W., 50(492), 106(491), 108(491, 492), *243*
Hooz, J., 127(833a, 834, 835, 836, 837, 837a, 838, 839, 840), *252*
Horeld, G., 57(1882), 151(1882), *282*
Horstschäfer, J. H., 21(962), 148(962), 149(962), 150(962), 151(962), 152(962), *256*
Horvitz, L., 24(1650), 49(1651), 55(1651), 76(1651), 77(1651), 78(1651), 138(1647), 164(1651), 175(1650), 176(1650), 183(1651), 184(1651), *276*
Hota, N. K., 21(579), 22(579), 66(579), 216(579), 220(579), *246*
Hough, W. V., 199(841), *253*
Hovey, M. M., 109(1816), *280*
Howarth, M., 12(361, 362), 22(346, 649), 65(346, 649), *239, 248*
Hubel, W., 21(154), 66(154), 220(154), *234*
Huber, H., 72(845), *253*
Hubert, A. J., 100(843), 120(842, 843, 844), *253*
Huemer, H., 37(1521), *272*
Hümpfner, K. H., 18(1747), *278*
Huisgen, R., 72(845), *253*
Humffray, A. A., 59(328), 92(327),

93(326), 104(327, 328), 108(846a), 144(846), 218(327), *238, 253*
Hummel, D., 46(680), 50(680), 56(680), *248*
Hunter, D. L., 21(1921), 59(1921), *283*
Hurd, D. T., 24(848), 26(848), 32(848), 138(847), 169(848), 174(848), 175(848), 179(848), *228, 253*
Husband, J. P. N., 14(452), 79(453), *242*
Hutchings, M. G., 141(1558), 143(1559, 1560), *273*
Hutchins, J. E. C., 141(998), *257*
Hutchins, R. O., 145(848a, 849, 850), *253*
Huttner, G., 3(850a, 850b), 4(850b), 90(850b), *253*

I

Ide, F., 224(851, 852), *253*
Ikeda, H., 29(1373), *268*
Imada, T., 224(631, 633), *247*
Imaki, N., 46(1022), 52(1022), *258*
Inatome, M., 50(853), 53(853), 86(1004), 107(853), *253, 257*
Ingold, K. U., 98(493, 854, 921), 99(493, 921), 100(921), 108(493, 854), 150(493), *243, 253, 255*
Inomata, K., 53(1395a), *269*
Inoue, N., 133(1595), *274*
Ioffe, S. L., *228*
Ione, S., 224(632), *247*
Iorns, T. V., 201(637), 206(637), *247*
Isagoulyants, G. V., 91(334b), 96(334b), *239*
Ishii, M., 224(855, 856), *253*
Isiyama, S., 115(1376), 117(1376), *268*
Ito, H., 13(1590), *274*
Ito, K., 77(857, 1854), *253, 282*
Itoh, M., 74(1752b), 97(1753), 115(1377, 1752, 1753a, 1755), 117(1377, 1752), 118(876, 1756), 128(1758), 129(1757), 131(1752b), 146(1752a), *254, 268, 279*
Ivanina, T. V., 224(34a), *231*
Ivanova, A. G., 224(34a), *231*
Ivanova, E. A., 62(919), 92(919), *255*
Iwabuchi, S., 224(907, 908, 909, 910), *254, 255*
Iwasaki, K., 24(963), 38(964), 50(964), 66(963), 68(963), 71(963), 72(963, 964), 73(963, 964), 77(964), 80(963), 87(963), 165(963), 175(963), 176(963), 177(963), 179(963), 181(963), 185(963), *256*
Iwata, Y., 224(908), *255*
Iyoda, J.,22(858), 46(858), *253*
Izatt, R. M., 211(734), *250*

J

Jackson, A. H., 21(441), 57(441), 79(441), *241*
Jackson, C. B., *228*
Jackson, R. C., 57(906), *254*
Jackson, W. R., 138(720), 145(36, 720), *231, 249*
Jacob, P., 115(1377, 1753a), 117(1377), *268, 279*
Jäger, H., 142(859), *253*
Jeffers, W., 174(816), *252*
Jefferson, R., 51(860), 56(860), 115(860), 116(860), *253*
Jehlička, V., 52(588a), 61(588a), 63(588a), 80(588a), 82(588a), *246*
Jenne, H., 22(1438), 23(1433), 24(1433), 49(1433, 1438), 50(1433, 1438), 51(120, 861, 1433, 1438), 56(120, 861, 1438, 1442), 57(1433), 115(120), 116(120), *233, 253, 270*
Jennings, J. R., 49(862), 50(862), 51(862), 56(862), 57(862), 219(862), *253*
Jensen, K. A., 1(20), *230*
Jensen, R. E., 138(862a), *253*
Jessop, G. N., 20(817), 22(817, 818), 46(817), 82(817), *252*
Johannesen, R. B., 148(218), 153(218), *235*
Johnson, F. A., 83(1364, 1365), 148(1365), *268*
Johnson, H. D., 206(166, 863, 863a), *234, 253*

Johnson, J. R., 21(93, 866, 1664, 1719), 23(865), 47(865), 53(865, 867), 54(865), 56(866, 1719), 57(93, 866, 1664, 1665), 58(93), 59(93), 61(867), 64(865), 79(1719), 92(865, 866, 1719), 94(866, 1719), 97(866), 98(865, 866, 867, 1719), 99(866, 867, 1719), 100(867, 1719), 102(866, 867), 103(867), 109(866, 1719), 133(866), 139(865), 222(1664), 223(866, 867, 1719), *232, 253, 276, 278*
Johnson, W. H., 16(864), *253*
Johnston, D. G., 56(778), 57(778), 62(778), 63(778), 130(778), *251*
Johnstone, R. A. W., 18(824), *252*
Jolicoeur, C., 138(868), *253*
Jolly, W. L., 140(519a), *244*
Joly, R., 22(338), 46(338), 47(338), *239*
Jones, J. R., 22(416), 46(416), 66(416), 67(416), 80(416), 92(416), *241*
Jones, P. R., 31(869), *253*
Jones, R., 7(510, 531), 217(511, 531), *243, 244*
Jones, W. J., 12(371, 869a), 62(869a), 190(371, 372, 373), 192(373), *240, 253*
Joshi, J. C., 18(1688), *277*
Jotham, R. W., 12(371), 190(371, 372, 373), 192(373), *240*
Joy, F., 35(871), 36(871), 43(870, 871), 57(871), 61(871), 64(871), 65(871), 69(871), 88(870), 89(870, 871), 108(870), 115(870), 116(879), *253*
Jugie, G., 148(641a, 872, 1062), 149(873), *247, 253, 259*
Jutzi, P., 22(874), 68(874, 874a), *253*
Juvinall, G. L., 40(1526), 41(1526), 125(1526), 153(1526), *272*

K

Kabalka, G. W., 95(283), 97(239), 108(219, 220), 115(219, 220, 222, 273, 282, 875, 1755), 117(273, 282), 118(2, 875, 876), 124(285), 126(283, 284, 285), *235, 236, 237, 254, 279*
Kachaeva, L. I., 143(876a), *254*
Kaesz, H. D., 22(77, 877), 62(77), 64(877), 97(77), 125(877), 148(878), 223(77), *226, 227, 232, 254*
Kaldor, A., 160(879), *254*
Kandasamy, D., 145(848a), *253*
Kaneko, C., 72(532), 73(532), *244*
Kang, S. Z., 25(1542), 28(1542), 35(1542), *273*
Kaplan, E., 223(793), *251*
Kapshtal, V. N., 190(879a, 879b, 879c), 192(879b), *254*
Karpeiskaya, E. I., 90(879d), *254*
Kasha, M., 13(1587), *274*
Kato, S., 21(880), 48(880), 49(880), 97(880), *254*
Katz, J. J., 28(268), 33(268), 46(1419), 60(268), 61(268), 62(268), 130(1419a), 148(268), 156(268), 177(1419), 186(1419), *237, 269*
Kaufman, J. J., 5(881), 16(881, 882), 218(882), *254*
Kawakami, J. H., 29(223), *236*
Kawasaki, A., 224(634), *247*
Kawazura, H., 224(1457, 1458), *271*
Kaye, S., 201(1028), *258*
Kazakova, E. B., *284*
Kazansky, B. A., 47(336), 92(336), 115(336, 1275b), 117(336), 208(336), *239, 264*
Keblys, K. A., 27(224, 225), 28(224), 29(224), 31(224, 225), 181(224, 225), *236*
Keicher, G., 48(1908), 137(1908), 138(1907, 1908), 140(1908), *283*
Keil, E., 145(569), *245*
Keller, H., 24(682), 77(683), 79(682, 683), *248*
Keller, R. N., 20(1564) 21(1564), 22(1564), 47(1564), 56(1564), 79(882a, 1564), 92(1564), 104(1564), 133(882a), *254, 273*
Kennerly, G. W., 12(1371), 56(1371), 85(1371), *268*
Kenson, R. E., 156(513), *244*
Kent, W. M., 12(116), 162(116, 117), *233*

Keough, A. H., 62(1012), 63(1012), *257*
Kerrigan, J., 20(390), 22(390, 1827), 38(390, 1827), 39(390, 1827), 80(390, 1827), 81(390), *240, 281*
Kessenikh, A. V., 10(141, 141a, 141b), *233*
Kettle, S. F. A., 30(181), 59(577), 65(577), *235, 246*
Khan, O. R., 49(485), 56(485), 97(485), 127(486), *243*
Khan, W. A., 138(720), 145(36, 720), *231, 249*
Khodabocus, M., 51(466), 77(466), 92(466), 140(466), 141(466), 151(466), 152(466), 153(466), *242*
Khol'tzanfel, K., 138(884), *254*
Khotinsky, E., 21(883), 48(883), 49(883), 56(883), 57(883), *254*
Kibby, C. L., 156(513), *244*
Kiefer, H. R., 20(381), 22(381), 46(381), 56(380), 71(379, 380), 127(379, 380), 219(380), *240*
Kilday, M. V., 16(864), *253*
Killip, K. A., 77(1779), 111(1779), *279*
King, C. S., 21(889), 77(889), 78(889), *254*
King, R. W., 148(1838), *281*
Kippur, P. R., 24(665, 666), 35(665, 666), 46(665, 666), 55(665, 666), 57(665), 64(666), *248*
Kirk, R. W., 159(885), 160(885), 212(885), *254*
Kirson, B., 77(1401), *269*
Kiselev, V. G., 28(1320, 1692), 29(1248, 1691), 47(333, 334, 1320), 51(1836c), 53(1836c), 55(1836c), 60(1320), 61(332), 62(332), 66(333, 334, 1252, 1692), 67(332, 333, 334, 1691, 1692), 68(1252), 69(1252, 1320), 74(1836c), 79(332, 334), 82(1320), 115(332, 333, 334, 1275b), 116(332, 333, 334), 118(1766c), 121(1272), 156(141c), 215(1691), *234, 238, 239, 263, 264, 266, 277, 279, 281*
Kiss-Eroess, K., 138(886), 212(886), *254*
Kitching, W., 138(887), 143(887), *254*
Klanberg, F., 136(1841), 140(1841), 212(790), *251, 281*
Klanica, A. J., 21(889), 22(888), 77(889), 78(888, 889), *254*
Klein, F. M., 29(1543), *273*
Klein, J., 29(891), 30(891), 31(891), *254*
Klein, M. J., 88(1723), 115(1146), 165(1723), 175(1723), 176(1723), 194(741, 1724), 195(741), 205(1146, 1724), 206(1724), *250, 261, 278*
Klein, R., 131(890), 174(890, *254*
Klender, G. J., 27(226), 28(226), 53(226), 60(226), 148(226), 149(226), 170(226), 172(226), 173(226), 177(226), 186(226), *236*
Klimentova, N. V., 48(913), 53(913), 92(913), 224(916), *255*
Knickel, B., 138(1776), *279*
Knights, E. F., 27(893), 28(893, 894), 29(227, 892), 31(227), 33(892, 893, 894), 66(892), 111(892), 122(894), 123(228), 172(892), 177(892), 179(892), 181(227, 893), *236, 254*
Knoth, W. H., 209(899), 210(792, 895, 896, 898, 900, 902), 211(791, 898, 901, 902), 212(791, 792, 896, 897, 898, 899, 900, 901), *227, 251, 254*
Kobetz, P., 143(624), *247*
Kobrich, G., 127(904), 130(903, 904), *254*
Kochi, J. K., 98(1002), 101(1002), 108(1002), *257*
Kodama, G., 199(905), 201(905), *254*
Kögler, H. P., 77(1673, 1674), *276*
Koehler, K. A., 57(906), *254*
König, W., 21(920), 51(920), 57(920), 58(920), 59(920), 79(920), 109(920), *255*
Köster, R., 5(690), 13(779), 20(926), 21(926, 946, 947, 959, 962), 22(129, 926, 928, 932, 933, 935, 936, 948, 951, 980), 24(130, 935, 936, 938, 943, 945, 947, 958, 963, 977), 25(130, 938, 956, 958, 959, 979), 26(977), 28(129, 130, 929, 937, 958, 959, 975), 29(929, 936, 937, 943, 979, 1617, 1618), 33(929, 936, 937), 38(947, 960, 964, 965, 970, 974, 977, 978), 41(946),

45(130), 46(131, 131a, 932, 942, 943, 947, 951, 959, 972), 47(130, 131a, 932, 951, 952, 966, 980), 48(932, 951, 952, 959, 971), 49(931, 932, 951, 959, 972), 50(931, 942, 943, 944, 952, 964), 51(597, 943, 976), 52(131a, 976), 53(951, 956, 967, 976), 54(956), 55(956), 57(932), 58(948), 59(932), 62(690, 955, 956, 976), 64(956), 65(956), 66(936, 937, 943, 944, 949, 956, 958, 959, 963, 965, 976, 977, 978, 979), 67(697, 956, 959, 965, 970, 974, 976, 977, 979), 68(947, 956, 963, 975), 69(929, 943, 945, 946, 947, 949, 958, 959, 965, 976, 978, 1617, 1618), 71(963), 72(963, 964), 73(963, 964), 74(949), 75(979), 77(931, 952, 960, 964), 80(963), 81(131), 82(946), 84(108), 86(976), 87(963), 92(108, 597, 940, 952, 976, 1618), 97(940), 107(967, 968), 109(129, 941, 944, 946, 949, 950, 951), 111(930, 932, 978), 113(129, 966), 114(129, 130, 956), 115(932), 116(932), 119(928, 930, 932, 936), 120(936), 124(129), 131(927, 928, 934, 952), 132(951), 133(941), 137(131), 138(130, 131, 972), 139(131), 141(131), 143(130), 145(130), 148(952, 962, 969, 972), 149(962), 150(951, 962), 151(929, 962, 972), 152(962), 165(963), 166(937), 168(945), 170(937), 172(937), 173(932, 937, 978), 174(928, 942, 970), 175(937, 963), 176(937, 945, 949, 952, 954, 957, 961, 963, 973), 177(932, 937, 952, 963, 970), 178(947), 949, 952, 970), 179(937, 963), 180, 181(963), 182(130, 958, 959), 185(942, 963), 187(947), 188(939, 965, 969, 970, 971, 974, 977, 978), 215(974, 979, 1617), 223(967, 968), *227, 233, 246, 249, 251, 255, 256, 257, 275*

Kohl, H., *241*

Kojima, K., 224(907, 908, 909, 910, 911), *254, 255*

Koksharova, A. A., 48(982), 88(1598), 138(912a, 1954b), 143(912, 982), *255, 257, 274, 284*

Kolb, G., 132(1909), 139(1909), *283*

Kolesnikov, G. S., 48(913), 53(913), 92(913), 224(914, 915, 916, 917), *255*

Kollonitsch, J., 48(918), *255*

Kolobova, N. E., 89(1419b), 90(1419b), *269*

Kolodkina, I. I., 62(919), 92(919), *255*

Konecky, M. S., 79(1718), *278*

Konrod, N. P., 21(1640), 57(1640), 109(1640), 139(1640), *275*

Korcek, S., 98(921), 99(921), 100(921), *255*

Koreshkov, Y. D., 216(1839), *281*

Korner, V. A., 22(1565), 48(1565), *273*

Korobeinikova, S. A., 47(1273, 1275), 49(1273, 1275), 60(1275), 61(1271), 66(1273a, 1274), 67(1271), 91(334b), 92(1271, 1273), 96(334b), 115(1271, 1273, 1273a, 1274, 1275, 1275a), 116(1271, 1273, 1273a, 1274, 1275, 1275a), 121(1275), *239, 264*

Korshak, V. V., 56(922), 58(923), 59(923), 77(922, 922a, 925), 78(922), 148(922), 185(922, 923), *255*

Korytnyk, W., 28(229), 153(229), *236*

Koski, W. S., 160(595), 163(595), *246*

Kostroma, T. V., 22(1301), 23(1301), 48(1301), 51(1301), 52(1301), 54(1301), 63(1299, 1300), 77(1300), 139(1301), *265*

Kotz, J. C., 22(984), 41(983, 984), 64(983, 984), 77(981), 152(984), *257*

Kovaleva, L. S., 139(1642), *275*

Kovredov, A. I., 21(1931), 24(985, 1931, 1932, 1933, 1934, 1935, 1936), 25(985), 28(1931, 1935, 1936), 45(1931, 1932, 1933, 1934, 1935), 66(985, 1932, 1934), 67(985, 1931), 69(985), 80(985, 1931, 1932, 1933, 1934, 1935, 1936), 81(1936), 124(1935), 168(1932), 176(1932), 180(985), *257, 284*

Kow, R., 222(1594a), *274*
Kozeschkow, K. A., 109(621), *247*
Kozima, S., 21(579), 22(579), 66(579), 216(579), 220(579), *246*
Kozitskii, V. P., 138(986, 987), *257*
Kozminskaya, T. K., 24(1307, 1309), 25(1309), 51(1311), 54(1311), 55(1302, 1304, 1305, 1311), 56(1311), 57(1305), 61(1302), 63(1303, 1305, 1306, 1311), 64(1302, 1304, 1305, 1306), 66(1255a, 1309), 67(1255a, 1307), 77(1302, 1304, 1311), 80(1309), 81(1309), 83(1308), 140(1303, 1308, 1310), 158(1307), *263, 266*
Kraus, C. A., 40(57), 47(57), 50(144), 55(144), 64(144), 91(144), 92(144), 109(57, 144), 139(1714), 140(612, 613, 991, 1714), *231, 234, 247, 257, 278*
Krause, D., 21(993), 48(993), 57(993), *257*
Krause, E., 21(992, 994, 996, 997), 47(992, 994), 48(990, 992, 994), 49(990, 992, 994, 996, 997), 56(992), 79(990), 131(989, 995), 140(988), 150(990, 996, 997), 151(988, 990), *257*
Kreevoy, M. M., 141(998), *257*
Krieg, B., 3(850a, 850b), 4(850b), 90(850b), *253*
Kriner, W. A., 22(588), 47(588), 92(588), *246*
Krishnamurthy, S., 138(230, 231, 231a), 145(230, 231a), *236*
Krohmer, P., 69(998a), 80(998a), *257*
Krongauz, E. S., *284*
Krüerke, U., 21(999, 1001), 48(999, 1000, 1001), 51(1881), 59(1881), 92(999, 1000), 139(999, 1000, 1001), 151(1000), *257, 282*
Krusic, P. J., 98(1002), 101(1002), 108(1002), *257*
Kryakova, I. V., 77(591), *246*
Kubba, V. P., 16(535), 56(537), 57(537), 62(537), 63(537), 72(524, 533, 534, 535, 537), 73(524), 217(533, 535, 536), 218(535), 222(534, 535), *244*
Kubo, M., 8(1494), 10(1494), 77(857, 1854, 1855), *253, 272, 282*
Kuchen, W., 83(1003), *257*
Kuck, J. A., 21(1719), 56(1719), 79(1719), 92(1719), 94(1719), 98(1719), 99(1719), 100(1719), 109(1719), 223(1719), *278*
Kudryavtsev, L. F., 143(876a), *254*
Kuhn, L. P., 50(853), 53(853), 86(1004), 107(853), *253, 257*
Kuimova, M. E., 28(1312), 53(1313), 54(1313), 65(1313), 66(1313), 120(1313), 121(1312, 1314), 176(1312), *266*
Kuivila, H. G., 57(1011), 58(1009, 1011, 1015), 59(1008, 1011, 1019), 62(1012), 63(1012), 79(1404), 92(1014, 1015, 1016, 1018, 1404), 93(1014, 1015, 1016), 94(1017, 1018, 1404), 100(1005, 1006, 1007), 101(1006, 1020), 104(1008, 1010, 1011, 1019, 1021), 110(1013), 222(1009, 1011), *257, 269*
Kukushkin, Yu, N., 90(879d), *254*
Kulakov, S. I., 100(1698b), *277*
Kumada, M., 46(1022), 52(1022), *258*
Kurashev, M. V., 21(1784, 1785), 47(1783, 1785), 48(1784), 53(1782, 1783, 1785), 60(1782, 1783, 1785), 92(1782, 1784, 1785), 121(1782, 1785), 149(1783, 1785), 151(1784), 224(1786), *280*
Kurita, Y., 8(1494), 10(1494), *272*
Kursanov, D. N., 216(1839), *281*
Kuwada, D., 56(380), 71(380), 127(380), 219(380), *240*
Kuznesof, R. M., 15(1023), *258*

L

Labarre, J. F., 162(1023a), 163(1023a), *258*
Lafferty, W. J., 4(1024, 1025), *258*
Lagowski, J. J., 21(22), 22(1217), 49(587), 50(587), 77(22, 429, 1217, 1233), 78(22, 1217), 124(1026), 125(1026), *230, 241, 246, 258, 262, 263*

Lalor, F. J., 149(1027), 154(1027), *258*
Lambert, L., 159(1027a), *258*
Lamneck, L., 201(1028), *258*
Landesman, H., 7(1032, 1503), 22(1030), 191(1503), 196(1029, 1030, 1031), *258, 272*
Lane, C. F., 49(233), 66(1035), 102(1034), 103(233a, 311a, 1034, 1035, 1036), 105(232, 233), 108(311a, 1034), 119(1037), 127(233a, 311a), 133(1033), *236, 238, 258*
Lange, K. R., 46(91), *232*
Lapkin, I. I., 6(1038), 49(1038), 100(1038a), 150(1038a), 153(1038), *258*
Lappert, M. F., 8(71, 72), 9(72), 12(56, 107), 21(171, 427, 650, 1040, 1043, 1046), 23(1043), 24(4, 6, 1044a), 35(871, 1046), 36(340, 871, 1040), 43(870, 871, 1046), 48(1046), 49(8), 50(71, 427, 555, 556, 648, 650), 51(427, 455, 648, 860, 1043, 1048, 1049), 52(2, 3, 4, 5, 6, 572, 651, 652, 1049), 53(572, 650, 651, 652), 54(5, 650), 55(2, 5, 454, 650, 1044a, 1046), 56(70, 71, 454, 455, 456, 860, 1044, 1045, 1048), 57(70, 71, 871, 1042, 1043), 58(470), 59(56), 60(1046), 61(171, 871, 1046), 62(6, 171, 172, 173, 469, 572, 652, 1049), 63(172, 173, 456, 468, 469, 1047), 64(871, 1044a, 1046), 65(2, 6, 468, 871, 1046), 69(871), 78(1043), 79(2, 5, 6, 56), 80(71), 88(456, 870), 89(870, 871), 91(172), 92(4, 6, 650, 652, 1040, 1046), 97(4), 99(651), 102(6), 108(870), 109(648), 110(648), 113(1046), 115(455, 555, 860, 870), 116(455, 860, 870, 1046), 124(1046), 140(1040), 151(8, 650), 152(8, 173), 153(8, 1039, 1041), 158(650), 209(107, 572), 219(427, 555, 556), *226, 227, 230, 231, 232, 233, 234, 239, 241, 242, 245, 247, 248, 253, 258*
LaPrades, M. D., 195(1050), *258*
Larbig, W., 21(959), 25(959), 28(959), 38(965), 46(959), 47(966), 48(959), 49(959), 66(959, 965), 67(959, 965), 69(959, 965), 113(966), 182(959), 188(965), *227, 256*
Lardicci, L., 109(1051), *258*
Larkworthy, L. F., 11(484), 107(15), *230, 243*
Larock, R. C., 109(1052, 1053, 1054), 110(1053, 1054), *258*
Larsson, E. N. V., 21(1799), 56(1799), 57(1799), *280*
Laubengayer, A. W., 6(1055), 21(1379), 78(1379), *258, 268*
Laurent, J.-L., 148(1056), *258*
Laurent, J. P., 24(1059, 1812), 25(1811), 47(1059), 48(1059), 49(1059), 50(1058, 1059, 1060), 51(1060), 52(1059), 53(1059, 1064), 54(1059, 1061), 55(1058, 1059, 1060, 1063), 59(1057), 60(1057, 1812), 61(1057), 62(1064, 1811), 63(1061, 1063), 64(1043), 65(1063), 92(1059), 148(641a, 872, 1056, 1062), 149(873), 151(1812, 1813), 152(1812, 1813), *247, 253, 258, 259, 280*
Lavrinovich, L. I., 50(556b), 86(556a), 92(556a, 556b, 1283a), *245, 265*
Law, R. W., 13(1065), 157(1065), *259*
Lawesson, S. O., 21(1065a), 23(1065a), 56(1065a, 1065b), 59(1065c), *259*
Lawrance, R. B., 162(1867), *282*
Laws, E. A., 207(1065d), *259*
Lawson, G., 13(457), 14(452), *242*
Layton, R. B., 127(837a), *252*
Leach, J. B., 8(1066, 1068, 1815), 21(1066), 55(1066), 56(1066), 188(706), 190(1068), 196(1815), 198(1815), 199(1815), 201(1067, 1815), 202(706), 204(1067, 1815), 216(706), *226, 249, 259, 280*
Leatham, M. J., 22(164), *234*
Ledwith, A., *232*
Lee, D. A., 139(1069), *259*
Lee, J., 8(71, 72), 9(72), 50(71), 56(70, 71), 57(70, 71), 80(71), *232*
Lee, L. T. C., 42(1818), 94(1818), 125(1818, 1819), *280, 281*
Leffler, A. J., 79(1070), *259*
Leffler, J. E., 13(1591), 49(1074),

51(1073), 109(1071), 127(1072), 129(1073), 132(1071, 1074), *259, 274*

Lehmann, H., 51(684) 59(684), 70(677), 91(677, 684), 92(684, 1075), 93(677), 135(677, 1075), *248, 259*

Lehmann, W. J., 12(1078, 1080, 1081, 1084), 13(1082, 1083, 1089), 33(1866), 36(1076), 46(1080, 1084, 1683), 47(1683), 79(1089), 176(1083, 1086, 1087, 1088, 1683, 1866), 177(1866), 190(1076, 1077, 1078, 1079, 1083, 1085, 1086, 1087, 1088, 1685, 1866), 192(1086, 1683), *259, 276, 277, 282*

Leibovici, C., 162(1023a), 163(1023a), *258*

Lengyel, B., 64(1090), 65(1090), *259*

Lennarz, W. J., 21(1091), 49(748), 51(748), 53(748), 56(748), 57(1720), 58(1091, 1720), 59(748), 72(1720), 74(748, 1092, 1720), 76(1720), 79(748, 1718), 222(748), *250, 259, 278*

Lenz, H., 36(1519), *272*

Lepeska, B., 31(1544, 1545), 166(1544), 186(1544), *273*

Lepley, A. R., 16(538), 72(524), 73(524), *244*

Letsinger, R. L., 11(1105), 21(1094, 1096, 1102, 1103, 1104, 1105, 1106, 1107, 1108, 1110), 23(1102, 1107, 1109), 25(1102, 1107, 1109), 36(1100), 50(1106), 51(1104), 52(1104, 1107, 1109), 53(1102, 1104, 1106, 1107, 1109), 55(1097), 56(1095, 1097, 1105), 57(1096, 1099, 1110), 58(1096), 61(1106), 62(1099, 1107), 63(1107), 68(1108), 75(1103), 84(1103), 86(1100), 87(1093, 1094, 1103), 92(1104), 96(1104), 97(1105), 100(1105), 151(1107), 224(1095, 1098, 1101), *259, 260*

Leusink, A. J., 22(1111), 67(1111), 214(1111), 216(1111), *260*

Levens, E., 21(1850, 1851), 23(1850, 1851), 51(1850), 57(1850, 1851), 62(1850), 79(1852), 92(1850), *281, 282*

Levina, I. S., 115(24a), *230*

Levison, K. A., *260*

Levy, A. B., 28(240), 64(234), 65(234), 127(241, 833a), 128(240, 241a, 1112a), *236, 252, 260*

Levy, H. A., 4(1113), *260*

Levy, L. A., 124(1114), 131(1114), *260*

Lewis, E. S., 155(772), 156(772, 1115), *251, 260*

Lewis, R. N., *228*

Lhomme, J., 97(1763a), *279*

Liberman, G. S., 31(1426), 48(1426), 49(1426), 138(1426), 139(1426), *270*

Lide, D. R., Jr., 3(1116, 1117), *260*

Liedtke, J. D., 59(1179), 60(1178), 61(1178, 1180), 92(1179), 113(1180), 219(1178), 221(1179), *261*

Lienhard, G. E., 57(906), *254*

Likussar, W., 141(1583a), *274*

Lindner, H. H., 8(1119), 28(1118), 32(1118), 169(1118), 170(1118), 176(1118), 177(1118), 190(1118, 1119), *260*

Lindstrom, K., 51(644a), *247*

Linke, S., 127(838, 839), *252*

Lippard, S. J., 138(1120), 146(1120), *260*

Lippincott, W. T., 20(1575), 21(1575), 51(1575), 78(1575), *274*

Lipscomb, W. N., 4(625, 1389a, 1501), 197(1759), 199(1121, 1501), 201(1501), 203(625, 1501), 207(1065d), 216(1120a), *226, 227, 247, 259, 260, 269, 272, 279*

Lissi, E. A., 17(712), 98(713), 99(713), 103(712), 104(712), 108(16, 584, 709, 710, 711, 712, 714, 1122, 1123), 111(17), 160(715, 716), 188(1123), 194(714), 224(431), *230, 241, 246, 249, 260*

Livingstone, J. G., 21(422), 23(422), 51(420, 422), 52(422), 53(420, 422), 54(420), 55(422), 152(419), 219(420, 422), *241*

Lloyd, D. R., 77(1125), 162(1124, 1126), *260*

Lloyd, H. H., 156(513), *244*
Lloyd, J. E., 34(1127), 83(1127), 169(1127), *260*
Lockhart, J. C., 11(1129), 12(1130), 23(601), 57(133), 61(601), 62(601, 804), 64(606), 140(133), 152(133, 1128, 1131), 153(1131), *227, 233, 246, 252, 260*
Lockman, B., 59(1696), 63(1696), 65(1131a), *260, 277*
Logan, T. J., 28(1133), 46(1133), 111(1132), 170(1133), *260*
Logodzinskaya, G. V., 134(1255), *263*
Long, J. R., 59(1696), 63(1696), *277*
Long, L. H., 12(371), 16(1135), 24(1137), 47(1134, 1138), 54(1138), 55(553, 1134, 1138), 92(1135), 108(1134), 167(1136, 1137, 1138, 1139), 173(1139), 176(1137, 1138, 1139), 190(371, 372, 373), 192(373), *240, 245, 260*
Loomis, C. C., 162(1867), *282*
Lopatin, B. V., 51(1696a), 52(1696a), 53(1696a), *277*
Lory, E. R., 194(514), *244*
Love, P., 3(1117), 18(1140), 49(1141), 148(1141), 153(1141), *260*
Lovejoy, R. W., 12(1600), 62(1600), 79(1600), *274*
Lowe, J. U., 208(1709), 209(1709), *277*
Lowman, D. W., 10(580), 201(1141a), 204(580), *246, 260*
Lown, J. W., 182(547a), *244*
Lucchesi, C. A., *226*
Luft, R., 27(592), 28(592, 593a), 181(592), 182(593), *246*
Lukas, S., 21(1470), 57(1470), 158(1470), *271*
Luna, I. G., 223(793), *251*
Lutsenko, G. A., 119(1775), 121(1775), *279*
Lutz, C. A., 193(1142), 194(1142), 205(1142), 206(1142), *260*
Lyle, R. E., 11(1143), 21(1143), 47(1143), 48(1143), 49(1143), 219(1143), *260*
Lynaugh, N., 77(1125), 162(1124, 1126), *260*
Lynds, L., 28(1144), 64(1144), 65(1144), *260*
Lyons, A. R., 109(1145), *261*

M

McClendon, J. H., 224(1800), *280*
McCloskey, A. L., 22(1922), 24(1431), 44(179), 47(1922), 55(585), 57(179), 77(180), 78(1431), 80(179), 135(1430), 152(179), 159(1922), 218(180), *228, 234, 246, 270, 283*
McCormack, C. G., 15(182), *235*
McCoy, R. E., 163(1200), 164(1200), 187(1200), *262*
McCusker, P. A., 5(465), 21(781, 782, 783, 784, 1207, 1383, 1616), 22(781, 782, 784, 1383), 23(784), 24(780, 782, 783, 784, 1203, 1205, 1207), 25(1204, 1205), 47(781, 1207), 48(781, 782, 784, 1207), 49(782, 1616), 53(784), 54(1207), 55(1205), 56(1201), 62(1204, 1206), 63(1206), 64(1202), 65(1202, 1206), 77(1383), 79(780, 1201, 1203), 98(1204), 111(780, 781, 1209, 1615), 112(1615), 173(781), *242, 251, 262, 268, 275*
McDaniel, D. H., 218(1210), *262*
MacDiarmid, A. G., 22(588), 47(588), 92(588), *246*
MacDonald, A. A., 212(738, 739), 213(739), *250*
McEwen, W. E., 21(1447), 48(1448), 51(1448), 53(1448), 63(1448), 65(1447), 81(1447), 82(1447), 151(1448), 168(1448), 174(1448), 179(1448), *270*
McGee, H. A., 160(642), 163(643), *247*
McGinnis, G. D., 62(1682), 224(1682), *276*
Mack, J. L., 208(1708, 1709), 209(1709), *228, 277*
McKenna, J., 30(181), *235*
McKenna, J. M., 30(181), *235*
Mackie, R. K., 77(401), *240*
McKinley, I. R., 62(1211), *262*
McLafferty, J. J., 21(385, 1389), 139(385, 1389), *240, 269*

McLaughlin, D. E., 148(1212), *262*
Maclean, D. B., 36(1100), 86(1100), *259*
McMullen, J. C., 24(1213), 42(1213), 46(1213), 47(1213), 141(1213), 148(1213), 149(1213), 158(1213), 175(1213), 177(1213), 180(1213), *262*
McNamara, P. M., 61(602), 62(602), *246*
McReynolds, K., 60(1545a), 103(1545a), 108(1545a), 220(1545a), *273*
Maddox, M. L., *227*
Maguire, R. G., 115(1146), 194(1724), 205(1146, 1724), 206(1724), *261, 278*
Mah, R. W. H., 21(1181), 51(1181), 52(1181), 59(1183), 60(1183), 61(1182, 1183), 220(1181), 221(1183), *261*
Maier, D. P., 44(1893, 1894), 145(698), *249, 283*
Maina, G., 28(367), 92(367), *239*
Maisch, H., 36(1519), *272*
Maitlis, P. M., 16(1147), 21(1147, 1148), 37(404), 56(405), 70(1147), 72(404, 540, 541, 1147), 75(404, 539), 79(1148), 109(541), 217(1147), *227, 241, 244, 261*
Majert, H., 51(1214), 92(1214), 93(1214), 121(1214), *262*
Majumdar, M. K., 21(1043), 23(1043), 51(1043), 56(1044), 57(1042, 1043), 78(1043), *258*
Makaeva, S. Z., 28(1954), 47(1954), 48(1954), *284*
Makarov, Yu. V., 89(1419b), 90(1419b), *269*
Makin, G. I., 100(1698b), 101(1148a), 102(1698a), *261, 277*
Makosza, M., 21(1668, 1712), 58(1667, 1668, 1712), 106(1667), *276, 278*
Makowski, H. S., 5(465), 56(1201), 62(1206), 63(1206), 64(1202), 65(1202, 1206), 79(1201), *242, 262*
Malhotra, S. C., 43(1149), 83(1149), *261*
Malone, L. J., 12(1152), 56(1151), 138(811, 1150, 1152, 1952), 159(1152), 161(811, 1150, 1151, 1152, 1952), *252, 261, 284*
Maltesson, A., 71(703a), 76(703a), *249*
Manabe, O., 53(1928), 107(1928), *283*
Manasevit, H. M., 44(179), 57(179), 80(179), 152(179), *234*
Mangravite, J. A., 92(1018), 94(1017, 1018), *257*
Manley, M. R., 56(1151), 161(1151), *261*
Manley, R. M., 138(811), 161(811), *252*
Mappes, G. W., 160(1153, 1154), 163(596), *246, 261*
Maraschini, F., 16(148), *234*
Marecek, H., 21(1232), 22(1220), 24(1220), 78(1220, 1232), 84(1220), *262, 263*
Margerison, D., 224(89), *232*
Margoshes, M., 162(1155), 163(1155), *261*
Margrave, J. L., 13(1065), 157(1065), 209(1156), *259, 261*
Maringgele, W., 21(1223), 22(1223), 49(1225, 1226), 50(1225), 51(1225), 57(1226), 83(1221, 1222, 1223, 1224), 151(1224, 1226), 152(1224, 1226), 219(1221, 1222, 1223, 1224, 1225), *262, 263*
Marquez, L. A., 28(794, 796), 182(794), *251*
Marr, G., 89(1157), *261*
Marr, P. A., 71(463, 542), 72(463), 75(463), 76(463), 215(463), 217(463, 542), *242, 244*
Marra, J. V., 21(782, 783, 1207), 22(782), 24(782, 783, 1207), 28(1207), 47(1207), 48(783, 1207), 49(782), 54(1207), *251, 262*
Marsh, J. F., 28(132), 32(132), 46(132), 47(132), 53(132), 62(132), 113(132), 124(132), 182(132), *233*
Marshall, A. S., 8(165), 49(165), 50(165), 51(165), *234*
Marshall, J. A., 114(1158), *261*
Marsico, W. E., 92(1864), 94(1864), 95(1864), *282*
Martin, D. R., 74(1159), 224(1159, 1160), *227, 261*
Martinek, K., 224(34a), *231*

Maryanoff, B. E., 145(849, 850), *253*
Marynick, D., 204(1161), 205(1161), *261*
Maseles, F., 13(430), 14(430), *241*
Maslennikov, V. P., 100(1698b), 101(1148a), 102(1698a), *261, 277*
Maslov, P. G., 16(1162), 53(1162), 54(1162), 55(1162), 61(1162), 62(1162), 63(1162), 64(1162), 65(1162), *261*
Massey, A. G., 8(1168), 9(1168a), 10(1168, 1168a), 21(1164, 1166, 1167), 22(1166), 38(821), 49(1164, 1165, 1167), 77(1166), 80(819, 820), 81(819), 91(461), 92(461), 115(819), 138(1168a), 139(1164, 1167), 150(1164, 1165, 1167), 153(1165), 223(461, 1163), *227, 242, 252, 261*
Matsui, Y., 147(1169), *261*
Matsumoto, H., 115(1752), 117(1752), *279*
Matsumoto, T., 224(634), *247*
Matteson, D. S., 20(1184, 1186), 21(1171, 1181, 1184, 1185, 1186, 1187), 22(1177), 23(1171, 1186, 1187), 29(1189), 31(1190), 40(386), 44(386), 45(1177, 1189, 1190), 51(1181), 52(1181, 1187), 55(1170), 56(1177, 1185, 1188, 1191), 57(386), 59(1170, 1171, 1172, 1174, 1177, 1179, 1183, 1184, 1185, 1188), 60(1170, 1178, 1183, 1184, 1185, 1186, 1188, 1194), 61(1176, 1177, 1178, 1180, 1182, 1183, 1184, 1185, 1193, 1196, 1197), 70(1171), 76(1191), 79(1172), 80(386, 1177, 1183a, 1189, 1190), 82(386, 1183a, 1190), 87(1173), 89(1177a), 92(386, 1170, 1173, 1179, 1184, 1185), 94(1184, 1185), 95(386), 108(1174), 109(1175, 1176, 1190, 1192, 1193, 1195, 1196, 1197), 110(1175), 113(1180), 115(1190), 219(1178, 1188, 1193, 1194), 220(1181, 1184, 1185, 1191), 221(1179, 1183, 1184, 1188), 222(1177), *227, 240, 261, 262*
Matthies, P., *232*
Mattschei, P. K., 80(1183a), 82(1183a), 202(1504), *261, 272*
Mauchamp, J., 223(1198), *262*
Maurel, J., 208(721), *249*
Mayer, E., 159(1199), *262*
Ma Zhui-Zhan, *255*
Mazur, Y., 28(1492, 1917), 30(1917), *272, 283*
Medvedera, V. G., 89(1425), *270*
Meerwein, H., 50(1215), 51(1214, 1215), 92(1214, 1215), 93(1214), 121(1214), *262*
Mehrotra, I., 93(1216), *262*
Meissner, B., 6(1215a), 52(1732b), 58(1732b), 59(1732b), 63(1732b), *262, 278*
Meister, W., 50(1469), *271*
Melamed, M., 21(883), 48(883), 49(883), 56(883), 57(883), *254*
Melcher, L. A., 22(1217), 77(1217), 78(1217), *262*
Meller, A., 21(725, 1223, 1231, 1232), 22(724, 1218, 1220, 1223, 1230), 24(1220), 49(1225, 1226), 50(1225), 51(1225), 57(1226), 77(724, 725, 1227, 1230, 1231), 78(724, 725, 726, 727, 1218, 1220, 1227, 1232), 83(1221, 1222, 1223, 1224, 1228, 1229), 84(728, 1219, 1220), 106(1227), 151(1224, 1226), 152(1224, 1226), 219(1221, 1222, 1223, 1224, 1225, 1229), *227, 250, 262, 263*
Mellon, E. K., 20(1768), 22(1768), 77(1233, 1768), *263, 279*
Merkle, H. R., 127(904), 130(903, 904), *254*
Meyer, H., 57(1801), 58(1801), 59(1801), *280*
Mezey, E. J., 196(1628), 197(1628, 1630), 201(1628, 1630), 202(1628), *275*
Michael, V., 51(703b), 76(703b), *249*
Michaelis, A., 20(1236), 22(1234, 1236), 51(1234, 1237), 54(1237), 55(1234), 57(1234, 1235), 58(1237), 59(1234), 62(1235), 63(1234), 64(1234, 1237), 65(1234, 1235, 1236, 1237),

79(1234, 1235, 1237), 92(1234), 106(1236), *263*

Middleditch, B. S., 224(178), *234*

Midland, M. M., 28(240), 97(238, 239, 1238, 1238a, 1754), 98(237, 1238), 99(237), 100(235), 102(236), 108(236), 115(1377), 117(1377), 127(241, 833a), 128(240, 241a, 1758), *236, 252, 263, 268, 279*

Miesel, J. L., 32(1546, 1547), *273*

Miginiac, L., 21(1239), 22(1239), 62(1239), *263*

Mikhailov, B. M., 3(1339), 13(1450), 16(639, 640), 18(1332), 21(1241, 1242, 1244, 1249a, 1250, 1276, 1292, 1297, 1323, 1325, 1328, 1330, 1337, 1339, 1352a, 1353, 1354, 1356, 1357, 1359, 1450), 22(1301), 23(1244, 1301, 1323, 1354), 24(331, 1277, 1279, 1307, 1309, 1316, 1318, 1329, 1332a, 1333d, 1343, 1344, 1345, 1346, 1347, 1348, 1352, 1390), 25(1309, 1344, 1347, 1348), 26(1329), 28(331, 331a, 1254, 1278, 1284, 1285, 1286, 1312, 1318, 1320, 1321, 1327, 1333d, 1390, 1692), 29(330, 554, 1248, 1280, 1285, 1330, 1331, 1333, 1691), 31(1254, 1327), 32(1333d), 35(1284, 1315), 36(1315), 44(1330a), 45(1247), 46(331, 331a, 1253, 1254, 1280, 1284, 1323, 1327, 1330a, 1390), 47(333, 334, 336, 1256, 1273, 1275, 1320, 1323, 1327, 1337, 1338, 1359, 1390), 48(140a, 1278, 1279, 1301, 1315, 1355, 1450), 49(335a, 1273, 1275, 1315, 1359), 50(556b, 558, 1240, 1245, 1253, 1257, 1258, 1259, 1260, 1261, 1264, 1265, 1280, 1284, 1328, 1337, 1341, 1345, 1346, 1361, 1690), 51(1243, 1244, 1245, 1254, 1261, 1264, 1280, 1281, 1282, 1289, 1291, 1292, 1293, 1295, 1301, 1311, 1328, 1351, 1352a, 1357, 1360, 1450, 1836b, 1836c), 52(331, 1240, 1241, 1242, 1244, 1245, 1253, 1254, 1258, 1261, 1263, 1277, 1282a, 1286, 1292, 1301, 1322, 1323, 1328, 1345, 1346, 1347, 1348, 1352a, 1356, 1357, 1359, 1361, 1390, 1450), 53(1240, 1241, 1263, 1264, 1267, 1277, 1278, 1313, 1322, 1347, 1348, 1349, 1353, 1358, 1359, 1361, 1450, 1836a, 1836b, 1836c), 54(1240, 1244, 1253, 1256, 1257, 1258, 1263, 1278, 1288, 1292, 1301, 1311, 1313, 1315, 1331, 1340, 1342, 1349, 1361, 1390, 1450), 55(1244, 1288, 1302, 1304, 1305, 1311, 1324, 1342, 1345, 1346, 1349, 1836c), 56(1254, 1281), 1282, 1311), 57(1282, 1305, 1342, 1450), 59(331, 331a, 1246, 1253, 1254, 1325, 1339), 60(628, 1246, 1249a, 1256, 1267, 1275, 1320, 1324, 1325, 1337, 1342, 1345, 1346, 1347, 1348, 1349, 1390), 61(332, 628, 1271, 1302, 1345, 1347, 1348, 1360, 1836a), 62(332, 1242, 1290, 1329, 1450), 63(1250, 1266, 1290, 1299, 1300, 1303, 1305, 1306, 1311, 1324, 1325, 1331, 1332a, 1339, 1340, 1342, 1344, 1349), 64(1302, 1304, 1305, 1306, 1325, 1326), 65(1250, 1313), 66(333, 334, 334a, 1249, 1252, 1255a, 1268, 1270, 1271a, 1273a, 1274, 1276, 1276a, 1276c, 1278, 1309, 1313, 1330, 1333, 1333a, 1333b, 1333c, 1350, 1352, 1692), 67(332, 333, 334, 628, 1249, 1255a, 1271, 1276b, 1276c, 1307, 1330, 1333, 1333b, 1352, 1692), 68(1252, 1276b), 69(1252, 1316, 1320, 1338, 1352), 71(554, 554b, 557, 1283, 1285, 1286a, 1319, 1351, 1836b), 72(1350), 74(1249, 1836a, 1836c), 75(554a, 1286), 76(554a), 77(1243, 1265, 1300, 1302, 1304, 1311, 1342, 1450), 78(1297, 1298), 79(332, 334, 628, 1250, 1292, 1337, 1342), 80(1247, 1309, 1318, 1319, 1321, 1350, 1352), 81(1309, 1316, 1318, 1319, 1321), 82(1317, 1320, 1352), 83(46, 140, 330, 335, 559, 1256, 1287, 1308), 84(559a, 1286b), 85(1283), 86(556a, 1262), 87(557), 90(1326), 91(334b), 92(331, 335a, 336, 556a, 556b, 559b, 1240, 1245, 1246, 1251, 1253, 1259, 1260, 1261,

1262, 1263, 1264, 1265, 1266, 1271, 1273, 1278, 1282b, 1283a, 1285, 1290, 1293, 1315, 1337, 1340, 1341, 1342, 1345, 1359, 1360, 1361), 93(1251, 1278, 1340, 1342), 95(1359), 96(334b, 1261, 1263, 1264, 1266, 1278, 1345), 102(1836a, 1836b), 103(1836a), 107(1257, 1258, 1326), 108(1263, 1266, 1342), 113(1246, 1269a), 114(1254, 1269a), 115(24a, 330, 332, 333, 334, 335, 336, 628, 1246, 1249, 1268, 1269, 1270, 1271, 1271a, 1273, 1273a, 1274, 1275, 1275a, 1275b, 1276a, 1333d, 1335, 1766b), 116(332, 333, 334, 334a, 628, 1249, 1267, 1268, 1269, 1269a, 1270, 1271, 1273, 1273a, 1274, 1275, 1275a, 1276a, 1333a, 1333c), 117(336), 118(1334, 1335, 1336, 1766c), 120(1313, 1327, 1335, 1338), 121(1272, 1275, 1312, 1314), 134(140a, 142, 1255), 137(46), 138(1354), 139(1301, 1354), 140(1242, 1291, 1293, 1303, 1308, 1310, 1328, 1343, 1352a, 1353, 1354, 1357, 1358, 1359), 148(1256, 1280, 1281, 1282, 1286), 149(1260, 1281, 1282, 1341), 150(1260, 1354, 1355), 151(1289, 1292, 1295, 1354, 1355, 1450), 152(1329), 155(1355), 156(1836a, 1836b), 157(1294), 158(1294, 1296, 1307, 1690), 165(1240, 1279), 172(1277, 1279, 1281, 1344, 1345), 175(1240, 1277, 1278), 176(1240, 1312, 1345, 1346), 177(1277, 1278, 1279, 1281), 179(1277), 180(1277, 1279, 1343, 1344, 1345, 1346, 1347), 182(1240, 1278), 183(1278), 185(1240, 1280, 1281, 1282), 187(1277), 208(336), 215(1691), 219(330, 335, 1246, 1287), 223(1323), *227, 230, 231, 233 234, 238, 239, 245, 247, 263, 264, 265, 266, 267, 268, 269, 271, 277, 279, 281*

Mikhoilova, L. N., 224(1362), *268*

Mikulaschek, G., 50(1472), 151(1471), *271*

Milewski, C. A., 145(849, 850), *253*

Milks, J. E., 12(1371), 56(1371), 85(1371), *268*

Miller, B., 21(1619), 52(1619), 53(1619), 86(1620), *275*

Miller, C. D., 51(1445), 84(108), 92(108), *233, 270*

Miller, D. S., 49(1074), 132(1074), *259*

Miller, F. M., 193(1363), 194(1363), *268*

Miller, H. C., 7(1569), 83(23), 137(23), 199(23), 209(23), 210(900, 902), 211(902), 212(897, 900), *230, 254, 273*

Miller, J. J., 83(1364, 1365), 148(1365), *268*

Miller, J. T., Jr., 51(1530), 97(1530), 99(1530), 179(1530), *273*

Miller, N. E., 24(1213), 42(1213, 1366, 1368, 1369), 46(1213), 47(1213), 71(1366, 1367, 1368, 1369), 141(1213), 148(1213), 149(1213), 157(1367), 158(1213, 1366, 1601), 159(1601), 175(1213), 177(1213), 180(1213), 210(898), 211(898), 212(898), 215(1367), 219(1366, 1369), *254, 262, 268, 274*

Millero, F. J., 138(1370), *268*

Mills, J. L., 148(444), 154(444), *242*

Minato, H., 100(1372), 101(1372), *268*

Minoura, Y., 29(1373), *268*

Mirviss, S. B., 98(1374, 1375), 99(1374), *268*

Mischke, P., 72(1902a), 127(1902a), *283*

Mitsuomi, I., 97(1754), *279*

Miyamoto, N., 115(1376), 117(1376), *268*

Miyaura, N., 74(1752b), 97(1753), 115(1377, 1753a), 117(1377), 129(1757), 131(1756b), *268, 279*

Moedritzer, K., *227*

Möhrle, H., 52(1382a), *268*

Moeller, C. W., 132(1378), *268*

Moerikofer, A. W., 27(242, 243, 244), 28(242, 244), 29(243), 30(242), 33(244), 49(243), 53(243 111(243), 170(242, 243), 173(243), 177(243), 181(242, 244), 182(242), *236*

Moews, P. C., Jr., 21(1379), 78(1379), 149(1380), 158(1380), 183(1380), 184(1380), *268*
Mogel, L. G., 62(1381), *268*
Mohanty, R. K., 138(1382), *268*
Mohler, F. L., 88(1844), 89(1844), *281*
Molinari, M. A., 21(1383), 22(1383), 77(376, 1383), *240, 268*
Moll, R. E., 47(26), 49(26), *231*
Mongeot, H., 194(513a), *244*
Montalbano, D., 57(1636), 58(1636), 79(1636), 92(1636), *275*
Moodie, R. B., 100(495), *243*
Mooney, E. F., 8(1384), 9(1384), 12(344, 361, 362), 21(653), 22(346, 649, 655), 50(653), 51(653, 655), 54(655, 656), 55(655), 56(347), 60(655), 61(655), 62(73, 344, 1388), 63(73), 64(655), 65(346, 649, 655), 77(345, 347, 654, 655, 656), 79(656), 148(1385), 152(1386, 1387), 218(654), *232, 239, 248, 268, 269*
Moore, C. E., 10(1832), 12(1832), 21(385, 1389, 1831), 139(385, 1389, 1831), *240, 269, 281*
Moore, C. J., 138(887), 143(887), *254*
Moore, E. B., 4(1389a), *269*
Moore, L. O., 20(661), 22(661), 57(661), *248*
Moore, R. E., 89(1157), *261*
Moore, W. R., 182(1389b), *269*
Moormeier, L. F., 21(1451), 48(1451), *271*
Moreland, C. G., 83(1732), 92(1732), *278*
Morita, Y., 38(960, 964), 50(964), 53(967), 72(964), 73(964), 77(960, 964), 107(967, 968), 148(969), 188(969), 223(967, 968), *256*
Moritani, I., 20(1763), 22(1763), 65(1763), 80(1763), *279*
Morris, J. H., 8(1066), 21(1066), 28(693), 29(692, 693), 55(803, 1066), 56(803, 1066), 69(691, 692), 83(693), 148(589), 151(692), 152(589), 186(589), 214(693), *226, 246, 249, 252, 259*
Morrison, G. F., 127(840), *252*
Morrison, J. D., 56(1095), 224(1095, 1101), *259, 260*
Mostovoi, N. N., 24(1390), 28(1284, 1285, 1286, 1390), 29(1285), 35(1284), 46(1284, 1390), 47(1390), 50(1284), 52(1286, 1390), 54(1390), 60(1390), 71(1285), 75(1286), 92(1285), 148(1286), *265, 269*
Mousseron, M., 176(1391), 181(1391), *269*
Müllbauer, R., 50(1514), 54(1514), 55(1514), 79(1515), 129(1515), 150(1514), 151(1514), *272*
Mueller, J., 90(786), 134(786), *251*
Müller, K. H., 48(971), 188(971), *256*
Muetterties, E. L., 7(1569), 36(1393, 1394, 1395), 41(1394), 64(1394, 1395), 65(1393, 1394), 71(1367), 83(23), 92(1392), 137(23), 157(1367), 168(1394), 169(1394), 199(23), 209(23, 899), 210(792, 900, 902), 211(791, 901, 902), 212(790, 791, 792, 897, 899, 900, 901), *227, 230, 251, 254, 268, 269, 273*
Mukaiyama, T., 53(1395a), 126(1395b), *269*
Muller, H., 38(247), 48(787), 67(247), 89(787), 134(787), *236, 251*
Muller, T. C., 110(1013), *257*
Mulligan, B. W., 12(134), *233*
Mulliken, R. S., 6(1396), *269*
Munekata, T., 28(1967), 29(193), 32(1967), 111(193), 119(193), 182(1967), *235, 285*
Murakami, K., 224(910), *255*
Muraki, M., 53(1395a), *269*
Murphy, C. B., 204(1397, 1398), *269*
Murphy, C. J., 58(1009), 222(1009), *257*
Murphy, M. D., 42(1368), 71(1368), 219(1368), *268*
Murphy, W. S., 106(217), 133(217), *235*
Murray, K. J., 28(248), 29(245, 248), 38(247), 67(247), 91(245, 246), 92(245, 246), 93(245, 246), 96(245), *236*
Murray, L. J., 28(248), 29(248), *236*

Musgrave, O. C., 21(438, 441), 43(150), 57(441, 1399), 79(441), 80(438), 81(1399), 82(438), 86(1400), 88(438), *234, 241, 269*
Muszkat, K. A., 77(1401), *269*

N

Naar-Colin, C., 209(805), *252*
Nadeau, H. G., 131(890), 174(890), *254*
Nagasawa, K., 49(1403, 1856), 50(1403, 1856, 1857), 77(1402), 218(1402), *269, 282*
Nagase, K., 33(1973, 1974), 170(1973), 181(1974), *285*
Nagayama, M., 224(908), *255*
Nahabedian, K. V., 58(1015), 79(1404), 92(1014, 1015, 1016, 1404), 93(1014, 1015, 1016), 94(1404), *257, 269*
Nahm, F. C., 8(1405), 12(1405), 51(1445), *269, 270*
Nakagawa, T., 8(1494), 10(1494), 49(1856), 50(1856, 1857), 77(1858), *272, 282*
Nakane, R., *269*
Nakayama, Y., 224(1407, 1408), *269*
Nam, M. H., 120(1409), *269*
Nambu, H., 95(250, 251), 126(249, 250, 251, 252, 280, 1410, 1411), *236, 237, 269*
Namtvedt, J., 71(1413), 76(704, 705, 1412), *249, 269*
Narisada, M., 77(1858), *282*
Naruse, M., 134(1414), *269*
Naumov, A. D., 71(1283), 83(140), 85(1283), *233, 265*
Naylor, R. E., Jr., 5(1415), *269*
Nazarova, I. I., 146(1641), *275*
Nazy, J. R., 21(1096, 1102, 1103), 23(1102), 25(1102), 53(1102), 57(1096), 58(1096), 75(1103), 84(1103), 87(1103), *259, 260*
Neece, G. A., 219(1439a), *270*
Negishi, E., 24(263, 264), 28(260, 261, 262, 268, 1418), 29(253, 254, 260, 264, 265, 266, 1418), 33(268), 46(1419), 60(268), 61(268), 62(268, 1417), 66(262, 360a, 1418), 67(257, 260, 261, 262, 263, 265, 267, 360a, 1418), 80(263, 264), 92(263, 266), 93(263, 266), 97(259), 111(1418), 118(258, 1756), 122(253, 254, 255, 256, 257, 267, 1417), 123(255, 1416), 130(1419a), 138(259), 148(268), 156(268), 166(263, 264, 265, 1418), 170(259, 265), 171(260, 261), 172(260), 176(263, 264), 177(259, 260, 261, 263, 1418, 1419), 178(1418), 180(259, 1418), 182(262), 186(1419), 215(261, 360a, 1418), *225, 236, 237, 239, 269, 279*
Negrebetsky, V. V., 10(141, 141a, 141b), *233*
Nesmeyanov, A. N., 21(1421, 1422, 1423, 1426), 48(1426), 49(1426), 89(1419b, 1420a, 1423, 1424, 1425, 1426a), 90(1419b), 92(1424), 97(1423, 1424), 109(621, 1424), 133(1424), 138(1420, 1426), 139(1421, 1422, 1426), 150(1422), *247, 269, 270*
Nesmeyanova, O. A., 47(336), 92(336), 115(336, 1275b), 117(336), 208(336), *239, 264*
Neu, R., 21(1427), 52(1427), 56(1429), 92(1428, 1428a, 1429), 141(1428a, 1428b), *270*
Neville, A. F., 18(824), *252*
Newitt, D. M., 97(67), 99(67), *232*
Newsom, H. C., 24(1431), 78(1431), 135(1430), *270*
Nichols, L. D., 15(1737), *278*
Niedenzu, K., 8(1405), 10(1446), 11(1446), 12(1405), 20(1434), 22(626, 1434, 1436, 1438, 1446), 23(1433, 1434, 1440), 24(1432, 1433, 1444), 37(1435), 49(1433, 1434, 1438, 1440, 1443), 50(62, 1433, 1434, 1436, 1438, 1440, 1443), 51(120, 861, 1432, 1433, 1434, 1436, 1438, 1444, 1445, 1846), 55(626, 1432, 1441, 1444, 1860), 56(120, 432, 861, 1432, 1438, 1441, 1442, 1860), 57(626,

1433, 1436, 1439), 64(626, 1437, 1446), 77(626, 627, 1435), 79(1437, 1444), 92(1444), 115(120), 116(120), 219(626, 1439a, 1440), 222(1437), *227, 231, 233, 241, 247, 253, 269, 270, 281, 282*

Nielsen, D. R., 21(1447), 48(1448), 51(1448), 53(1448), 63(1448), 65(1448), 81(1447), 82(1447), 151(1448), 168(1448), 174(1448), 179(1448), *270*

Nikitina, A. N., 13(1449, 1450), 21(1450), 48(1449, 1450), 51(1449, 1450), 52(1449, 1450), 53(1449, 1450), 54(1450), 57(1450), 59(1449), 62(1449, 1450), 77(1450), 151(1449, 1450), *270, 271*

Nikolaeva, M. E., 35(1315), 36(1315), 48(1315), 49(1315), 54(1315), 92(1315), *266*

Nikolaeva, N. A., 115(1335, 1766b), 118(1335, 1336, 1766c), 120(1335), *267, 279*

Nishida, S., 20(1763), 22(1763), 65(1763), 80(1763), *279*

Nishimura, A., 224(55), *231*

Nitsche, R., 21(992, 993), 47(992), 48(992, 993), 49(992), 56(992), 57(993), *257*

Nixon, G. A., 56(1191), 76(1191), 220(1191), *262*

Nobbe, P., 21(994, 997), 47(994), 48(994), 49(994, 997), 150(997), *257*

Nobis, J. F., 21(1451), 48(1451), *271*

Nölle, D., 55(1452), 56(1452), *271*

Nöth, H., 7(1488), 8(1491), 19(1487), 21(1466, 1468, 1470), 22(1473, 1487, 1490), 23(1465, 1466, 1468), 24(1876), 40(1656), 44(1467), 46(1468, 1490), 47(1487, 1490), 48(1656), 49(1459, 1460, 1462, 1468, 1489), 50(1459, 1460, 1461), 1469, 1472, 1475, 1489, 1491), 51(1460, 1462, 1475, 1476, 1479, 1491), 52(1459, 1481), 53(1484, 1490), 54(1487, 1490), 55(646, 1452, 1459, 1463, 1466, 1473, 1477, 1478, 1487, 1490, 1491), 56(1452, 1462, 1474, 1475, 1476, 1477, 1478, 1486, 1491), 57(1465, 1466, 1467, 1468, 1470, 1474, 1485, 1491, 1876), 59(1491), 60(1467), 61(1467), 62(1490), 64(1487, 1490), 65(1473, 1487, 1490), 68(1483), 77(1459, 1474, 1486), 80(1465, 1467), 83(1475), 89(1482, 1483), 90(464, 1480, 1481), 92(1460), 109(1484), 136(1877), 138(1464, 1876, 1877), 140(1466), 148(1876), 151(1471), 155(1460), 158(1460, 1466, 1470, 1475), 165(1876), 174(1876), 175(1876), 177(1876), 179(1876), 180(1876), 181(1475), 185(1876), 187(1876), 219(1468, 1489), *227, 242, 247, 271, 276, 282*

Nolte, M. J., 3(1453), 90(1453), 146(1453), *271*

Noltes, J. G., 22(1111), 67(1111), 140(343a), 214(1111), 216(1111), *239, 260*

Nordman, C. E., 138(1525), 161(1525), 194(1525), 196(1050), 1525), *258, 272*

Norman, A. D., 196(1454), *271*

Norman, J. H., 208(1523), 209(1523), *272*

Norman, R. O. C., 58(743), 222(743), *250*

Normant, H., 21(1455, 1456), 56(1455), 58(1455), 59(1455), 60(1455), 61(1455), 63(1455), 100(1455), *271*

Noro, K., 224(1457, 1458), *271*

Norrish, R. W., 16(1135), 92(1135), *260*

Novikov, S. S., *228*

Nozaki, H., 86(1827a), 115(1376), 117(1376), 134(1414), *268, 269, 281*

Nozawa, S., 97(1754), 115(1755), 118(1756), 129(1757), *279*

Nussim, M., 28(1917, 1492), 30(1917), *272, 283*

Nystrom, R. F., 120(1409), *269*

O

O'Brien, R. J., 9(1493), 11(1493), 12(1493), *272*
Odom, J. D., 10(580), 47(26), 49(26), 201(1141a), 204(580), *231, 246, 260*
Oganesyan, R. M., 77(925), *255*
Ogushi, T., 53(1928), 107(1928), *283*
Ohashi, O., 8(1494), 10(1494), 50(1857), *272, 282*
Ohkubo, K., 15(1494a), *272*
Okada, M., 15(1494a), *272*
Okamoto, Y., 77(1495), *272*
Okhlobystin, O. Y., 20(1939), 21(1939), 22(1937, 1938, 1939, 1940, 1941, 1943), 23(1940), 46(1939, 1941), 47(1937, 1938, 1939, 1941), 48(1937, 1942, 1943), 64(1940), 109(1939, 1941), *284*
Olin, A. D., 47(1626), 77(1626), *275*
Oliver, A. J., 10(58a), *231*
Oliver, J., 24(1830), 165(1830), 179(1830), 202(1830), *281*
Oliver, J. P., 165(1665), *276*
Onak, T. P., 3(1505), 4(1501), 7(1503), 8(1068, 1119, 1815), 10(1067), 28(1118), 32(1118), 65(1131a), 169(1118), 170(1118), 176(1118), 177(1118), 188(706, 1506), 190(1068, 1118, 1119), 191(1503), 196(1500, 1502, 1815), 197(1505), 198(1496, 1815), 199(1500, 1501, 1502, 1502a, 1815), 201(1067, 1496, 1500, 1501, 1502, 1814, 1815), 202(706, 1497, 1498, 1504), 203(1499, 1501), 204(1067, 1161, 1496, 1500, 1814, 1815), 205(1161), 216(706, 1506), *227, 249, 259, 260, 261, 272, 280*
Onishi, A., 224(633, 634), *247*
Onyszchuk, M., 92(1553), 148(1553), 154(1553), *273*
Opperman, M., 29(152), 196(152), *234*
Orlandini, A. B., 3(550), 138(550), *245*
Osborn, J. A., 146(1662), *276*
Ospici, A., 22(1702), 54(1702), 64(1702), 90(1702, 1702a), *277*
Ossko, A., 77(1227), 78(1227), 83(1228, 1229), 106(1227), 219(1229), *263*
Ottley, R. P., 24(822), 38(823), 39(822, 823), 55(823), 81(823), 82(822, 823), 109(823), 111(823), *252*
Otto, M. M., 56(1507), 57(1507), *272*
Ourisson, G., 97(1763a), *279*
Owen, A. J., 208(583), 209(582), *246*
Oyama, T., *269*
Ozin, G. A., 9(1493), 11(1493), 12(1493), *272*

P

Pace, E., 174(1507a), *272*
Pace-Asciak, C., 223(1508), *272*
Paetzold, P. I., 36(1518, 1519), 46(719, 719a), 50(1510, 1514), 51(1509, 1516), 54(1514), 55(1514), 56(1510, 1512), 60(719), 67(1518), 77(1509, 1511, 1512), 79(1515), 83(1517), 128(1509, 1512), 129(1513, 1515), 148(719, 719a), 150(1509, 1511, 1514), 151(1514), 152(1511), *227, 249, 272*
Pailer, M., 37(1521), 56(1520), 57(1520), 62(1520), *272*
Paine, R. T., 193(1522), *272*
Painter, W. J., 77(981), *257*
Pake, G. E., 132(1868), *282*
Palchak, R. J. F., 208(1523), 209(1523), *272*
Panaiotov, I. M., 224(1524), *272*
Paraschiv, M., 62(64), 86(64), 92(64), 140(64), *232*
Park, A. J., 21(1164, 1166, 1167), 22(1166), 49(1164, 1165, 1167), 77(1166), 99(1165), 139(1164, 1167), 150(1164, 1165, 1167), 153(1165), *261*
Park, T. O., 86(1400), *269*
Parry, R. W., 12(1152), 138(378, 1152, 1380, 1525), 149(1380), 158(1380), 159(1152), 161(378, 1152, 1525), 183(1380), 184(1380), 193(1522), 194(1525), 195(1766), 196(1525), *227, 240, 261, 268, 272, 279*
Parshall, G. W., 212(897), *254*

Parsons, T. D., 5(1529), 21(1527, 1529), 20(390), 22(390, 1827), 24(1527), 25(1527, 1529), 38(390, 1827), 39(390, 1827), 39(1827), 40(1526, 1528), 41(1526), 46(1529), 47(1527, 1528, 1529), 62(1528), 80(390, 1827), 81(390, 1527), 97(1527), 98(1529), 125(1526), 153(1526), *240, 272, 273, 281*
Partridge, J. A., 211(734), *250*
Parts, L., 51(1530), 97(1530), 99(1530), 179(1530), *273*
Pasdelorys, M., 55(1063), 63(1063), 64(1063), 65(1063), *259*
Pasto, D. J., 15(1533, 1534), 25(1542), 28(1533, 1538, 1542), 29(1538, 1543), 31(1535, 1536, 1540, 1544, 1545, 1548), 32(1546, 1547, 1548), 33(1531, 1533), 35(1537, 1542), 47(1551), 52(1550), 56(1541), 57(1535), 60(1545a), 61(1535, 1536), 80(1535, 1536), 81(1535, 1536), 90(1536), 102(1532), 103(1545a), 108(1536, 1545a), 109(1551), 113(1536), 1537, 1548, 1549), 114(1537, 1549), 124(1536, 1539, 1540, 1541, 1548), 127(1550), 139(1552), 149(1541), 153(1533), 165(1534), 166(1534, 1544), 176(1541), 186(1534, 1544), 188(1539, 1540, 1541), 219(1536), 220(1545a), *273*
Pastukhova, I. V., 122(1582a), *274*
Paterson, W. G., 92(1553), 148(1553), 154(1553), *273*
Pattison, I., 49(862, 1554), 50(862), 51(862, 1555), 56(862), 57(862), 92(1554, 1555), 148(1554), 219(862, 1554), *253, 273*
Pattison, I. C., 11(1143), 21(1143), 47(1143), 48(1143), 49(1143), 219(1143), *260*
Paul, D. E., 132(1868), *282*
Paul, I. C., 4(1769a), *279*
Paushkin, Ya. M., 21(1579, 1781, 1785), 22(1781), 43(623), 46(1781), 47(1781, 1783, 1785), 49(1579), 53(1782, 1783, 1785), 60(1782, 1783, 1785), 62(1579), 65(623), 92(1782, 1785), 121(1782, 1785), 149(1783, 1785), 150(1579), 224(1786), *247, 274, 280*
Paxson, T., 149(1027), 154(1027), *258*
Payling, D. W., 92(125), 92(125), 102(123, 124), *233*
Peacock, K., 20(1184, 1186), 21(1184, 1185, 1186, 1187), 23(1186, 1187), 52(1187), 56(1185), 59(1184, 1185), 60(1184, 1185, 1186), 61(1184, 1185), 92(1184, 1185), 94(1184, 1185), 220(1184, 1185), 221(1184), *261, 262*
Pearn, E. J., 16(603), 22(603), 23(601), 48(603), 49(603), 54(603), 55(603), 63(607), 65(603), 66(605), 176(605), *246*
Pearsall, C. S., 162(1744), *278*
Pearson, R., 56(1848), 169(1848), 175(1848), 176(1848), *281*
Pearson, R. J., 224(174), *234*
Pease, R. N., 28(176, 177), 30(176), 43(176), 97(175), 169(1873), 179(1873), 224(177), *234, 282*
Peat, I. R., 62(598), *246*
Pecile, C., 71(160), 72(160), 127(160), *234*
Pedley, J. B., 24(1044a), 55(1044a), 64(1044a), *258*
Pellon, J., 21(1557), 58(1557), 77(1556), 224(1556, 1557), *273*
Pelter, A., 28(1561), 48(1561), 49(1561), 141(1558), 143(1559, 1560), *273*
Pennartz, P. L., *262*
Pepin, C., 159(1027a), *258*
Perkins, P. G., 5(39), 12(37, 44), 13(38), 16(39, 41, 43, 44, 1562), 55(803), 56(803), 157(40), 163(42), *231, 252, 260, 273*
Perloff, A., 3(1563), 208(1563), 209(1563), *273*
Perrin, C., *232*
Perrine, J. C., 20(1564), 21(1564), 22(1564), 47(1564), 56(1564), 79(1564), 92(1564), 104(1564), *273*
Persianova, I. V., 139(1954a), 141(1954a), *284*
Peters, M. D., 13(147), *234*

Peterson, L. K., 148(60), *231*
Peterson, S. W., 61(1197), 109(1197), *262*
Petree, H. E., 131(1573), 156(1573), *274*
Petropoulos, J. C., 77(1693), *277*
Petrov, A. A., 21(1945), 22(1565), 23(1945), 48(1565), 61(1945), *273, 284*
Petry, R. C., 51(1566), 98(1566), *273*
Pettit, R., 56(537), 57(537), 62(537), 63(537), 72(537), 217(536), *244*
Petukhov, G. G., 48(982), 88(1598), 138(912a, 1954b), 143(876a, 912, 982), *254, 255, 257, 274, 284*
Petz, W., 22(1473), 55(1473), 65(1473), *271*
Pfaffenberger, C. D., 28(269), 66(269), 67(269), 182(269), *237*
Phillips, C. R., 195(149), *234*
Phillips, C. S. G., 21(1567, 1576), 77(1576), 78(1567, 1576), *237, 274*
Phillips, J. R., 28(1568), 31(1568), 32(1568), 62(1568), 92(1568), 113(1568), 124(1568), 125(1568), 152(1568), *273*
Phillips, W. D., 7(1569), *227, 273*
Pierpont, C. G., 3(1569a), *274*
Pierson, R. H., 223(1570), *274*
Pilger, R. C., *262*
Pirozhkov, S. D., 122(1582a), *274*
Pitzer, K. S., 6(1571), *274*
Platt, A. E., 46(106a), 108(106a), *233*
Platt, J. R., 77(1599), *274*
Plešek, J., 208(733a), *250*
Ploquin, J., 62(1572), 63(1572), 85(1572), *274*
Podall, H. E., 131(1573), 156(1573), *274*
Poesche, W. H., 22(543), 37(543), 73(543), 75(543), *244*
Poholsky, F. D., 27(21), 29(21), 33(21), 46(21), *230*
Polack, H., 21(996), 49(996), 131(995), 150(996), *257*
Polevy, J. H., 12(1371), 56(1371), 58(1009), 85(1371), 222(1009), *257, 268*
Polston, N. L., 33(1970), 104(1976), 107(1976), 181(1975), 228(1976), *285*
Pope, A. E., 16(1574), *274*
Porter, R. F., 77(1853), 159(1853), 160(879), *232, 254, 282*
Post, E. W., 22(984), 41(983, 984), 64(983, 984), 152(948), *257*
Posvic, H., 21(1389, 1831), 139(1389, 1831), *269, 281*
Pouyanne, J. P., 149(873), *253*
Povlock, T. P., 20(1575), 21(1575), 51(1575), 78(1575), *274*
Powell, H. M., 3(138), *233*
Powell, P., 21(1567, 1576), 77(1576), 78(1567, 1576), *273, 274*
Powell, R. E., 92(434), 137(434), *241*
Pozdnev, V. F., 24(1316, 1318), 28(1318, 1320), 47(1320), 60(1320), 69(1316, 1320), 71(1319) 80(1318, 1319), 81(1316, 1318, 1319), 82(1317, 1320), 134(142, 1255), *234, 263, 266*
Prager, R. H., 127(197), *235*
Pratt, D. E., 22(649), 65(649), 77(345, 654), 218(654), *239, 248*
Pressley, G. A., 15(788, 1736), 163(788), *251, 278*
Price, A. D., 58(1009), 222(1009), *257*
Price, W. C., 138(1577), 147(1577), *274*
Prinz, R., 88(1578, 1872), *274, 282*
Prokai, B., 21(1046), 35(871, 1046), 36(871), 43(871, 1046), 48(1046), 51(860), 55(1046), 56(860, 1045), 57(871), 60(1046), 61(871, 1046), 63(1047), 64(871, 1046), 65(871, 1046), 69(871), 89(871), 92(1046), 113(1046), 115(860), 116(860, 1046), 124(1046), 129(1675), 130(1675), *253, 258, 276*
Prokhorova, A. A., 21(1579, 1781, 1784, 1785), 22(1781), 43(623), 46(1781), 47(1781, 1783, 1785), 48(1784), 49(1579), 53(1782, 1783, 1785), 60(1782, 1783, 1785), 62(1579, 65(623), 92(1782, 1784, 1785), 121(1782, 1785), 149(1783, 1785), 150(1579), 151(1784), 224(1786), *247, 274, 280*
Prokof'ev, E. P., 66(1333c), 116(1333c), *267*
Prosen, E. J., 16(864), *253*

Puill, A., 28(593a), 182(593), *246*
Pungor, E., 138(1709b, 1709c), *277*
Puosi, G., 43(156), 72(156), *234*
Purcell, K. F., 10(114), 12(113, 114), 83(1840), 137(1840), 138(1840), 147(1580), 162(1580), *233, 274, 281*
Purser, J. M., 8(1581, 1582), *274*
Puzitskii, K. V., 122(1582a), *274*
Pyszora, H., 12(56), 51(1048, 1049), 52(572, 1049), 53(572), 56(1048), 59(56), 62(572, 1049), 79(56), 209(572), *227, 231, 245, 258*

Q

Qaseem, M. A., 148(1385), 152(1386, 1387), *268, 269*
Quayle, A., 208(1583), 209(1583), *274*

R

Raab, G., 140(1675a), *276*
Raber, H., 141(1583a), *274*
Raff, P., 48(1908, 1910), 137(1908, 1911), 138(1908, 1910, 1911), 139(1910, 1911), 140(1908, 1910), 146(1910), *283*
Ramp, F. L., 174(1584), *274*
Ramsay, B. G., 11(1585, 1586), 12(1588, 1589), 13(1585, 1587, 1590, 1591), 49(1585, 1589), 127(1072), *259, 274*
Ranck, R. O., 21(662, 663), 79(662), *248*
Randall, E. W., 8(1168), 9(1168a), 10(1168, 1168a), 138(1168a), *261*
Rasmussen, R. S., 5(622), *247*
Rathke, J., 46(1592), 48(1592), 49(1592), 50(1592), 51(1592), 53(1592), 54(1592), 126(1592), 219(1592), 221(1592), *274*
Rathke, M. W., 29(271), 44(1593), 79(1593), 95(283), 97(281), 105(274), 115(222, 273, 282, 1752), 117(222, 273, 282), 122(222, 270, 271), 123(199, 270, 271, 272, 1594), 124(285), 126(280, 281, 283, 284, 285), 133(1595), 222(1594a), *235, 236, 237, 274, 279*
Ravindran, N., 28(275, 275a, 275b, 275c), 33(275), 53(275), 55(275), 60(275c), 104(214a), *235, 237*
Razuvaev, G. A., 88(1598), 138(1596), 139(1596, 1597), 143(1597), 144(1597), 145(1597), 146(1596), 151(1596, 1597), *274*
Reade, W., 18(824), *252*
Rector, C. W., 77(1599), *274*
Reed, P. R., 5(1918), 12(1600), 62(1600), 79(1600), *274, 283*
Reedy, A. J., 57(1720), 58(1720), 72(1720), 74(1720), 76(1720), *278*
Rees, R. G., 22(655), 51(655), 54(655), 55(655), 60(655), 61(655), 64(655), 65(655), 77(655), *248*
Reeves, L. W., 140(398), *240*
Regan, T. H., 144(1892, 1893, 1894), 145(698), *249, 282, 283*
Regnet, W., 50(1475), 51(1475, 1476, 1479), 55(1477, 1478), 56(1474, 1475, 1476, 1477, 1478), 57(1474), 77(1474), 83(1475), 158(1475), 181(1475), *271*
Reigel, F., 90(1702a), *277*
Reinert, K., 38(970), 48(971), 67(697, 970), 174(970), 177(970), 178(970), 188(970, 971), *227, 249, 256*
Reinhart, P. B., 5(1918), *283*
Reintjes, M., 50(775), 86(774), 92(774), 96(774), 115(773, 775), 118(773, 775), *251*
Remes, N., 21(1104), 23(1109), 25(1109), 51(1104), 52(1104, 1109), 53(1104), 1109), 92(1104), 96(1104), *260*
Reppe, H. G., 138(1912), *283*
Resenbaum, J., 11(1611, 1612), *275*
Reuwer, J. F., 92(1018), 94(1017, 1018), *257*
Reynard, K. A., 50(815), 51(815), 55(815), *252*
Reznicek, D. L., 42(1368, 1369), 71(1368, 1369), 158(1601), 159(1601), 219(1368), 1369), *268, 274*

Rhine, W., 147(717), *249*
Rhodes, S. P., 28(276), 114(276, 277), *237*
Rice, B., 12(1602), 17(1603), 176(1602, 1603), *274*
Richards, O. V., 109(393), *240*
Richardson, N. M., 224(89), *232*
Richter, E., 51(1237), 54(1237), 58(1237), 64(1237), 65(1237), 79(1237), *238*
Richter, K., 138(884), *254*
Rickborn, B., 42(1604), 46(972), 49(972), 138(972), 148(972), 151(972), *256, 274*
Riddle, C., 140(519a), *244*
Riddle, J. M., 28(833), 109(831, 832), 110(831), 136(833), 138(833), 139(833), 141(833), 145(833), 147(833), *252*
Rieber, M., 51(1049), 52(1049), 62(1049), *258*
Rihl, H., 51(1479), *271*
Riley, P. N. K., 24(1044a), 55(1044a), 64(1044a), *258*
Rinehart, E. A., 5(1918), *283*
Ring, H., 4(671, 672, 673), 162(671, 672, 673), 163(673), *248*
Ritter, D. M., 5(1529), 7(670), 8(515, 670), 12(670), 16(670), 21(1527, 1529), 22(670), 24(1527, 1652, 1830), 25(1527, 1529), 46(670, 1529), 47(670, 1527, 1529), 54(670), 77(1652), 78(1652), 81(1527), 97(1527), 98(1529), 136(287), 138(287), 149(670), 165(1665, 1830), 179(1830), 193(515, 1142, 1363), 194(514, 515, 1142, 1363), 202(1830), 205(1142), 206(1142), *237, 244, 248, 260, 268, 272, 273, 276, 281*
Ritter, J. J., 4(1024, 1025), 22(1606), 38(445, 446), 39(445, 1607), 62(1606), 64(1605, 1606, 1607), 80(445), 81(445, 446), 82(445, 1607), 92(1606), 113(445), 124(445), 221(1606), 222(446), *226, 242, 258, 274*
Rittig, F. R., 22(1660), 43(1660, 1703), 54(1704), 63(1657), 65(1660, 1703), 88(1657b, 1660), 90(1660), 152(1657a), *276, 277*
Rittner, R.C., 223(1608), *275*
Roberts, B. P., 21(505), 47(505), 48(505), 50(492), 53(609), 57(497), 61(497), 92(500, 609), 93(500), 95(500), 97(502), 98(493, 497, 499, 501, 502), 99(493), 101(480, 496), 102(494, 503), 106(491), 108(480, 481, 482, 491, 492, 493, 498, 502, 503), 121(479, 504, 609), 129(500), 150(493), 223(500), *243, 246*
Roberts, J. D., 10(1862a, 1862b), *282*
Robinson, B. P., 16(582), 29(582), 72(528), 73(528), 218(528), *244*
Robinson, G. C., 143(624), *247*
Robinson, R. E., 21(1451), 48(1451), *271*
Rochow, E. G., 22(660), 55(660), 57(659), 77(659), *228, 248*
Rockett, B. W., 89(1157), *261*
Römer, R., 57(32), *231*
Rogers, H., 16(544), 218(544), *244*
Rogic, M. M., 43(279), 95(250, 251, 283), 97(281), 105(274), 115(222, 273, 282, 1752), 117(222, 273, 282, 1752), 124(285), 126(250, 251, 252, 278, 279, 280, 281, 283, 284, 285), *225, 236, 237, 279*
Rohwedder, K. H., 62(685), 127(685), *249*
Roiger, P. S., 138(862a), *253*
Roman, M., 62(64), 86(64), 92(64), 140(64), *232*
Romas, E., 62(64), 86(64), 92(64), 140(64), *232*
Rona, P., 8(545), 13(430), 14(430), 21(512), 29(512), 71(512), 215(512), *241, 243, 244*
Rondestvedt, C. S., 52(1609), 59(1609), 79(1609), 92(1609), 96(1609), *275*
Roques, B., 76(1609a), *275*
Rosen, A., 39(1610), 40(1949), 80(1610, 1949), 81(1610, 1949), 82(1949), *275, 284*
Rosenblum, L., 16(1614), 21(1614), 47(1614), 48(1614), 111(1613), 173(1613), *275*
Rossetto, F., 43(156), 72(156), *234*

Rossi, F. M., 21(1616), 49(1616), 111(1209, 1615), 112(1615), *262, 275*
Rotermund, G., 24(977), 26(977), 28(975), 29(1617, 1618), 38(965, 977, 978), 51(976), 52(976), 53(976), 56(978), 62(976), 66(965, 976, 977, 978), 67(965, 974, 976, 977), 68(975), 69(965, 976, 978, 1617, 1618), 86(976), 92(976, 1618), 111(978), 173(978), 176(957, 973), 188(965, 974, 977, 978), 215(974, 1617), *227, 256, 257, 275*
Roth, H. J., 21(1619), 52(1619), 53(1619), 86(1620), *275*
Rothergy, E. F., 8(1405), 12(1405), *269*
Rothstein, E., 53(1621), 179(1621), *275*
Roushdy, M. I., 24(581), 111(581), *246*
Rozas, R., 224(431), *241*
Rudashevskaya, T. Yu., 47(336), 92(336), 115(336, 1275b), 117(336), 208(336), *239, 264*
Rudolph, R. W., 38(1622), 39(1622), 81(1622), *275*
Rückert, A., 48(1908), 137(1908), 138(1908, 1913), 139(1913), 140(1908), 145(1913), *283*
Rüdorff, W., 138(1623), 139(729a), *250, 275*
Ruff, J. K., 46(1624), 50(1624), 51(1624), 55(1624), 56(1624), 109(1624), *275*
Ruigh, W. L., 47(1626), 77(1625, 1626), *275*
Rundle, R. E., 162(1155), 163(1155), *261*
Ruschmann, W., 22(1883), 24(1883), 45(1883), 54(1883), 64(1883), 147(1883), 179(1883), *282*
Russo, A. J., 120(1409), *269*
Rutkowski, A. J., 21(781, 784), 22(781, 784), 23(784), 24(784), 47(781), 48(781, 784), 53(784), 111(780, 781), 173(781), *251*
Ryabova, K. G., 122(1582a), *274*
Ryazanova, O. D., 24(1333d), 28(1333d), 32(1333d), 115(1333d), *267*
Ryschkewitsch, G. E., 8(165, 1627), 22(1629), 49(165), 50(165), 51(165), 77(1629), 78(1629), 196(1628), 197(31, 1628, 1630), 201(31, 1628, 1630), 202(1628), *231, 234, 275*

S

Sacconi, L., 3(114a), *233*
Saegebarth, K. A., 28(1631), 66(1631), 67(1631), 215(1631), *275*
Safonova, E. N., 24(1352), 28(1321), 66(1352), 67(1352), 69(1352), 80(1321, 1352), 81(1321), 82(1352), *266, 268*
Sagusa, T., 224(633, 634), *247*
Saha, H. K., 22(1632), 39(1632), 81(1632), 82(1632), *275*
Saji, A., 8(1627), *275*
Samuel, B., 51(1633), *275*
Sanderson, R. T., 20(1653), 22(1653), 46(1563), 187(1634), *275, 276*
Sandhu, J. S., 198(355), 199(355), 201(355), *239*
Sandvik, E. I., 179(1634a), *275*
Sanhueza, A. C., 167(1136), *260*
Sanhueza, E., 108(16, 1122, 1123), 188(1123), *230, 260*
Santucci, L., 16(1637), 21(662, 1635), 57(1636), 58(1635, 1636), 59(1635), 79(662, 1636), 92(1636), *248, 275*
Sarma, T. S., 138(1382), *268*
Sato, P., 31(1681), 47(1681), 59(1681), 77(1674, 1676, 1681), *276*
Saturnino, D. T., 3(416a), *241*
Sauer, J. C., 210(900, 902), 211(901, 902), 212(900, 901), *254*
Saunders, V. R., 15(722), *250*
Savereide, T. J., 21(1096), 57(1096), 58(1096), *259*
Saville, R. W., 53(1621), 179(1621), *275*
Sawodny, W., 10(1446), 11(1446), 22(1446), 64(1446), 219(1439a), *270*
Sayre, R., 18(1638), *275*
Sazonova, V. A., 21(1421, 1422, 1423, 1426, 1640), 48(1426), 49(1426),

57(1640), 89(1420a, 1423, 1424, 1425), 92(1424), 97(1423, 1424), 109(1424, 1640), 115(1639), 133(1424), 138(1420, 1426), 139(1421, 1422, 1426, 1640), 146(1641), 150(1422), *270, 275*
Scaiano, J. C., 17(712), 98(713), 102(503), 103(712), 104(712), 108(481, 482, 503, 506, 712, 714), 121(479, 504), 194(714), *243, 249*
Schaad, L. H., 40(1528), 47(1528), 62(1528), *272*
Schabacter, W., 11(1643), 24(1643), 54(1643), 64(1643), *275*
Schaefer, H., 52(1481), 90(1480, 1481), *271*
Schaeffer, G. W., 77(1599), 173(1648), 176(1648), *274, 276*
Schaeffer, R., 22(159), 25(1780), 41(552), 42(746a), 46(746a, 1592, 1644), 48(1592), 49(746a, 1592), 50(1592), 51(1592), 53(1592), 54(1592), 77(159), 80(552), 81(552), 82(552), 126(1592), 129(1644), 180(1780), 183(551), 194(155), 195(155, 1643a), 196(1454), 219(1592, 1644), 221(1592, 1644), *234, 245, 250, 271, 274, 275, 280*
Schaper, K. J., 43(1705), 63(1705), 64(1705), 65(1705), *277*
Scharrnbeck, W., 21(920), 51(920), 57(920), 58(920), 59(920), 79(920), 109(920), *255*
Schaschel, E., 21(725), 22(724), 77(724, 725), 78(724, 725), *250*
Schaumberg, G. D., 21(1645), 56(1188), 59(1188), 60(1188), 61(1645), 79(1645), 219(1188), 221(1188), *262, 275*
Schechter, W. H., *228*
Schenker, E., *228*
Schick, H., 50(1469), *271*
Schiemenz, G. P., 138(1645b), 147(1645a), *276*
Schlegel, R., 78(726, 727), 84(728), *250*
Schlesinger, H. I., 20(390, 1653), 22(390, 1653, 1827), 24(1650, 1652, 1654), 38(390, 1827), 39(390, 1827), 46(1653), 49(1651), 50(1654), 55(1651), 76(1651), 77(1649, 1651, 1652), 78(1651, 1652), 80(390, 1826, 1827), 81(390), 136(287, 1646), 138(287, 1647, 1849), 139(1646), 148(286, 1649), 149(1649), 151(1649), 153(286, 1649), 155(1649), 159(356), 160(356), 164(1651, 1654), 165(1649), 173(1646, 1648), 175(1650, 1654), 176(1648, 1649, 1650, 1654), 179(1654), 183(1649, 1651), 184(1649, 1651), *228, 237, 239, 240, 276, 281*
Schleyerbach, C., 51(1824), 52(1824, 1825), *281*
Schmid, G., 40(1656), 48(1656), 52(1481), 68(1483), 89(1482, 1483, 1655, 1655a), 90(1480, 1481, 1655a), *228, 271, 276*
Schmidt, G., 50(1469), *271*
Schmidt, M., 22(1660), 24(1706), 43(1660, 1703), 54(1701, 1704, 1706), 55(1706), 63(1657, 1705), 64(1705), 65(1660, 1703, 1705), 88(1657b, 1658, 1659, 1660), 90(1659, 1660), 143(1706), *276, 277*
Schmulbach, C. D., 140(1695), *277*
Schneider, H., 47(1751), *278*
Schoen, L. J., 131(890), 174(890), *254*
Scholer, F. R., *228*
Schomburg, G., 5(690), 25(979), 29(979), 62(690), 66(979), 67(979), 75(979), 215(979), *249, 257*
Schrägle, W., 53(1484), 57(1485), 109(1484), *271*
Schrauzer, G. N., 92(1661), 140(1661), *276*
Schrock, R. R., 146(1662), *276*
Schrotter, H. W., 109(950), *256*
Schug, J. C., 6(1865), *282*
Schuster, R. E., 10(382), 127(382), 136(382), 139(382), 141(382), 147(382), 219(382), *240*
Schweizer, P., 21(1470), 57(1470), 158(1470), *271*
Schwind, L. H., 21(1557), 58(1557), 224(1557), *273*

Scott, K. N., 8(1663), 9(1663), 79(1663), *276*
Scribner, R. M., 52(1609), 59(1609), 79(1609), 92(1609), 96(1609), *275*
Seaman, W., 21(1664), 57(1664), 58(1664), 222(1664), *276*
Searcy, I. W., 196(1500), 199(1500), 201(1500), 203(1499), 204(1500), *272*
Searles, S., Jr., 148(1212), *262*
Sedlak, M., 47(1626), 77(1626), *275*
Sedova, N. N., 89(1420a, 1426a), *270*
Seely, G. R., 24(1830), 165(1665, 1830), 179(1830), 202(1830), *276, 281*
Seigel, B., *228*
Sejnowski, P., 224(588b), *246*
Self, J. M., 40(1528), 47(1528), 62(1528), *272*
Selva, A., 28(367), 92(367), *239*
Semlyen, J. A., 21(1567, 1576), 77(1576, 1666), 78(1567, 1576), *273, 274, 276*
Serafin, B., 21(1668), 58(1667, 1668), 106(1667), *276*
Serebryakov, E. P., 139(1642), *275*
Seredenko, V. I., 84(1286b), *265*
Sergienko, T. Y., 100(1038a), 150(1038a), *258*
Seyferth, D., 21(1680), 27(1670), 29(1669), 31(1669, 1681), 46(1669, 1670), 47(1670, 1681), 59(1681), 77(1671, 1673, 1674, 1676, 1677, 1678, 1681), 78(1671), 85(1671), 113(1678), 129(1675), 130(1675), 138(1672), 139(1679, 1680), 140(1672, 1675a), 218(1671), *276*
Shafferman, R., 21(650), 49(8), 50(650), 52(572, 651, 652), 53(572, 650, 651, 652), 54(650), 55(650), 62(572, 652), 92(650, 652), 99(651), 151(8, 650), 152(8), 153(8), 158(650), 209(572), *230, 245, 248*
Shafizadeh, F., 62(1682), 224(1682), *276*
Shagova, E. A., 53(1313), 54(1313), 65(1313), 66(1313), 120(1313), 121(1314), *266*
Shapiro, I., 7(1503), 12(1078, 1080, 1081, 1084), 13(1082, 1083, 1089), 33(1866), 46(1080, 1084, 1683), 47(1683), 79(1089), 170(1866), 176(1083, 1086, 1087, 1088, 1683, 1866, 1901a), 177(1866), 190(1078, 1079, 1083, 1085, 1086, 1087, 1088, 1685, 1866), 191(1503), 192(1086, 1683, 1901a), *228, 259, 272, 276, 277, 282, 283*
Sharefkin, J. G., 106(1686), *277*
Sharma, A., 18(1688), *277*
Sharma, R. K., 29(1687), 66(1687), *277*
Sharp, R. L., 29(288), 31(288, 289), *237*
Sharrocks, D. N., 28(1561), 48(1561), 49(1561), *273*
Shashkova, E. M., 24(1332a), 28(1692), 29(1691), 63(1332a), 66(1692), 67(1691, 1692), 215(1691), *267, 277*
Shaver, A., 140(368), *240*
Shaw, D., 8(1168), 9(1168a), 10(1168), 1168a), 138(1168a), *261*
Shchegoleva, T. A., 21(1323, 1325, 1328, 1330), 22(1323), 24(1329, 1332a), 26(1329, 1332a), 28(1327, 1692), 29(1330, 1331, 1333, 1689, 1691), 44(1330a), 46(1323, 1330a), 47(1323), 50(1328, 1690), 51(1328), 52(1322, 1323, 1328), 53(1322), 54(1331, 1689), 55(1324), 59(1325), 60(1324, 1325), 62(1329), 63(1324, 1325, 1331, 1332a, 1689), 64(1325, 1326), 66(1330, 1333, 1692), 67(1330, 1333, 1691, 1692), 90(1326), 107(1326), 120(1327), 140(1328), 152(1329), 158(1296, 1690), 215(1691), 223(1323), *265, 266, 267, 277*
Shchteinshneider, A. Ya., 10(141b), *233*
Shdo, J. G., 29(1189), 31(1190), 45(1189), 1190), 80(1189, 1190), 82(1190), 109(1190), 115(1190), *262*
Sheehan, D., 77(1693), *277*
Sheft, I., 136(287), 138(287), *237*
Sheldon, J. C., *228*
Sheludyakov, V. D., 29(1333),

66(1333), 67(1333), 158(1296), *265, 267*
Shepp, A., 163(1694), *277*
Sheppard, N., 11(1924), 12(869a, 1924), 62(869a), *253, 283*
Sherif, F. G., 140(1695), *277*
Shevtsova, O. N., 62(1930), *284*
Shiihara, I., 22(858), 46(858), *253*
Shiller, N. A., 115(1639), *275*
Shimada, H., 15(1494a), *272*
Shiono, M., 126(1395b), *269*
Shiotani, S., 224(635), *247*
Shore, S. G., 3(416a), 36(1870), 38(1870), 39(1870), 59(1696), 63(1696), 81(1870), 170(1929), 176(1929), 177(1929), 183(1929), 184(1929), 199(167), 202(167), 206(166, 863, 863a), *234, 241, 253, 277, 282, 284*
Shorygin, P. P., 51(1696a), 52(1696a), 53(1696a), *277*
Shoulders, B. A., 29(1687), 66(1687), *277*
Shriver, D. F., 15(1023), 66(1697), 77(1698), 78(1698), 89(1697), *258, 277*
Shtov, O. P., *228*
Shu, P., 7(50), 22(50), 67(50), 216(50), *231*
Shuikin, N. I., 119(1775), 121(1775), *279*
Shushunov, V. A., 100(1698b), 101(1148a), 102(1698a), *261, 277*
Siebert, H., 11(1699), 12(1699), *277*
Siebert, W., 22(1660, 1702), 24(1706), 43(1660, 1703, 1705), 47(1700), 49(1700), 54(1701, 1702, 1704, 1706), 55(1706), 63(1705), 64(1702, 1705), 65(1660, 1700, 1703, 1705), 88(1658, 1659, 1660), 90(1659, 1660, 1702, 1702a), 140(1699a), 143(1706), *276, 277*
Siedle, A. R., 212(1707), *277*
Siegel, B., 208(641, 1708, 1709), 209(1709), *277*
Sikora, J., 127(486), *243*
Silverman, M. B., 5(1529), 21(1529), 25(1529), 46(1529), 47(1529), 98(1529), *273*
Sime, R. J., 3(1709a), *277*
Simon, Z., 66(65), *232*
Simpson, W., 83(1517), *272*
Singh, A., 52(10), 63(10), *230*
Singh, G., 151(1957), *284*
Singh, K. P., 27(290), 28(290), 181(290), *237*
Siryatskaya, V. N., 28(1954), 47(1954), 48(1954), *284*
Siska, E., 138(1709b, 1709c), *277*
Sisler, H. H., 16(745), 22(1629), 77(1629), 78(1629), 196(1628), 197(31, 1628, 1630), 201(31, 1628, 1630), 202(1628), *231, 250, 275*
Sistrunk, T. O., 143(624), *247*
Skinner, H. A., 16(1574, 1709d, 1710), 17(1709d), 32(109), 92(1710), *228, 233, 274, 277*
Skoog, I. H., 11(1105), 21(1105, 1106, 1107, 1108), 23(1107, 1109), 25(1107, 1109), 50(1106), 52(1107, 1109, 1711), 53(1106, 1107, 1109), 56(1105), 61(1106), 62(1107), 63(1107), 68(1108), 97(1105), 100(1105), 151(1107), 220(1105), *260, 277*
Skowronska-Serafin, B., 21(1712), 58(1712), *278*
Smalley, J. H., 21(1713), 77(1713), 78(1713), 219(1713), *278*
Smirnov, R. F., 224(1713a), *278*
Smirnov, V. N., 24(1333d), 28(1333d), 32(1333d), 66(1333a, 1333b, 1333c), 67(1333b), 115(1333d), 116(1333a, 1333c), *267*
Smith, B. C., *228*
Smith, D. E., 77(1698), 78(1698), *277*
Smith, D. L., 159(885), 160(885), 212(885), *254*
Smith, E. B., 139(1714), 140(1714), *278*
Smith, H. D., Jr., 56(806), *252*
Smith, J. A. S., 18(1715, 1716), *278*
Smith, J. D., 154(25), *231*
Smith, K., 141(1558), 143(1559, 1560), *273*
Smith, P., 77(1698), 78(1698), *277*
Smith, R. L., 113(29), *231*
Smolinsky, G., 60(1717), 62(1717), *278*
Smolka, H. G., 36(1518), 67(1518), *272*
Snaith, R., 21(427), 50(427), 51(427), 219(427), *241*

Sneeden, W., *228*
Snover, J. A., 28(248), 29(248), *236*
Snow, J. T., 28(1971), 92(1971), 93(1971), 104(1971), 141(1972), *285*
Snyder, C. H., 109(215, 291, 308), *235, 237, 238*
Snyder, H. R., 21(1091, 1719, 1722), 23(865), 46(388), 47(865), 49(748), 51(748), 53(748, 865), 54(865), 56(748, 1719), 57(1720), 58(387, 1091, 1720, 1721), 59(748, 749, 1722), 63(388), 64(865), 72(749, 1720), 73(387, 388), 74(387, 748, 1092, 1720), 75(387), 76(1720), 79(749, 1718, 1719), 92(865, 1719, 1722), 94(1719), 98(865, 1719), 99(1719), 100(1719), 109(1719), 139(865), 222(748), 223(1719), *240, 250, 253, 259, 278*
Snyder, R., 31(1548), 32(1548), 113(1548, 1549), 114(1549), 124(1548), *273*
Soboczenski, E. J., 59(1019), 62(1012), 63(1012), 104(1019), *257*
Soboleva, T. A., 224(917), *255*
Sololev, E. S., 139(1954a), 141(1954a), *284*
Solomatina, A. I., 77(922a), *255*
Solomon, I. J., 88(1723), 115(1146), 165(1723), 175(1723), 176(1723), 194(741), 195(741), 205(146, 1724), 206(1724), *250, 261, 278*
Soloway, A. H., 27(363), 56(363, 1191), 58(1725, 1727), 59(363, 1725), 76(1191), 81(1726), 87(1726), 211(745a), 212(745a), 220(1191), 222(1725, 1726), *228, 239, 250, 262, 278*
Somers, G. F., 224(1800), *280*
Sondheimer, F., 28(1492, 1728, 1917), 30(1917), *272, 278, 283*
Sonke, H., 50(1215), 51(1214, 1215), 92(1214, 1215), 93(1214), 121(1214), *262*
Sono, S., 128(1758), *279*
Soulié, J., 21(667, 1729, 1730), 22(667, 1729), 52(667, 1729), 65(1729), 79(1729), 149(667, 1729, 1730), 150(667, 1729, 1730), 151(667, 1729, 1730), 220(1729), *248, 278*
Sowinski, F. A., 56(1927), 57(1927), 71(1927), *283*
Sparfel, L., 62(1572), 63(1572), 85(1572), *274*
Spauninger, P. A., 55(529), 56(529), *244*
Spielman, J. R., 20(358), 22(358), 53(358), 159(1731), 160(1731), 170(1731), 194(357, 1731), 195(357, 1731), 196(1500, 1731), 199(1500), 201(549, 1500), 204(1500), *239, 245, 272, 278*
Spielvogel, B. F., 8(1581, 1582), 83(1732), 92(1732), *274, 278*
Sprague, M. J., 56(1486), 77(1486), *271*
Sprecher, R. F., 208(1732a), *278*
Squire, D. R., 219(1439a), *270*
Staab, H. A., 6(1215a), 52(1732b), 58(1732b), 59(1732b), 63(1732b), *262, 278*
Stafford, F. E., 15(788, 1023, 1736), 163(788), 196(827), *251, 252, 258, 278*
Stafford, S. L., 5(448), 12(1733, 1734), 15(448), 22(877, 1735), 23(1735), 25(448), 46(447, 1735), 55(1735), 62(447, 1735), 64(447, 877, 1735), 92(1735), 125(877), 140(1733), 141(1733), 218(448), *227, 242, 254, 278*
Stafiej, S. F., 21(1713), 22(707), 77(707, 1713), 78(1713), 219(1713), *249, 278*
Standfest, R., 51(1479), *271*
Stang, A. F., 199(841), *253*
Stange, H., 22(416), 46(416), 66(416), 67(416), 80(416), 92(416), *241*
Stanko, V. I., 21(1944), 22(1944), 24(1944), 47(1944), 60(1944), 106(1944), 109(1944), *228, 284*
Statton, G. L., 28(364, 365, 366), 35(364, 365, 366), 71(365, 366), 72(365), 75(364), 83(366), *239*
Steck, S. J., 15(1736), *278*
Steele, W. C., 15(1737), *278*
Steinberg, H., 55(585), *225, 228, 246*
Steinberg, N. G., 47(1626), 77(1626), *275*

Sterlin, R. N., 12(1738), 46(1738), 92(1738), *278*
Stern, D. R., 28(1144), 64(1144), 65(1144), *260*
Steudel, O. W., 47(1956), 143(1956), *284*
Stevens, R. M., 197(1759), 207(1065d), *259, 279*
Stewart, J. E., 11(1739), 12(1739), *278*
Stibr, B., 208(733a), *250*
Stilz, W., 138(1914), *283*
Stock, A., 22(1740), 46(1740), 47(1740), 148(1740), *228, 278*
Stohr, G., 36(1519), *272*
Stone, F. G. A., 5(448, 449), 9(450), 12(1734), 15(448, 1737), 20(169), 21(1167), 22(77, 168, 169, 877, 1735), 23(169, 1735), 25(79, 169, 448), 28(79, 689, 1568, 1741), 31(1568), 32(1568, 1741, 1742), 41(552), 46(447, 1735), 47(689), 48(1741), 49(1167), 53(169, 1742), 55(169, 1735), 62(77, 79, 168, 169, 447, 1568, 1735, 1742), 64(168, 169, 447, 877, 1735), 80(552), 81(552), 82(552), 92(1568, 1735), 94(169), 97(77), 113(79, 1568, 1742), 124(79, 1568, 1742), 125(79, 877, 1568, 1742), 139(1167), 140(168), 147(449), 148(688, 878), 150(1167), 151(79), 152(79, 168, 449, 1568, 1742), 174(689), 176(689), 218(448, 449), 223(77), *226, 228,* 232, 234, *242, 245, 249, 254, 261, 273, 278*
Storch, W., 19(1487), 22(1487), 47(1487), 54(1487), 55(1487), 64(1487), 65(1487), *271*
Strafford, R. G., 13(669), *248*
Strahm, R. D., 223(1743), *278*
Strandberg, M. W. P., 162(1744, 1867), *278, 282*
Straus, S., 88(1844), 89(1844), *281*
Strebel, P., 21(1884), *282*
Strecker, W., 22(1745), 57(1745), *278*
Strehmeier, W., 18(1747), *278*
Streib, W. E., 3(1810a), 71(1810), *280*
Streitwieser, A., Jr., 31(1746), 32(1746), 182(1746), *278*
Strong, P. L., 21(1923), 22(1923a), 29(1923a), 59(1923), 60(1923), 61(1923), 220(1923), *283*
Strunin, B. N., 22(1943), 48(1942, 1943), *284*
Stucky, G. D., 147(717), *249*
Sturm, C. L., 224(610), *246*
Sturm, W., 24(1890), 63(1886, 1890), 88(1885, 1888, 1889, 1890), *282*
Subba Rao, B. C., 26(293, 294, 296, 297), 27(297), 28(294, 296), 29(292, 293, 294, 296, 297), 48(298), 49(294, 298), 100(293, 294), 111(295, 296, 298), 119(295, 298), 120(295, 298), 169(294, 297), 170(296, 297), 173(298), *237, 238*
Subramanian, S., 138(1382), *268*
Sujishi, S., 6(299), 49(299), 148(300, 302, 303), 150(299), 153(299, 300, 303), *238*
Sullivan, E. A., 83(1840), 137(1840), 138(1840), *281*
Summerford, C., 49(862), 50(862), 51(862, 1748, 1479), 56(862, 1749), 57(862), 219(862), *253, 278*
Sundaram, S., 10(1750), 12(1750), 159(1750), *278*
Suprun, A. P., 224(917), *255*
Sutton, N., 47(1751), *278*
Suyama, H., 224(856), *253*
Suzuki, A., 74(1752b), 97(1753, 1754), 115(1377, 1752, 1753a, 1755), 117(1377, 1752), 118(876, 1756), 128(1758), 129(1757), 131(1752b), 146(1752a), *254, 268, 279*
Suzuki, T., 224(911), *255*
Svec, H. J., 25(1780), 180(1780), *280*
Sverdlov, L. M., 190(879b, 879c), 192(879b), *254*
Svitsyn, R. A., 139(1954a), 141(1954a), *284*
Swayampati, D. R., 21(662, 663), 79(662), *248*
Switkes, E., 197(1759), *279*
Sykes, A., 223(106), *233*
Symons, M. C. R., 11(1611, 1612), 109(1145), *261, 275*
Szabady, P., 58(1727), *278*
Szász, G., 138(687), 223(687), *249*

T

Taft, R. W., Jr., 3(1117), 49(1141), 148(1141), 153(1141), *260*
Takamizawa, M., 77(1671, 1674, 1676, 1677, 1678), 78(1671), 85(1671), 113(1678), 218(1671), *276*
Takayama, Y., 224(851, 852), *253*
Talamini, G., 224(1760, 1761), *279*
Talbot, M. L., 61(1193), 109(1192, 1193), 219(1193), *262*
Tamres, M., 148(1212), *262*
Tanahashi, Y., 97(1763a), *279*
Tanaka, J., 71(1762), *279*
Tanaka, T., 86(1827a), *281*
Tanigaki, T., 49(1074), 109(1071), 132(1071, 1074), *259*
Tanigawa, P., 20(1763), 22(1763), 65(1763), 80(1763), *279*
Tartakovski, V. A., *228*
Tarumi, N., 224(909, 911), *255*
Tatani, T., 6(1859), 22(1859), 77(1859), *282*
Taylor, F. B., 24(825), 62(825), *252*
Taylor, M. D., 46(190), 148(190, 301, 302, 303), 153(190, 301, 302, 303), *235, 238*
Taylor, R. C., 10(1764, 1765), 147(1169), 162(1764, 1765), *261, 279*
Tebbe, F. N., 36(1395), 64(1395), *269*
Tees, T. F. S., 16(1710), 92(1710), *277*
Templeton, D. H., 3(1709a), *277*
Terhaar, G., 138(1525), 161(1525), 194(1525), 195(1766), 196(1525), *272, 279*
Tersac, G., 138(1766a), *279*
Ter-Sarkisyan, G. S., 115(1335, 1766b), 118(1334, 1335, 1336, 1766c), 120(1335), *267, 279*
Tevebaugh, A. D., 89(1767), *279*
The, N. D., 138(868), *253*
Therrell, B. L., 20(1768), 22(1768), 77(1768), *279*
Thevénot, G., 51(1237), 54(1237), 58(1237), 64(1237), 65(1237), 79(1237), *263*
Thiele, K. H., 22(1946), 46(1946), 83(329), *238, 284*
Thielens, G., 21(1769), 59(1769), *279*
Thierig, D., 92(1823), 140(1823), *281*
Thoburn, J. M., *226*
Thomas, P. C., 4(1769a), *279*
Thomas, R. I., 24(343), 53(343), 55(343), 62(343), 63(343), 64(343), *239*
Thomas, W. M., 21(1557), 58(1557), 77(1556), 224(1556, 1557), *273*
Thompson, C. C., 16(392), 21(392), 219(392), *240*
Thompson, N. R., 89(1770), 91(461), 92(461, 826), 137(826, 1770), 139(826), 143(826, 1770), 223(461), *242, 252, 279*
Thompson, P. G., 124(1026), 125(1026), *258*
Thompson, R. J., 7(1771), 138(1771), 139(1771), *279*
Thorpe, F. G., 22(164), *234*
Tikhomirov, B. I., 224(1362, 1713a), *268, 278*
Tilley, B. P., 51(455, 860), 56(455, 456, 860, 1044), 63(456), 88(456), 115(455, 860), 116(455, 860), *242, 253, 258*
Timms, P. L., 21(1567), 42(1774), 69(1774), 78(1567), 80(1773), 82(1774), 159(885, 1772), 160(885), 212(885), 214(1774), 216(1774), *254, 273, 279*
Timofeeva, E. A., 119(1775), 121(1775), *279*
Titov, Yu. A., 115(24a), *230*
Tochtermann, W., 137(1777), 138(1776), 139(1777), *228, 279*
Todd, J. F. J., 13(457, 458, 460), 14(452, 459), *242*
Todd, L. J., 46(1644), 51(1073), 129(1073, 1644), 219(1644), 221(1644), *228, 259, 275*
Toeniskoetter, R. H., 22(1778), 77(1779, 1916), 78(1778), 111(1779), *279, 283*
Tollin, B. C., 25(1780), 180(1780), *280*
Tomlinson, D. W., 56(1191), 76(1191), 220(1191), *262*
Tong, D. A., 18(1715, 1716), *278*
Topchiev, A. V., 21(1781, 1784, 1785), 22(1781), 43(623), 46(1781), 47(1781, 1783, 1785), 48(1784),

53(1782, 1783, 1785), 60(1782, 1783, 1785), 65(623), 92(1782, 1784, 1785), 121(1782, 1785), 149(1783, 1785), 151(1784), 224(1786), *247, 280*

Toporcer, L. H., 86(1788), 87(1787, 1788), 91(1787), 92(1787), 96(1788), *280*

Torssell, K., 21(1789, 1790, 1791, 1794, 1798, 1799), 22(1789), 48(1789), 51(1791), 52(1790, 1791, 1798), 53(1798), 56(1799), 57(1789), 1793, 1799, 1801), 58(1792, 1793, 1794, 1801), 59(1792, 1794, 1801), 62(1789, 1796), 66(1789), 67(1789), 68(1798), 69(1789), 98(1789, 1790), 159(1792), 222(1792, 1794), 223(1795), 224(1796, 1800), *280*

Tossell, J. A., 197(1759), *279*

Totani, T., 8(1494), 10(1494), 49(1856), 50(1856, 1857), *272, 282*

Townsend, J., 132(1868), *282*

Trapasso, L. E., 174(1584), *274*

Trautwein, H. D., 137(1777), 139(1777), *279*

Travoletti, L., 57(1636), 58(1636), 79(1636), 92(1636), *275*

Traylor, T. G., 100(1372), 101(1372), 107(1847), *268, 281*

Treichel, P. M., 22(1802), 55(1802), 65(1802), *228, 280*

Triboulet, C., 16(1637), *275*

Trofimenko, S., 51(1807), 85(1803, 1804, 1806, 1809), 92(1804, 1806, 1807, 1808, 1809), *228, 280*

Trofimov, V. A., 90(879d), *254*

Tsai, C. C., 3(1810a), 71(1810), *280*

Tsukamoto, A., 24(304), 29(304), 53(304, 305), 60(305), 61(304), 305), 148(305), 165(304, 305), 169(304, 305), 175(304, 305), *238*

Tsuruta, T., 224(630, 631, 632, 633, 634, 635, 1407, 1408), *247, 269*

Tsuzuki, Y., 21(880), 48(880), 49(880), 97(880), *254*

Tuchagues, J. P., 24(1812), 25(1811), 53(1064), 60(1812), 62(1064, 1811), 151(1812, 1813), 152(1812, 1813), *259, 280*

Tuck, L. D., 138(1647), *276*

Tucker, P. M., 8(1815), 41(552), 80(552), 81(552), 82(552), 196(1815), 198(1815), 199(1815), 201(1814, 1815), 204(1814, 1815), *245, 280*

Tudor, R., 21(505), 47(505), 48(505), 98(493), 99(493), 102(507), 108(482, 493), 150(493), *243*

Tufariello, J. J., 42(1818), 94(1818), 109(1816), 125(1817, 1818, 1819), *280, 281*

Tully, T. J., 18(409), *241*

Turco, A., 71(160), 72(160), 127(160), *234*

Turner, H. S., 77(1820), *281*

Turner, R. B., 14(459), *242*

Tutorskaya, F. B., 3(1339), 21(1337, 1339), 47(1337, 1338), 50(1337, 1341), 54(1340, 1342), 55(1342), 57(1342), 59(1339), 60(1337, 1342), 63(1339, 1340, 1342), 77(1342), 79(1337, 1342), 92(1337, 1340, 1341, 1342), 93(1340, 1342), 108(1342), 120(1338), 149(1341), *267*

Tweedale, A., 2(1044a), 55(1044a), 64(1044a), *258*

U

Uden, P. C., 62(73), 63(73), *232*

Uemura, E., 224(1458), *271*

Ugi, I., 72(845), *253*

Ulmschneider, D., 50(1821, 1822), 51(1822), 55(1822), 59(1822), 70(677), 79(1822), 91(677), 92(1075, 1822), 93(677), 95(1822), 135(677, 1075), *248, 259, 281*

Umland, F., 51(1824), 52(1824, 1825), 92(1823), 140(1823), *281*

Ungermann, C. B., 8(1068), 188(706), 190(1068), 202(706), 216(706), *249, 259*

Unni, M. K., 27(306), 28(306), 29(306), 31(306), *238*

Untch, K. G., 21(1833), 68(1833), 214(1833), *281*

Uriarte, A. K., 41(733), 50(733), 57(733), *250*

Urry, G., 20(390), 22(390, 1632, 1827), 38(390, 1827), 39(390, 1632, 1827), 80(390, 1826, 1827), 81(390, 1632), 82(1632), *240, 275, 281*

Utimoto, K., 86(1827a), 115(1376), 117(1376), 134(1414), *268, 269, 281*

V

Vahrenkamp, H., 7(1488), 8(1491), 22(1490), 46(1490), 47(1490), 50(1489, 1491), 51(1491), 53(1490), 54(1490, 1828, 1829), 55(1490, 1491), 56(1491), 57(1491), 59(1491), 62(1490), 63(1829), 64(1490), 65(1490), 219(1489), *271, 281*

Van Alten, L., 24(1830), 165(1830), 179(1830), 202(1830), *281*

Van Artsdalen, E. R., 61(576), *245*

Van Campen, M. G., Jr., 21(866), 23(865), 47(865), 53(865, 867), 54(865), 56(866), 57(866), 61(867), 64(865), 92(865, 866), 94(866), 97(866), 98(865, 866, 867), 99(866, 867), 100(867), 102(866, 867), 103(867), 109(866), 133(866), 139(865), 223(866, 867), *253*

Vandeberg, J. T., 10(1832), 12(1832), 21(1831), 139(1831), *281*

Van der Kelen, G. P., 9(518), 18(519), 153(519), 217(518), *244*

van der Kerk, G. J. M., 22(1111), 67(1111), 214(1111), 216(1111), *260*

Vanderwerf, C. A., 48(1448), 51(1448), 53(1448), 63(1448), 151(1448), 168(1448), 174(1448), 179(1448), *270*

Van Tamelen, E. E., 21(1833), 68(1833), 214(1833), *281*

Van Veen, R., 38(1835a), 68(1834, 1835), 69(1835), 216(1834, 1835), *281*

Van Wazer, J. R., 24(814), *252*

Van-Willigen, H., 109(1869), 131(1869), 132(696), *249, 282*

Vara, M. C., 111(17), *230*

Varma, K. R., 30(1836), 133(1595), 182(384, 1836), *240, 274, 281*

Varma, V., 133(307), 182(307), *238*

Varshavskara, L. S., 62(919), 92(919), *255*

Vartanyan, M. M., 51(1836c), 53(1836c), 55(1836c), 74(1836c), *281*

Varushchenko, R. M., 16(639, 640, 640a), *247*

Vasil'ev, L. S., 24(1343, 1344, 1345, 1346, 1347, 1348, 1352), 25(1344, 1347, 1348), 50(1240, 1345, 1346), 51(1351, 1836b, 1836c), 52(1240, 1345, 1346, 1347, 1348), 53(1240, 1347, 1348, 1349, 1836a, 1836b, 1836c), 54(1240, 1349), 55(1345, 1346, 1349, 1836c), 60(1345, 1346, 1347, 1348, 1349), 61(1345, 1347, 1348, 1836a), 63(1349), 66(1252, 1350, 1352), 67(1352), 68(1252), 69(1252, 1352), 71(1351, 1836b), 72(1350), 74(1836a, 1836c), 80(1350, 1352), 82(1352), 92(1240, 1345), 96(1345), 102(1836a, 1836b), 103(1836a), 140(1343), 156(141c, 1836a, 1836b), 165(1240), 172(1344, 1345), 175(1240), 176(1240, 1345, 1346), 180(1343, 1344, 1345, 1346, 1347), 182(1240), 185(1240), *234, 263, 267, 268, 281*

Vaver, V. A., 13(1450), 21(1352a, 1353, 1354, 1356, 1357, 1359, 1450), 23(1354), 47(1359), 48(1355, 1450), 49(1359), 50(1361), 51(1352a, 1357, 1450), 52(1352a, 1356, 1357, 1359, 1361, 1450), 53(1353, 1358, 1359, 1361, 1450), 54(1361, 1450), 57(1450), 62(1450), 77(1450), 92(1359, 1361, 95(1359), 138(1354), 139(1354), 140(1352a, 1353, 1354, 1357, 1358, 1359), 150(1354, 1355), 151(1354, 1355, 1450), 155(1355), *268, 271*

Végh, A., 138(687), 223(687), *249*

Verbit, L., 31(1746), 32(1746), 182(1746), *278*
Verbrugge, C., 109(308), *238*
Verhoek, F. H., 51(1566), 98(1566), *273*
Verkade, J. G., 148(777, 1837, 1838), *251, 281*
Vernon, C. C., 21(664), *248*
Vetter, H. J., 138(1464), *271*
Vevashov, A. V., 224(34a), *231*
Videla, G. J., 77(376), *240*
Vidotto, G., 224(1760, 1761), *279*
Villa, A. E., 160(715, 716), *249*
Vitek, A., 208(733a), *250*
Vogel, H., 86(1904), *283*
Vogt, F., 51(35), 92(35), 151(35), *231*
Volpin, M. E., 216(1839), *281*
von Rosenberg, J. L., 16(546), 73(546), 218(546), *244*
Vulle, W. J., 56(1095), 224(1095), *259*

W

Wada, M., 21(880), 48(880), 49(880), 97(880), *254*
Waddington, T. C., 136(1841), 140(1841), *281*
Wade, K., 3(342), 21(427), 34(1127), 49(862, 1554), 50(427, 862), 51(427, 862, 1555, 1633, 1748, 1749), 56(862, 1749), 57(862), 83(1127), 92(1554, 1555), 148(1554), 157(694), 169(1127), 219(427, 862, 1554), *226, 239, 241, 249, 253, 260, 273, 275, 278*
Wade, R. C., 83(1840), 137(1840), 138(1840), *281*
Wagner, R. I., 24(1843), 46(359), 51(359), 53(359), 54(360), 56(383), 77(1842), 78(1842), 92(1842), 148(359), 149(359), 155(359), 175(360), 184(359), *239, 240, 281*
Wagner, R. P., 49(829), 53(829), 133(829), *252*
Wahab, M. A., 24(581), 111(581), *246*
Waldbillig, J. O., 60(1194), 61(1196, 1197), 109(1195, 1196, 1197), 219(1194), *262*
Walker, A. O., 24(1654), 50(1654), 138(1647), 164(1654), 175(1654), 176(1654), 179(1654), *276*
Walker, D. J., 52(9, 11), 62(9, 11), *230*
Walker, S. M., 38(394), 82(394), *240*
Wall, D. H., 16(1562), *273*
Wall, E. M. V., 79(882a), 133(882a), *254*
Wall, L. A., 88(1844), 89(1844), *281*
Wallbridge, M. G. H., 24(1137), 47(1138), 54(1138), 55(1138), 167(1137, 1138, 1139), 176(1137, 1138, 1139), *260*
Walling, C., *228*
Walmsley, D. E., 148(1845), 149(1845), 154(770, 1845), *251, 281*
Walter, M. K., *227*
Wang, T. T., 51(1846), *281*
Ware, J. C., 100(1372), 101(1372), 107(1847), *268, 281*
Wartik, T., 22(74), 38(1948, 1951), 39(1948), 40(614, 1951), 42(74), 49(1141), 54(74), 56(1848), 64(74), 65(614), 82(614, 1950, 1951), 92(74), 138(1849), 148(1141), 153(1141), 169(1848), 175(1848), 176(1848), 186(1950), *232, 247, 260, 281, 284*
Washburn, R. M., 21(1850, 1851), 23(1850, 1851), 51(1850), 57(1850, 1851), 62(1850), 79(1852), 92(1850), *281, 282*
Wason, S. K., 77(1853), 159(1853), *282*
Watanabe, H., 6(1859), 8(1494), 10(1494), 22(1859), 49(1403, 1856), 50(1403, 1856, 1857), 77(857, 1854, 1855, 1858, 1859), *253, 269, 272, 282*
Watanabe, T., *269*
Watts, G. B., 16(603, 604), 22(603), 48(603), 49(603, 1074), 54(603), 55(603), 65(603), 98(921), 99(921), 100(921), 132(1074), *246, 255, 259*
Way, G. M., 80(325a), 135(325a), *238*
Weaver, C., 58(1721), *278*
Weber, L., 24(943), 29(943), 46(943), 50(943), 51(943), 66(943), 69(943), *256*
Weber, W., 49(1443), 50(1443),

55(1860), 56(1860), 57(1439), 219(1439a), *270, 282*
Wechsberg, M., 21(1231), 22(1230), 77(1230, 1231), 78(1232), *263*
Wechter, W. J., 30(1861), 100(1861), *282*
Weidmann, H., 51(407), 52(407), *241*
Weigel, H., 62(1211), *262*
Weigert, F. J., 10(1862a, 1862b), *282*
Weilmuenster, E. A., 196(1628), 197(1628), 201(1628), 202(1628), 224(1863), *275, 282*
Weiner, M. A., 21(1680), 139(1679, 1680), *276*
Weinheimer, A. J., 92(1864), 94(1864), 95(1864), *282*
Weismann, T. J., 6(1865), 131(413), 132(411, 412, 413), *241, 282*
Weiss, H. G., 33(1866), 170(1866), 170(1866), 176(1866), 177(1866), 190(1866), *282*
Weiss, M. T., 162(1744, 1867), *278, 282*
Weissman, S. I., 109(1869), 131(1869), 132(1868), *282*
Welch, C. N., 36(1870), 38(1870), 39(1870), 81(1870), *282*
Welch, F. J., 99(1961), 224(1871, 1960), *282, 284*
Welcker, P. S., 138(1120), 146(1120), *260*
Wellum, G. R., 61(602), 62(602), *246*
Wendlandt, W. W., 139(1871a), *282*
Werner, H., 88(1578, 1872), 89(513b), *244, 274, 282*
Weston, A. F., 13(460), 14(459), *242*
Whatley, A. T., 169(1873), 179(1873), *282*
Whitcombe, R. A., 153(423), *241*
White, D. G., 28(1874), 35(1874), 57(1874), 71(1874), *282*
White, R. F. M., 21(489), 48(487, 489, 490), 53(488, 489), 56(489), 61(487, 488, 489, 490), 79(487, 489), 97(488), 98(489), 99(487, 489, 490), *243*
Whitney, C. C., 28(1971), 92(1971), 93(1971), 104(1964, 1971, 1976), 107(1976), 223(1976), *285*
Wiberg, F., 21(1884), 22(1883), 24(1876, 1883, 1890), 45(1883), 49(1878, 1879, 1880), 51(1881), 54(1883), 56(1879, 1880), 57(1876, 1879, 1882), 59(1881), 63(1886, 1890), 64(1883), 77(1875, 1878, 1879, 1880), 78(1875), 79(1879), 1880), 88(1885, 1887, 1889, 1890), 92(1875, 1878, 1879, 1880), 136(1877), 138(1876, 1877), 147(1883), 148(1876, 1878, 1879, 1880), 151(1882), 152(1879), 165(1876), 174(1876, 1878), 175(1876), 177(1876), 179(1876, 1883), 180(1876), 185(1876), 187(1876), *228, 282*
Wiedmann, H., 18(1862), *282*
Wiles, R. A., 101(1020), *257*
Wilke, G., 24(1891), 48(1891), 49(1891), 150(1891), *282*
Willcockson, G. W., 22(145), 84(145), *234*
Wille, D., 56(560), *245*
Wille, H., 28(1888), 57(1888), 71(1888), 77(1888), *282*
Willemart, A., 21(1730), 149(1730), 150(1730), 151(1730), *278*
Willey, G. R., 56(1), 77(1), *230*
Williams, J., 199(137), 204(137), 208(137), 209(137), *233*
Williams, J. L. R., 13(561), 21(700), 49(561), 66(700), 72(699, 701), 75(701), 134(1895), 144(562, 1892, 1893, 1894, 1895), 145(698, 702), 146(700), 220(700), *245, 249, 272, 283*
Williams, L. F. G., 108(846a), 144(846), *253*
Williams, R. E., 3(1505), 7(1032, 1503), 20(381), 22(381), 46(381), 112(1897), 190(1898), 191(1503), 194(1684, 1899), 195(1899), 196(1896), 197(1505), 199(1502a), 201(549), 204(548), 208(1523), 209(1523), *240, 245, 258, 272, 277, 283*
Williams, R. L., 12(107), 85(575), 196(135, 599), 197(599), 199(137), 201(135, 599), 204(137), 207(135, 573, 575,

1900), 208(135, 136, 137, 573, 575, 1900), 209(107, 135, 137, 573, 575, 1900), *233, 245, 246, 283*
Williams, R. M., 104(1021), *257*
Willis, C. J., 23(399, 400), 136(399, 400), 140(398, 399, 400), 147(399, 400), *240*
Willis, H. A., 12(344, 361, 362), 54(656), 62(344), 77(345, 656), 79(656), *239, 248*
Willsek, R. J., 89(1177a), *261*
Wilmarth, W. K., 132(1378), *268*
Wilson, C. O., 12(1080, 1081, 1084), 13(1082, 1083, 1089), 46(1080, 1084, 1683), 47(1683), 79(1089), 176(1083, 1086, 1087, 1088, 1683, 1901a), 190(1079, 1083, 1085, 1086, 1087, 1088, 1685, 1898), 192(1086, 1683, 190a), *259, 276, 277, 283*
Wilson, E. B., Jr., 5(1415), *269*
Wingfield, J. N., 52(9, 11), 62(9, 11), *230*
Winson, P. H., 62(1388), *269*
Winternitz, P. F., 38(1901), 66(1901), 67(1901), 111(1901), 173(1901), *283*
Witte, A., 76(158), 143(158), *234*
Witte, H., 71(799, 800, 801, 802, 1902), 72(799, 800, 1902a), 127(799, 800, 801, 1902, 1902a), 139(801), 141(802), *252, 283*
Wittenberg, D., 81(1915), 138(1915), *283*
Wittig, G., 21(1906), 48(1908, 1910), 49(1906), 69(1906), 81(1906, 1915), 86(1904), 132(1909), 137(1908, 1911), 138(1903, 1907, 1908, 1910, 1911, 1912, 1913, 1914, 1915), 139(1906, 1909, 1910, 1911, 1913), 140(1908, 1910), 145(1913), 146(1910), 151(1905), *229, 283*
Wojnowsko, M., 21(1232), 78(1232), *263*
Wojtkowski, P., 42(1818), 94(1818), 125(1818, 1819), *280, 281*
Wojtkowski, P. W., 15(1534), 47(1551), 52(1550), 109(1551), 127(1550), 139(1552), 165(1534), 166(1534), 186(1534), *273*
Wolf, C. J., 77(1916), *283*
Wolfe, L. S., 223(1508), *272*
Wolfe, S., 28(1728, 1917), 30(1917), *278, 283*
Wolff, M. A., 29(891), 30(891), 31(891), *254*
Wollrab, J. E., 5(1918), *283*
Wong, G. T. F., 188(706, 1506), 202(706), 216(706, 1506), *249, 272*
Wood, S. E., 42(1604), *274*
Wood, W. S., *283*
Woods, W. G., 21(1921, 1923), 22(1922, 1923a), 24(1431), 29(1923a), 47(1922), 59(1920, 1921, 1923), 60(1920, 1923), 61(1920, 1923), 78(1431), 88(1920), 135(1430), 219(1920), 220(1923), *270, 283*
Woodward, L. A., 11(1924), 12(1924), *283*
Worley, S. D., 217(511), *243*
Worsley, M., 182(547a), *244*
Wright, J. C., 12(695), 28(693, 695), 29(693), 66(695), 83(693), 214(693), *249*
Wulfman, C. E., 52(1609), 59(1609), 79(1609), 92(1609), 96(1609), *275*
Wyatt, B. K., 49(862), 50(862), 51(862), 56(862), 57(862), 219(862), *253*
Wyman, F. W., 21(1722), 59(1722), 92(1722), *278*
Wysocki, A. J., 21(1110), 57(1110), *260*

Y

Yabroff, D. L., 5(119, 151), 57(119, 1925, 1926), 58(119), 59(119, 151, 1926), 218(119, 1926), *233, 234, 283*
Yakovlev, I. P., 28(331a), 46(331a), 50(556b), 59(331a), 71(1283), 83(1287), 85(1283), 219(1287), *238, 245, 265*
Yakubchik, A. I., 224(1713a), *278*
Yale, H. L., 56(1926a, 1927), 57(1927), 71(1927), *283*
Yamamoto, K., 46(1022), 52(1022), *258*

Yamamoto, N., 224(634), *247*
Yamamoto, S., 126(1395b), *269*
Yamamoto, Y., 46(309), 50(309, 311), 52(309), 54(309), 103(309, 311a), 108(309, 311a), 126(311), 127(311a), 151(309), *238*
Yamazaki, H., 31(1681), 47(1681), 59(1681), 77(1674, 1681), *276*
Yampol'skay, M. A., 48(913), 53(913), 92(913), *255*
Yamuchi, J., 224(1407), *269*
Yanez, A., 224(18), *230*
Yoon, N. M., 182(196, 216), *235*
Yoshida, Z., 53(1928), 107(1928), *283*
Yoshikuni, M., 224(907), *254*
Yoshizaki, T., 6(1859), 22(1859), 49(1403, 1856), 50(1403, 1856, 1857), 77(1859), *269, 282*
Young, D. E., 170(1929), 176(1929), 177(1929), 183(1929), 184(1929), *284*
Young, T. F., 12(1602), 17(1603), 176(1602, 1603), *274*
Yurkevich, A. M., 62(919, 1381, 1930), 92(919), *255, 268, 284*
Yuzhakova, G. A., 6(1038), 49(1038), 100(1038a), 150(1038a), 153(1038), *258*

Z

Zacharias, B., 57(1801), 58(1801), 59(1801), *280*
Zakharkin, L. I., 20(1939), 21(1931, 1939, 1944), 22(1937, 1938, 1939, 1940, 1941, 1943, 1944), 23(1940), 24(1931, 1932, 1933, 1934, 1935, 1936, 1944), 24(985), 25(985), 28(1931, 1935, 1936), 45(1931, 1932, 1933, 1934, 1935), 46(1939, 1941), 47(1937, 1938, 1939, 1941, 1944), 48(1937, 1942, 1943), 49(1937), 60(1944), 64(1940), 66(985, 1932, 1934), 67(985, 1931), 69(985), 80(985, 1931, 1932, 1933, 1934, 1935), 81(1936), 106(1944), 109(1939, 1941, 1944), 124(1935), 168(1932), 176(1932), 180(985), *228, 257, 284*
Zalkin, A., 3(1709a), *277*
Zamyatina, V. A., 56(922), 58(923), 59(923), 77(922, 922a, 925), 78(922), 148(922), 185(922, 923), *229, 255*
Zannier, H., 138(1628), *275*
Zappel, A., 49(686), 55(686), 92(686), *249*
Zavgorodnii, V. S., 21(1945), 22(1565), 23(1945), 48(1565), 61(1945), *273, 284*
Zdunneck, P., 22(1946), 46(1946), *284*
Zeidler, F., 22(1740), 46(1740), 47(1740), 148(1740), *278*
Zeigler, I., 72(845), *253*
Zeldin, M., 38(1948, 1951), 39(1610, 1948), 40(1949, 1951), 80(1610, 1949), 81(1610, 1949), 82(1947, 1949, 1951), 186(1950), *275, 284*
Zetlmeisl, M. J., 138(1952), 161(1952), *284*
Zhigach, A. F., 28(1954), 47(1954), 48(1954), 139(1954a), 141(1954a), *284*
Zhil'tsov, S. F., 48(982), 88(1598), 138(912a, 1954b), 143(876a, 912, 982), *254, 255, 257, 274, 284*
Ziegler, E., 50(944), 109(941, 944), 133(941), *256*
Ziegler, K., 22(980), 47(980, 1956); 109(1955), 140(1955), 143(1956), *257, 284*
Zietz, J. R., 131(1573), 156(1573), *274*
Zimmer, H., 151(1957), *284*
Zimmerman, H. K., 18(1862), 51(407), 52(407, 1959), 62(1949), 92(1958), *241, 282, 284*
Zubiani, G., 21(367a), 45(367a), *240*
Zuckerman, J. J., 56(473, 474), 62(473), *242*
Zutty, N. L., 99(1961), 224(1960), *284*
Zvonkova, Z. V., 4(1961a, 1961b), *284*
Zweifel, G., 6(320), 27(317, 318, 321, 1966, 1968, 1969), 28(185, 186, 187, 248, 314, 315, 317, 318, 320, 321, 322, 1962, 1965, 1966, 1967, 1968, 1969, 1971), 29(186, 187, 193, 248, 314, 317, 319, 321, 323, 325, 1969), 30(312, 319, 322, 1969), 31(319), 32(185), 186, 187,

1965, 1967), 33(315, 321, 323, 1962, 1968, 1970, 1973, 1974), 38(247), 45(323, 1962a, 1963), 49(186), 56(1962), 67(247), 91(323), 92(1971), 93(1971), 100(322, 1969), 101(323), 104(1964, 1971, 1976), 107(1976), 111(193, 316, 324, 325, 1968), 112(324, 325), 119(193), 130(1970a), 134(415), 141(1972), 170(186, 319, 322, 1966, 1968, 1969, 1973), 173(187), 177(186, 187, 1968), 181(185, 186, 313, 317, 318, 321, 323, 1966, 1968 1974, 1975), 182(185, 186, 187, 320, 1965, 1967, 1969), 223(319, 1976), *229, 235, 236, 238, 241, 285*

Subject Index

A

Acceptor strength, boron, 25, 98–100, 153
Acetate ion, oxidation of, 107
Acetates, formation of, 127
Acetic acid, 91
Acetic anhydride, 222
Acetone, triplet state, 108
Acetophenone, triplet state, 108
Acetyl chloride, 142
Acetylene(s)
 formation of, 113
 hydroboration of, 26, 33, 45, 91, 101, 171, 229
 reaction with decaborane, 209
 thioboration of, 44
Acetylene boronates, 220
Acetylenylboron compounds, 4, 20, 52, 60, 62, 64, 94, 221
Acidity, of polyborane carboxylic acids, 211
Activation energies
 adduct displacement reactions, 154
 $Al(BH_4)_3$, olefins, 30
 allylic rearrangement, 134
 autoxidation of organodiboranes, 179
 2-ethyldecaborane decomposition, 209
 radical reactions, 101
 for S_H2 reactions at boron, 101
 THF·BH_3, olefins, 31
Acylation reactions, 143
Addition compounds, *see* Lewis base adducts
Addition reactions, *see also* Haloboration, Hydroboration
 1,4-addition, 117, 118
 diboration with diboron(4) compounds, 38–40
 organoboranes with unsaturated compounds, 115–119, 122–124
Adduct stabilities, 147–153
Adducts, *see* Lewis base adducts
Adiponitrile, polymerization of, 224
Alcohols
 from organoboron compounds, 34, 42, 97–98, 100–101, 107, 122–123, 127, 181
 as protodeboronation agents, 92, 95–96, 161
Alcoholysis
 2,4-dimethylenetetraborane, 195
 of organodiboranes, 175
 sodium triphenylborane, 137
Aldehydes, 127
 formation of, 100, 123
 reaction with organoboron compounds, 108, 115, 117–118, 121, 124
 reduction by cyanoborohydride ion, 145
Alkali, 125, 130
Alkali metals, 109, 131, 132, 223
Alkaline hydrogen peroxide, reaction mechanism, 100, 101
Alkaline hydrolysis of the C—B bond, 94–95
Alkaloids, tetraphenylborate derivatives, 223
Alkanes, formation of, 91–97, 109, 131, 141
Alkenes
 diboron tetrahalide addition, 38–39
 formation of, 91, 97, 110–115, 119–121, 124, 130, 173
 haloboration, 35–36, 41
 hydroboration of, 26–27, 30–35, 95, 97–98, 120–121, 169–172

reaction with organoboron compounds, 115–121
Alkenylboron compounds, *see* Vinylboron compounds, Allylboron compounds
Alkenylboronic acids, iododeboronation, 104
Alkenylboronic esters, 33
Alkoxide ion, 92, 95, 113, 125–127, 221
Alkoxy radicals, 101, 102, 108
Alkoxyboron compounds, 6, 11–12, 20–25, 41, 51–53, 56–57, 59–63, 80–82, 88, 95–103, 105, 107, 117, 119–122, 125–126, 129, 134, 137, 140, 151–153, 167–168, 180, 185–186, 221, 227
2-Alkoxyvinylboranes, 46, 52, 59, 96
Alkyl radicals, 98–99, 108–109, 133
Alkylating agents, organoboranes, 115–119, 124–131
Alkylborates, *see* Alkoxyboron compounds
Alkylboron group stabilities, 111
Alkylchlorides, formation of, 106
Alkyldiboranes, *see* Diboranes, organo
Alkyl(hydro)dipinan-3α-ylborate anion, 145
Alkylideneamine-*t*-butylboranes, 34
Alkylperoxy radical, 98, 108
Alkylperoxyboron compounds, 98
Alkyne–organoborane displacement, 120
Alkynes
  B_2X_4 addition, 38
  formation of, 124
  haloboration of, 35, 36
  hydroboration of, 26, 33, 91, 101, 181, 229
  reaction with BF, 42
  reaction with organoboron compounds, 115–116
  reaction with R_3BH^-, 137
Alkynylboron compounds, 46, 48, 52–53, 57, 60–62, 64, 77–78, 81, 86, 107, 141–142, 146
  base cleavage of C—B bond, 94
  formation of, 20, 137
  physical properties, 2, 4, 7
Allenes, hydroboration of, 32, 33, 170
Allylamine, hydroboration of, 35
Allylboron compounds, 227
  addition to unsaturated compounds, 116, 117
  protodeboronation, 93
Allyl bromide, 221
Allyl compounds, hydroboration of, 181
Allylic rearrangements, 93, 134
Aluminum, organic derivatives, 5–6, 9, 13, 15, 19, 20–23, 153, 164, 167
Aluminum borohydride
  exchange with Me_4Pb, Me_4Sn, 174
  as hydroborating agent, 28, 30
  reaction with ethylene–oxygen, 224
Aluminum compounds, 109
Aluminum halides, 37, 126–127, 157
Aluminum metal, 41, 169
Aluminum methoxide, 153
Amides, as protodeboronating agent, 92
Amine adducts of organoboron compounds, 8, 13, 28, 40, 126, 147–159
Amineboranes, as hydroborating agents, 28, 29
Amines, 113, 126–128
  formation of, 106, 133
  as inhibitor of autoxidation, 99
  as protodeboronating agent, 92
α-Amino acids, synthesis of, 182
Aminoboranes, 5, 8–9, 12, 16, 20, 29, 49–51, 55–57, 143, 166, 181, 184, 185, 227
Amino radicals, 108
Ammonia
  as protodeboronating agent, 92
  reaction with B_4H_8CO, 196
  reaction with borane carbonyl, 161
  reaction with Lewis base adducts, 154
  reactions with organodiboranes, 183
Ammoniacal silver oxide, 97
Ammonium ion, as proton source, 141, 153
Ammonium tetraphenylborate, thermal decomposition of, 146
Analytical chemistry of boron, 226
Analytical determinations, organoboron compounds, 222
Anchimeric assistance, 129
Anodic oxidation, 143
Antimony compounds, 109, 133
Appearance potentials, organopentaboranes, 204

Applications of organoboron compounds, 223, 224
Aromatic substitution, arylboron compounds, 222
Arsenic compounds, 42, 125
Arsenic–boron compounds, 54, 219
Arylborane–amine adducts, hydrolysis of, 156
Arylboranes, acceptor strengths, 153
Arylborate anions, 10, 12, 136–141, 143–146
Arylboron bond, angle of twist, 6, 9, 13, 132, 218–219
Arylboron compounds, 16, 41, 175, 214–220
  addition reactions, 115–116
  as arylating agents, 125–130, 146
  autoxidation, 99
  biochemical applications, 223, 224
  C—B bond distances, 2–4
  π-bonding, 2, 4, 6–7, 9, 11, 17
  borohydrides, 136–141, 143–146
  electrochemical oxidation, 144
  electronic transitions, 12–13
  exchange reactions, 24–25, 164–165, 172, 179, 182
  formation of, 20–24, 36–38, 43, 44, 113, 136, 143
  halodeboronation, 104–106, 146
  hydrogenolysis, 131
  hydrolysis of, 156
  Lewis base adducts, 154–155
  mass spectrometry, 13–14
  nuclear magnetic resonance, 6–10
  peroxidation of, 100, 101
  phenylboroles, 220
  photolysis, 145
  protodeboronation of, 93–94, 96
  reaction with alkali metals, 131–132
  reaction with trimethylamine oxide, 107
  restricted rotation of aryl group, 153
  thermodynamic properties, 17–18
  transmetallation, 110
  vibrational spectroscopy, 10–12
Arylmetallic reagents, 20–22
Asymmetric hydroboration, 32
Asymmetric organoboron compounds, *see* Optically active organoboron compounds
Autoxidation, 18, 97–100, 134–135, 175, 179
Azides, 128, 227
Azobenzene, reaction with organoboranes, 121

B

Back-bonding to boron, 2, 25–26, 98, 100, 153, 216–217
Base catalysis, bromodeboronation, 104
Base-catalyzed rearrangement, 126, 198
Benzaldehyde, 121
Benzene
  formation of, 94, 97, 144–145
  reaction with B_2Cl_4, 40
  reaction with elemental boron and bromine, 43
  reaction with organodiboranes, 188
Benzeneboronic acid, *see* Phenylboronic acid
Benzilic acid, as chelating agent, 101
1,3,2-Benzodioxaborole, as hydroborating agent, 29, 33
1,4-Benzoquinone, reaction with trialkyl-boranes, 118, 119
Benzoyl derivatives, of $B_{10}H_{10}^{2-}$, 211, 212
Benzyl chloride, reaction with decaborane(14), 208
Benzylboron group, uv, 12
Beryllium compounds, 109
  organo, 21
Biacetyl triplet, 108
Biaryls, 144
Biferrocene, formation of, 97
Biochemistry, of organoboron compounds, 223, 224
Biphenyl, formation of, 144
Biphenyl anion radical, formation from triphenylboron anion radical, 132
Bis-9-borabicyclo[3.3.1]nonane, 9-BBN, 172, 179
Bisboracyclane, 180
Bis(boracyclopentane), 170, 175, 181
Bis(9-borafluorene), synthesis of, 180
Bisborepane, 171
Bisborolane, rearrangement of, 180
Bis-2-chlorovinylchloroborane, 35
Bis(cyclopentadienyl)cobalt, 134

1,2-Bis(dimethoxybora)ethane, 195
Bis(dimethoxyboryl)methane, 95
Bis(2,3-dimethyl-2-butyl)borane, 186
Bis-3-methyl-2-butylborane (dimer), 27, 30
Bis(1,2-tetramethylene)diborane, 175, 179
Boiling points, 18
Bond angles
  C—B—C, 179
  H—B—H, 162
Bond dissociation energies, B—C bond, 160
Bond hybridization energy, boron, 153
Bond lengths
  B—B, 5, 189–190
  C—B, 2–4, 162, 189, 203, 209
  C—C, 209
  C—H, 189
Bond orders, borane carbonyl, 162
Bond strength, boron-donor compounds, 153
Bonding
  borane carbonyl, 162
  C—B bond, 4–6
Borabenzimidazolines, 155
9-Borabicyclo[3.3.1]nonane, 9-BBN, as hydroborating agent, 33
1-Bora-1-*t*-butyl-4,4-dimethyl-4-silacyclohexane, 34
Boracyclanes, *see* Cyclic organoboranes
Boracyclopropene, 216
9-Borafluorene, 187
  cation, 157
Borane, BH_3, 5, 15, 187, 226
  displacement of from adducts, 156
  as hydroborating agent, 28, 30, 33
  adducts
    as hydroborating agents, 26–29, 31, 33
Borane carbonyl(s), 1, 4, 10, 12, 15, 159–163, 170, 195
  organo substituted, 122
Borates, reaction with pentaborane(9), 196
2,1-Borazaronaphthalene, 218
10,9-Borazarophenanthrene, 222
Borazines, 1, 3, 4, 6, 9, 15, 43, 77, 78, 88, 89, 106, 128, 135, 154, 184, 185, 218, 224, 227
Borenylide intermediate, 142
Boric acid, 97, 107, 115, 141, 143–144, 161, 179, 223
Borinanes, 171
Borinates, reduction of, 186
Borinic acids, *see also* Hydroxyboron compounds
  bromination, 127
  formation of, 107, 133, 141, 144, 156, 175
  hydrolysis, 96
  peroxidation of, 100
  reaction with diborane, 172
Borinic esters, exchange with organodiboranes, 180
Borohydride ion(s), 44, 141, 228
  organo, 7, 10, 83, 95, 136–147, 174
Borolanes, 66, 93
Boroles, 220
Boron
  acceptor ability, 25, 98–100
  analytical determination of, 223
  charge density, 12, 15, 126, 141, 162, 186, 197, 204, 207
  back-bonding to, 2, 25–26, 98, 100, 153, 216–217
  electron density of, 156
  π-electron density, 12
  elemental, reaction with benzene, 43
Boron atom
  π-charge, 16
  empty *p*-orbital, 17, 104, 108–109, 156–157, 162, 179, 187
Boron–boron bond, formation of, 44, 45
Boron–boron exchange, 24–26, 45, 133, 143, 203, 214
Boron–carbon bond, *see* Carbon–boron bond
Boron cations, *see* Boronium ions
Boron halides, 15, 19–29, 35, 36, 39, 41–43, 102
Boron–metal exchange, *see* Transmetallation
Boron monofluoride, reaction with acetylene, 42
Boron tribromide, 227
  as haloborating agent, 35, 36
  as transmetallation agent, 20
Boron trichloride, 15, 39, 226
  as haloborating agent, 35–36

reaction with aluminum carbide, 42
reaction with benzene, 36
reaction with carbon vapor, 41
as transmetallation agent, 19, 20, 23
Boron trifluoride, 20, 23, 28–29, 35, 136
Boron trifluoride etherate, 26–29, 174
Boron trihalides, reaction with olefins, 35–36, 41
Boron triiodide, 43, 227
Boron trimethyl, *see* Trimethylborane
Boronates, *see* Alkoxyboron compounds
Boronic acids, 1, 4–5, 7, 56–59, 94, 100, 105, 107, 109, 118, 144, 156, 175, 218, 222, *see also* Hydroxyboron compounds
Boronium ions, 13–15, 42, 144, 155, 157–159, 192, 228
Boronous acids, *see* Hydroxyboron compounds
Boroxines, 4, 12–13, 16, 20, 44, 79, 101, 104, 133, 159
Bridge bonding, exchange reactions, 25
Bridge deprotonation, organopentaboranes, 202
Bromide ion, displacement of, 221
Bromine, 103–106, 127, 156, 219, 220, *see also* Halodeboronation
Bromine atoms, 108
reaction with α-C—H bonds, 103
Brominolysis, *see* Halodeboronation
Bromobenzene, formation of, 133
Bromodeboronation, 103–106, 146
Bromotrichloromethane, addition to vinylboronates, 220
Butadiene
hydroboration of, 170
reaction with triphenylborane, and tritylsodium, 137
t-Butanol, 101
1-Butene, hydroboration of, 31
Butenes, formation of, 107, 117, 143
t-Butoxide ion, 126
1-Butoxy-3-methylboracyclopentane, 168
n-Butylboronate anion, anodic oxidation of, 143
4-*t*-Butylethoxycyclohexene, hydroboration of, 35
t-Butylhydroperoxide, 101
t-Butylhypochlorite, 102
t-Butylhyponitrite, as radical generator, 98
Butyl lithium, reaction with boronium ions, 42
n-Butyl nitrite, 107

C

Cadmium compounds, 94
Cadmium, organo, 20
Cage polyboron hydrides, 210
Carbanions, 42, 92, 94, 95, 126
boron-stabilized, 222
Carbenes, 31, 129, 130
Carbon
analytical determination of, 223
orbital hybridization, 9, 204
Carbon–boron bond
barrier to rotation, 5, 11, 189, 192, 203–204
bond distances, 2–4, 147, 162, 189, 203, 209
bond order, 12, 162
π-bonding, 2, 4–7, 9–11, 15–18, 25, 218
dissociation energies, 16–17
force constants, 9–11, 163
ligand exchange, 24–26
stretching frequencies, 9–12
Carbon–boron bond construction, 19–45, 136–137, 159–161, 169–173, 193–197, 199, 200, 207–211, 213–215
cyclization by loss of H_2, 38, 188
using diboron(4) compounds, 38–40
Friedel-Crafts reaction, 36–37
haloboration, 35–36
hydroboration, 26–35
metalboron compounds with halides, 40–41
redistribution and exchange reactions, 24–26
transmetallation, 19–24, 173–174, 214
Carbon–boron bond strength, 15–17
Carbon–boron cleavage reactions, 91–135, 137, 141–146, 160–162, 179–181
with alkali, 94
with antimony pentachloride, 133
using *n*-butyl nitrite, 107

carbonylation, 122–124
with *N*-chloroamines, 106
dehydroboration, 111–113
halodeboronation, 102–106, 146, 156
homolytic reactions, 98–99, 135
hydrogenolysis, 131
using nitric oxide, 107
using nitrosocyclohexane, 107
oxidation, 97–100, 107, 108
with peroxides, 100–102
protodeboronation, 36, 91–97, 103, 126, 141, 223
radical reactions, 109
with selenium, 107
with sulfur, 107
transmetallation, 109, 143
Carbon–boron coupling, nmr, 10, 204
Carbon–hydrogen bond
bond energy, 17, 103
hydrogen abstraction, 100
radical attack, 103, 108, 109
spin coupling, 9
Carbon monoxide, 225
reaction with B_2H_6, 44, 159
reaction with B_4H_{10}, 195
reaction with B_5H_{11}, 195
reaction with boroxine, 159
reaction with 1,10-$(N_2)_2B_{10}H_8$, 210
reaction with organoboranes, 122–124
Carbon monoxide(C—B)borane, *see* Borane carbonyl
Carbon monoxide–triborane, 193
Carbon–oxygen bond, 19
Carbon tetrachloride
reaction with boron halides and metals, 40
reaction with diboron(4) compounds, 41
Carbon vapor, reaction with boron compounds, 41–42
Carbonium ions, 1, 142, 211, 231
Carbonyl compounds, reaction with organoboron compounds, 115, 117–119, 121–124, 146
Carbonyl derivatives
$B_{10}H_{10}^{2-}$, 210
$B_{12}H_{12}^{2-}$, 210
Carbonyl group
B_4H_8CO, 195
in borane carbonyl, 163
Carbonylation, 44, 122–124, 210
Carboranes, 107, 109, 135, 188, 189, 202, 203, 209, 216, 226, 227
Carboxy group, 222
Carboxylic acids
acidity of, 218
for organoborane analysis, 223
of polyboranes, 210–212
use as protodeboronating agent, 91–93, 141
reduction of, 182
Cationic boron complexes, 228, *see also* Boronium ions
Charcoal, activated, as catalyst in haloboration reaction, 35
Charge densities, heterocyclic compounds, 217, 218
Charge distribution
decaborane, 207
pentaborane(9), 197
Charge transfer, intramolecular, 12, 13
Chelate complexes, protodeboronation reaction, 96
Chelating agents, as catalysts, 101
Chelation, autoxidation inhibition, 99
Chlorination, of polyboranes, 211
Chlorine, 102, 104, *see also* Halodeboronation
Chlorine atoms, 210
Chloroalcohols, hydrolysis of, 224
N-Chloroamines, 106
Chlorobenzene, formation of, 133
Chloroborane, as hydroborating agent, 28, 35
3-Chlorocyclohexene, hydroboration of, 113
Chlorodifluoromethane, reaction with organoboranes, 127
Chloroform, reactions of, 40, 146
Chloromethyldimethylborane, 46, 129
γ-Chloropropylboranes, 95
2-Chlorovinyldichloroborane, 35, 64
Chromic acid, 107
Chromium compounds, 3, 88–89
Cineole, 209
Cis addition, to alkynes, 116
Cis-1-butene-1-*d*, hydroboration of, 30
$(CO)_2B_{10}H_8$, 210
$(CO)_2B_{12}H_{10}$, 210, 211
Cobalt compounds, 3–4, 89, 134

Combustion, 16
Coordinate bond, 99, *see also* Lewis base adducts
Copper compounds, 109, 133
Copper tetraphenylboron, 146
Corticosteroid boronates, 223
Coupling, nmr, 9, 10, 190, 191, 204
Coupling of radicals, 97, 109
Cyanide ion
  reaction with decaborane, 209
  reaction with pentaborane(9), 199
Cyanides, organo, reaction with organodecaboranes, 209
Cyano compounds, 83
Cyanoboranes, 211
Cyanoborohydride ion, 10, 12, 137–138, 141, 145–146, 224
Cyclanones, reduction of, 182
Cyclic ketones, alkylation of, 117, 118, 119
Cyclic organoboranes, 3, 7, 14, 20, 34, 36, 66–70, 114, 118, 120, 122–123, 130, 134, 145–146, 165–168, 174, 176–181, 183, 187, 190, 214–220, *see also* Borazines, Boroxines, Heterocyclic compounds
  aromatic character, 216, 217
  cleavage of, 118, 156, 166
  formation of 36–38, 41, 116, 123, 137, 188, 215
  protodeboronation of, 93, 96, 175
  stability of, 93, 96, 173, 180, 215
Cyclic organodiboranes, 165–168, 170–172, 176–178, 180–181, 184, 185, 189, 190
Cyclobutene, reaction with B_2Cl_4, 39
Cyclohexene
  hydroboration of, 172
  reaction with BBr_3, 36
  reaction with B_2Cl_4, 39
  reaction with Et_3B, 117
1,5-Cyclooctadiene, hydroboration of, 172
Cyclopentadienyl group, 20
Cyclopentene
  hydroboration of, 30
  reaction with B_2Cl_4, 39
Cyclopropane
  cleavage with diborane, tetraborane, 42, 174
  reaction with B_2Cl_4, 40
  synthesis of, 95
Cyclopropanes, formation of, 114
Cyclopropene, reaction with B_2Cl_4, 39
Cyclopropene, 1-methyl, reaction with trially borane, 117

D

Deboronation, *see* Carbon–boron cleavage reactions
Decaborane(14), 226, 228
  reaction with acetylenes, 209
  compounds, organo, 3, 199, 207–209
Decaboranylmagnesium iodide, 207, 208
1-Decene, 120
Dehydroboration, 1, 24, 38, 111–113, 120, 157, 173, 174, 188, 215
Dehydrobromination, 221
Dehydrogenation of cyclic compounds, 215
Deprotonating agents, 126
Deprotonation, of decaborane(14), 208
Deuteriodeboronation, 94
Deuteriumisotope studies, 195, 222
Deuterium oxide
  exchange with decaborane, 209
  exchange with $NCBH_3^-$, 141
Diacylperoxides, 118
Dialkylborane (dimer), synthesis of, by hydroboration, 27, 31
Dialkylboranes, 38, *see also* Diboranes, organo
Diallylphenylphosphine, hydroboration of, 35
Diamines, 185
Diammoniates, 183
Diazadiboretanes, 129
Diazoacetate, ethyl, 127
Diazoalkanes, polymerization of, 224
Diazoketones, reaction with organoboranes, 127
Diazomethane
  reaction with *B*-chloroborazine, 43
  reaction with trialkylboranes, 127
Diazonium salts, 210
1,3-Dibora-2,4-diazarobenzene, 218
Diborane, 227, 228
  B—B bond distance, 5, 190

exchange reactions, 173, 180
as an exchange reagent, 38
formation of, 131, 168
as hydroborating agent, 26, 28, 30, 32, 33, 169, 179, 204, 208
reaction with acetylene, 45
reaction with B_4H_8CO, 195
reaction with carbanions, 136
reaction with carbon monoxide, 159
reaction with $Me_3B \cdot N_2H_4$, 154
reaction with organoboranes, 164–165, 169, 172, 174–175, 181
reaction with PhLi, 136
reaction with trimethylborane, 199
reaction with triphenylborane, 165
reaction with ylides, 42, 174, 175
transmetallation, 20
tritiated, 120
Diborane–borane equilibrium, 186, 187
Diborane(4) compounds, 43, 80, 226
formation of, 44–45
insertion reactions, 194
reaction with alkenes, alkynes, 38–40, 169
reaction with CCl_4, 41
reaction with ferrocene, 41
reaction with HCN, 43
transmetallation, 24
Diboranes, organo, 164–192
alcoholysis of, 175
autoxidation of, 99
bond distances, 3, 5, 189
coordination compounds, 183
disproportionation and exchange reactions, 172, 179–181
dissociation energy, 187
equilibria, 169, 172
geometrical isomerism, 187
as hydroborating agents, 27, 28, 30, 33, 34, 183
hydrolysis of, 175, 179
monomer–dimer interconversion, 179
nmr, 8, 190–191
oxidation of, 175, 179
physical properties, 189–192
reaction with fluoroolefins, 113
reaction with hydrazine, 155
reaction with LiH, 136
reaction with pentaborane(11), 204–205
reaction with tetraborane, 193, 205
synthesis, 27, 42, 112, 120, 154, 164–175
vibrational spectra, 12, 190, 192
Diboration with diboron(4) compounds
mechanism of, 39
stereochemistry of, 38
Dibutylboron azide, 50, 129
Di-*t*-butylethylene, hydroboration of, 170
Di-*t*-butyl peroxide, 100, 101
2,6-Di-*t*-butylphenoxide, 126
Dichloroborane, as hydroborating agent, 28
Dichloromethyllithium, 127
Diels-Alder reactions, 146, 219, 220
Dienes
formation of, 104
hydroboration of, 32, 33, 181, 229
reaction with B_2Cl_4, 39–40
Dienophiles, 146
Dienylboranes, 134
Diethylphenylboronate, reduction of, 168
Difunctional molecules, 219
Diglyme, as hydroboration solvent, 27
Dihydroxy(methyl)borane, 1, 56, 175
Diisopinocamphenylborane (dimer), 30, 32, 182
Dimer-monomer equilibria, difunctional molecules, 219
Dimesitylboron azide, 128, 129
Dimethylboron chloride, 54, 173, 194
Dimethylboron fluoride, 12, 53
Dimethylborinic acid, 50, 175, 189
2,3-Dimethyl-2-butene, hydroboration of, 169
2,3-Dimethyl-2-butylborane, 27, 31
2,4-Dimethylenetetraborane, 194, 205
Dimethylphosphine, 184
Dimethyl sulfide, formation of, 42
3,5-Dimethyl-1,2,4-trioxa-3,5-diborolane, 100
1,3-Diones, 96
Diperoxyborane, 100
Diphenylborane, 172, 187
Diphenylborinic acid, 13, 51, 97, 141, 144
Diphenylboron chloride, 13, 17, 55, 137, 172
2,3-Diphenylbut-2-enes, 95
1,2-Diphenyldiborane, 179, 180

Diphenylmercury, 24
Dipolar organoboron compounds, 217
Dipole moments, 6, 18, 162, 203, 216
Displacement reactions, 115, 119–121, 154
Displacement-elimination reactions, 113, 114
Disproportionation reactions, 31, 125, 155, 156, 168, 172, 179–181, 183
Dissociation energy, organodiboranes, 187
Dissociative mechanisms, 154
Disulfides, organic, 102
Divinylmethylborane, 12, 16, 47
Diynes, hydroboration of, 33
Donor-acceptor combinations, 147–152
Donor compounds, *see* Lewis base adducts

E

Electric discharge, organopentaboranes, 202
Electron affinities, triarylboranes, 132
Electron delocalization, 154
Electron density, boron, 6, 16, 156, 186
Electron diffraction, 3, 4, 162, 189
Electron impact studies, 15
Electron transfer, 133, 144
Electronegativity, boron, 7, 18
Electronic effects, protodeboronation, 93
Electronic factors, adduct stability, 153
Electronic spectra, 11–13, 16, 216
Electro-oxidation, 108, 143, 144
Electrophilic substitution
  of $B_{10}H_{10}^{2-}$, 211
  BI_3 and ArI, 43
  of decaborane, 207
  heterocyclic compounds, 217–218, 222
  of pentaborane, 196
Electrophilic reactions
  displacement, 104
  halodeboronation, 104, 105
  protodeboronation, 94
  transmetallation, 110, 111
α-Elimination, 129–130
Elimination-displacement sequence, 117
Elimination reactions
  acid catalyzed, 114
  base catalyzed, 113, 114, 115
  dehaloboronation, 113
  dehydrohalogenation, 221
  dehydroboration, 111–113
Entropies of activation, adduct displacement reactions, 154
Entropy, borane carbonyl, 163
Equilibrium constants, monomer–dimer boranes, 165–167
Esr spectra, 101
  $Me_2BCH_2\cdot$, 109
  $Ph_3B\cdot^-$, 132
Esters, 42, 125–127
Ethers
  as catalysts, 26, 32, 97, 111, 159, 170, 173
  reaction with pentaborane, 196
Ether cleavage, 1-bromopentaborane(9), 199
Ethylacetoacetyldi-*n*-butylboronite, 99
Ethylborane, use as hydroboration reagent, 30
Ethyl bromide, reaction with $B_{10}H_{14}$, 208
Ethylene
  force constant, 10
  formation of, 111, 114
  hydroboration of, 26, 30
  and Me_3B, 6
  reaction with B_4H_8CO, 196
  reaction with diborane, 199, 204, 208
  reaction with pentaborane(11), 205
  reaction with tetraborane, 194
Ethyltriphenylborate, 143
Ethynylboron compounds, 52, 60, 62, 64, 138–139, 149
  B—C bond cleavage by base, 94
  C—B bond length, 4
  formation of, 20
Exchange reactions, 24–26, 143, 164–167, 179–181, 203, 214

F

Fenton's reagent, 102
Ferric ion oxidation, 144
Ferrocene
  formation of, 97
  reaction with B_2Cl_4, 41

Ferrocenylboron compounds, 97, 133, 152
Flash thermolysis, 188, 202
Fluorescence, 18
Fluorine–boron π-bonding, 5
Fluorine transfer reactions, 113
Fluoroolefins, hydroboration of, 113
Fluoroorganic compounds, 127, 208
Fluoroorganoboron compounds, 12, 23, 40–41, 94–95, 99, 125, 139, 147, 153, 226
Force constants, 9–11, 163
Formaldehyde, reaction with organoboron compounds, 117
Four-coordinate organoboranes, 136–163
  carbon monoxide–borane, 159–163
  Lewis-base adducts, 147–157
  organoborohydride ions, 136–147
  organoboronium ions, 157–159
Free-electron model, 204
Friedel-Crafts reactions, 36–37, 196, 204, 214
Furan, 131, 133

G

Gallium, organic derivatives, 6, 9, 13, 153, 173
Galvinoxyl, as radical scavanger, 98, 118
Germanium compounds, 81, 140
Glycinate ion, reaction with borane carbonyl, 161
Grignard reagents, 19–23, 113, 166, 207, 208

H

Halide displacement, 125, 126
Halides of boron, 227
Halides, organic, 102–106, 126, 127, 143, 145, 196–197, 207–208, 221
Halides, organoboron, *see* Organoboron halides
Haloboranes
  used for hydroboration, 28
  reduction by triethylborohydride ion, 145
Haloboration, 35, 36, 146
Halodeboronation, 102–106
Halogen–boron bond, ligand exchange, 24–26
Halogen migration from carbon to boron, 124–126, 130
Halogenation, of arylboron compounds, 222
α-Haloorganoboranes, 31, 32, 113, 124–126, 129, 130, 188, 221, 222
β-Haloorganoboranes, 32, 113, 114
Halopolyboranes, 201, 212
Hammett σ-constants, 7
Heats of formation, 16, 17, 163
Heats of hydroboration, 16
Heteroaromatic compounds, aromaticity, 217
Heteroaromatic organoboron compounds, 226
  ^{11}B nmr, 7
Heterocyclic organoboranes, *see also* Cyclic organoboranes
  nitrogen–boron, 1, 3, 15–16, 20, 27, 34–38, 43, 70–78, 83–87, 96, 106, 113, 128, 130, 142, 154, 155, 157, 214, 215, 217, 222
  oxygen–boron, 4, 12, 13, 16, 20, 29, 33, 37, 68, 70, 74–76, 79, 85–89, 96, 99–100, 102–103, 113, 122, 124, 131, 142–143, 159, 161, 220
  phosphorus–boron, 75, 88, 89
  review, 225
  selenium–boron, 90, 107
  silicon–boron, 34, 70
  sulfur–boron, 25, 37, 70, 75–76, 88, 217
  tin, 20
Hexaborane(10) compounds, organo, 206
Hexamethylenetetramine, 198
Homologation of organic groups, 42, 115–117, 125–130
Homolytic bond cleavage, 98, 99, 101–103, 108, 135
Hybrid orbital character
  boron, 2, 18, 93, 147, 204
  carbon, 9, 204
Hydrazines, as protodeboronating agent, 92
Hydrazinoboranes, 155

Hydride abstraction reactions, 142
Hydride shifts, 175
Hydride transfer reactions, 42, 124, 125, 188
Hydridic reducing agents, organoborohydrides, 145
Hydroborating agents, 26–35, *see also* individual compounds
Hydroboration, *see also* Carbon–boron bond construction
of alkynes, 33, 45, 101
alkyne–alkene competition, 33
catalysis of by ethers, 26, 32
in conjunction with alkaline hydrolysis, 95
in conjunction with autoxidation, 97, 98
in conjunction with carbonylation, 122–123
in conjunction with peroxidation, 100, 101
in conjunction with protodeboronation, 91
of haloolefins, 31–32, 113
mechanism, 30, 31, 35
of olefins, 26–35, 111–113, 120–121, 160, 169–172, 182, 214
using organodiboranes, 181–183
reviews, 225, 229
stereochemistry of, 30
steric effects, 27, 30, 31, 33
substituent effect on, 31–34
of vinylboronic esters, 45
Hydroboration–carbonylation–oxidation sequence, 122, 123
Hydroboration–dehydroboration sequence, 215
Hydroboration-displacement sequence, 120
Hydrocarbon autoxidation, 98
Hydrocarbons, formation of, 91, 145, 167
Hydrogen
formation of, 38, 137, 141, 155–157, 161, 167, 175, 184–185, 214
reaction with B_4H_8CO, 195
Hydrogen abstraction, 144, 145
Hydrogen, analytical determination of, 223
Hydrogen–boron bond
as catalyst in alkylborane isomerization, 111–112
force constant, 10
as hydroborating agent, 26–35
ligand exchange, 24–26
polarizability, 30
protolysis, 155, 175
Hydrogen, bridge
proton magnetic resonance, 190–191, 204
symmetrical cleavage, 195
Hydrogen, bridge bond, 147, 164–209
dissociation energy, 187
vibrational spectroscopy, 190
Hydrogen bromide, 91, 103, 219
Hydrogen chloride, 91, 104, 126, 155, 167, 173, 206
Hydrogen cyanide, 43
Hydrogen halides
addition to vinylboronates, 219
formation of, 43
metal hydride reactions, 167
as protodeboronating agents, 91–92
Hydrogen isotope effects, 18
Hydrogen peroxide, 100–102, 107, 125, 179
Hydrogen sulfide, as protodeboronating agent, 92
Hydrogen tautomerism, 112
Hydrogenation
of trialkylboranes, 174
of vinyl and acetylenic boron compounds, 220
Hydrogenolysis, 131
borane–amine adducts, 156
Hydrolysis, 27, 91–92, 97–98, 101, 127–129, 156, 175, 195, 222–223, *see also* Protodeboronation
Hydroperoxide radicals, 102
Hydroquinone, 118, 119
Hydroxide ion, use as deboronating reagent, 92, 94, 95, 113, 114
Hydroxyboron compounds, 1, 4, 5, 7, 45, 50–51, 56–59, 80–82, 84, 86–90, 94, 96, 99–101, 103, 105, 107, 109, 118, 127, 133, 140–141, 144, 156, 172, 175, 189, 217–218, 222
Hydroxydeboronation, *see* Peroxidation
ω-Hydroxyketones, 118

Hydroxylamine-*o*-sulfonic acid, reaction with organoboron compounds, 133
α-Hydroxyorganoboranes, 221, 222
8-Hydroxyquinoline, 92
Hyperconjugation, 2, 3, 11, 16, 18, 187, 190, 219

I

Imines, stereospecific reduction of, 145
Iminoboranes, 227
Imino esters, reduction by cyanoborohydride ion, 145
Indium, organic derivatives, 6, 9, 13, 153
Induction period, autoxidation, 99
Inductive effects, 95, 187, 219
Infrared spectroscopy, *see* Vibrational spectroscopy
Insertion reactions, 173, 193, 194, 206
Internal rotation, 5, 192
Iodide, as nucleophile, 221
Iodine
  as radical scavanger, 98, 102, 118
  reaction with adducts, 155, 157
Iodine atoms, 108
Iododeboronation, 104, 105, 146
Ionization energies, organopentaboranes, 204
Ionization potential, ethyldecaborane, 209
Iridium complex with tetraphenylborate anion, 146
Iron compounds, 77, 89, 134–135
Isocyanates, reaction with organoboron compounds, 115
Isocyanide–trimethylboron complex, proton nmr, 147
Isocyanides, 43, 127, 128
Isomerization, alkyl group, 23–24, 111–113, *see also* Rearrangement reactions
Isonitriles, reaction with organoboranes, 127, 128, 226
Isotope studies, 94, 141, 156, 163, 195, 209, 222

K

Ketoboryl compounds, 40
Ketones
  formation of, 100, 101, 107, 122, 123, 141
  as protodeboronating reagent, 92, 96
  reaction with organoboron compounds, 108
  reduction of, 145, 182
Kinetic isotope effects, 156
Kinney boiling point constants, 18

L

Lactones, reduction of, 182
Lead compounds, 20, 22, 109, 143
Lead metal, reaction with $NaBEt_4$ and ethyl halides, 143
Lewis acids, 93, 118, 124, 153, 157, 188, 196, 197
Lewis base, sterically hindered, 222
Lewis base adducts, 9–10, 40, 103, 108, 147–157, 214, 222
  dissociative bond strength, 154
  elimination reactions, 113
  exchange reactions, 25–26
  hydroboration, 26–29, 31, 33–35
  hydrolysis of, 155
  hydrogenolysis, 131, 156
  of organoboranes
    B—C bond cleavage reactions, 155
  organoboron radicals, 109
  pyrolysis of, 155
  reaction with mineral acids, 155
  stabilities, 147, 153, 183
  synthesis, 147, 186
  transmetallation, 20, 23
Lewis bases, reactions with organodiboranes, 183
Ligand exchange, 1, 24–26, 143, 179
Lithium aluminum hydride, 28, 167, 168, 186
Lithium borohydride, 28, 29, 123, 137, 167
Lithium compounds, 109, 207, 214
Lithium hydride, 28, 136
Lithium metal, 41
Lithium octahydropentaborate(1—), reaction with Me_2BCl, 206
Lithium, organo, 19–21, 23, 129, 131, 173

Lithium perhydro-9*b*-boraphenalylhydride, as stereospecific reducing agent, 145
Lithium triethylcarboxide, reaction with organoboranes, 127
Lithium triphenylborohydride, use in the reduction of acyl halides, 145
2,6-Lutidine, 198

M

Macrocyclic alkadienes, 120
Magnesium compounds, 109, 136
Magnesium, organo, *see* Grignard reagents
Manganese compounds, 89
Mannich base, 117, 118
Mannitol, 223
Mass spectrometry, 13–15, 100, 157, 162, 163, 192, 196, 204, 209, 224, 228
Mechanisms of reaction, *see* specific reaction type
Medicinal chemistry of organoboron compounds, 223, 224
Mercuric acetate, 109, 110
Mercurideboronation, 109–110
Mercury compounds, 109
Mercury compounds, organo, 20, 22–24, 130, 143, 193
Mercury, elemental, 110, 146
Mesidine, 129
Meta directors, boron containing, 222
Metaboric acid, formation of, 97
Metal–boron exchange, *see* Carbon–boron cleavage, Transmetallation, and Carbon–boron bond construction, transmetallation
Metal hydride reduction, 167–169
Metal organoboron compounds, 109
Metalloboron compounds, 225
Metals
  as catalysts, 131
  reaction with organoboranes, 131, 132
Metastable ions, 13
Methaneboronate ion, anodic oxidation, 107
Migration reactions, organoboranes as alkylating agents, 103, 104, 125–131, 141, 146
Migratory aptitudes of organo groups, 116, 129
Molecular orbital calculations, 5, 12, 15, 16, 157, 162, 164, 186, 197, 204, 207, 217, 218
Molybdenum compounds, 90
Monoalkylborane–tetrahydrofuran complex, 31
Monalkylboranes, 27, 31
Monomer–dimer equilibria, boranes, 186, 187
N-Methylanilinophenylalkylboranes, photocleavage of, 108
O-Methylboranocarbonate, 161
Methylborohydride ion, 138, 147
Methylboron dichloride, 12, 42, 64
Methylboron difluoride, 12, 17, 62, 147
Methyl–boron rotation, 5, 11, 189, 192, 203–204
Methylboronic acid, 1, 56, 175
Methylboronic ester, 147
2-Methyl-2-butene, hydroboration of, 169
1-Methylcyclopropene, reaction with triallylborane, 117
Methyl(dimethylamino)borane, 181
Methylene chloride, reaction of, 40
Methylenetriphenylphosphorane, 154
2-Methyl-2-nitrosopropane, reaction with organoboranes, 121
2-Methyl-1-propene-1-boronic acid, 11, 56
Methyl radicals, 108
Methyl sulfate, reaction with $B_{10}H_{13}^-$, 208
Methyl sulfide, formation of, 175
Methyltrinaphthylborate anion, 143
Microwave spectroscopy, 3, 4, 162, 203, 204

N

Na_2HBMe_2, 189
Naphthalene, 40, 96, 217
Napthylene boronic acid, 59
  ododeboronation, 104
Nickel catalyst, 43
Nickel compounds, 90
Nitration, of arylboron compounds, 222
Nitric oxide, 107, 162
Nitrile oxides, 220

Nitriles, 126, 127
hydroboration of, 34
reaction with organoboron compounds, 115
reduction of, 182
Nitrogen–boron bond
π-bonding, 5, 16
ligand exchange, 24, 181, 184–185
rotational barriers, 8–9
transmetallation, 22–23
Nitrogen–boron bond order, 219
Nitrogen–boron compounds, 3–6, 8–9, 12, 15–16, 20, 22–24, 27, 29, 34–38, 43, 49–51, 55–57, 70–78, 80–81, 83–89, 96, 106–107, 113, 128–130, 133, 135, 147–159, 181, 184–185, 209, 217, 226, *see also* Aminoboranes, Borazines, and Heterocyclic compounds
Nitrogen–boron polymers, 225
Nitrogen–hydrogen bond, 217
Nitrosocyclohexane, 107
Nomenclature, 1
Norbornyl radical, 102
Nuclear magnetic resonance, 226, 227
boron-11, 4, 6–8, 147, 190, 191, 193, 194, 196, 197, 204, 209
carbon-13, 8
fluorine-19, 5, 218
proton, 6, 8, 9, 147, 190, 191, 194, 197, 202, 204, 205, 216
spin couping, 9, 10, 190, 191, 204
Nuclear quadrupole resonance, ^{11}B, ^{10}B, ^{35}Cl, 18
Nucleophilic reactions, 94, 125–131, 197, 207, 221
Nucleophilic reagents, 113, 114, 127

O

Olefin-BF_3 complexes, 6
Olefin elimination reactions, 111
Olefin–organoborane displacement, 119–121
Olefin polymerization, 224
Olefinic halides, reduction with $NaBH_4$, 169
Olefins
formation of, 91, 97, 110–115, 119–121, 124, 130, 173, 188
hydroboration of, 26, 27, 30–35, 97, 98, 121, 145, 169–172, 181, 229
reaction with Al, BX_3, 41
reaction with boron halides, 35–36
reaction with decaborane(14), 207
reaction with diborane, 179
reaction with pentaborane(9), 196, 197
Optically active organoboron compounds, 6, 32, 94, 129, 181
Orbital symmetry, hydroboration reaction, 31
Organic iodides, formation of, 105
Organoborane (monomer), 46, 186–187
Organoboranes
as alkylating agents, 115–119, 122–131
as Lewis acids, 118
mixed, formation of, 119
optically active, 6, 32, 94, 129, 181
Organoborate anions, as hydridic reducing agents, 145
Organoborohydrides, *see* Four-coordinate boron compounds
Organoboron bromides, 54, 57, 62, 64–65, 80, 90, 151–152, 156
Organoboron chlorides, 45, 54–55, 57, 63–65, 80–82, 89, 99, 140, 151–152, 158, 167–168, 172
Organoboron fluorides, 53, 62–63, 80–82, 140, 147, 151–152, 214
Organoboron halides, 1, 3–5, 9, 11, 12, 15, 17, 19–10, 23–26, 40–42, 44–45 53–55, 57, 62–65, 80–82, 89–90, 102, 106, 125–126, 130, 140, 146, 151–153
as alkylating agents, 127, 128
autoxidation of, 99
bromodeboronation, 103
formation of, 133, 146
Lewis acidity, 128
peroxidation of, 100
protodeboronation, 94
reaction with alkali metals, 109
reaction with diborane, 172
reaction with trimethylamine oxide, 107
reduction of, 167, 168
Organodecaboranes, *see* Decaboranes, organo
Organodiazo compounds, 127

Organodiboranes, *see* Diboranes, organo
Organohexaboranes, *see* Hexaborane, organo
Organopentaboranes, *see* Pentaborane, organo
Organoperoxy radical, 108
Organopolyboranes, *see* Diboranes, Tetraboranes, and other polyboranes
Oxaboroles, formation of, 142
Oxalyl chloride, as carbonylation agent, 210
Oxidation, 1, 42, 127, 228, *see also* Autoxidation, Peroxides
  as an analytical procedure, 222, 223
  of borane carbonyl, 162
  of organoboron compounds, used in conjunction with hydroboration, 27, 98
  of organodiboranes, 175, 179
  organopentaboranes, 202
  two-electron process, 144
Oxidative hydrolysis, 122
Oxidative polymerization, 224
Oxidative reactions
  organoborohydrides, 143–145
  using trimethylamine oxide, 107
Oximes, as protodeboronating agent, 92
Oxygen, 18, 97–99, 102, 117–118, 134, 144, 162, 175, 179
Oxygen–boron bond
  ligand exchange, 24–25, 172–173
  transmetallation, 20–23
Oxygen–boron compounds, *see also* Autoxidation, Alkoxyboron compounds, Carbon–boron cleavage reactions, Hydroxyboron compounds, Heterocyclic organoboranes, Peroxidation
  formation of, 97, 129
  reduction of, 167, 185, 186

P

Palladium-charcoal catalyst, 215
Paraffinic compounds, formation of, 91
Paramagnetic organoboranes, 109, 132, 134
Pentaborane(9)
  alkylation, 196
  organo derivatives, 4, 8, 196–204
    physical properties, 203
    reactions, 202
    synthesis, 196–199
Pentaborane(11)
  reaction with carbon monoxide, 195
  reaction with ethylene, 205
  reaction with organodiboranes, 204, 205
  organo derivatives, 204–206
1,4-Pentadiene, hydroboration of, 171
Pentafluorophenylboronic acid, 59, 95
Perbenzoic acid, 100
Perchloric acid, as coreagent for peroxidation, 101
Perfluorotriphenylborane, stability toward autoxidation, 99
Peroxidation
  in conjunction with hydroboration, 100–101
  organoborane analysis, 223
  reaction mechanism, 100, 101
  stereospecificity of, 100
Peroxide, diacyl, 118
Peroxides, 141, 226
  as catalysts in the cleavage of C—B bonds with thiols, 96
  reaction with organoboron compounds, 100–102
Peroxyboranes, 98–100, 226
Peroxy radicals, 108
Phenanthrene, 217
Phenols, formation of, 100, 134, 144
Phenoxide, ion, 126
Phenylborane, 168, 174, 187
Phenylborinic acid, 51, 218
Phenylboron compounds, *see also* Arylboron compounds
  perfluoro-, 125
Phenylboron dibromide, 43, 65, 134
Phenylboron dichloride, 10, 17, 35, 43, 65, 106, 167–168, 214
Phenylboron difluoride, 24, 62
Phenylboron halides, 18
Phenylboronic acid, 57, 218
  formation of, 96, 134, 144
  halodeboronation, 104
  mass spectra, 13
  mercurideboronation, 110

M.O. calculations, 16
nitration of, 222
peroxidation of, 100–101
protodeboronation, 93–94, 97
reaction with CuX_2, 133
Phenyldibromoborane, *see* Phenylboron dibromide
Phenyldioxaborolane, 13
o-Phenylenediamine, 155
1,1-Phenylethylene, interaction of π-electrons with boron of R_3B, 11
Phenyl group migration, 116, 128
Phenyl radical, 144–146
Phenyl sulfide anion, 126
Phosphoric acid
as coreagent for peroxidation, 101
reaction with $K_2H_3BCO_2$, 161
Phosphorus compounds, 136–137, 153–154, 188, 195
Phosphorus–boron compounds, 53, 57, 63, 75, 88–89, 137, 143, 194–195, 219
Phosphorus halides, exchange with trialkylboranes, 133
Photochemical rearrangements, 144, 145
Photocyclization, 134
Photoelectron spectra, 18, 162, 217
Photolytic reactions
benzene and BX_3, 43
cyclohexene, BEt_3, 117
dehydrohalogenation, 221
diboron(4) compounds, 41
1,2-diborylolefin isomerization, 222
halogenation, 103, 106, 127
α-hydrogen abstraction, 103, 109
iron–organoborane complex, 134
organoborohydride ions, 143–145
S_H2 reactions, 101–102, 108
trimethylborane, 135
α,β-unsaturated carbonyl compounds with BR_3, 118
Pinacol, as chelating agent, 101
α-Pinene, 172–173
Pivolic acid, as a protodeboronation catalyst, 97
Polar effects, organodiboranes, 186, 187
Polyborane carbonyl compounds, 210, 211
Polyboranes, organo, *see* Diboranes, Tetraboranes, Pentaboranes, Decaboranes, and other individual polyboranes
Polyboron compounds, 41, 43, 38–41, 43–45, 68–90, 95, 110, 195
Polyhedral boranes, 227
Polymers, 170–171, 174, 190, 224–225, 229
Polypyrazolylborates, 228
Poly(μ-trimethylene)diborane, 170, 171
Potassium
reaction with organoboron halides, 168
reaction with trimethylborane, 137
Potassium aminotrimethylborate, 137
Potassium borohydride, 28
Potassium compounds, organo, 20, 22
Potassium dimesityldiphenylborate, photolysis of, 145
Potassium ion, analytical determination of, 223
Potential barriers, B—C rotation, 6
Propane, 17
Propargly chloride, reduction of, 182
Propellants, 224
Propionic acid, 92
Propylene, 10
Protic agents, 91–97
Protodeboronation, 36, 91–97, 103, 126, 141, 223
Proton ionization, of polyborane carboxylic acids, 211
Protonation, of 3-MeB_6H_9, 206
Pyridine adducts, 33–34, 153, 155, 157, 168, 180, 185
Pyridinium ions, 147
Pyrolysis reactions, 188–189, 202–203, 208

Q

Quadrupole coupling constants, 6, 18, 162
Quadrupole moment, boron, 9
Quaternary boron compounds, 2, 6, 7, 10, 12, 95, 104, 125–131, 136–163, *see also* Lewis base adducts, Four-coordinate boron compounds
Quinolineboronic acid, 224

Quinones, reactions with trialkylboranes, 118

R

Racemization, 6
Radical reactions
  addition reactions, 220
  AgOH, NH_3, BR_3, 97
  alkylboranes with carbonyl compounds, 117–119
  autoxidation, 98–100
  Br_2, BR_3, 103
  N-chloroamines with organoboranes, 106
  copper tetraphenylboron with carbonyl compounds, 146
  $CuBr_2$, BR_3, 133
  general, 108–109
  halogenation of α-carbon, 103, 220
  oxidation, 100–102
  polymerization reactions, 224
Radical scavangers, 98, 102, 117
Radical stabilization by boron, 17, 104, 109
Radicals
  alkyl, 143–144
  boron containing, 104, 108, 109, 144
Raman spectroscopy, 10, 162–163, 190, *see also* Vibrational spectroscopy
Reactivity trends
  autoxidation, 98
  displacement reactions, 120
Rearrangement reactions, 1, 42, 103, 106, 124–131, 137, 141, 145, 173, 198, 211, 221
Redistribution reactions, *see* Carbon-boron bond construction
Redox reaction intermolecular, 98
Reduction, with hydrogen, 174
Reduction reactions, 172–173
Reductions with metal hydrides, 167
Reductive alkylation, 118
Refractivities, bond and molar, 18
Relative rates of reaction, hydroboration, 182
Rhenium compounds, 90
Rhodium compounds, 3, 90, 146
Ring currents, 147, 204
Ring strain
  cyclic organoboranes, 93, 96
  organodiboranes, 170
Rose Bengal-generated singlet oxygen, 144
Rotational barriers, methylboron compounds, 5, 11, 189, 192, 203–204
Ruthenium compounds, 90, 146

S

Selenium compounds, 22, 139
Selenium–boron compounds, 54, 90, 107
S_H2 reactions, 98–102, 106, 108, 109
1,3-Shifts, 145
Silane, formation of, 189
Silicates, organo, reaction with pentaborane(9), 196
Silicon compounds, 22, 70, 81, 85, 109, 158, 180
Silicon–boron compounds, 57, 140, 160
Silver, 97, 155
Silver compounds, organo, 22
Silver cyanide, reaction of, 43
Silver hydroxide, ammoniacal, 97
Silver ion, ammoniacal, organoborane analysis, 223
Silver nitrate, alkaline, 109
S_N1 mechanisms, adduct reactions, 154
Sodium, liquid ammonia, reaction with organodiboranes, 189
Sodium borohydride
  as catalyst in carbonylation reactions, 123
  reaction with Me_2BCl, 173
  as reducing agent, 167
  used in hydroboration, 26, 27, 29
Sodium compounds, organo, 20, 21, 137
Sodium hydride, 28, 136, 208, 215
Sodium hydroxide, organoborane analysis, 223
Sodium metal, 131, 132
Sodium methoxide
  reaction with organodiboranes, 180
  use with bromine, 105
Sodium/potassium alloy, 40
Sodium triethyl-1-propynylborate, 142

Sodium trimesitylborane, 132
Sodium tri-α-naphthylborane, 132
Sodium triphenylborane, 132
Solvolysis, of α-haloorganoboranes, 221
Stereochemistry, 225
of chromic acid oxidation, 107
halogen migration from carbon to boron, 124
1,2-migrations, 126
of reaction between *N*-chloroamines with organoboranes, 106
reaction of hydroxylamine-*o*-sulfonic acid with organoboranes, 133
transmetallation reaction, 109
trimethylamine oxide with organoboranes, 129
Stereoisomeric groups, 98
Stereospecific reactions, 104
carbonylation, 123, 124
displacement, 121
hydroboration, 145
peroxidation, 100
reduction of cyclic ketones, 145
Steric effects, 5
autoxidation, 99
borane-donor complex formation, 153
exchange reactions, 165
hydroboration, 27, 30, 31, 33, 169, 181
hydrogenolysis of amine-organoborane adducts, 156
isomerization of alkyl group, 111
Lewis base adducts, 154
ligand exchange reaction, 25
organodiboranes, 186, 190
organopentaborane rearrangement, 198
protodeboronation, 93
S_H2 reactions, 108
sodium triarylborane monomer–dimer equilibrium, 132
Styrene, polymerization of, 224
Styreneboronic acid, hydroboration of, 31
Sulfonic acids, as protodeboronating agent, 92
Sulfur, 107
Sulfur compounds, 108, 125–126, 141, 221
Sulfur–boron compounds, 15, 24–25, 28, 54, 57, 63–64, 70, 75–76, 80, 88, 102 107, 108, 146, 152, 185, 226
Sulfur halides, exchange with trialkylboranes, 133
Sulfuric acid, organoborane analysis, 223

T

Tetrakis(dimethylamino)diboron, 43
Tetraborane
organo derivatives, 193–196
reaction with dimethylmercury, 193
reaction with ethylene, 194
reaction with organodiboranes, 193, 205
Tetraborane(8)carbonyl, 195, 196
Tetracovalent boron anions, 221
Tetraethylborate anion, 139, 143
Tetraethyllead, 143
Tetraethynylborate anion, 139, 141
Tetrahydrofuran, 97, 106, 186
Tetrahydrofuran–borane, 26–29, 31, 166
Tetrakis(dimethoxyboryl)methane, 40, 82–95
Tetramethylborane(1−) ion, 8, 10, 138, 147
1,2-Tetramethylenediborane
conversion to carborane, 188–189
formation of, 165, 170
Tetramethylethylene, hydroboration of, 27, 31
Tetraphenylborate(1−) ion, 225, 226
analytical reagent, 223
^{11}B nmr, 7
biochemical applications, 223
bond distance, 3
electrochemical oxidation, 144
ferric ion oxidation, 144
metal complexes, 146
nmr shift reagent, 147
nomenclature, 1
oxidation by hexachloroiridate, 144
photochemical decomposition, 144
protodeboronation, 141, 153
reactions of, 137–138
spin coupling, 10
transmetallation, 143
Thallium, organic derivatives, 13, 153
Thermal decomposition reactions, 128, 129, 135

Thermal isomerization, organopentaboranes, 199
Thermal reactions, organoborohydrides, 146
Thermodynamic properties, 16, 17, 164, 204
Thermolysis reactions, 188
Thexylborane(dimer), 27, 33
β-Thioalkenylboranes, 44
Thioboration, 44
Thioborinic esters, 107, 175
Thioboronate esters, 155
Thiocyanate ion, 221
Thiols, 155
  as catalysts in the alcoholysis of C—B bonds, 96
  reaction with organodiboranes, 175
Thioperoxides, 102
Thiopheneboronic acids, 218
Tin compounds, 19–20, 22–23, 90, 109, 136, 143, 157, 214
Tolylboronic acid, 223
Tosylates, reduction by cyanoborohydride ion, 145
Trans-cinnamyl group, 20
Transition states
  cyclic six-membered, 116, 121
  cyclization by loss of H_2, 38
  diboration with diboron(4) compounds, 39
  exchange reactions, 165
  four-center, 112, 114, 188
  hydroboration, 30–31
  hydrogenolysis, 131, 156
  hydrolysis of organoborane adducts, 156
  ligand exchange reaction, 25
  protodeboronation, 91, 93
Transmetallation, *see* Carbon–boron cleavage, Carbon–boron bond construction
Triallylborane, 47
  addition to alkenes, 117
  displacement reaction, 120
  oxidative polymerization, 224
  rearrangement, 134
Triallylboron, halodeboronation, 106
Tribenzylborane, 49
  autoxidation of, 99
  reaction with alkali metals, 131, 132
Triborane(7)-ether complex, 193
Tri-3-butenylborane, 11, 48
Tri-*n*-butylborane, 11, 47, 48
  pyrolysis of, 38
  reaction with nitric oxide, 107
Tri-*n*-butylphosphine, 154
Tri-*s*-butylborane, 48
  autoxidation of, 99
  isomerization of, 111
Trichloromethyllithium, 127
Tricyclohexylborane, 49
  autoxidation of, 99
  C—B bond strength, 17
  reaction with nitrosocyclohexane, 107
Tricyclopropylborane, acceptor strength, 153
Triethylaminephenylborane, as hydroborating agent, 35
Triethylborane, 47
  C—B bond strength, 17
  C—H bond strength, 17
  catalyst, 120
  dehydroboration of, 111
  formation of, 146, 160, 168
  heat of formation, 17
  mass spectroscopy, 13–14
  nmr, 7–8, 10
  photoelectron spectrum, 18
  protodeboronation, 91
  reaction with atomic oxygen, 18
  reaction with cyclohexene, 117
  reaction with NaH, 136
  reaction with Na-K, 132
  transmetallation, 110
  vibrational spectrum, 12
Triethylborohydride anion
  as hydroborating agent, 145
  hydrolysis of, 141
  infrared of, 147
Trifluoromethyl radicals, 108
Trifluorophosphine, reaction with B_4H_8CO, 195
Tri-*n*-hexylborane, 49
  reaction with hydrogen peroxide, 102
Triisobutylborane, 48
  autoxidation of, 99
  isomerization of, 111
  olefin-displacement, 120
Triisopropylborane, acceptor strength, 153

Trimesitylborane, 49
autoxidation of, 99
protodeboronation of, 93
Trimesitylboron anion radical, 132
Trimesitylboron-sodium adduct, 132
Trimethylaluminum, 5, 153, 164
Trimethylamine
reaction with borane carbonyl, 160
reaction with organoboranes, 147, 153
reaction with organopentaboranes, 198, 199
Trimethylamine borane
formation of, 160
as hydroborating agent, 33, 34
Trimethylamine-*t*-butylborane, as hydroborating agent, 34
Trimethylamine oxide, 107, 129, 223
Trimethylborane, 46
acceptor strength, 153
ammonia adduct, 109
autoxidation of, 99, 179
C—B strength, 17
C—B bond distance, 4, 189
exchange reactions, 166
force constants, 11
heat of formation, 17
γ-irradiated, 109
Lewis-base adducts, 153, 154
mass spectroscopy, 13–14
methyl rotation, 5, 11
M.O. calculations, 15
nmr, 7
nqr, 18
photolysis of, 135
radical, spin density on boron, 109
reaction with diborane, 199
reaction with isonitriles, 136
reaction with $KPMe_2$, 137
reaction with LiEt, 136
reaction with Na_2HBMe_2, 189
reaction with potassium metal in ammonia, 137
reaction with ylides, 42
thermal decomposition, 135
vibrational spectroscopy, 9, 12, 192
Trimethylborazine, 4, 77, 106
Trimethylboroxine, 4, 13, 44, 79
1,2-Trimethylenediborane, 190
formation of, 170
Trimethylgallium, 153, 173
Trimethylindium, 153
2,4,4-Trimethyl-2-pentene, hydroboration of, 169
Trimethylphosphine
Lewis base adducts, 153–154
reaction with 2,4-dimethylenetetraborane, 195
Trimethylthallium, 153
Tri-α-naphthylborane, 49
amine adducts, 153
autoxidation of, 99
protodeboronation of, 93
reaction with sodium, 132
Tri-2-norbornylboranes, 102
Tri-4-pentenylborane, 11, 49
Triphenylborane, 48
autoxidation of, 99
C—B bond strength, 17
electronic spectra, 13
formation of, 146, 168, 173, 179, 182
nmr, 7
oxidation with organic hydroperoxides, 102
photolysis of, 134, 144
protodeboronation of, 93
pyridine adduct, 145
reaction with diborane, 165
reaction with NaH, 137
reaction with sodium, 132
reaction with tritylsodium and butadiene, 137
transmetallation, 110
Triphenylboron anion radical, 132
Triphenylboroxine, 16, 79
Triphenylphosphine, 188
Triphenylphosphine oxide, 188
Triplet ketones, 102
Tri-*n*-propylborane, 47
exchange with borane, 165
formation of, 174
reaction with dizoketones, 127
reaction with Na-K, 132
reaction with olefins, 120
Tri-*i*-propylborane, 47
exchange with borane, 165
1,1,2-Tris(dichloroboryl)ethane, 39, 82
Tritiated olefins, 120
Tritylsodium, 137
Trivalent boron, vacant orbital of, 99
Trivalent boron compounds, 2–135

Trivinylborane, 16, 46
nmr, 7–8
photoelectron spectra, 18
protodeboronation, 97
reaction with B_2Cl_4, 39
synthesis of, 33
Tropenylium ion, 213
Tropilium ion, 13

U

Ultraviolet spectroscopy, 11–13, 216
Unsymmetrical trialkylboranes, 33
Uses of organoboron compounds, 223, 224
uv radiation, 101, *see also* Photolytic reactions

V

Vacuum thermolysis, 113
Vibrational spectroscopy, 4, 5, 9–11, 147, 155, 162–163, 190, 192, 194, 195, 205, 209, 217, 219
Vinylborohydride anion, B-trifluoro derivatives, 141
Vinylboron compounds, 24, 107, 214, 219–221
alcoholysis of C—B bond, 96
autoxidation of, 99
B—C twist angle, 18
bond hybridization, 18
^{11}B nmr, 6–8
π-delocalization, 5–12, 15–16, 18, 25
electron impact, 15
electronic transitions, 11–12
elimination reactions, 113
formation of, 20, 23, 33, 35–36, 38–39, 41, 44, 141
heats of formation, 17
hydroboration of, 31, 33, 45
Lewis acidity, 218
perfluoro-, 12, 23, 46, 62, 94, 125
peroxidation of, 100, 102
photolysis, 134
protodeboronation of, 91, 94, 97
proton nmr, 8–9
reaction with iodine, 104
rearrangement, 130
spin coupling, 10
transmetallation, 110
vibrational spectroscopy, 9–11
Vinylboron dibromide, 9, 11, 64
Vinylboron dichloride, 39, 64
Vinylboron difluoride, 15, 23, 62
Vinylboron halides, 16, 20, 23
Vinylboronates, 219
Vinylboronic esters, hydroboration of, 45
Vinyl chloride
hydroboration of, 113
reaction with B_2Cl_4, 39
Vinyldimethylborane, 12, 16, 47
Vinylesters, 126
Vinyl ethers, reaction with organoboron compounds, 115
Vinyl halides, hydroboration of, 31, 169
Vinylmetallics, 20–23
Vinylmethylchloroborane, 12, 54
Vinylthioesters, hydroboration of, 32
Volume, molar, 18

W

Water, 127
as inhibitor of autoxidation, 99
Wurtz reaction, 44

X

X-ray diffraction, 3, 4, 203, 209

Y

Ylides, 42, 125, 136, 154, 174–175

Z

Zinc compounds, 109
Zinc, organo, 19, 20, 22
Zwitterions, 210

A 5
B 6
C 7
D 8
E 9
F 0
G 1
H 2
I 3
J 4